AF710821

VOLUME ONE HUNDRED AND TEN

CURRENT TOPICS IN DEVELOPMENTAL BIOLOGY

bHLH Transcription Factors in Development and Disease

CURRENT TOPICS IN DEVELOPMENTAL BIOLOGY

"A meeting-ground for critical review and discussion of developmental processes"
A.A. Moscona and Alberto Monroy (Volume 1, 1966)

SERIES EDITOR

Paul M. Wassarman
Developmental and Regenerative Biology
Mount Sinai School of Medicine
New York, NY, USA

FOUNDING EDITORS

A.A. Moscona and Alberto Monroy

CURRENT ADVISORY BOARD

Blanche Capel	Susan Mango
B. Denis Duboule	Yoshiki Sasai
Anne Ephrussi	Philippe Soriano
Julian Lewis	Cliff Tabin

FOUNDING ADVISORY BOARD

Vincent G. Allfrey	Dame Honor B. Fell
Jean Brachet	John C. Kendrew
Seymour S. Cohen	S. Spiegelman
Bernard D. Davis	Hewson W. Swift
James D. Ebert	E.N. Willmer
Mac V. Edds, Jr.	Etienne Wolff

VOLUME ONE HUNDRED AND TEN

CURRENT TOPICS IN DEVELOPMENTAL BIOLOGY

bHLH Transcription Factors in Development and Disease

Edited by

RESHMA TANEJA
Department of Physiology,
Yong Loo Lin School of Medicine, and
National University of Singapore Graduate
School for Integrative Sciences and Engineering,
National University of Singapore,
Singapore, Singapore

ELSEVIER

AMSTERDAM • BOSTON • HEIDELBERG • LONDON
NEW YORK • OXFORD • PARIS • SAN DIEGO
SAN FRANCISCO • SINGAPORE • SYDNEY • TOKYO
Academic Press is an imprint of Elsevier

Academic Press is an imprint of Elsevier
225 Wyman Street, Waltham, MA 02451, USA
525 B Street, Suite 1800, San Diego, CA 92101-4495, USA
32 Jamestown Road, London NW1 7BY, UK
The Boulevard, Langford Lane, Kidlington, Oxford OX5 1GB, UK

First edition 2014

Copyright © 2014 Elsevier Inc. All rights reserved.

No part of this publication may be reproduced or transmitted in any form or by any means, electronic or mechanical, including photocopying, recording, or any information storage and retrieval system, without permission in writing from the publisher. Details on how to seek permission, further information about the Publisher's permissions policies and our arrangements with organizations such as the Copyright Clearance Center and the Copyright Licensing Agency, can be found at our website: www.elsevier.com/permissions.

This book and the individual contributions contained in it are protected under copyright by the Publisher (other than as may be noted herein).

Notices

Knowledge and best practice in this field are constantly changing. As new research and experience broaden our understanding, changes in research methods, professional practices, or medical treatment may become necessary.

Practitioners and researchers must always rely on their own experience and knowledge in evaluating and using any information, methods, compounds, or experiments described herein. In using such information or methods they should be mindful of their own safety and the safety of others, including parties for whom they have a professional responsibility.

To the fullest extent of the law, neither the Publisher nor the authors, contributors, or editors, assume any liability for any injury and/or damage to persons or property as a matter of products liability, negligence or otherwise, or from any use or operation of any methods, products, instructions, or ideas contained in the material herein.

For information on all Academic Press publications visit our website at store.elsevier.com

ISBN: 978-0-12-405943-6
ISSN: 0070-2153

Working together to grow libraries in developing countries

ELSEVIER Book Aid International

www.elsevier.com • www.bookaid.org

CONTENTS

CONTRIBUTORS

Takanori Amano
Mammalian Genetics Laboratory, Genetic Strains Research Center, National Institute of Genetics, Mishima, Shizuoka, Japan

Avinash G. Bahirvani
Department of Physiology, Yong Loo Lin School of Medicine, National University of Singapore, Singapore, Singapore

Ian Belle
Department of Immunology, Duke University Medical Center, Durham North Carolina, USA

Jennifer A. Chan
Department of Pathology & Laboratory Medicine, Southern Alberta Cancer Research Institute, University of Calgary, Calgary, Alberta, Canada

Glenda Comai
Stem Cells and Development, CNRS URA 2578, Department of Developmental & Stem Cell Biology, Institut Pasteur, Paris, France

Christos Delidakis
Institute of Molecular Biology & Biotechnology, Foundation for Research & Technology Hellas, and Department of Biology, University of Crete, Heraklion, Crete, Greece

Katsumi Fujimoto
Department of Dental and Medical Biochemistry, Basic Life Sciences, Institute of Biomedical and Health Sciences, Hiroshima University, Hiroshima, Japan

Manfred Gessler
Developmental Biochemistry, Theodor-Boveri-Institute/Biocenter, and Comprehensive Cancer Center Mainfranken, Wuerzburg University, Wuerzburg, Germany

Carol Huang
Department of Pediatrics, Alberta Children's Hospital Research Institute, University of Calgary, Calgary, Alberta, Canada

Yu Jin
Department of Physiology, Yong Loo Lin School of Medicine, National University of Singapore, Singapore, Singapore

Ryoichiro Kageyama
Institute for Virus Research; Japan Science and Technology Agency, Core Research for Evolutional Science and Technology (CREST), Saitama, and World Premier International Research Initiative–Institute for Integrated Cell-Material Sciences (WPI-iCeMS), Kyoto University, Kyoto, Japan

Bin Kang
Immunobiology Cancer Research Program, Oklahoma Medical Research Foundation, Oklahoma City, Oklahoma, USA

Yukio Kato
Department of Dental and Medical Biochemistry, Basic Life Sciences, Institute of Biomedical and Health Sciences, Hiroshima University, Hiroshima, Japan

Takeshi Kawamoto
Department of Dental and Medical Biochemistry, Basic Life Sciences, Institute of Biomedical and Health Sciences, Hiroshima University, Hiroshima, Japan

Taeko Kobayashi
Institute for Virus Research, Kyoto University, Kyoto, and Japan Science and Technology Agency, Core Research for Evolutional Science and Technology (CREST), Saitama, Japan

Flora Ling
Immunobiology Cancer Research Program, Oklahoma Medical Research Foundation, Oklahoma City, Oklahoma, USA

Srivathsa S. Magadi
Institute of Molecular Biology & Biotechnology, Foundation for Research & Technology Hellas, and Department of Biology, University of Crete, Heraklion, Crete, Greece

Maria Monastirioti
Institute of Molecular Biology & Biotechnology, Foundation for Research & Technology Hellas, Heraklion, Greece

Mitsuhide Noshiro
Department of Dental and Medical Biochemistry, Basic Life Sciences, Institute of Biomedical and Health Sciences, Hiroshima University, Hiroshima, Japan

Jin Rong Ow
Department of Physiology, Yong Loo Lin School of Medicine, and National University of Singapore Graduate School for Integrative Sciences and Engineering, National University of Singapore, Singapore, Singapore

Carol Schuurmans
Department of Biochemistry and Molecular Biology, Alberta Children's Hospital Research Institute, Hotchkiss Brain Institute, University of Calgary, Calgary, Alberta, Canada

Toshihiko Shiroishi
Mammalian Genetics Laboratory, Genetic Strains Research Center, National Institute of Genetics, Mishima, Shizuoka, Japan

Xiao-Hong Sun
Immunobiology Cancer Research Program, Oklahoma Medical Research Foundation, Oklahoma City, Oklahoma, USA

Shahragim Tajbakhsh
Stem Cells and Development, CNRS URA 2578, Department of Developmental & Stem Cell Biology, Institut Pasteur, Paris, France

Masaru Tamura
Technology and Development Team for Mouse Phenotyping Analysis, RIKEN BioResource Center, Tsukuba, Ibaraki, Japan

Yong Hua Tan
Department of Physiology, Yong Loo Lin School of Medicine, National University of Singapore, Singapore, Singapore

Reshma Taneja
Department of Physiology, Yong Loo Lin School of Medicine, and National University of Singapore Graduate School for Integrative Sciences and Engineering, National University of Singapore, Singapore, Singapore

David Weber
Developmental Biochemistry, Theodor-Boveri-Institute/Biocenter, Wuerzburg University, Wuerzburg, Germany

Cornelia Wiese
Developmental Biochemistry, Theodor-Boveri-Institute/Biocenter, Wuerzburg University, Wuerzburg, Germany

Yuan Zhuang
Department of Immunology, Duke University Medical Center, Durham North Carolina, USA

PREFACE

The development of various tissues and cell types each with a specific structure and function is a highly complex process that requires precisely regulated gene expression. Among the various transcription factors that regulate developmental processes, arguably, the basic helix-loop-helix (bHLH) superfamily plays a prominent role in the establishment of tissue-specific gene expression programs. Moreover, these factors also play a role in tissue maintenance and homeostasis.

The bHLH transcription factor superfamily is highly conserved over evolution and has been identified in organisms ranging from yeast to humans. The bHLH motif was first recognized in the E-proteins E12 and E47, and now over 200 factors all of which share a bHLH domain have been identified in this superfamily. In addition to the conserved bHLH domain, some family members contain a Per-Arnt-Sim [PAS] domain or a leucine zipper [LZ] domain, whereas the Id proteins lack a basic region. Elegant studies over the years in various species and using different model systems have identified these factors as key regulators of lineage specification, sex determination, and cellular differentiation along multiple lineages. In addition, bHLH factors regulate immune cell function, circadian rhythms, hypoxia responses, toxin metabolism, cell proliferation, apoptosis, and many other biological processes. Given their requirement for growth, development, and differentiation of several cell types, it is not surprising that deregulated expression or activity of numerous bHLH factors is linked to human pathologies. This volume brings together our current knowledge on the structure, function, and involvement of bHLH transcription factors in development and in human disease.

In Chapter 1, Tajbakhsh and colleagues provide an in-depth review on the myogenic regulatory factors MyoD, Myf5, Myogenin, and Mrf4 that coordinate developmental myogenesis. They provide a perspective on regulation and function of these factors, as well their function in regulating chromatin structure through recruitment of chromatin-modifying and remodeling complexes.

Proneural genes that confer neural identity have been well studied both in *Drosophila* and in vertebrates by regulating Notch signaling. Chapter 2 by Huang et al. reviews proneural gene function and discusses the deregulated expression of these genes in various human disease including cancer and psychiatric disorders.

In Chapter 3, Tamura et al. review *Hand* genes that have been genetically proven to play a critical role in heart, craniofacial, and limb development. The human *Hand2* gene resides on 4q34.1 and its expression is reduced in 4q− syndrome suggesting it may be causally involved. The impact of *Hand2* gene dosage in developmental disorders and in the human chromosomal disorder partial trisomy distal 4q is discussed.

E-proteins that include E2A, HEB, and E2-2 have been widely studied as central regulators of T and B lymphocyte development. In addition, E-proteins heterodimerize with various tissue-restricted bHLH factors such as MyoD to regulate cellular differentiation. In Chapter 4, Belle and Zhuang provide an overview of the role of E-protein function in development of the immune system. They also discuss how deregulation of these factors leads to autoimmune disease, as well as T and B cell leukemias and lymphomas.

Id proteins function predominantly, but not exclusively, as dominant-negative inhibitors of E-proteins. Unsurprisingly, gain of Id proteins mimic loss of E-proteins and also block differentiation along multiple lineages. In Chapter 5, Ling et al. discuss the role of Id proteins as antagonists of E-proteins in the immune system, as well as in blocking differentiation and promoting tumorigenesis.

The *Enhancer of Split [E(Spl)]* locus which consists of seven bHLH genes was identified as spontaneous loss-of-function mutant resulting in a neurogenic phenotype in *Drosophila.* These genes are perhaps among the best-characterized targets of the Notch signaling pathway. In Chapter 6, Delidakis and colleagues provide a comprehensive review on the evolution, structure, and function of these genes in lateral inhibition that serves to restrain neuroblast commitment.

The vertebrate counterparts of E(Spl) are the *Hes* and *Hey* genes. In Chapter 7, Kobayashi and Kageyama provide a comprehensive review on the role of *Hes* genes as downstream effectors of Notch signaling. They discuss their function in the nervous system, in the maintenance of stem cells, as well as the oscillatory expression of *Hes* genes, which is important in somite segmentation.

In Chapter 8, Weber et al. describe the striking defects in heart and vascular development as evidenced by genetic disruption of *Hey1*, *Hey 2*, and *HeyL* in mice. They also discuss the role of Hey2 in cardiac function and its association with Brugada syndrome.

Stra13/Dec1 and Sharp-1/Dec2 bHLH factors exhibit domain structure similarity with the Hes and Hey proteins. However, these factors exhibit several distinctive transcriptional properties and functions in various tissues.

In Chapter 9, we review the function of these proteins in tissue repair and regeneration especially in the musculoskeletal system, as well as their function in homeostasis in the immune system. Growing evidence documenting deregulated expression of both genes in various pathologies is reviewed. In Chapter 10, Kato and colleagues discuss the role of these factors in circadian rhythms and adaptive responses.

Given the very large number of bHLH factors, it would be impossible to cover the entire superfamily in a single volume. Nonetheless, this volume attempts to bring together for the first time a compilation of in-depth reviews on several subfamilies that positively and negatively regulate development and differentiation of many cell types. These reviews reflect a continuum of progress that has been made toward understanding the biological functions of this superfamily.

Significant advances have been made in our understanding of bHLH factor function in developmental processes through loss-of-function and gain-of-function studies in mice and other model systems. Numerous cascades involved in cell fate determination during development and governing cell proliferation, differentiation, apoptosis, and migration are regulated by bHLH family members that are highly conserved from invertebrates to humans. Moreover, deregulated expression or activity of several factors is seen in human pathologies. In some instances, there is strong evidence of causal involvement in various disease conditions, whereas in others, direct evidence of mutations in human pathologies awaits further investigation. In spite of these rapid strides in linking bHLH factors to many human diseases, much remains to be understood with regard to the molecular mechanisms underlying their function. For instance, while most bHLH proteins bind to highly similar DNA-binding sites (E-box or N-box motifs), they interact with distinct genomic regions *in vivo* and regulate different sets of target genes. The mechanisms by which specificity and selectivity of DNA binding is achieved in the genome are not completely understood. The advent of ChIP-seq technology along with transcriptome analysis has started to shed light on genome-wide binding sites and identification of direct target genes. Posttranslational modifications that can alter stability, subcellular localization, DNA binding, dimerization, and association with cofactors remain to be deciphered in many cases. Epigenetic reprogramming is a hallmark of terminal differentiation during which cells acquire specialized functions and undergo dramatic alterations in gene expression. The identity of chromatin-modifying and remodeling complexes that influence tissue-specific differentiation programs by distinct bHLH transcription factors is unclear in most tissues. Association with cofactors can influence transcriptional activity and function of bHLH proteins and may be relevant in pathologies where genetic mutations have not been identified in disease. Such studies aimed at understanding of the molecular and cellular basis of function will undoubtedly clarify and put into perspective how bHLH factors function in normal developmental programs and how their deregulation results in disease pathogenesis and progression.

This book is dedicated to the loving memory of my father who passed away as I was working on this book in December 2013. He led a life of honesty, hard work, perseverance, and goodwill to all that will be an everlasting source of inspiration.

Reshma Taneja

CHAPTER ONE

Molecular and Cellular Regulation of Skeletal Myogenesis

Glenda Comai, Shahragim Tajbakhsh[1]
Stem Cells and Development, CNRS URA 2578, Department of Developmental & Stem Cell Biology, Institut Pasteur, Paris, France
[1]Corresponding author: e-mail address: shahragim.tajbakhsh@pasteur.fr

Contents

Current Topics in Developmental Biology, Volume 110
ISSN 0070-2153
http://dx.doi.org/10.1016/B978-0-12-405943-6.00001-4
© 2014 Elsevier Inc.
All rights reserved.

Abstract

Since the seminal discovery of the cell-fate regulator Myod, studies in skeletal myogenesis have inspired the search for cell-fate regulators of similar potential in other tissues and organs. It was perplexing that a similar transcription factor for other tissues was not found; however, it was later discovered that combinations of molecular regulators can divert somatic cell fates to other cell types. With the new era of reprogramming to induce pluripotent cells, the myogenesis paradigm can now be viewed under a different light. Here, we provide a short historical perspective and focus on how the regulation of skeletal myogenesis occurs distinctly in different scenarios and anatomical locations. In addition, some interesting features of this tissue underscore the importance of reconsidering the simple-minded view that a single stem cell population emerges after gastrulation to assure tissuegenesis. Notably, a self-renewing long-term Pax7+ myogenic stem cell population emerges during development only after a first wave of terminal differentiation occurs to establish a tissue anlagen in the mouse. How the future stem cell population is selected in this unusual scenario will be discussed. Recently, a wealth of information has emerged from epigenetic and genome-wide studies in myogenic cells. Although key transcription factors such as Pax3, Pax7, and Myod regulate only a small subset of genes, in some cases their genomic distribution and binding are considerably more promiscuous. This apparent nonspecificity can be reconciled in part by the permissivity of the cell for myogenic commitment, and also by new roles for some of these regulators as pioneer transcription factors acting on chromatin state.

1. INTRODUCTION

The discovery of Myod in 1987 was a landmark step in our understanding of the processes leading to the acquisition of cell fates—the ectopic expression of a single transcription factor converted many nonmuscle cells to skeletal muscle (Davis, Weintraub, & Lassar, 1987). What followed was perplexing, as researchers failed to identify a similar single potent transcription factor in other tissues, until combinations of transcription factors were identified that had similar cell-fate transforming capabilities from fibroblasts or other adult cell types (Han et al., 2012; Huang et al., 2011; Ieda et al., 2010; Marro et al., 2011; Pang et al., 2011; Sekiya & Suzuki, 2011; Szabo et al., 2010; Vierbuchen et al., 2010; Zhou et al., 2008). More recently, this historical loop was closed following the seminal discovery by Yamanaka that a handful of transcription factors can reprogram adult committed cells to the pluripotent embryonic state (Takahashi & Yamanaka, 2006). These findings raised other issues regarding the permissivity of some cells to conversion, as well as efficacy. It is now clear that the state of the chromatin plays a critical role in defining cell-fate conversion, particularly in those cells that were refractory to myogenic conversion by

Myod. Although Myod proved to be potent to initiate the myogenic program, strangely, one such transcription factor was not enough to assure skeletal myogenesis in vertebrates. The closely related genes *Myf5*, *Mrf4* were identified subsequently as determination genes acting in the progeny of stem cells—the muscle progenitor cells—to establish myogenic fate, whereas the related myogenic regulatory factor (Mrf) *Myogenin* (*Myog*) drives myogenic differentiation along with *Mrf4*. Muscle stem cell specification is driven by upstream factors, *Pax3*/*Pax7* in the trunk, and *Tbx1*/*Pitx2* in the head.

This review will highlight recent findings and attempt to link them with initial studies on the functional role of the Mrfs. Specifically, four major axes will be developed: (1) similarities and differences between the Mrfs and speculations on why four potent transcription factors are necessary to govern myogenesis; (2) the link between stem cells and their progeny where Mrfs effect their functions in different anatomical locations; (3) challenges in the fields of transdifferentiation and reprogramming seeking to treat muscle disorders; and (4) new mechanistic insights emerging from genome-wide studies into the spatiotemporal control of myogenesis.

2. MOLECULAR FEATURES OF THE MRFS

2.1. Mrfs at the nodal point of cell identity—similarities and differences

Mrfs are class II bHLH transcription factors (Murre et al., 1994) that are structurally highly similar, each containing three conserved domains: a transactivation domain in the amino terminal region including a histidine/cysteine (H/C)-rich domain, a bHLH in the central region, and an amphipathic α-helix domain (helix III) in the carboxy terminal transactivation domain (see Singh & Dilworth, 2013).

The HLH domains (helix I and II) are required for heterodimerization with ubiquitously expressed class I bHLH E-proteins (HEB/HTF4, E2-2/ITF-2, E12 and E47) (Conway, Pin, Kiernan, & Merrifield, 2004; Henthorn, Kiledjian, & Kadesch, 1990; Lassar et al., 1991; Murre, McCaw, & Baltimore, 1989; Murre, McCaw, Vaessin, et al., 1989; Parker, Perry, Fauteux, Berkes, & Rudnicki, 2006). The basic domain, which is also α-helical, acts as a sequence-specific DNA-binding domain that recognizes the E-box consensus sequence (CANNTG) present in the regulatory region of muscle-specific genes (Blackwell & Weintraub, 1990; Davis, Cheng, Lassar, & Weintraub, 1990; Lassar et al., 1989). A conserved "muscle recognition motif" or "myogenic code" composed of the amino acids AT and K (alanine–threonine within the basic domain, followed by lysine in the junction with helix-I) distinguishes the Mrfs from all the other bHLH transcription factors. These three amino acids confer the ability to activate muscle-specific genes by either interacting with cofactors, or inducing conformational change, or both (Black, Molkentin, & Olson, 1998; Brennan, Chakraborty, & Olson, 1991; Davis et al., 1990; Molkentin, Black, Martin, & Olson, 1995). These amino acid motifs, when substituted into the corresponding sequence of the nonmyogenic bHLH protein E12 are sufficient to confer myogenic potential (Davis & Weintraub, 1992).

Myod requires a minimum of two paired E-boxes or an E-box plus a site for a cooperative coactivator (such as Mef2, Sp1, or Pbx and Meis) for activation of reporters *in vitro* (Biesiada, Hamamori, Kedes, & Sartorelli, 1999; Knoepfler et al., 1999; Molkentin et al., 1995; Sartorelli, Webster, & Kedes, 1990; Weintraub, Davis, Lockshon, & Lassar, 1990). Therefore, dimer formations (Myod–Myod or Myod–coactivator) are crucial for establishing a stable complex with DNA, possibly through induced conformational changes (see Tapscott, 2005). Mef2 family members are expressed in most tissues but are not capable of directing skeletal muscle differentiation in the absence of Mrfs (Molkentin et al., 1995). However, in the promoters and enhancers of muscle-specific genes, E-boxes and Mef2-binding sites are often located in close proximity to one another (Wasserman & Fickett, 1998), and Mef2 proteins and Myod can synergistically accelerate skeletal muscle differentiation in culture (Black et al., 1998; Molkentin et al., 1995; Penn, Bergstrom, Dilworth, Bengal, & Tapscott, 2004). The AT residues appear to be required to mediate the cooperativity between Mrfs and Mef2 proteins to activate muscle genes (Black et al., 1998; Molkentin et al., 1995).

Having presumably evolved from a single ancestral invertebrate gene following two rounds of duplication (Atchley, Fitch, & Bronner-Fraser, 1994; Yuan, Zhang, Liu, Luan, & Hu, 2003), the question is why are four different Mrfs required to establish vertebrate muscles. Within their bHLH domains, amino acid sequences have remained largely unchanged during evolution from nematodes to human (Atchley et al., 1994; Zhao, Yu, Huang, & Liu, 2014). In contrast, upstream and downstream of this highly conserved DNA-binding region, considerable differences in sequence become apparent, and they seem to have diverged under positive selection during evolution (Zhao et al., 2014). Therefore, functional differences between the individual Mrfs can be explained in part by their different structural characteristics as explained below.

First, generation of chimeric Mrfs by domain swapping experiments *in vitro* have demonstrated that the N- and C-terminal domains of Myf5 and Myod are interchangeable for the activation of gene expression during myoblast proliferation but not in differentiation culture conditions. In this case, cooperation of the N- and C-terminal regions of Myod was required to activate a variety of differentiation markers (Ishibashi et al., 2005). These results support the notion that Myf5 (and Myod) function(s) toward myoblast proliferation, whereas Myod is inclined to promote differentiation.

Second, differences within the H/C and helix III domains seem to account for their differential ability to activate expression from silent loci, an activity that is dependent on SWI/SNF enzymes and is one of the hallmarks of establishing a lineage (see Section 5.2). This property is shared by the determination genes *Myod*, *Myf5*, and *Mrf4*, but not *Myog* (Bergstrom & Tapscott, 2001; Berkes et al., 2004; de la Serna, Carlson, & Imbalzano, 2001; Gerber et al., 1997; Roy, de la Serna, & Imbalzano, 2002). This explains to some extent the reduced efficiency of myogenic conversion by *Myog in vitro* (Section 4.1). Substitution of the Myod helix III domain into Myog increases its ability to activate muscle-specific gene expression, and replacement of both helix III and H/C domains with the corresponding Myod sequences enhanced the activity of the chimeric molecule, to be nearly as active as Myf5 and Myod (Bergstrom & Tapscott, 2001). Therefore, sequence divergence between individual Mrfs outside the bHLH domain appears to account for the differential activities and developmental roles of these highly related family members. Although domain swap experiments have provided mechanistic insights, it remains unclear from these *ex vivo* studies what distinguishes these Mrfs functionally. These chimeric constructs would need to be tested during tissuegenesis *in vivo* where more subtle roles can be unmasked.

2.2. Negative regulation of Mrf potency: establishing the onset of differentiation

Potent transcription factors can lead to cell commitment at an inappropriate time, thus fail-safe mechanisms that limit premature activity of the Mrfs should be operational. For example, Myod in a homodimer conformation binds preferentially to quadruplex structures of guanine-rich sequences in the promoters of muscle specific genes, thereby precluding productive binding to E-boxes (Davis et al., 1990; Etzioni et al., 2005; Lassar et al., 1991; Murre, McCaw, Vaessin, et al., 1989; Shklover, Weisman-Shomer, Yafe, & Fry, 2010). Thus, the preferential binding of relative inactive Myod homodimers to quadruplex DNA structures or transcriptionally active Myod/E-protein heterodimers to E-boxes might contribute to the timed regulation of gene expression during myogenesis (Etzioni et al., 2005; Shklover et al., 2010). Posttranslational modifications associated with corepressor proteins and with proteins that titrate Myod away from its DNA binding have also been reported (Puri & Sartorelli, 2000). Many of these antagonists are themselves HLH proteins, such as Inhibitor of DNA-binding (Id) proteins (Id1–4), Twist, and MyoR.

Twist heterodimerizes with E-proteins thereby sequestering them and preventing their interaction with the Mrfs (Spicer, Rhee, Cheung, & Lassar, 1996). Twist can also inhibit myogenesis by direct interaction with Mef2 proteins, or by binding to the basic region of the Mrfs (Hamamori, Wu, Sartorelli, & Kedes, 1997; Spicer et al., 1996). MyoR and Mist-1 form heterodimers with the Mrfs and bind to E-boxes; however, these dimers are unable to activate transcription (Lemercier, To, Carrasco, & Konieczny, 1998; Lu, Webb, Richardson, & Olson, 1999).

Similarly, Id proteins heterodimerize with E-proteins (Benezra, Davis, Lockshon, Turner, & Weintraub, 1990), or with Mrfs directly to prevent transcriptional activity (Neuhold & Wold, 1993). Consistent with this view, Id mRNAs are detected in proliferating skeletal muscle/high serum conditions and then downregulated in differentiating muscle cultures (Benezra et al., 1990; Chen, Han, Sun, & Lim, 1997; Wu & Lim, 2005). Although the mechanisms regulating the inhibition of Id protein expression are still unclear, using a combined systems approach, the transcriptional repressor RP58 (mouse Zfp238), was found to directly repress Id2/Id3 expression, thereby allowing the progression of myogenic differentiation (Yokoyama et al., 2009). The study by Yokoyama and colleagues revealed a regulatory loop by which Myod positively regulates Id expression in myoblasts (Id3 is a growth phase target of Myod; Wyzykowski, Winata, Mitin, Taparowsky, & Konieczny, 2002) and then at the initiation of differentiation, it activates *RP58* (Yokoyama et al., 2009) which in turn represses Id2/3 expression, thereby allowing the progression of myogenic differentiation. Interestingly, Id3 is also a direct transcriptional target for Pax7 in quiescent satellite cells; hence, induction of Id2/3 in these cells might act to block premature differentiation by the Mrfs (Kumar et al., 2009).

Snail1/2 are zinc finger repressor proteins that bind and compete for the same DNA motif (E-box) as the bHLH transcription factors and they recruit the histone deacetylaces HDAC1 and HDAC2, which render chromatin less accessible (Bolos et al., 2003; Hajra, Chen, & Fearon, 2002; Peinado, Ballestar, Esteller, & Cano, 2004). At the onset of differentiation, Snail1/2 must be removed to allow Myod access to differentiation gene sequences. Snail was reported to bind E-box motifs that are G/C rich in their central dinucleotides, and such sites are almost exclusively associated with genes expressed during differentiation (Soleimani, Yin, et al., 2012). By contrast, Snail does not bind the A/T-rich E-boxes associated with Myod targets in myoblasts. In addition, Snail1 and 2 are targets of miR-30a and miR-206, which are in turn Mrf targets (Soleimani, Yin, et al., 2012; Sweetman et al., 2008). Therefore, Soleimani and colleagues proposed a regulatory loop involving these factors for entry into differentiation. When cells receive a differentiation signal, Mrfs activate the miRNAs that prevent Snail mRNA translation. As Snail proteins turn over, Myod/E47 can gain access to G/C-rich differentiation-specific E-boxes (Soleimani, Yin, et al., 2012). Consistent with this, overexpression of Snail blocks differentiation, whereas siRNA knockdown induces precocious differentiation.

ZEB1 is another zinc finger protein that, similarly to Snail proteins, binds to G/C-centered E-boxes in muscle differentiation genes at the myoblast stage but not in myotubes (Siles et al., 2013). Inhibition of muscle genes by ZEB1 occurs via transcriptional repression and involves recruitment of the CtBP corepressor. *ZEB1* knockdown results in precocious expression of muscle differentiation genes and acceleration of myotube formation. Similarly, the pattern of gene expression associated with muscle differentiation is accelerated in *ZEB1* null mouse embryos.

3. SKELETAL MUSCLE DEVELOPMENT

3.1. Giving birth to muscle—some unusual twists

Skeletal muscle development is a multistep process where founder stem cells are specified and allocated to the lineage. This is followed by the proliferation of committed myoblasts, their differentiation, and fusion to generate multinucleated, contractile myofibers (see Biressi, Molinaro, & Cossu, 2007; Sambasivan & Tajbakhsh, 2007). However, there are some features of this process that are counterintuitive. The long-term stem cell population (Pax3+/Pax7+ in body; Pax7+ in head; see below) emerges relatively late, only after an anlage of differentiated cells has already been established during mid-embryogenesis (Fig. 1.1). Therefore, founder stem cells that initiate skeletal myogenesis do not all persist to adulthood (Fig. 1.2). Second, although cell identity is established by the Mrfs throughout the embryo, distinct signatures of transcription factors program founder stem cell fates in different regions of the embryo.

Myf5$^{nlacZ/+}$

Origin	Core genetic networks*	Comments
Extra ocular muscles	Pitx2 → Myf5, Mrf4 → Myod	• *Myod* does not rescue myogenesis in *Myf5:Mrf4* KO
Branchial arches	Tbx1, Pitx2 → Msc, Tcf21, Lhx2 → Myf5, Myod	• *Msc/Tcf21* act on BA1-derived muscles
Limb	Pax3 → Myf5, Mrf4 → Myod	• Delayed myogenesis in *Myod* null mutants
Trunk	Pax3, Myf5, Mrf4 → Myod	• *Myod does not* rescue myogenesis in *Pax3:Myf5(Mrf4)* KO

* For initiation of myogenic program

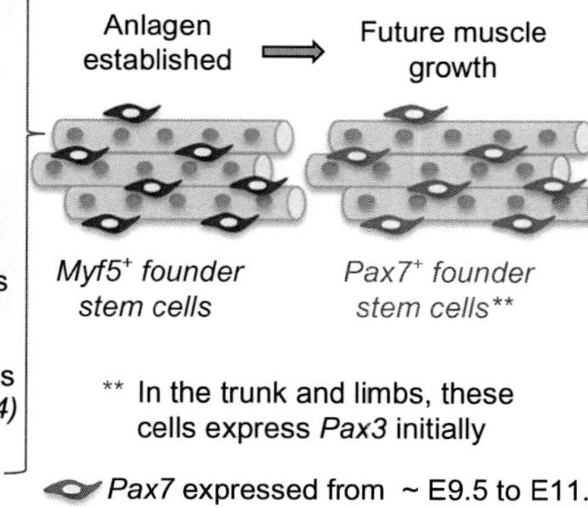

Figure 1.1 Tissuegenesis is initiated with an anlagen prior to the emergence of the self-renewing stem cell population. At anatomical locations throughout the embryo (*Myf5*^{nlacZ} E11.5, Tajbakhsh et al., 1996), an anlagen of differentiated cells is established before the long-term myogenic stem cell (Pax7+) population emerges. In the head and limb, myogenic cells of the anlagen are specified by the upstream regulators *Tbx1/Pitx2* or *Pax3*, respectively, prior to activation of the core set of myogenic factors *Myf5/Mrf4/Myod*. In the trunk, *Pax3* and also *Myf5/Mrf4* act as upstream regulators, given that in *Pax3* null mice, muscle is still specified by these factors in the epaxial domain of the somite. (See the color plate.)

3.2. Diversity in origin and modular design

Trunk and limb muscles originate from somites, transient epithelial metameric segments of paraxial mesoderm (Tajbakhsh & Buckingham, 2000). In contrast, head muscles originate from cranial mesoderm that is unsegmented (cranial paraxial (CPM), cranial splanchnic (SpM), and precordal mesoderm (PCM); see Noden & Francis-West, 2006; Sambasivan, Kuratani, & Tajbakhsh, 2011). The emergence of a Pax7+ myogenic stem cell population by mid-embryogenesis occurs differently in trunk, limb, and head muscle masses. These differences, which are highlighted below, point to an intriguing plasticity in how cell fate is established from myogenic founder stem cells.

The ventral somite contains sclerotomal precursors that will give rise to the axial skeleton and ribs, whereas the dorsal somite gives rise to the epithelial dermomyotome (DM) that is the source of all skeletal muscles in the body, tongue (from the hypoglossal cord), some neck muscles, as well as dermal, brown fat, smooth muscle, and endothelial progenitors (Atit et al., 2006; Ben-Yair & Kalcheim, 2008; Seale et al., 2008). Myogenesis is initiated by delamination of cells from the borders of the DM in a rostrocaudal developmental gradient (Fig. 1.2B). At the interlimb level, cells from the dorsomedial (epaxial) and ventrolateral (hypaxial) part of the DM expressing *Myf5* (founder stem cell-1, FSC1) are displaced underneath the DM to form the primary myotome (from E8 in rostral-most somites). These cells become rapidly postmitotic and constitute the first skeletal muscle (anlagen) of the embryo. In the chick, the rostral and caudal lips of the DM have also been reported to directly input progenitors into the myotome in a biphasic manner (Nitzan & Kalcheim, 2013; Relaix & Marcelle, 2009). Once these anlagen are in place, *Pax3*/*Pax7*+ stem cells from the central portion of the DM (FSC2) "parachute" into this underlying myotome to assure continued muscle growth. The epaxial part of the DM and the myotome will give rise to the deep back muscles. At the limb level, the scenario is somewhat different. Cells from the ventrolateral (hypaxial) lip of the DM expressing *Pax3* but none of the *Mrfs* (FSC3) undergo epithelial–mesenchymal transition (EMT) and migrate into the forming limb bud. Notably, unlike the trunk, *Pax7* expression emerges in the majority of *Myf5*-expressing cells to mark the future myogenic stem cell population (Kassar-Duchossoy et al., 2005). The key molecules involved in EMT of these myogenic precursor cells are hepatocyte growth factor/scatter factor and its receptor c-Met (Tajbakhsh & Buckingham, 2000). Apart from *Pax3* and *Met*, migrating myogenic precursor cells express *Lbx1*, *CXCR4*, *Sp5*, and *Pitx2* (see

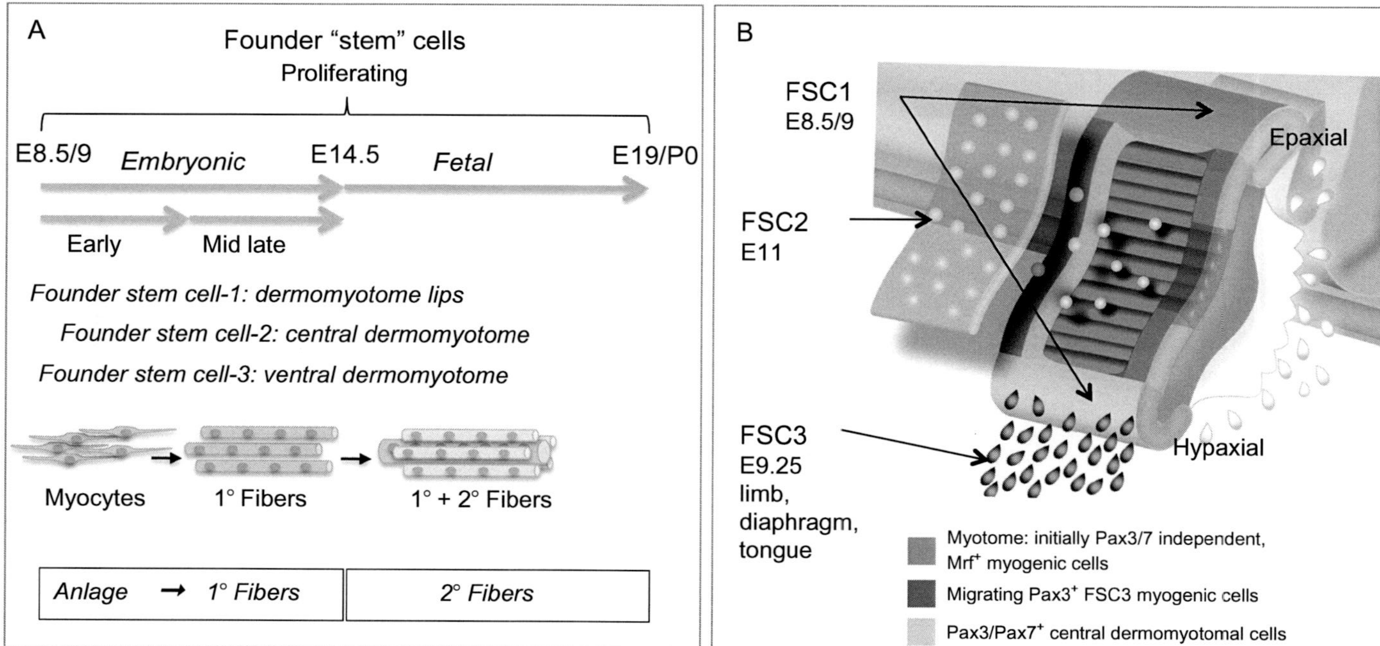

Figure 1.2 See legend on the opposite page.

Vasyutina & Birchmeier, 2006; Yokoyama & Asahara, 2011). The diaphragm muscles also originate from migrating cells from the DM, at the level of the cervical somites. The primary target of their migration is the pleuroperitoneal folds; subsequently, the folds and the developing diaphragm descend caudally, ultimately to lie at the thoracic/lumbar boundary (see Merrell & Kardon, 2013).

In the head, the extraocular muscles (EOM) and branchiomeric (branchial arch)-derived (BA) muscles are the most studied muscle groups. EOM are reported to have a dual origin, at least in the chick, arising from CPM and PCM (Noden & Francis-West, 2006), but detailed information is lacking in the mouse. Retrospective clonal analysis in the mouse has demonstrated a common lineage relationship between EOM and first arch-derived muscles, confirming the contribution of CPM to both muscle groups (Lescroart et al., 2010). Branchimeric muscles are involved in opening and closing of the jaw, facial expression (derived from first and second BA, respectively) and pharyngeal and laryngeal function (derived from the lower arches) (Kelly, 2013). In a developmental context, the CPM

Figure 1.2 Emergence of distinct founder stem cell populations from the somite with different genetic requirements. (A) Embryonic development can be divided into early and late phases. During the early phase, progenitors are specified and determined as myoblasts and an anlage is established in the somite consisting of a primary myotome with differentiated/postmitotic mononuleated myocytes. Myocytes fuse to form multinucleated myofibers (primary fibers). Fetal myogenesis is characterized by growth and maturation of each muscle anlage and by the onset of innervation. This second wave of myogenesis involves fusion of fetal myoblasts to form secondary fibers (initially smaller and surrounding primary fibers). Cells that establish muscles prenatally are referred to as founder stem cells (FSC1, FSC2, FSC3 in somites) to distinguish them from the juvenile and adult satellite (stem) cell populations. FSC1 is largely exhausted in the embryo, while FSC2/FSC3 will give rise to the majority of the adult skeletal muscle stem cells. (B) Localization of the founder stem cell populations highlighted in (A). FSC1 (founder stem cell population-1) emerges first (E8.5/E9.0) in the dorsomedial (epaxial) and later in the ventrolateral (hypaxial) dermomyotome (DM). These are initially Pax3+ then Myf5/Mrf4+. They are reduced in number but not lost in *Pax3* mutants. FSC3 cells, that emerge from ≈E9.25 onwards, are Pax3+ and this migratory population (from the somites to limbs, diaphragm, and tongue) is lost in *Pax3* null embryos. Note, *Pax7* expression is initiated from E9.5, therefore it is not implicated in the establishment of FSC1 and FSC3. FSC2 cells emerge from the central dermomyotome between E10.5 and E11.5 starting from anteriormost somites. FSC2 cells are initially *Mrf*-negative, present in the *Pax3* null, but lost in the *Pax3:Pax7* double mutant (Kassar-Duchossoy et al., 2005; Relaix, Rocancourt, Mansouri, & Buckingham, 2005). Each of these populations can be distinguished by their distinct genetic requirements, as they are spared, or lost in *Pax3* or *Pax3:Pax7* mutants. (See the color plate.)

contributes mainly to the proximal region of the myogenic core in the first branchial arch, whereas splanchnic mesodermal (SpM) cells contribute to its distal region (see Tzahor, 2009). Less information is available on the neck muscles. Notably though, some muscles such as the cucullaris group (including the trapezius and the sternocleidomastoid muscles) have been shown to originate from caudal arches, therefore of cranial mesodermal origin (Tajbakhsh, Rocancourt, Cossu, & Buckingham, 1997; Theis et al., 2010). However, the precise origin of their progenitors has been contested (Kelly, 2013; Sambasivan, Kuratani, et al., 2011; Theis et al., 2010).

Recent findings using mouse genetics and manipulations of the avian embryo demonstrated that heart and craniofacial muscle developmental programs are tightly linked, given that cranial mesoderm also contributes to the formation of the arterial pole of the heart (Kelly, 2013; Tzahor, 2009). Retrospective clonal analysis has also shown the existence of a common progenitor for these tissues (Lescroart et al., 2010). In addition, although the regulation of skeletal and cardiac muscle is quite different, progenitors share common upstream factors. The term "cardio-craniofacial morphogenetic field" reflects the intimate links between development of the head, face, and heart, which is manifested in numerous cardiac and craniofacial birth defects (Hutson & Kirby, 2003).

Despite the differences in their embryological origins, skeletal myogenic progenitor cells at all anatomical locations are regulated by the Mrfs. However, the upstream regulators that initiate this cascade in myogenic stem cells differ depending on the location. In the trunk, *Pax3* is expressed in proliferating stem cells upstream of the Mrfs. By contrast, *Pax3* is not expressed in head myogenic stem/progenitors, a role enacted by a suite of transcription factors, including *Tbx1*, *Musculin*, *Tcf21*, *Isl1*, *Lhx2*, and *Pitx2* (Bothe & Dietrich, 2006; Hacker & Guthrie, 1998; Harel et al., 2009, 2012; Mootoosamy et al., 2002; Nathan et al., 2008; Sambasivan et al., 2009). Not surprisingly, this set of transcription factors is also shared with cardiac progenitor cells (Kelly, 2013; Tzahor, 2009). In the head, *Pax7* expression ensues after *Myf5* and *Myod* expression (Horst et al., 2006; Tajbakhsh et al., 1997), and like the limb, often in cells that have already expressed *Myf5* (R. Sambasivan, G. Comai, S. Tajbakhsh, unpublished observations). The differences in genetic programs between trunk, limbs, and head persist in adulthood (Sambasivan et al., 2009). In addition, satellite cells of head and trunk muscles exhibit different molecular signatures and also different functional properties in terms of proliferative capacity and differentiation efficiency (Harel et al., 2009; Ono, Boldrin, Knopp, Morgan, & Zammit, 2010; Sambasivan et al., 2009). Finally, it is also curious that head and trunk myogenic programs differ in the response to different signaling inputs from adjacent tissues. For example, whereas trunk myogenesis is promoted by the action of Wnt and Shh, Wnts were reported to inhibit head myogenesis (Tzahor et al., 2003).

3.3. Lessons from gene deletions

The generation of mouse knock-out (KO) lines for regulatory genes has been used to define a hierarchical relationship between the four Mrfs and to demonstrate that embryonic and fetal myoblasts have different genetic requirements for their determination (see Murphy & Kardon, 2011). A summary of the individual and compound Mrf mutants is shown in Table 1.1, and the main features of these mutants are highlighted below.

As indicated above, myogenic cell fate and lineage progression to myoblasts and differentiated myofibers are directed by the determination genes *Myf5*, *Myod*, and *Mrf4*. Accordingly, triple mutants in mouse lack myoblasts and differentiated cells (Kassar-Duchossoy et al., 2004; Rudnicki et al., 1993). When the initial experiments were reported, the logical assumption was that these determination genes act in muscle stem cells; however, as indicated above, this view has been revised—*Myf5*, *Myod*, and *Mrf4* act in the progeny of stem cells—muscle progenitors. Analysis of the role played by Mrfs *in vivo* has been complicated by the fact that *Mrf4* and *Myf5* are located in tandem and perturbations of one affect expression of the other *in cis* (Kassar-Duchossoy et al., 2004; Olson et al., 1996). Therefore, many of the original *Myf5* and *Mrf4* loss-of-function alleles also compromised the expression of the neighboring gene. Notably, the original *Myf5:Myod* double KO (Rudnicki et al., 1993) was found to be a *Myf5:(Mrf4):Myod* triple null (Kassar-Duchossoy et al., 2004). *Myf5(Mrf4)* KO mice (where *Mrf4* expression affected *in cis*) has compromised epaxial muscles and a 1- to 2-day delay in the development of trunk muscles due to delayed expression of *Myod* (Braun & Arnold, 1995; Kassar-Duchossoy et al., 2004; Tajbakhsh et al., 1997). While these double mutant embryos show normal limb muscle, Myod fails to rescue EOM development, suggesting that either *Myf5* or *Mrf4* is required for muscle formation and that *Myod* activation is dependent on one of this factors. In the absence of *Myf5* and *Mrf4*, EOM primordia undergo cell death, demonstrating that *Mrf* activity is also required for progenitor cell survival (Sambasivan et al., 2009). Forced expression of *Myod* from the *Myf5* locus rescues progenitor cell survival and EOM development in *Myf5:Mrf4* double-mutant embryos (Sambasivan et al., 2009).

In *Myod:Myf5* null animals where *Mrf4* expression is not affected by *Myf5* gene disruption, muscle progenitors are determined by *Mrf4* in the first wave of embryonic myogenesis particularly in the trunk (E8–E12.5), but not as robustly in the head or limbs (Kassar-Duchossoy et al., 2004). Although *Myod* null mutants can be obtained alive, many die at birth, depending on the genetic background of the mice. The most dramatic prenatal phenotype is a delay in limb muscle differentiation until about E12.5 (Kablar et al., 1997).

Other genetic studies in mice have shown that *Myod*, *Mrf4*, and *Myog* regulate myogenic differentiation (Hasty et al., 1993; Nabeshima et al., 1993; Rawls et al., 1995, 1998; Venuti et al., 1995). *Myog* acts downstream of the three other *Mrfs*, and it has a unique nonredundant function which cannot be compensated by the other *Mrfs* during embryonic development. In mice lacking *Myog*, myoblasts are present, but skeletal muscle tissue is poorly developed (complete absence of fetal myofibers), pointing to its critical role in terminal differentiation of myoblasts (Hasty et al., 1993; Nabeshima et al., 1993; Venuti et al., 1995). Normal levels of *Myod* cannot compensate for the absence of *Myog* in this context (Davie et al., 2007). Unusually, *Mrf4* expression is restricted to progenitors in the embryo and thus it fails to drive myogenesis autonomously beyond the embryonic period in mice lacking *Myf5* and *Myod* (embryonic day E14.5; Kassar-Duchossoy et al., 2004). However, it is the most abundant Mrf expressed in differentiated myonuclei postnatally (Hinterberger, Sassoon, Rhodes, & Konieczny, 1991). Finally, while in triple *Myog:Myod:Mrf4* mutant animals no embryonic or fetal myofibers differentiate (Valdez et al., 2000),compound *Myog:Mrf4*, *Myog:Myod*, or *Myog:Myf5* show embryonic but not fetal muscle similar to the *Myog* null (Rawls et al., 1995, 1998; Valdez et al., 2000). The knockin of *Myog* into the *Myf5* locus in a *Myod* null background is not sufficient to establish the skeletal muscle lineage in mouse embryos (Wang & Jaenisch, 1997), indicating that *Myod* and *Myf5* have different instrinsic functions to *Myog*, rather than simply different temporal expression patterns. Interestingly, conditional inactivation of *Myog* at late fetal and perinatal stages does not result in a major loss of muscles; however, it is required to allow sufficient muscle development for survival until birth (Knapp et al., 2006; Meadows et al., 2008). Taken together, these studies argue for distinct roles for the *Mrfs* although they can functionally compensate for each other in some cases.

3.4. Upstream regulators

As seen in Section 3.3, the *Mrfs* control the entry into the myogenic program, which culminates in the formation of skeletal muscle. However, upstream of these critical regulators, other transcription factors direct cells toward myogenesis. We will summarize here the respective roles of the main upstream regulators in the head and trunk (see also Table 1.1).

Pax3 and *Pax7* belong to the paired/homeodomain family of transcription factors that play key roles during development (see Buckingham & Relaix, 2007). Unlike the Mrfs, their expression is restricted not only to the skeletal muscle but also to the neural crest and central nervous system; however, they act as critical upstream regulators of muscle. *Pax3* marks migrating progenitor cells from the ventrolateral lip of the DM to the limbs, the diaphragm, and the hypoglossal cord (Bober et al., 1994; Franz, Kothary, Surani, Halata, & Grim, 1993; Goulding, Lumsden, & Paquette, 1994; Tajbakhsh et al., 1997; Tremblay et al., 1998). In *Pax3* null (Splotch; spontaneous mutation in *Pax3*) mice, the hypaxial somite is lost by apoptosis and as a result the formation of limb and some trunk muscles is impaired. Epaxial-derived muscles are much less affected showing that epaxial and hypaxial muscles do not have the same requirements for *Pax3* (Franz et al., 1993; Relaix et al., 2004; Tajbakhsh et al., 1997). In these mice, initial muscle formation occurs in the myotome by *Myf5*-expressing cells, and head muscles are not affected.

Pax3 transcription is downregulated during fetal myogenesis when the *Pax3* paralog *Pax7* becomes a common signature of all myogenic stem/progenitor cells. Notably, *Pax7* expression identifies these cells throughout development starting from mid-embryogenesis through adulthood as it marks all adult satellite cells irrespective of the anatomical location (see Relaix & Zammit, 2012; Tajbakhsh, 2009). Mutants of *Pax7* do not show overt muscle defects during embryonic development (Mansouri et al., 1996), and the consequence of *Pax7* absence is only revealed at late fetal and postnatal stages (Hutcheson, Zhao, Merrell, Haldar, & Kardon, 2009; Seale et al., 2000). Gene replacement studies have demonstrated that *Pax7* can replace *Pax3* function in neural crest cells, dorsal neural tube, and trunk muscles. On the contrary, *Pax7* cannot compensate for *Pax3* function in the delamination and migration of limb muscle progenitors (Relaix et al., 2004); however, this point would need to be re-evaluated after removal of extraneous sequences that could affect *Pax7* expression levels from the targeted *Pax3* locus. *Pax3:Pax7* double KO mice display relatively normal early embryonic muscles in the early myotome and the head. This is consistent with the view that the first muscles to differentiate are *Pax3:Pax7* independent (see Fig. 1.2). However, *Pax3:Pax7* double KO mice completely lack further muscle development in the trunk as muscle progenitor cells assume other cell fates or die by apoptosis (Kassar-Duchossoy et al., 2005; Relaix et al., 2005).

Myf5(Mrf4):Pax3 compound mutants fail to make somite-derived muscles (Tajbakhsh et al., 1997); however, in these embryos, branchial-arch myogenesis proceeds normally. This was the first evidence suggesting that *Myod* is activated by Pax3-independent pathways in the head. *Pax3* can directly regulate *Myf5* expression in the limb, but this regulation is not a general feature further underscoring the notion that genetic hierarchies vary depending on anatomical location (Bajard et al., 2006; Sambasivan et al., 2009; Tajbakhsh & Cossu, 1997).

Six proteins also play important upstream roles in myogenesis. They form complexes with their transcriptional coactivators, eyes absent (Eya) proteins to stimulate transcription (Kawakami, Sato, Ozaki, & Ikeda, 2000). Double mutants for *Six1/4* or *Eya1/2* do not express *Pax3* in the hypaxial DM, indicating that these genes lie upstream of *Pax3* in the genetic hierarchy of hypaxial myogenesis. As a result, double mutants display loss of all muscles derived from the hypaxial DM including limb and a subset of trunk muscles (Grifone et al., 2005, 2007). *Six/Eya* directly regulate enhancer elements of *Myf5* and *Myod*. Specifically, *Six1/4* directly regulate the −57.5 kbp enhancer element that controls *Myf5* activation in the limbs and mature hypaxial DM (Giordani et al., 2007). However, the early activation of *Myf5* in the epaxial somite does occur and deep back muscles can be detected, in these double-mutant mice (Grifone et al., 2005). Key enhancer sequences of the *Myod* gene (CE and DRR, see below) control *Myod* expression through the recruitment of Six1/4 and also Six2 transcription factors at all sites of myogenesis *in vivo* (Relaix et al., 2013). Mutation of these Six (Mef3)-binding sites severely compromised expression of a transgene driven by the two enhancer sequences and the *Myod* proximal promoter sequences, demonstrating direct regulation of *Myod* by Six/Eya complexes in the embryo, both in the trunk and in the head. In the absence of *Six1/4*, *Myf5 (Mrf4)* is required for *Myod* activation in the trunk (Relaix et al., 2013). However, *Myod* expression and head myogenesis is preserved in *Six1/Six4:Myf5* triple mutant embryos, illustrating that upstream regulators of *Myod* are distinct in different embryonic locations. Finally, unlike *Pax3*, *Six/Eya* are also expressed at the onset of differentiation where they directly regulate the activation of *Myog* (Grifone et al., 2005; Spitz et al., 1998) and later where they regulate the expression of sarcomeric proteins (Niro et al., 2010; Richard et al., 2011). Recent studies based on genomic approaches evoke a model whereby Six proteins act not only upstream of the *Mrfs* during embryonic development but also parallel to them in a cooperative manner for terminal differentiation of myoblasts *in vitro* (Liu, Chu, Chakroun, Islam, & Blais, 2010). This effect on muscle differentiation distinguish Six/Eya from Pax regulation of myogenesis; *Pax3* and *Pax7* play a critical role in controlling the upstream events leading to myogenesis, and *Pax7* appears to directly repress genes required for differentiation (Soleimani, Punch, et al., 2012).

As *Pax3* does not regulate head myogenesis, the focus had been on other transcription factors. Several factors are expressed in regionalized domains of cranial mesoderm before Mrf expression (Bothe, Tenin, Oseni, & Dietrich, 2011; Bothe and Dietrich, 2006), and some will be discussed here.

Pitx2 is a bicoid-related homeodomain transcription factor expressed early in cranial mesoderm and it is a primary regulator of EOM specification (Gage, Suh, & Camper, 1999; Kitamura et al., 1999; Zacharias et al., 2011). EOMs are absent in *Pitx2*-mutant embryos as *Pitx2* is required cell autonomously for cell survival in the premyogenic mesoderm. After this initial requirement for cell survival, inducible mesodermal inactivation *Pitx2* showed that this gene is required for activation of *Myf5* and *Mrf4* expression, and therefore *Myod*, in the forming EOM (Zacharias et al., 2011). Nevertheless, in the absence of both Myf5 and Mrf4, Pitx2 cannot ensure survival and activation of Myod expression (Sambasivan et al., 2009). Being a regulator of cell survival and myogenic specification, *Pitx2* assumes the role that *Pax3* has at sites of myogenesis in the body. Interestingly, extraocular myogenesis is sensitive to *Pitx2* dosage, as *Pitx2* heterozygous mutant mice also exhibit defects in EOM morphogenesis and *Mrf* expression (Diehl et al., 2006). It has been recently shown that EOM express higher levels of *Pitx2* than limb muscles and that high levels of *Pitx2* were retained in dystrophic and aging mouse EOM, which could help explain the sparing of the EOM in muscular dystrophies (Hebert, Daniel, & McLoon, 2013).

In the first BA, inactivation of *Pitx2* results in increased cell death in the mesodermal core and loss of the premyoblast specification markers *Tbx1*, *Musculin*, *Tcf21* and *Six2*. As a result, expression of *Myf5*, *Myod*, and *Myog* are severely compromised in the first arch (Dong et al., 2006; Shih, Gross, & Kioussi, 2007). These observations suggest that *Pitx2* acts genetically upstream of these factors in the BA and also that has a survival role in both EOM and BA-derived muscles. At the trunk level, in the absence of *Pitx2*, the onset of *Myod* expression is affected in the limb but not in the myotome, where *Myf5* and *Mrf4* play a predominant role. The initial downregulation of *Myod* in the limb is later compensated by *Myf5* and *Pitx2* has been shown to directly activate *Myod* in the limb through the CE enhancer element (L'Honore et al., 2010). Nonetheless, *Pitx2*, *Myf5*, and *Mrf4* together contribute to *Myod* expression in both the limb and myotome as revealed by their compound mutants. Skeletal muscle does not form in *Myf5(Mrf4):Pitx2* compound mutant similar to the phenotype of *Myf5(Mrf4):Pax3* embryos, suggesting that *Pitx2* lies genetically downstream of *Pax3*.

The transcriptional repressors *Musculin* (*Msc/MyoR*) and *Tcf21* (*Epicardin/Capsulin/POD-1*) are also bHLH transcription factors, known to be markers of undifferentiated muscle precursor cells in the first arch (von Scheven et al., 2006). Although mice lacking either *Musculin* or *Tcf21* show normal branchyomeric muscle development, *Musculin:Tcf21* double-mutant mice fail to activate *Mrf* expression in the first branchial arch. This results in elevated cell death in the cells that failed to enter the myogenic program and absence of the major mastication muscles (Lu et al., 2002). This phenotype mirrors *Myf5:Myod* KO mice where branchiomeric myogenic progenitors are not determined and skeletal muscle is not formed (Kassar-Duchossoy et al., 2004). More recently, *Musculin* and *Tcf21* were shown to control the transcriptional levels of *Myf5* and *Myod* in the branchial arches through binding to specific enhancers (Moncaut et al., 2012).

Tbx1 is a T-box factor expressed in the BA before the activation of the *Mrfs* (Kelly et al., 2004) and required for the formation of BA-derived muscles. It is a major candidate gene for Di George Syndrome, a disorder characterized by craniofacial and cardiovascular defects (Jerome & Papaioannou, 2001). Its inactivation results in severely atrophied second BA and the sporadic unilateral development of first BA-derived muscles (Kelly et al., 2004). In addition, all caudal arch-derived muscles including the trapezius and sternocleidomastoid muscle are missing in *Tbx1* null embryos, while somite-derived muscles such as those in the tongue are present (Kelly et al., 2004; Theis et al., 2010). A mesodermal-specific KO of *Tbx1* shows that the requirement for *Tbx1* in BA myogenesis is cell autonomous (Aggarwal et al., 2010; Dastjerdi et al., 2007; Kong et al., 2014); however, it is not clear whether the phenotype of the mutant mice results from increased cell death (Grifone et al., 2008; Kong et al., 2014). In the absence of *Tbx1*, stochastic and low-level activation of the *Mrfs* appear to result in the phenotypically normal, although hypoplastic, unilateral first arch-derived muscles observed at late fetal stages. Therefore, *Tbx1* ensures robust bilateral specification at the earliest stages of myogenesis in the BAs, and muscles that escape this requirement subsequently develop normally (Grifone et al., 2008; Kelly, 2013). The sporadic myogenesis observed in *Tbx1*-null embryos is virtually lost in *Tbx1:Myf5* double KO embryos, suggesting that *Tbx1* acts synergistically with *Myf5* to assure robust *Myod* expression and myogenic cell fate (Sambasivan et al., 2009).

Isl1 is a marker of the splanchnic mesoderm. Tracing studies using *Isl1*^{Cre} labeled facial and BA-derived muscles to varying extents, with a stronger labeling in distal first arch-derived muscles (Harel et al., 2009). Interestingly, *Isl1*^{Cre} also labeled neck muscles derived from occipital-level somites, suggesting a shared transcriptional program with BA-derived muscles (Harel et al., 2009; Theis et al., 2010). Nevertheless, as *Isl1* mutants die before head muscle specification (at around E10), the direct role of *Isl1* in pharyngeal muscle development still needs to be resolved (Cai et al., 2003).

Lhx2 was identified recently as another regulator of BA muscle development (Harel et al., 2012). The *Lhx2* KO results in defects in pharyngeal muscle specification and head muscle patterning. *Lhx2* acts genetically downstream of *Tbx1*. *Tbx1:Lhx2* compound mutants completely lack pharyngeal arch muscles and *Tbx1* and *Lhx2* synergistically regulate heart morphogenesis.

3.5. Temporal specification of myogenesis

As tissuegenesis proceeds, temporal classes of precursors lead to differentiated cells characteristic of developmental stages, as with skeletal muscle where embryonic and fetal myoblasts make primary and secondary myofibres (Fig. 1.2A). A major finding that demarcated embryonic from fetal myogenesis was the discovery of *Nfix* as a critical regulator of the fetal myogenic program (Messina et al., 2010). *Nfix* mutants lack fetal myogenesis, in spite of its low level of expression also in embryonic myogenic cells. In other studies, using a cell ablation approach in which cells that express *Pax3* were specifically deleted, it was shown that *Pax3*-expressing progenitors give rise to all Pax7+ cells and are required for embryonic and fetal myogenesis in the limb (Hutcheson et al., 2009). This same approach demonstrated that Pax7+ progenitors are required to establish fetal but not embryonic myogenesis in the limb and give rise to all satellite cells.

Therefore, *Nfix* and to some extent *Pax3* and *Pax7* can define the temporal specification of embryonic and fetal myogenesis. The signaling pathways that allow a continuum of stem cells to generate temporally different progeny were only recently uncovered. Interestingly, the myogenic stem cell population requires the presence of differentiating cells for their maintenance, where a lack of differentiated cells results in the loss of upstream Pax7+ cells in the fetus (Kassar-Duchossoy et al., 2005). We recently showed that constitutive Notch activity, via expression of the transcription factor NICD (Notch intracellular domain, cleaved receptor) is sufficient to autonomously maintain and self-renew muscle stem/progenitor cells, in spite of the complete abrogation of committed and differentiated cells (Mourikis, Gopalakrishnan, Sambasivan, & Tajbakhsh, 2012). Remarkably, this myogenic population undergoes correct temporal specification—when isolated from the embryo or fetus, as myogenic cells express markers specific for those stages. They subsequently leave the cell cycle, in spite of continued NICD expression, and are surrounded by laminin, in the absence of differentiated myofibers. These findings indicate that Notch is permissive, allowing the stem cell population to autonomously interpret developmental stage signals and mature accordingly. During normal development, the Notch ligand delta-like 1 (Dll1) is upregulated in differentiating myogenic cells, thereby signaling back to upstream cells via Notch activity to assure their maintenance (Delfini, Hirsinger, Pourquie, & Duprez, 2000; Hirsinger et al., 2001; Mourikis, Sambasivan, et al., 2012; Schuster-Gossler, Cordes, & Gossler, 2007), similar to progenitors in the nervous system (Dong, Yang, Yeo, Chitnis, & Guo, 2012; Henrique et al., 1997). During myogenesis, therefore, parental upstream cells are maintained by their committed daughters through Notch signaling allowing them to adopt their respective stage-specific phenotypes.

Pax7 is used extensively as a marker of adult satellite cells, the major adult stem cell population responsible for muscle regeneration (Lepper & Fan, 2010; Lepper et al., 2011; Murphy, Lawson, Mathew, Hutcheson, & Kardon, 2011; Sambasivan, Yao et al., 2011; Seale et al., 2000; Zammit, 2008), and it marks their ancestors from mid-embryogenesis (Hutcheson et al., 2009; Kassar-Duchossoy et al., 2005; Lepper & Fan, 2010; Relaix et al., 2004, 2005). In contrast, *Pax3* is expressed in satellite cells in a subset of muscles such as the diaphragm and some limb muscles (Relaix et al., 2006), but not in the head (Tajbakhsh et al., 1997). Prior to birth, *Pax7* does not appear to be essential for myogenesis, presumably due to compensation by *Pax3* in the limbs and trunk. However, after birth, *Pax7* null mice exhibit loss of satellite cells leading to progressive muscle wasting even in muscles where Pax3 is normally expressed such as the diaphragm (Kuang et al., 2006; Oustanina et al., 2004; Relaix et al., 2006; Seale et al., 2000). These differences could result from divergent functions between *Pax3* and *Pax7* postnatally (Soleimani, Punch, et al., 2012). Thus, Pax7 plays a critical role in specifying juvenile satellite cells perinatally. The role of Pax7 in quiescent satellite cells has been less clear. Although it was initially believed that *Pax7* does not play a role in adult satellite cells (Lepper et al., 2009), since the satellite cell population was still present and muscle could regenerate even in absence of both *Pax3* and

Pax7, more detailed analysis showed that conditional inactivation of *Pax7* leads to the gradual loss of satellite cells (Gunther et al., 2013; von Maltzahn et al., 2013). This could result from premature differentiation of the satellite cell pool at the expense of proliferation (Gunther et al., 2013; von Maltzahn et al., 2013) or perhaps through the pioneer functions already suggested for this transcription factor in other tissues (Budry et al., 2012). Therefore, temporal specification of postnatal myogenic cells can also be associated with Pax7 function.

3.6. Enhancer interplays in regulation of the Mrfs

Expression of the Mrf proteins during development is under strict temporal and spatial control. In the last two decades, transgenic analysis was used to identify the *cis*-regulatory elements of the *Myf5/Mrf4*, *Myod*, and *Myog* loci. These studies showed that *Myf5* regulation is remarkably complex, driven by a combination of promoter and enhancer elements that span 150 kbp, buried in intronic sequences and adjacent genes (reviewed in Carvajal & Rigby, 2010). Myf5 activation during the first wave of myogenesis depends on an early epaxial enhancer (EEE), directly controlled by Wnt and Shh signaling from the neural tube and the midline through Tcf- and Gli-binding sites. Later embryonic expression depends on a region at −58/−48 kbp. Within this sequence, a 145-bp element alone directs expression in myogenic cells of mature somites and limb buds. This activity depends directly on *Pax3* and is modulated by *Six1/4* (see Lagha, Sato, et al., 2008). *Mrf4* and *Myf5* coding sequences are separated by only 9 kbp, but they are regulated in a remarkably specific fashion. A new type of element, termed TRABS (transcription balancing sequences), acts to regulate the equilibrium between the enhancers and promoters of *Myf5* and *Mrf4* in order to establish specificity (Carvajal, Keith, & Rigby, 2008). Finally, using genome alignment to compare noncoding evolutionary conserved regions in vertebrate species, two recent studies identified new regulatory elements within the *Myf5/Mrf4* locus responsible for driving specific *Myf5* expression in the ventral somite and in the branchial arches during development (Moncaut et al., 2012; Ribas et al., 2011). In addition, a −111 kbp element upstream of *Myf5* (ECR111) is a conserved *Pax7*-dependent enhancer, which is required for *Myf5* expression in adult quiescent satellite cells (Ribas et al., 2011; Soleimani, Punch, et al., 2012).

Myod expression is largely regulated by two enhancers, the core enhancer located at −20 kbp (CE) and the distal regulatory region (DRR) located at −4 kbp (Asakura, Lyons, & Tapscott, 1995; Chen & Goldhamer, 2004; Faerman, Goldhamer, Puzis, Emerson, & Shani, 1995; Goldhamer, Faerman, Shani, & Emerson, 1992; Tapscott, Lassar, & Weintraub, 1992). The CE enhancer directs gene expression in somitic precursors during embryogenesis and the DRR enhancer does so in early differentiated myogenic cells (see Pownall, Gustafsson, & Emerson, 2002). While genetic studies suggest that *Pax3* and *Pax7* are potential upstream regulators of *Myod* during both embryonic and postnatal myogenesis (Bober et al., 1994; Goulding et al., 1994; Maroto et al., 1997; Relaix et al., 2005; Seale et al., 2000), the direct activation of *Myod* transcription by *Pax3* and *Pax7* has not been demonstrated. Recently though, two *Pax7*-binding sites were found in conserved intergenic regions upstream of *Myod* (Soleimani, Punch, et al., 2012). In addition, Pax3/7 and FoxO3 appear to work cooperatively on the Myod promoter for regulating cell-type-specific transcriptional activation in postnatal muscle cells (Hu, Geles, Paik, DePinho, & Tjian, 2008).

Little is known about the regulation of *Pax3/7* during myogenesis. Apart from a number of *Pax3* regulatory elements driving expression in neural crest and dorsal neural tube (Degenhardt et al., 2010; Li, Chen, & Epstein, 2000; Milewski et al., 2004; Moore et al., 2013), an enhancer element driving hypaxial somite expression has been identified upstream of *Pax3* (Brown, Engleka, Wenning, Min Lu, & Epstein, 2005). In addition, a *Pax3-Cre* transgenic mouse using the proximal 1.6 kbp of the mouse *Pax3* promoter combined with an ubiquitous Rosa-reporter showed a rostrocaudal gradient of skeletal muscle expression, although this did not recapitulate complete endogenous *Pax3* expression. (Jarad & Miner, 2009). In the case of *Pax7*, distinct upstream and intronic enhancer elements have been identified that direct expression in CNC, facial mesenchyme, mesencephalon (Lang et al., 2003); however, the regulatory enhancers for the myogenesis in the embryo and satellite cells have not yet been identified.

3.7. Lessons from lineage tracing and cell ablation studies—a cautionary tale

Cell ablation studies are a powerful approach to identify lineage relationships of subpopulations of cells within a tissue. For example, DTA-mediated ablation of adult stem cell populations has been informative for skeletal muscle (Lepper et al., 2011; Murphy et al., 2011; Sambasivan, Yao, et al., 2011) and the small intestine (Tian et al., 2011). The Cre/loxP system has been used to follow the genetic lineage of Pax3+, Pax7+, or Mrf+ cells and to test the requirement of particular cell populations for myogenesis (reviewed extensively in Murphy & Kardon, 2011). This is achieved by crossing the *Cre* lines of the gene of interest with either Cre-responsive reporter mice (such as the $R26^{YFP}$, $R26^{lacZ}$, $R26^{mTmG}$ reporters) or with Cre-responsive ablator lines that kill autonomously *Cre*-expressing cells (see Hutcheson & Kardon, 2009). It is worth highlighting that whereas *Cre* is expressed in temporally and spatially defined patterns, recombination of a Cre-dependent reporter is an on/off readout in which only the progenitors having reached an appropriate threshold of *Cre* expression can be activated (see Ma, Zhou, & Pu, 2008).

Ablation of *Myf5*-expressing cells resulted in virtually normal myogenesis, leading to a model where a Myf5-independent/Myod+ subpopulation of myogenic cells rescues the myogenic program by mid-embryonic and fetal stages (Gensch, Borchardt, Schneider, Riethmacher, & Braun, 2008; Haldar, Karan, Tvrdik, & Capecchi, 2008). For this model to be valid, both the $Myf5^{Cre}$ and $R26^{DTA}$ lines need to be working with virtual 100% efficacy to eliminate the possibility of *Myf5*-expressing escaper cells contributing to myogenesis (see below).

In contrast to the ablation of *Myf5* expressing progenitors, a recent report showed that DTA-mediated ablation of *Myod*-expressing cells results in complete loss of differentiated muscle at all anatomical locations by E12.5 (Wood, Etemad, Yamamoto, & Goldhamer, 2013). Concomitant loss of Myf5+ progenitors in $MyoD^{iCre}$:$R26^{DTA}$ embryos indicates that the vast majority of Myf5+ progenitors also expressed Myod during their developmental history. Loss of Pax7+ progenitors in *Myod* ablated embryos can be explained by the fact that *Myod* expression precedes that of *Pax7* in limb muscle progenitors (Chen & Goldhamer, 2004; Relaix et al., 2004; Tajbakhsh et al., 1997) and also non-cell autonomously due to loss of Notch ligand Dll1 from the committed progeny (Kassar-Duchossoy et al., 2005; Mourikis, Gopalakrishnan, et al., 2012). Thus, the vast majority of myogenic progenitors transit through a Myod+ stage and myogenesis cannot be rescued by a *Myod*-negative population of myogenic progenitors. This suggests that if a Myod−/Myf5+ and/or a Myod−/Mrf4+ progenitor population existed, it would be incapable of supporting robust myogenesis in *Myod*-lineage ablated embryos. The results obtained from *Myf5* and *Myod*-lineage ablation are difficult to reconcile; while these results do not formally exclude the existence of a *Myod+/Myf5*-independent lineage, they underscore the importance of clearly demonstrating that the genetic tools employed perform as advertised. In this respect, whereas the loss of *Cre*-dependent reporter gene expression was shown in *Myf5*^{Cre}-ablated embryos (Gensch et al., 2008; Haldar et al., 2008), as both DTA and reporter gene expression are dependent on *Cre*, these observations do not prove that all myogenic cells that expressed endogenous *Myf5* were truly ablated. Indeed, we showed recently with new generation reporter mice that Myf5-escaper cells masquerade as a Myf5-independent/Myod+ population following DTA-mediated ablation of *Myf5*-expressing cells (Comai et al., in press).

Ablation of *Myog*-expressing cells leads to complete loss of muscle by birth, consistent with the finding that all fetal muscle has expressed *Myog* (Gensch et al., 2008; Li et al., 2005). Ablation of the *Mrf4* lineage does not significantly compromise early embryonic myogenesis; however by birth, differentiated myofibers show signs of apoptosis and altered histology (Haldar et al., 2008). Haldar and colleagues assumed a primary role for *Mrf4* in differentiated cells of the myogenic lineage, although *Mrf4* is also clearly a determination gene acting in myogenic progenitors in the embryo (Kassar-Duchossoy et al., 2004). Furthermore, Pax7+ muscle progenitors should have been also ablated in *Mrf4*^{Cre}*:R26*^{DTA} mice as all of these progenitors have been normally primed by *Mrf4* expression by the end of the embryonic phase (Sambasivan et al., 2013). Inadequate expression of the Cre/reporter combination might explain some of these discrepancies.

In conclusion, lineage tracing and ablation studies are commonly used tools to reveal the lineage relationships between progenitors and their descendants within a tissue. However, for proper interpretation of results, one must ensure that: (i) *Cre* expression faithfully follows that of the endogenous gene, (ii) the reporter line is responsive in the tissue of interest, and (iii) the reporter line reflects the full spectrum of *Cre* and endogenous gene expression. Common issues with both *Cre* and reporter lines have been raised elsewhere (Liu et al., 2013; Sambasivan et al., 2013; Tajbakhsh, 2009).

4. MRFS AS TRANSDIFFERENTIATION FACTORS

4.1. *In vitro* studies provide a new view on transdifferentiation

Myogenesis is an excellent paradigm for examining how cell fate is induced. It has long been hypothesized that prior to the acquisition of a differentiated phenotype, the process of cell commitment relies on the expression of a small number "master" regulatory or master "switch" genes. Pioneering studies by Harold Weintraub and colleagues demonstrated that the constitutive expression of a single transcription factor, *Myod*, is sufficient to convert fibroblast cells into skeletal muscle cells, bypassing normal developmental lineage progression (Davis et al., 1987). It was subsequently shown that a number of differentiated cell types, including fat, liver, chondroblasts, retinal pigmented cells, smooth muscle, melanoma, and neuroblastoma lines, could also be diverted to myogenesis following forced expression of *Myod* (Choi et al., 1990; Weintraub et al., 1989). Overexpression of *Myod* was not only able to promote the expression of myofibrillar proteins but also to inhibit, in most but not in all cases, the ongoing cellular program. This property was extended to the other *Mrfs* (*Myf5*, Braun, Buschhausen-Denker, Bober, Tannich, & Arnold, 1989; *Mrf4*, Braun, Bober, Winter, Rosenthal, & Arnold, 1990; Miner & Wold, 1990; Rhodes & Konieczny, 1989) and, to a certain extent, to *Myog* (Edmondson & Olson, 1989; Wright, Sassoon, & Lin, 1989). Remarkably, these studies provided the first demonstration of cell-fate "reprogramming" or "transdifferentiation" by a single transcription factor; however, the usefulness of these approaches for regenerative medicine had remained largely neglected until the generation of induced pluripotent stem (iPS) cells (see below). While *Myod* was capable of cell-fate reprogramming in a variety of "permissive" cell lines derived from the three embryonic layers, it is intriguing that a number of differentiated and undifferentiated cell types resisted myogenic conversion (Fig. 1.3A). These "refractory" cell lines include endodermal HepG2 cells (Schafer, Blakely, Darlington, & Blau, 1990; Weintraub et al., 1989), HeLa cells (Weintraub et al., 1989), the P19 pluripotent mouse embryonal carcinoma line (Armour, Garson, & McBurney, 1999; Pari, Jardine, & McBurney, 1991; Skerjanc, Slack, & McBurney, 1994), and undifferentiated human and mouse embryonic stem (ES) cell lines (Albini et al., 2013; Dekel et al., 1992). With novel perspectives provided by the post-Yamanaka reprogramming era, it is worthwhile revisiting some of these studies.

First, in the original transdifferentiation studies, overexpression levels were not assessed in detail, and the readout was often transcript analysis of myogenic markers or immunofluorescence of differentiated cells using a limited number of markers. With our current knowledge on how partial cell-fate changes can occur in some circumstances (Ferrari et al., 1998; Lapidos et al., 2004; Tedesco, Dellavalle, Diaz-Manera, Messina, & Cossu, 2010),

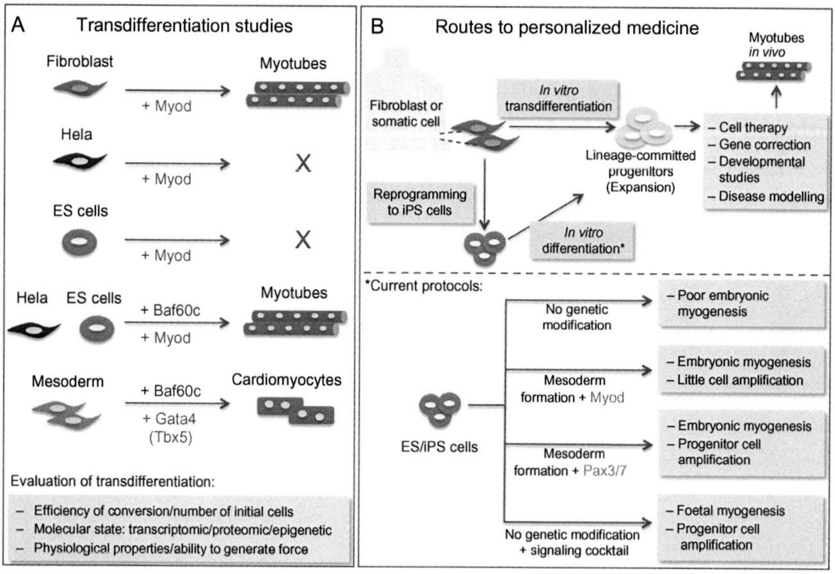

Figure 1.3 Myogenic conversion and routes to personalized medicine. (A) The hallmark of the reprogramming era began with the transdifferentiation of multiple cell types to skeletal myogenesis and cell-fate conversion with *Myod* (Davis et al., 1987). Some cells remained refractory to myogenic conversion (HeLa, ES cells) (Albini et al., 2013; Dekel, Magal, Pearson-White, Emerson, & Shani, 1992; Weintraub et al., 1989). More recently, chromatin modifiers (e.g., Baf60c) were found to render some of these cells permissive to conversion (Albini et al., 2013; Forcales et al., 2012; Takeuchi & Bruneau, 2009). This underscores the notion that rendering the chromatin accessible should be considered in concert with the choice of transcription factor(s) for directing the desired cell fate. In addition, a combination of parameters should be routinely evaluated to assure full cell-fate conversion. (B) Schematic representation of some of the approaches under investigation to produce new muscle tissue for therapeutic use (upper panel). Patient fibroblasts can be directed to transdifferentiate *in vitro* into lineage-committed myogenic cells directly or via a reprogramming step through iPS cells. The generation of lineage-committed progenitors has a number of advantages: reduced chance of teratoma formation, minimization of alternative cell fates and ideally, the potential to specifically generate the desired type of muscle progenitors (location specific, slow vs. fast fibers). The conversion of pluripotent ES or iPS cells to myogenesis has been a major obstacle (lower panel). Although different strategies have now made it possible to amplify skeletal muscle stem/progenitor cells, this is achieved by genetic manipulation of the ES/iPS cells. Furthermore, fetal/adult myogenesis has not been systematically generated in these *in vitro* differentiation assays (see Section 4.3 for reference). (See the color plate.)

it remains possible that some of the cell types that were phenotypically scored for myogenesis were not fully transdifferentiated. Therefore, those experiments need to be revisited using genome-wide analyses to assess whether full or partial reprogramming was achieved by *Mrf* overexpression and to identify potential regulators that might promote or act as barriers to myogenesis. Second, a potential caveat in those studies was their reliance on cell lines as opposed to primary cells. It has been suggested that continuous maintenance of cell lines in culture could result in changes in lineage identity, for example, epithelial-to-mesenchymal transition (EMT), and this could greatly potentiate the ability of *Myod* to induce myogenic differentiation in some cell lines (Fong & Tapscott, 2013). Similarly, while a few reports showed apparent direct conversion of ES cells or iPS cells into the myogenic lineage after the conditional expression of Myod (Ozasa et al., 2007; Tanaka et al., 2013; Thoma, Maurus, Wagner, & Schartl, 2012), most studies reported the generation of myoblasts from ES cells only after transition through a "mesodermal stage" (e.g., embryoid body derivation or mesenchymal derivatives of ES cells). This was most efficiently achieved by induced expression of *Pax3/Pax7* (Darabi et al., 2012, 2008), *Myf5* (Iacovino et al., 2011), or *Myod* (Goudenege et al., 2012; Rao et al., 2012) (see Section 4.3). A major breakthrough was reported recently, where high efficiency of paraxial mesoderm and skeletal myogenesis was reported from ES cells, even in the absence of genetic modifications (J. Chal and O. Pourquie, personal communication) and where the developmental program *in vivo* was recapitulated *in vitro*.

Finally, the chromatin signature of a cell can no longer be ignored in transdifferentiation studies (Zhou & Melton, 2008). The observation that simple overexpression of *Myod* has limited ability to induce muscle differentiation in undifferentiated ES cells and other differentiated cell types could be the result of a nonpermissive "nuclear landscape." Early studies using heterokaryons via cell–cell fusion showed that despite the high level of *Myod* overexpression achievable in hepatocytes, these cells would not convert into myoblasts except after fusion with fibroblasts (Schafer et al., 1990). Cell complementation proved to be a useful model to assess how the differentiated state and associated nuclear landscape determine cell fate. In keeping with this notion, an elegant study showed that the selective absence of *BAF60C* is responsible for the resistance of human ES cells to *Myod*-mediated activation of skeletal myogenesis (Albini et al., 2013). *BAF60C* is a key structural component of the SWI/SNF chromatin remodeling complex that is known to enable *Myod*-dependent gene expression (de la Serna et al., 2001, 2005) (see also Section 5.2). While all three variants of the BAF60 subunits (i.e., BAF60A, BAF60B, and BAF60C) are expressed in human skeletal myoblasts and fibroblasts, BAF60C was the only member missing in human ES cells. Forced expression of the *BAF60C2* isoform prior to *Myod* overexpression enabled the direct conversion of human ES cells into skeletal myoblasts without the need of a transition through the mesodermal stage (Albini et al., 2013) (Fig. 1.3A). Accordingly, the *BAF60C2* isoform is upregulated during embryoid body formation (Albini et al., 2013), a permissive stage for the activation of skeletal myogenesis (Iacovino et al., 2011). These findings suggest that repression of *BAF60C2* in undifferentiated human ES cells might be a necessary event to maintain pluripotency, presumably to prevent a permissive chromatin state for cell commitment (Albini et al., 2013). Similarly, BAF60C deficiency was shown to be responsible for the resistance of HeLa cells to *Myod*-mediated myogenic conversion (Forcales et al., 2012). These observations provide a link between resistance to myogenic conversion and chromatin state and underscore the importance of remodeling chromatin for transcription factor accessibility (see Section 5.2).

Experimentally, induced exit from the myogenic state has also provided insights into the acquisition of cell fate. Here, in the context of reprogramming myogenic cells to the pluripotent state, suppression of *Myod* expression by Oct4 is required to initiate the reprogramming step for the generation of iPS cells (Watanabe et al., 2011). The chromatin status and associated factors remain to be explored in this paradigm. It would be interesting to assess the expression of chromatin remodelers such as *BAF60C* during the generation of iPS cells, and in other cell lines that are "permissive" or "refractory" to myogenic conversion by *Myod* overexpression.

4.2. Transdifferentiation *in vivo*

Mrfs can also initiate ectopic myogenesis *in vivo*, though with variable success rates depending on the organ or cell type. Studies in *Caenorhabditis elegans* showed that induction of CeMyoD (HLH-1) is sufficient to convert virtually all somatic cells into a body wall muscle-like fate; the capacity of myogenic conversion by *hlh-1* declines rapidly after the first hours of development and is completely lost in differentiated cells (Fukushige & Krause, 2005). Earlier studies with *Xenopus* and mouse embryos reported that overexpression of *Myf5* and *Myod* in ectodermal and mesodermal lineages, including the heart, leads to expression of early muscle differentiation markers; however, full skeletal muscle differentiation was not achieved (Faerman, Pearson-White, Emerson, & Shani, 1993; Hopwood & Gurdon, 1990; Hopwood, Pluck, & Gurdon, 1991; Miner, Miller, & Wold, 1992; Santerre et al., 1993). Nevertheless, ectopic expression of *Myf5* and *Myod* lead to apparently complete skeletal myogenesis (including classical sarcomeric organization of normal muscle fibers) in *Xenopus*, the mouse brain (Ludolph et al., 1994; Santerre et al., 1993) and chick neural tube (Delfini & Duprez, 2004). In this context, the cell of origin (mesenchymal, neuronal, etc.) was not clearly defined in those studies. As discussed for the *in vitro* experiments, these studies also underscore the importance of cell history and context in determining whether, and to what extent, ectopic expression of a transcription factor, such as a myogenic regulator, can fully reprogram cell types to the desired fate (Zhou & Melton, 2008).

Several explanations can account for incomplete dominant myogenic activities of *Myf5* and *Myod* in embryos. *Mrf* function might be inhibited in nonmyogenic cells by the absence of essential mesodermal coregulators, or by a transcriptional regulatory network that maintains Mrf target genes in a repressed state. For example, it has been shown that ectopic induction of *hlh-1* is able to convert mitotic germ cells into muscle only upon removal of the histone 3 lysine 27 (H3K27) methyltransferase Polycomb repressor complex 2 (PRC2), suggesting that PRC2 normally protects the germline from the aberrant execution of somatic differentiation programs (Patel, Tursun, Rahe, & Hobert, 2012). In addition, as indicated above, the selection of specific structural variants of the SWI/SNF chromatin remodeling complex is important to promote chromatin accessibility at lineage-specific loci and determine specific lineages (reviewed in Ho & Crabtree, 2010; Lessard & Crabtree, 2010). Examples of this are the requirement of specific SWI/SNF variants during neuronal development (Lessard et al., 2007; Wu et al., 2007; Yoo, Staahl, Chen, & Crabtree, 2009) as for the induction of ectopic cardiogenic activity in embryonic mesoderm in conjunction with *Gata4* (Takeuchi & Bruneau, 2009). Therefore, it is also possible that the availability of the chromatin remodeling protein BAF60C or other specific SWI/SNF variants limit the efficiency of *Mrf*-mediated transdifferentiation *in vivo*.

4.3. From transdifferentiation studies to in-dish models of disease

Building on the discovery that *Myod* expression can reprogram fibroblasts to myogenesis, other groups succeeded in using different combinations of transcription factors to directly convert fibroblasts into neurons, cardiomyocytes, blood cell progenitors, and hepatocytes *in vitro* (Han et al., 2012; Huang et al., 2011; Ieda et al., 2010; Pang et al., 2011; Sekiya & Suzuki, 2011; Szabo et al., 2010; Vierbuchen et al., 2010) and even reprogramming of endogenous fibroblasts into functional cardiomyocytes *in vivo* (reviewed in Addis & Epstein, 2013). In addition, transcription factors can induce direct conversions of one specialized cell type into another, such as conversion of hepatocytes into neurons (Marro et al., 2011) or pancreatic exocrine cells into beta cells (Zhou et al., 2008). However, for cell-based therapies, a major aim has been the generation of stem/progenitors that can be amplified prior to differentiation *in vitro* or *in vivo* (Fig. 1.3B). ES cells can address this limitation by allowing the generation of large quantities of specific cell populations for regenerative purposes. The establishment of human iPS cells with unlimited self-renewal potential from somatic cells (Takahashi et al., 2007; Yu et al., 2007) extents this model to patient-specific cell types and diseases. Nevertheless, a key requirement for clinical studies is the development of protocols for large-scale expansion of specific cell types (lineage-committed progenitors) from ES or iPS cells and the elimination of residual pluripotent cells, as they have the potential to form teratomas after transplantation (Blum & Benvenisty, 2008).

Several groups have demonstrated myogenic differentiation with murine and human ES and iPS cells with different protocols (reviewed in Salani et al., 2012). Possibilities include the use of a dissociation culture method from human ES/iPS-derived embryoid bodies for the selective expansion of myogenic cells (Awaya et al., 2012) and also the isolation of cells expressing a mesodermal marker, such as platelet-derived growth factor receptor-α (PDGFRA), following embryoid body formation (Hwang et al., 2013; Sakurai, Okawa, Inami, Nishio, & Isobe, 2008). While these methods do not require previous genetic manipulations, inefficient conversion and long culture times for differentiation are required.

As indicated above, myogenic differentiation of human ES or iPS cells can also be achieved by overexpressing *Mrfs*. Here, differentiation of human ES cells by inducible expression of *Myod* using a Tet-On system, led to progressive differentiation into mesodermal precursors, myogenic precursors, and finally terminal differentiated myoblasts (Rao et al., 2012). Although this differentiation method passes through a transient myogenic progenitor phase, as early as 4 days after induction, the majority of the cells stop proliferating likely due to induction of the cell cycle exit gene *p21* by *Myod* (Guo, Wang, Andres, Smith, & Walsh, 1995). This could explain the low transplantation efficiency of *Myod*-expressing iPS cells in other studies, due to their advanced commitment state (Tanaka et al., 2013). In another report (Goudenege et al., 2012), human ES and iPS cells derived from a DMD patient were maintained in a highly proliferative mesenchyme-like state (called MB1-ES/iPS cells). Subsequently, upon infection of the MB1-ES/iPS cells with an adenovirus-expressing *Myod* under a ubiquitous promoter (MB1-Myod cells), the cells expressed several myogenic markers including the surface maker CD56, and they fused into abundant multinucleated primary myotubes when exposed to differentiation media *in vitro*. When transplanted into cardiotoxin-injured muscle of $Rag^{-/-}$: *mdx* mice, MB1-Myod cells participated in muscle regeneration with a higher efficiency than non-*Myod*-expressing cells or control myoblasts as assessed by human dystrophin or spectrin staining. MB1-Myod cells are probably a mixed population of early and late myogenic cells, given the concomitant expression of endogenous *Myod*, *Myog*, and *MyHC* in the cell population even in proliferation media. Although regeneration was reported to be efficient and teratoma formation was not observed, it would be desirable for clinical studies to have access to a homogenous self-renewable population of upstream myogenic progenitors (i.e., Pax3+/Pax7+). This has been already achieved for the neuronal lineage, where self-renewable oligopotent neuronal progenitors can be generated from ES cells with chemically defined media or from mouse and human fibroblasts by direct reprogramming with Sox2 (Han et al., 2012; Ruggieri et al., 2013).

In other studies, mouse and human ES/iPS cells were directed to the myogenic lineage using doxycycline inducible expression of *Pax3* or *Pax7* in embryoid bodies (Darabi et al., 2012, 2008; Darabi, Pan, et al., 2011; Darabi, Santos, et al., 2011). Fractionation by PDGFRA, a paraxial mesoderm marker, and absence of Flk-1, a lateral plate mesoderm marker, resulted in the isolation of cell expressing *Myf5* and *Myod*, but also *Myog* and *MyHC* at low levels. In the case of Pax7-induced human ES/iPS cells, purification of the Pax7-hi (GFP-hi) fraction of induced cells resulted in cells with notable expansion potential and quasi-absence of Myogenin and MyHC expression as expected (Darabi et al., 2012). In all cases, upon removal of dox and exposure to differentiation media, Pax3/Pax7-induced ES/iPS cells expressed MyHC and exhibited a typical and uniform morphology of multinucleated myotubes. Myogenic progenitors derived from Pax3/Pax7-induced ES /iPS cells contributed to regenerating fibers after transplantation, not only restoring Dystrophin expression but also improving the force generation of engrafted muscles when transplanted systemically or intramuscularly into cardiotoxin-injured immunodeficient or dystrophic (mdx) mice. Finally, the authors show that Pax3/Pax7-induced murine or human ES/iPS cells can replenish the muscle satellite cell compartment and allow durable engraftment without teratoma formation up to 46 weeks (Darabi et al., 2012).

While further investigations on the regulatory mechanisms governing the myogenic specification of ES and iPS cells need to be carried out, these results together with the possibility of correcting genetic defects (Filareto et al., 2013; Kazuki et al., 2010) and the use of integration-free methods for the generation of iPS cells and the skeletal muscle progenitor derivatives (Kaji et al., 2009; Kim et al., 2009; Okita, Nakagawa, Hyenjong, Ichisaka, & Yamanaka, 2008; Stadtfeld, Nagaya, Utikal, Weir, & Hochedlinger, 2008; Warren et al., 2010; Yu et al., 2009; Zhou et al., 2009) will make patient-specific iPS cells a feasible therapeutic approach for myogenic diseases.

Nevertheless, a few concerns remain. For all reprogramming strategies, it is important to clearly identify the cell type generated using gene expression signatures, cellular organization of structural proteins, morphology, epigenetic marks, and functional attributes (Addis & Epstein, 2013). In the studies mentioned above, the molecular signature of ES/iPS-derived myogenic cells was not clearly investigated. It is well established now by microarray and genetic studies that, depending on the developmental stage (Biressi, Tagliafico, et al., 2007; Messina et al., 2010) or anatomical location (Harel et al., 2009; Sambasivan et al., 2009), the intrinsic properties of myoblasts are different. Therefore, the phenotypic subtype of myogenic cells obtained has not been clarified. Notably also, generally only embryonic fibers were obtained, and their nature was generally not documented (slow, fast fibers). In addition, as culture environments are known to trigger macroscale chromatin state changes (Zhu et al., 2013), it will be important to obtain genome-wide maps during myogenic specification of ES and iPS cells. Another consideration is the composition of the extracellular matrix and substrate stiffness, which appear to be key factors in regulating stem cell self-renewal and stem cell fate in culture (Gilbert et al., 2010; Watt & Huck, 2013). Therefore, the evaluation of these parameters for myogenic differentiation of human ES or iPS cells will allow the expansion of more homogenous self-renewable populations of upstream myogenic stem/progenitors enabling future cell-based therapies for muscle-wasting diseases.

5. INDUCTION OF MUSCLE-SPECIFIC GENE EXPRESSION

5.1. Cooperation between the Mrfs and other transcription factors

Skeletal muscle genes are not all expressed simultaneously (Lin et al., 1994). For example, *Myod* upregulates numerous targets with different kinetics; *Cadherin-15* and *Myog* are considered to be early targets of *Myod*, and *Ckm* and *MyHC* are considered to be late muscle genes (Bergstrom et al., 2002; Blais et al., 2005; Cao et al., 2006; Chen et al., 1997; Di Padova, Caretti, Zhao, Hoffman, & Sartorelli, 2007). Microarray analysis has demonstrated that *Myod* activity alters the expression of ~5% of the assayed genes in a temporally specific, multistaged response, with some genes activated within 6 h, but others not activated until 2 days later (Bergstrom et al., 2002). Extensive evidence indicates that temporal control of muscle gene transcription by *Myod* is driven by a feedforward circuit, wherein *Myod*-dependent activation of downstream "early" targets will feedforward to regulate Myod activity and cooperatively enable transcription of "late" genes (e.g., Mef2D, Penn et al., 2004; Myog, Cao et al., 2006; and Meis/Pbx complex, Berkes et al., 2004; Maves et al., 2007; see Tapscott, 2005). Data from genome-wide ChIP-seq experiments have now provided a more detailed view of such cooperative interactions, demonstrating that Myod-binding sites associated with genes differentially expressed during differentiation are enriched for motifs of factors that are activated by Myod and function with Myod in a positive feedforward circuit (Cao, Yao et al., 2010). This notion was also extended to a human cell culture model of rhabdomyosarcomas, a pediatric tumor of skeletal muscle that resembles undifferentiated myogenic cells (Merlino & Helman, 1999). MacQuarrie and colleagues showed that while Myod binds in a similar genome-wide pattern in both normal and tumor (RD) cells, it binds poorly at a subset of myogenic genes that fail to activate in the tumor cells. Interestingly, binding differences are associated with Runx1, Mef2C, Jdp2, and Nfic motifs and terminal differentiation in RD cells can be induced by forced expression of these factors, all of which interact with and/or potentiate Myod-mediated activity in some manner (MacQuarrie et al., 2013).

Cooperativity between Myod and proteins of the Mef2 family, as originally proposed on the basis of biochemical and reporter gene experiments (Molkentin et al., 1995), is one of the best-characterized models available.

RNA-seq analysis during myotube formation in C2 cells identified the transcription factor *Mef2D*, a *Mef2* family member, among the genes that undergo a major isoform switch during myogenesis (Trapnell et al., 2010). Among Mef2 family members, *Mef2D* is unique, as it undergoes tissue-specific splicing to generate a muscle-specific isoform which is induced during muscle differentiation (Martin et al., 1994). ChIP-seq analysis showed that while the ubiquitously expressed (Mef2Dα1) and muscle-specific (Mef2Dα2) isoforms of Mef2D bind an overlapping set of genes, only Mef2Dα2 activates late muscle transcription (Sebastian et al., 2013). The biological role of both *Mef2D* isoforms was confirmed *in vitro* and *in vivo*. Conditional expression of Mef2Dα2 accelerates differentiation of C2 myoblasts and results in improved muscle regeneration in an *in vivo* model of muscle regeneration. Expression of the ubiquitous isoform had an antagonistic effect. Mechanistically, the differential ability of Mef2D isoforms to activate transcription depends on their susceptibility to phosphorylation by protein kinase A (PKA). During the early stages of differentiation, the ubiquitous isoform Mef2Dα1 is expressed almost exclusively and phosphorylation by active PKA signaling leads to a recruitment of class II HDAC4/9 onto muscle promoters to maintain a transcriptionally repressed state. At later stages of differentiation, alternative splicing of *Mef2D* transcripts gives rise to the Mef2Dα2 isoform, which cannot be phosphorylated by PKA and leads to recruitment of Ash2L complexes to muscle promoters to activate gene expression. Thus, regulated alternative splicing appears to have a major effect on the biological activity of one of the best-characterized transcription factors involved in myogenesis. These findings expose a novel mechanism in which a tissue-specific alternate splicing event has evolved to allow a ubiquitously expressed transcription factor to escape inhibitory signaling for temporal regulation of gene expression.

5.2. Mrfs and chromatin remodeling

A major question in the myogenic field has been how Myod, and the other Mrfs, gain access, in undifferentiated myoblasts, to canonical E-boxes located on promoters that are covered by nucleosomes prior to the activation of transcription. The ability of *Myod* to convert cells of different lineages and differentiation states to skeletal muscle suggests that it has the characteristics of a pioneer transcription factor: a factor that can access genes in a repressive chromatin context and actively remodel the appropriate loci independent of the previous lineage to allow the productive engagement of polymerase II (Pol II) (Cirillo et al., 2002). However, transdifferentiation by Myod is limited by the cellular context suggesting that Myod alone cannot act as a pioneer factor (Zaret & Carroll, 2011) or its role as a pioneer is insufficient for muscle differentiation in certain cell types (Fong & Tapscott, 2013). Therefore, Myod should be loaded to the chromatin through the interaction with a *bona fide* pioneer factor that is insensitive to repressive marks or that binds to adjacent target gene regulatory elements that do not carry repressive marks. These definitions will likely be revised as more detailed studies of chromatin-binding partners are examined.

While the temporal sequence of events that occurs at a muscle-specific promoter preceding and during activation is not fully understood, a two-step model for gene activation has been proposed (Albini & Puri, 2010). This model suggests that Myod is first recruited to inaccesible E-boxes of its target gene promoters, e.g., *Myog*, through interaction with the Pbx–Meis complex which appears to be constitutively bound to a Mef3 consensus site located upstream of the canonical E-box (Berkes et al., 2004). The Pbx1/Myod complex was suggested to recruit directly the histone acetyltransferases that induce early hyperacetylation of the distal region of the myogenic promoter, thereby relaxing chromatin (de la Serna et al., 2005). Accordingly, recent evidence by ChIP-seq demonstrated that Myod–chromatin interactions in myoblasts coincide with peaks of histone acetylation (Cao, Yao, et al., 2010; see below). Subsequently, Myod recruits SWI/SNF proteins that use ATP hydrolysis to disrupt the interaction between DNA and histone octamers (de la Serna et al., 2005). These events would then render the canonical E-box-binding sites more accessible to Myod to promote transcription (de la Serna et al., 2005; Dilworth, Seaver, Fishburn, Htet, & Tapscott, 2004; Puri et al., 1997), reviewed in Guasconi & Puri, 2009; Albini and Puri, 2010).

The SWI/SNF complexes act in virtually all cell processes, and they are essential for establishing and maintaining pluripotent and multipotent cell states (Lessard & Crabtree, 2010). In addition to the ATPase subunit (BRG1 or BRM), SWI/SNF complexes are composed of 10 other non-enzymatic structural subunits (BAFs), which are assembled in different combinations from 20 different gene products. This combinatorial can potentially yield hundreds of complexes with different functions in a variety of cell types

(Wu, Lessard, & Crabtree, 2009; Puri and Mercola, 2012). The selective activation of lineage-specific constellations of genes appears to be achieved through the combinatorial assembly of BAFs subunits, which are bridged to the chromatin through lineage-specific transcription factors (e.g., Myod for skeletal myogenesis; Gata4 for cardiac myogenesis) (Puri & Mercola, 2012). In undifferentiated myoblasts, BAF60C was proposed to colocalize with Myod at promoter regions without other SWI/SNF subunits (Forcales et al., 2012) prior to transcription. While the relationship between preassembled BAF60C–Myod complexes and the Myod/Pbx1 interaction mentioned above remains unclear, it is possible that BAF60C mediates the initial recruitment of Myod by facilitating interactions with Pbx (Forcales, 2012). Upon exposure to differentiation cues, p38 directly phosphorylates chromatin-bound BAF60C (Simone et al., 2004) and preassembled BAF60C–MyoD recruits a Brg1-based SWI/SNF complex that is competent to remodel the chromatin. This allows the exposure of the canonical-binding sites for the Mrfs and Mef2 factors and transcriptional activation from the target loci (Forcales et al., 2012).

Although Myog is not able to remodel chromatin efficiently (Bergstrom & Tapscott, 2001; Gerber et al., 1997; Roy et al., 2002), it is a strong activator of transcription at loci with an open chromatin structure (Cao et al., 2006). In both developing embryos as well as in a tissue culture model for differentiation, Myog was reported to cooperate with the Mef2D muscle-specific isoform and Brg1-based SWI/SNF chromatin remodeling enzymes to promote chromatin remodeling and transcription initiation at myogenic late gene promoters (Ohkawa, Marfella, & Imbalzano, 2006). However, at the regulatory sequences of these late genes, transcriptional activation requires the combined activity of Myod and Myog (Cao et al., 2006). These observations suggest that part of the distinction between early and late myogenic gene expression is due to the sequential recruitment of Brg1-based SWI/SNF chromatin remodeling enzymes to different genes by the Mrfs and their specific coregulators (Pbx for Myod, and Mef2D for Myog) (Ohkawa et al., 2006).

Global analysis of transcript levels has shown that the mammalian SWI/SNF proteins are required for the expression of only a fraction of all Myod targets. Another possible mechanism for facilitating muscle gene expression is the removal of repressive epigenetic marks. Spt6 is a histone chaperone that, besides its known role in facilitating PolII elongation, entails erasure of the repressive H3K27me3 epigenetic mark, deposited by the

PRC2, through recruitment of the H3K27 demethylase KDM6A (UTX) (Wang et al., 2013). ChIP-seq experiments in undifferentiated C2C12 myoblasts and myotubes revealed an extensive genome-wide overlap of Spt6, PolII, and KDM6A at transcribed regions that are devoid of H3K27me3. By regulating proper recruitment and function of KDM6A, Spt6 effectively promotes H3K27 demethylation, muscle gene expression, and cell differentiation. Consistent with this model, *Spt6*-mutant zebrafish embryos, which show disrupted somitogenesis and muscle differentiation defects (Kok et al., 2007), display elevated levels of H3K27me3 and repression of *Myod* expression (Wang et al., 2013).

These models of myogenic gene activation would need to be verified in primary myogenic cells. Notably, it is curious that culture of satellite cells does not result in the immediate activation of *Myog*, in spite of the presence of Myod for at least 2 days (Zammit, Partridge, & Yablonka-Reuveni, 2006; Zammit et al., 2004), raising the question of the functional potential of Myod protein for accessing chromatin. It is possible that E12, PBX, SWF activities are compromised or high amounts of a negative regulator (like ID/Twist) prevent Myod from remodeling chromatin during this period of transient cell amplification.

6. GENOME-WIDE STUDIES AND THE PARADOX OF EXCESS BINDING SITES FOR THE MRFS

It has been known for many years, that the Mrfs act as obligate heterodimers with E-proteins to activate the terminal muscle differentiation program. They do so by binding to E-boxes in the promoters and enhancers of their target genes and by recruiting coregulators and RNA Pol II. However, most of this knowledge is derived from experiments on individual targets. The major outstanding questions are: how do Mrfs bind to specific E-boxes that are in vast excess and distributed throughout the genome? Other bHLH proteins recognize similar canonical sequences, yet NeuroD, a neurogenic transcription factor, promotes neuronal fates, whereas Myod makes muscle. Therefore, other factors must limit promiscuity for non-specific sites. Second, although Myod binds a large number of E-boxes genome-wide (Cao, Yao, et al., 2010; Fong et al., 2012), why does it modify gene expression only at a fraction of its targets? Some key advances were made recently by genome-wide profiling of myogenic cells using ChIP-seq and RNA-seq to address these issues.

6.1. Myod binding through the genome: productivity or remodeling?

Cross-linking studies using promoter arrays identified around 100–200 proximal promoters bound by Myod in C2C12 or Myod-ER (MDER) cells (Blais et al., 2005; Cao et al., 2006). Using a more sensitive ChIP-seq assay to identify Myod-binding sites in the chromatin of undifferentiated (myoblast) and differentiated (myotubes) C2C12 cells, Cao et al. mapped a much larger number of sites (23k in C2C12 myoblasts and 26k in myotubes). Surprisingly, whereas a positive correlation between Myod binding and gene expression during differentiation was noted for 4% of the genes, the majority of the sites were constitutively bound by Myod in both myoblasts and myotubes, and these sites were not associated with Myod-mediated changes in gene expression (Cao, Yao, et al., 2010). In addition, the majority of the Myod-binding regions did not show standard enhancer activity *in vitro*, nor were they located in active promoter regions. The function of the additional binding sites is not clear. Binding of Myod correlated with regional acetylation of histones in both myoblasts and myotubes. Histone 4 acetylation is generally associated with tissue-specific transcriptionally active *cis*-regulatory regions (Heintzman et al., 2009). However, as microarray analysis revealed that of the genes showing Myod-induced acetylation only a few displayed significant changes in genes expression, most acetylation events were suggested to be transcriptionally nonfunctional and instead involved in chromatin remodeling.

Using a similar genome-wide approach, a pervasive binding of Myod and Myog was also found throughout the genome (18k for Myod in C2C12 myoblasts and 40k for Myod in myotubes; Mousavi et al., 2013). Remarkably, 70% of the Myog-binding sites (35k in myotubes) overlapped with the Myod sites. In addition, these authors showed that the majority of these Myod/Myog peaks corresponded to extragenic enhancer regions (marked by H3K4me1, H3K27ac, p300, and Pol II) coinciding with RNA synthesis (i.e., eRNAs). Among the enhancers transcribed in C2 myotubes, were the DRR and CE regions upstream of *Myod*. siRNA experiments demonstrated that transcription from the CE enhancer is required for efficient *Myod* expression, whereas transcription from the DRR enhancer is required for activation of *Myod* targets such as *Myog* and *MyHC*. In addition, the authors showed that both eRNAs promote chromatin accessibility and increase Pol II occupancy at their target loci. Thus, a novel regulatory mechanism was proposed where eRNAs contribute to the establishment of cell-type-specific transcriptional circuitry by directing chromatin remodeling.

Most of these studies have focused on Myod and there is no specific genome-wide study for Myf5 and Mrf4 to date. However, using ChIP-seq in primary mouse myoblasts expressing TAP-tagged Myf5 or Myod fusion proteins, it was reported that there is a 30% overlap of binding sites between Myod and Myf5 targets in proliferating myoblasts (Soleimani, Yin, et al., 2012). This finding is consistent with the notion that Myf5 and Myod share a role in defining myoblast identity. However, these authors reported a much smaller number of binding sites (1.4k in myoblasts and 9.3k in myotubes for Myod and 1k in myoblasts for Myf5) compared to the studies indicated above. This result is unexpected given that the proteins were overexpressed. Interestingly, a comparison of the DNA-binding profiles of endogenous and overexpressed *Myod* showed that overexpression of *Myod* (at approximately fourfold compared to endogenous levels) did not result in a significant increase in the number of binding sites when compared to the endogenous protein (Yao et al., 2013). Future comparisons will help clarify these discrepancies.

Using a different genome-wide approach, Myod was reported to play a prominent role in the acquisition of an enhancer state (Blum, Vethantham, Bowman, Rudnicki, & Dynlacht, 2012). Here, ChIP-seq in C2C12 myoblasts and myotubes on the basis of histone marks identified regions that could act as condition-specific (myoblast- or myotube-specific) or constitutive (marked in both conditions) enhancers. These authors observed a strong correlation between condition-specific enhancers and genes that are more highly expressed in myoblasts and myotubes. Comparison with RNA-seq data in C2C12 cells (Trapnell et al., 2010) indicated that condition-specific enhancers coincide with sites of active noncoding transcription (eRNA). Surprisingly, comparison with genome-wide Myod ChIP-seq data (Cao, Yao, et al., 2010) showed that only 20% of the predicted Myod-binding sites in these enhancers were actually bound by Myod, suggesting that there is highly selective binding of Myod to chromatin at regulatory sequences. Importantly, Myod binding to the condition-specific enhancers correlated with the recruitment of Pol II, p300, and Set7, a H3K4 monomethylase as well as a cohort of other transcription factors (c-Jun, Jdp2, Runx), suggesting a pivotal role of Myod in enhancer assembly (Blum et al., 2012). In primary myoblasts isolated from *Myod* null mice, there was a striking reduction in the recruitment of p300 and Pol II and expression of eRNAs at enhancers. Myod re-expression in primary *Myod* null myoblasts rescued some but not all hallmarks of these enhancers, indicating that the timely function of Myod is essential for the correct assembly of muscle enhancers. It is likely that chromatin plasticity at enhancer regions is temporally limited and irreversibly constrained once the appropriate window of enhancer assembly has passed (Blum et al., 2012). This might explain the differentiation defects, including the delayed expression of *Myog* in *Myod* null animals (Megeney et al., 1996).

In summary, Myod not only drives direct activation of a subset of genes in proliferating and differentiating myoblasts through binding to their promoters, it also plays a major role at specific loci to prime chromatin into a "muscle-specific state." In addition, as dynamic changes in nuclear organization are necessary to coordinate gene regulation during progenitor cell commitment and differentiation (see Rajapakse et al., 2009; Zhu et al., 2013), one can speculate that the genome-wide binding of Myod serves to initiate new patterns of gene looping and chromosome architecture necessary for these transitions (Cao, Yao, et al., 2010).

6.2. Private versus shared E-boxes and lineage-determined chromatin accessibility

As indicated above, while *Myod* expression can induce myogenic conversion in embryonic fibroblasts, it cannot do so in the P19 mouse carcinoma cell line (Skerjanc et al., 1994). Conversely, the bHLH transcription factor *Neurod2* can induce neurogenesis in P19 cells, but not in fibroblasts (Farah et al., 2000). These observations reveal a striking specificity of bHLH proteins to drive lineage-specific gene expression programs.

Chromatin-mediated repression can partly explain the paradox of lineage-specific gene activation among a family of factors that bind similar DNA sequences (Tapscott, 2005). An attractive model that was proposed is that the canonical E-box is hidden from Myod and other bHLH factors by chromatin in nonmuscle cells. The requirement for an interaction between Myod with the resident Pbx complex (and BAF60C) to expose the canonical E-boxes and binding sites to other factors, such as Mef2 and Six proteins, might explain why other bHLH transcription factors like E-proteins or Neurod do not bind and activate the *Myog* promoter *in vivo*.

Interestingly, Six1 ChIP-on-chip studies in C2C12 cells (Liu et al., 2010) showed that the E-box elements recognized by Myod and Myog were significantly enriched among the Six1-bound loci and that as much as 39% of Six1-bound loci overlapped with the reported Myod-bound regions

(Cao, Yao, et al., 2010). This suggests a concerted mode of action of Six factors and the Mrfs. While this study proposed synergy between Six and Mrfs at the transcriptional level based on reporter assays containing combinations of Mef3 (Six) or E-box (Mrf) elements, two other studies showed that Six proteins recruit factors that render the chromatin more permissive to transcription and to Mrf activity (Palacios, Summerbell, Rigby, & Boyes, 2010; Seenundun et al., 2010).

Using genome-wide binding and transcription prolifing of NeuroD2-mediated neurogenesis and Myod-mediated myogenesis, Fong and colleagues showed that the specificity to drive lineage-specific gene expression programs is determined at two levels: E-box sequence and chromatin accessibility (Fong et al., 2012). First, they showed that both Myod and NeuroD2 bind to the CAGCTG E-box, but that each factor also binds to a factor-specific motif (CAGATG for NeuroD2 and CAGGTG for Myod), and this specific binding greatly influences the outcome. While binding to the shared E-box sequence is associated with regional epigenetic modification, Myod binding to "private" E-boxes is associated with transcriptional activation of muscle-specific genes. Second, they showed that the binding sites accessible for each factor is constrained by chromatin accessibility and thus which sites are open is epigenetically determined in a lineage-specific fashion. These results support a model of cellular differentiation in which specificity of gene activation is determined genetically by E-box sequences and by a superimposed "epigenetic presetting" that determines the availability of E-box sequences for transcriptional initiation in a lineage-specific fashion (Giordani & Puri, 2013; Fong and Tapscott, 2013).

It should be noted that most of the genome-wide studies to date have been performed on cell lines (C2C12, myoblasts derived from MEFs and myotubes derived from them) or primary cells in culture. A major question for future work will be the study of the temporal evolution of Mrf binding and obtain the corresponding chromatin state maps during prenatal and postnatal development.

In the context of global changes in the epigenetic landscape, studies have focused primarily on ES differentiation. It has become clear that chromatin marks at enhancers are not static and they can be dynamically regulated during differentiation (Rada-Iglesias et al., 2011). In this context, temporal changes in the chromatin landscape during muscle differentiation were reported for C2C12 cells on a genome-scale by ChIP-seq analyses (Asp et al., 2011). The resulting histone modification maps suggested the existence of two classes of genes marked by H3K27me3, a well-known repressive mark

(Berger, 2007). Class II genes, which are associated with commitment to nonmuscle lineages or are inhibitors of myogenic differentiation (eg. Twist), exhibit the highest levels of H3K27me3 and retain this mark during differentiation. Class I genes, which are upregulated during differentiation, are less highly enriched for this mark and lose it during differentiation. Class I genes, such as *Myog*, are targeted by PRC2-mediated H3K27 methylation in myoblasts. During differentiation, the PRC2 methyltransferases Suz12 and Ezh2 are downregulated and the H3K27 methylation mark is lost (Asp et al., 2011; Caretti, Di Padova, Micales, Lyons, & Sartorelli, 2004). Accordingly, depletion of Suz12 *in vitro* markedly accelerated myogenic differentiation, in marked contrast to the situation in ES cells (Pasini, Bracken, Hansen, Capillo, & Helin, 2007). H3K27 methylation of class II genes is independent of PRC2 and dependent on PRC1. This regulatory mechanism ensures the retention of H3K27me3 on class II genes, which must remain silenced during differentiation. Among the *Mrfs*, *Myog* was the only gene marked by the repressive H3K27me3 mark, which was replaced by the ones correlated with activation (H3K27ac, H3K4me1, p300, Pol II) during differentiation. Therefore, H3K27me3 marks could portray part of a methylation-to-acetylation switch during differentiation, thereby determining the timing of expression of *Myog*, and consequently terminal differentiation.

Interestingly, a number of genes encoding critical regulators of adipogenesis (*C/EBP* family, *PPAR-γ*) and chrondrogenesis (*Runx2*, *Sox* family) were devoid of repressive chromatin marks and were expressed in C2C12 cells. Nevertheless, several downstream effectors of these pathways were solely marked by repressive H3K27me3, suggesting that although some key master regulatory genes are expressed, activation of essential downstream effectors awaits additional cues. This view is in agreement with the observation that C2C12 cells can give rise to myotubes upon depletion of mitogens, but they can also be diverted to adipocytes and osteoblasts upon exposure to the appropriate inducers (Hu, Tontonoz, & Spiegelman, 1995; Lee et al., 2000). Taken together, these observations suggest that different mesodermal fates (osteogenic, adipogenic, muscle) could be explored by a single cell type by modulation of distinct chromatin modifications.

6.3. Lessons learned from Pax3/7 genome-wide studies

As discussed earlier, *Pax3* and *Pax7* have overlapping functions in the myogenic developmental program (Relaix et al., 2004) and in the specification and progression of the adult satellite cell lineage, where Pax7 exerts a major role (Kuang et al., 2006; Lepper et al., 2009; Oustanina et al., 2004; Relaix et al., 2006; Seale et al., 2000). Pax3 and Pax7 can bind to DNA sequences containing either a consensus paired domain binding site (GTCAC) or a homeobox (hbox) binding site (ATTA) (Chalepakis & Gruss, 1995; Chalepakis, Wijnholds, & Gruss, 1994), and Pax3 and Pax7 proteins show 86% aminoacid sequence identity within the paired domain (Walther et al., 1991). Therefore, as with the Mrfs, the question of specificity for Pax3 and Pax7 binding sites is raised for skeletal muscle and neuronal lineages.

Until recently, with the exception of *Met* (Epstein, Shapiro, Cheng, Lam, & Maas, 1996), the limb enhancer in the *Myf5* promoter (Bajard et al., 2006), and *Fgfr4* (Lagha, Kormish, et al., 2008), very few *Pax3* targets had been identified for myogenesis *in vivo*. Loss-of-function screens in the embryo have been difficult, given that loss of *Pax3* function often results in apoptosis. Therefore, most of the information has been provided by overexpression of *Pax3/Pax7* or by expression of PAX3-FKHR or PAX7-FKHR fusion proteins, which act as strong transcriptional activators. These fusion proteins arise from gene rearrangements between the C-terminal TAD of FKHR (FOXO1A) to the N-terminal DNA-binding elements of Pax3/Pax7 in alveolar rhabdomyosarcomas (Robson, He, & Eccles, 2006). Affymetrix analysis of somites and limbs of PAX3-FKHR/GFP embryos showed that *Pax3*-regulated sequences modulate initial cell-fate decisions in multipotent *Pax3*-positive stem cells of the DM (Lagha et al., 2010). The Id gene family of myogenic inhibitors was positively regulated while muscle differentiation genes were mainly downregulated by Pax3. This suggests that Pax3 acts as a brake on muscle differentiation. Genome-wide mapping of PAX3–FKHR-binding sites by ChIP-seq in human rhabdomyosarcoma cell lines (Cao, Yu, et al., 2010) showed a bias toward genes involved in growth and oncogenesis, cell migration and metastasis but also developmental processes including the nervous and skeletal systems. This study identified binding sites in *Fgfr4*, *Met*, *Meox1/2*, *Myod*, and *Prrx1* genes controlling limb and axial skeleton development and myoblast proliferation. Interestingly, many of these sites were not bound by Pax3 in adult myoblast cells (see below, Soleimani, Punch, et al., 2012).

There is limited knowledge on how *Pax3* functions in regulating gene expression. It is a weak transactivator and might require other elements for regulating gene expression. For example, genetic analysis in the embryo revealed that *Myod* expression requires both *Pax3* and *Myf5* (Tajbakhsh et al., 1997). Interestingly, motif analysis showed a strong colocalization of Pax3 and E-box motifs, which would suggest a broad role for Pax3 and E-box-binding proteins in regulating the expression of many targets (Lagha, Kormish, et al., 2008).

New insights into potential Pax3 and Pax7 targets came from genome-wide ChIP-seq and trancriptome analysis in primary myoblast cell lines derived from satellite cells stably expressing tagged Pax3 or Pax7 proteins (Soleimani, Punch, et al., 2012). Whereas Pax3 was reported to bind a substantial set of targets not bound by Pax7, the majority of Pax3-binding sites identified overlaped with those of Pax7. Furthermore, Pax7 showed a more extensive genome coverage and its overexpression resulted in differential regulation of a much larger set of genes than Pax3, pointing to a more dominant role for *Pax7* in adult skeletal myoblasts. Overexpression of *Pax7* induced the expression of many genes involved in growth and proliferation of satellite cells (e.g., *Rspo1* and *Bmp4*) and repression of many muscle differentiation genes. However, as both upregulated and downregulated genes were significantly enriched for Pax7 binding sites, this raised the question whether *Pax7* acts not only as an activator but also as a repressor (Soleimani, Punch, et al., 2012). The vast majority of Pax3 and Pax7 binding sites were identified within intergenic or intronic regions away from transcription start sites (Budry et al., 2012; Cao, Yu et al., 2010; Soleimani, Punch, et al., 2012). Interestingly, in adult cells, Pax7 binds strongly to many sites that had been documented to be occupied by Pax3 in the embryo (e.g. *Fgfr4, Dmrt2* and *Spry1*) (Soleimani, Punch, et al., 2012). Together, these data support the notion that embryonic and adult myogenic programs represent distinct states in which *Pax3* and *Pax7* have critical nonredundant roles. This might explain why *Pax3* does not rescue the postnatal phenotypes of the *Pax7* null in muscles where both proteins are normally present in satellite cells (Relaix et al., 2004).

Pax7-binding sites were also found in regulatory regions of *Pbx1* (Soleimani, Punch, et al., 2012) and *Id2/3* in adult myoblast cell lines (Kumar et al., 2009). One can therefore speculate that Pax7 not only binds and upregulates key myogenic genes such as *Myf5* and *Myod* but also their coactivators and inhibitors for more refined control of cell-state transitions.

Interestingly, in spite of recognizing the same binding motifs, Pax3 and Pax7 have significantly different affinities for paired versus hbox motifs. Pax7 can activate target gene expression from combined paired/hbox or hbox motifs alone, whereas Pax3 is ineffective in inducing transcription from hbox only motifs (Soleimani, Punch, et al., 2012). This also suggests relevant differences between the functions of these factors. Additional mechanisms such as the presence of specific coactivators or posttranslational modifications might differentially influence the DNA-binding affinity for Pax3 and Pax7 (Mascarenhas et al., 2010; Miller, Dietz, & Hollenbach, 2008).

Comparison of genome-wide binding and expression data revealed that Pax7 binds to many more sites (50,000 genomic loci) than the number of genes it regulates (1300 genes), a scenario reminiscent of Myod. The functional relevance of the "nonproductive" binding sites is currently unknown. Previous studies showed that Pax7 is associated with active chromatin (McKinnell et al., 2008). Pax7 associates with the Wdr5–Ash2L–MLL2 histone methyltransferase complex that directs H3K4 trimethylation of surrounding chromatin. In addition, a recent study reported an essential function for Pax7 in maintenance of heterochromatin (Gunther et al., 2013). Interestingly, during pituitary gland development *Pax7* acts as a pioneer transcription factor, since its activity is associated with remodeling of chromatin, and genome-wide, this function appears to be preferentially associated with combined paired/hbox-binding sites (Budry et al., 2012). Taken together, these observations suggest a more general role in chromatin organization for Pax7, which has been also proposed for Pax3 (Bulut-Karslioglu et al., 2012).

In summary, the distinct scenarios for Pax3 and Pax7 binding provide a framework for understanding the differences in the organization of the transcriptional network between prenatal and adult myogenesis. Both factors may be sufficient to initiate myogenic programs, but different developmental contexts may use one or the other to fulfill the role (Soleimani, Punch, et al., 2012). As for the *Mrfs*, more refined analysis with cells isolated directly from the organism will be necessary to complement these studies.

7. MOLECULAR EVOLUTION AND CONSERVATION IN OTHER SPECIES

Phylogenetic analysis indicates that the four vertebrate *Mrfs* evolved from a single ancestral gene as result of two round gene duplication events prior to the divergence of chordates (Atchley et al., 1994). The complex musculature found in vertebrates might have arisen concomitantly with the duplication of these genes during evolution. The retention of high conservation of the DNA-binding domain (bHLH) while diversifying the transactivation domains has allowed the Mrfs to retain a certain degree of functional overlap while acquiring unique properties, as revealed by the molecular and genome-wide binding studies indicated above. Gene inactivation studies revealed distinct phenotypes for the Mrfs, thus underscoring overlapping and unique functions for each of these regulators. Unlike vertebrates, invertebrates such as *Caenorhadbitis elegans*, sea urchin, and *Drosophila melanogaster* are able to develop their musculature in the presence of a single myogenic factor (Krause & Weintraub, 1992; Michelson, Abmayr, Bate, Arias, & Maniatis, 1990; Venuti, Goldberg, Chakraborty, Olson, & Klein, 1991).

A comparison of vertebrate and invertebrate *Mrfs* reveal remarkable degrees of evolutionary conservation (Atchley et al., 1994; Zhao et al., 2014). For example, vertebrate and nematode myogenic proteins share similar temporal and spatial patterns of expression during development as well as functional similarities; *CeMyoD* can convert the murine cell line C3H10T1/2 to myoblasts and efficiently transactivate mouse muscle-specific promoters in transient transfection assays. Furthermore, mouse Myod appears to recognize, and activate, a *CeMyoD* autoregulatory site within the nematode *hlh-1* gene (Krause, Fire, White-Harrison, Weintraub, & Tapscott, 1992). In other studies, Lei and colleagues used ChIP-seq and ChIP-chip to determine the genome-wide distribution of CeMyod (HLH-1)-binding sites in the genome of *C. elegans* mixed-stage embryos (Lei et al., 2010). They found that CeMyod preferentially binds to E-boxes (CANNTG) in the promoter regions of genes known to be expressed in body wall muscle, consistent with its role as a direct transcriptional regulator. HLH-1 binding was also detected at numerous sites not associated with muscle gene expression, as has been previously described for its mouse homolog Myod (Cao, Yao, et al., 2010). Finally, from analysis of all of the E-boxes present in the *C. elegans* genome, HLH-1 was found to bind to about 4% of them, indicating that there is a high degree of selectivity for HLH-1 binding that restricts it to only a small fraction of the potential E-box sites present in the genome. The remarkably similar observations for Myod in mouse and *C. elegans* suggest evolutionary conservation in the roles for this master regulatory transcription factor, and likely also the cellular and molecular mechanisms that restrict its activity.

Finally, ascidians were reported to have two developmentally regulated splice forms of *Myod*: *CiMDFa* and *CiMDFb*. *CiMDFa* is expressed both maternally and zygotically and lacks a portion of the C-terminus that is present in *CiMDFb* (Meedel, Farmer, & Lee, 1997). Thus, ascidians might have solved the problem of having a single *Mrf* gene to accomplish myogenic determination and differentiation during development. In addition, the ascidian model has provided important insights into the cellular origins of distinct muscles. Notably, a single progenitor cell was shown to give rise to pharyngeal (siphon) and cardiac lineages, as described for vertebrates above (Razy-Krajka et al., 2014). Furthermore, many of the regulators such as the homologs for *Tbx1*, *Isl1*, etc., as well as *Notch* signaling for preserving progenitor cell populations are conserved. Intriguingly, many of these regulators are also conserved for *Drosophila* muscle development (Boukhatmi et al., 2012).

8. CONCLUSIONS AND PERSPECTIVES

Major advances have been made recently, particularly in the molecular analysis of myogenic regulatory gene function by genome-wide studies. In addition, the identification of chromatin modifiers that render myogenic conversion possible underscores the importance of assessing the epigenetic signatures of permissive and nonpermissive cells. Another area of major interest is the route to personalized medicine. Although genetic modifications of ES and iPS cells have made important advances in this field, obtaining fetal myofibers has not been achieved to any appreciable extent. The generation of *bona fide* adult-type satellite cells has also been an obstacle. In this context, the generation of muscle stem/progenitors by direct conversion of fibroblasts or other somatic cells has also been refractory, unlike advances in this direction for neural stem cells, for example. These will be major objectives for future research. Finally, the study of the origins of skeletal muscles from an evolutionary perspective will be critical to provide a more complete picture of how skeletal muscles from somitic, cranial, and the less explored neck/head region are established. The targeting of only a specific subset of muscles in numerous myopathies suggests that a developmental origin could be causative in the disease process, at least in part. Combined efforts to understand developmental and evolutionary aspects of skeletal muscle ontology will be critical also for generating region-specific muscles from ES/iPS cells.

Table 1.1 Summary of some genetic mouse mutants and their corresponding phenotypes

Transcription factor	Family	Expression/Role	Mutant	Phenotype: Trunk muscle	Phenotype: Limb muscle	Phenotype: Head muscle	Comments
MRFs (Myf5, Mrf4, Myod, Myog)	bHLH	Muscle specific expression ***Myf5, Mrf4 and Myod:*** expressed in muscle progenitors →muscle cell fate determinants (*Mrf4* only in the embryonic phase) ***Myog*, also *Myod* and *Mrf4*:** expressed in commited precursors →muscle differentiation factors **Myoblasts:** • Embryonic myoblasts require *Myf5, Mrf4* or *Myod* for their determination. • Fetal myoblasts require Myf5 or *Myod.* **Myofibers:** • Embryonic myofibers require *Myog*, *Myod* or *Mrf4*. *Myf5* alone cannot support myofiber differentiation. • Fetal myofiber differentiation requires *Myog* and either *Mrf4* or *Myod.* **Adult SC:** *Myf5* expressed in quiescent, *Myod* in activated SC; *Myog* during differentiation; *Mrf4* expression in differentiated myofibres.	**Myf5 null** • ***Myf5*** $^{loxP/loxP}$ (Kassar-Duchossoy et al., 2004) • ***Myf5*** $^{\Delta loxP/\Delta loxP}$ (Kaul et al., 2000)	Embryo: Delay in primary myotome formation till E10. MPCs are developmentally arrested until *Mrf4* and/or *Myod* programme muscle identity. Mislocalization of MPCs along the dorsal and ventral dermomyotomal lips. Fetus/Adult: Myogenesis proceeds normally. No significant deficit in major muscle masses. Lack of some epaxial muscles (deep back, tail).	Embryo/Fetus: No phenotype observed. Adult: regular muscle architecture with subtle progressive myopathy (fibrosis). SC number normal, reduced proliferation in vitro. Perturbed muscle regeneration upon injury (Gayraud-Morel et al., 2007; Ustanina et al., 2007). *Myf5* heterozygous SCs have greater self-renewal capacity and niche occupancy after transplantation (Gayraud-Morel et al., 2012).	No phenotype observed prenatally (Sambasivan et al., 2009).	Mice viable. *Mrf4* expression in cis is not affected. Two other alleles originally described as *Myf5* null mice (*Myf5* $^{nlacZ/nlacZ}$ (Tajbakhsh et al., 1996); *Myf5* ^{m1} (Braun et al., 1992)) also compromise *Mrf4* expression. See (Kassar-Duchossoy et al., 2004; Yoon et al., 1997; Kaul et al., 2000) Distal rib phenotype observed in previous alleles that also perturbed Mrf4 expression not observed.
			Mrf4 null (Zhang et al., 1995)	Embryo: grossly normal; decreased level of embryonic MyHC. Fetus/Adult: Myogenesis proceeds normally.	No phenotype observed (same as trunk).	No phenotype observed.	Mice viable. Myf5 expression in cis not significantly affected. Rib phenotype: mild anomalies that include bifurcations and fusions, good attachment to the sternum. Two other alleles described as *Mrf4* null mice (Braun and Arnold, 1995; Patapoutian et al., 1995) also compromise *Myf5* expression. See (Floss et al., 1996; Olson et al., 1996; Yoon et al., 1997).
			Myf5: Mrf4 null • ***Mrf4*** ^{nlacZ} **:*Myf5*** ^{loxP} (Kassar-Duchossoy et al., 2004) • ***Myf5 (Mrf4) null*** (***Myf5*** $^{nlacZ/nlacZ}$, *Mrf4* affected in cis, Tajbakhsh et al., 1996; Tajbakhsh et al., 1997) • ***Mrf4 (Myf5) null*** (***Myf6*** ^{m1}, *Myf5* affected in cis, Braun and Arnold, 1995)	Embryo: 1-2 day delay in primary myotome formation and no expression of *Myog* until *Myod* is expressed and programmes muscle identity (> E11). Mislocalization of MPCs along the dorsal and ventral dermomyotomal lips; ectopic migration of muscle progenitor cells; change in cell fate. Fetus: Lack of some epaxial muscles (deep back, tail). (Kablar et al., 1997; Kassar-Duchossoy et al., 2004; Tajbakhsh et al., 1996)	No phenotype observed.	EOM myogenesis fails. MPCs die by apoptosis; no rescue by *Myod* (Sambasivan et al., 2009). BA development normal.	Mice non viable. Rib phenotype: very severe, only residual proximal rib stubs present in *Myf5* $^{nlacZ/nlacZ}$ and *Myf6* $^{m1/m1}$; (See Yoon et al., 1997).

Abreviations: MPCs: muscle progenitor cells; SC: satellite cell; BA: branchial arches; EOM: extraocular muscles

Notes: • If adult phenotype is not mentioned either the mice are perinatal lethal or phenotypes have not been investigated

• Tongue included as 'trunk derived/somitic' muscle

Transcription factor	Family	Expression/Role	Mutant	Phenotype			Comments
				Trunk muscle	Limb muscle	Head muscle	
MRFs (Myf5, Mrf4, Myod, Myog	bHLH		**Myod null** ***Myod*** ^{m1} (Rudnicki et al., 1992)	No overt phenotype. Adult: mice/muscles reduced in size.	Embryo: Migration of *Pax3* + myogenic precursors and upregulation of *Myf5* occur normally, but 2.5 day delay in limb myogenesis. Normal embryonic myoblasts and myofibers (Kablar et al., 1997). Fetus: no overt phenotype observed; but reduced muscle mass. Adult: mice/muscles severely reduced in size. Increased absolute number of SC and myoblasts, severely reduced differentiation in vitro. Delayed regeneration upon injury due to sustained myoblast replication and fusion delay. Balance between self-renewal and differentiation is deregulated. *Myod* $^{-/-}$ myoblasts are resistant to apoptosis. (Gayraud-Morel et al., 2007; Megeney et al., 1996; Sabourin et al., 1999; White et al., 2000; Yablonka-Reuveni et al., 1999; Asakura et al., 2007; Macharia et al., 2010).	Embryo: delay in BA myogenesis (Kablar et al., 1997).	Although null adults can be obtained, many die at birth; genetic background dependent.
			Myf5: Myod null • ***Myf5*** $^{loxP/loxP}$ **:*Myod*** $^{m1/m1}$ (Kassar-Duchossoy et al., 2004; Kassar-Duchossoy et al., 2005)	Embryo: *Mrf4* can determine MPC identity and drive *Myog* expression cell autonomously in the embryo. Trunk muscle formation, but failure from mid-embryogenesis. Fetus: no fetal myogenesis, myoblasts or myofibers.	Embryo: absence of limb myogenesis, residual MyHC+ fibers. Fetus: no fetal myogenesis, myoblasts or myofibers.	EOM and BA myogenesis fails; some Desmin+ myoblasts via *Mrf4* (S. Tajbakhsh / R. Sambasivan unpublished observations)	Mice non viable.
			Myf5(Mrf4):Myod null • ***Myf5*** $^{nlacZ/nlacZ}$ **:*Myod*** $^{m1/m1}$ (Kassar-Duchossoy et al., 2004) • ***Myf5*** $^{m1/m1}$ **: *Myod*** $^{m1/m1}$ (Rudnicki et al., 1993)	Failure of embryonic and fetal myogenesis throughout the body. Myoblasts are not specified in the absence of *Myf5*, *Mrf4* and *Myod*, *Myog* is not expressed. Upstream stem cell population is present early, then lost due to lack of committed daughter cells.			Mice non viable. Severe rib defects as *Myf5:Mrf4* null.

(Continued)

Table 1.1 Summary of some genetic mouse mutants and their corresponding phenotypes—cont'd

Transcription factor	Family	Expression/Role	Mutant	Phenotype			Comments
				Trunk muscle	Limb muscle	Head muscle	
MRFs (Myf5, Mrf4, Myod, Myog	bHLH		**Myogenin null** (Hasty et al., 1993; Nabeshima et al., 1993; Venuti et al., 1995)	Embryo: Largely unaffected embryonic myogenesis. *Mrf4* and *Myod* can trigger embryonic fiber differentiation. Myofibers disorganized. Fetus: failure of fetal myogenesis *in vivo*. Presence of residual embryonic myofibers and myoblasts, no perinatal MyHC+ myofibers. *Myogenin* null myoblasts differentiate into seemingly normal myotubes when cultured *in vitro*.		Not reported.	Mice non viable. Rib phenotype: deformations in the rib cage, absence or reduction of intersternebral cartilage. *Myogenin:Myf5* null, *Myogenin:Myod* null and *Myogenin:Mrf4* null similar to *Myog* null (Rawls et al., 1995; Rawls et al., 1998). The severity of rib defect depends on the allelic combination.
			Myogenin CKO *Myog*$^{flox/flox}$: *CAGG*$^{Cre\text{-}ERTM}$ (Knapp et al., 2006; Meadows et al., 2008)(*CAGG*$^{Cre\text{-}ERTM}$, Hayashi and McMahon, 2002)	E17.5 deletion: no apparent defects in muscle morphology or histology; respiration muscles such as diaphragm, deep back, and intercostal muscles less developped. Postnatal deletion: Myog is not required for normal muscle development but is required for normal body growth. Adult deletion: No proliferation or differentiation defects of Myog-deleted adult myogenic cell; nonetheless found to regulate a distinct set of muscle expressed genes.		Not reported.	Deletion at E17.5, some viable mice. Mice 100% viable if deletion done at P1 or later.
			Myod:Mrf4 null (Rawls et al., 1998) Alleles: *Myod* : Rudnicki et al., 1992; *Mrf4* : Zhang et al., 1995	Indistinguishable from *Myog* null.	Embryo: 2.5 day delay in limb myogenesis as for *Myod* null. Fetus: failure of fetal myogenesis. No fetal myoblasts or myofibers.	Not reported.	Non viable. Rib defects as in *Mrf4* null (Zhang et al., 1995)
			Myod:Mrf4:Myog null (Valdez et al., 2000) Alleles: *Myod* : Rudnicki et al., 1992; *Mrf4* : Zhang et al., 1995; *Myog* : Hasty et al., 1993	Myoblasts are specified by *Myf5* but cannot differentiate in presence of *Myf5* alone. No Desmin or MyHC expression. Myoblasts do not differentiate even *in vitro*.		Not reported.	Severe rib defects. Increased amounts of cervical brown fat as in *Myogenin* null.

Transcription factor	Family	Expression/Role	Mutant	Phenotype			Comments
				Trunk muscle	Limb muscle	Head muscle	
Pax3/7	Paired box homeo-domain	Expressed in dorsal neural tube and neural crest cells. *Pax3* also expressed in presomitic mesoderm. *Pax3/Pax7* are expressed in muscle stem/progenitor cells and in adult SC (*Pax7* throughout the body, *Pax3* in subset of muscles).	**Pax3 null** (Auerbach, 1954; Bober et al., 1994; Franz et al., 1993; Goulding et al., 1993; Goulding et al., 1994; Relaix et al., 2003; Relaix et al., 2004; Tajbakhsh et al., 1997; Tremblay et al., 1998)	Somite defects, complete loss of hypaxial dermomyotome, epaxial domain less affected. Deficits in trunk, shoulder and tongue muscles (reduced or absent). Diaphragm muscles missing.	Migration and survival of muscle stem/progenitor cells from hypaxial dermomyotome severely compromised. Limb muscles missing. Complete lack of myoblasts and myofibers.	No phenotype observed.	Generally lethal before E15; but occasionally can be pushed to E18.5. Defects in neural tube closure. Skeletal malformations.
			Pax3:Myf5 (Mrf4) null (Tajbakhsh et al., 1997)	Trunk muscles missing at early stages. No *Myod* expression, muscle stem/progenitor cells almost entirely absent. *Pax3* and *Myf5* are genetically upstream of *Myod*.	Limb muscles missing.	No phenotype observed in BA. Trapezius muscle present (Theis et al., 2010) EOM missing as in *Myf5 (Mrf4)* null.	Non viable. Mislocated muscle stem/progenitor cells. Skeletal defects additive. Synergistic muscle phenotypes.
			Pax7 null (Kuang et al., 2006; Mansouri et al., 1996; Oustanina et al., 2004; Relaix et al., 2006; Seale et al., 2000)	No overt phenotype in embryo or fetus. Neonatal: muscle weakness, deficient growth. Postnatal (<P21): decreased muscle growth. Juvenile satellite cells are progressively lost because of apoptosis accompanied by cell cycle defects. *Pax3* cannot compensate. Adult: impaired regeneration upon injury. Extensive muscle wasting in aged mice.		No overt phenotype in embryo or fetus (Zacharias et al., 2011).	Facial malformations. Most *Pax7* null animals die within 3 weeks after birth in C57/B6 background. Better survival in 129Sv/J background.
			Pax3:Pax7 null (Relaix et al., 2005)	Only form primary myotome. Majority of muscle stem/progenitor cells die or do not enter the myogenic programme. Trunk muscles missing. Only a few embryonic muscle fibres are detectable.	Limb muscles missing.	Not reported.	Mice die at midgestation. Mutant cells assume non-myogenic fates.
			Pax7 or Pax3:Pax7 CKO (Gunther et al., 2013; Lepper et al., 2009; von Maltzahn et al., 2013)	Adult: apparent normal muscle regeneration if injury done after P21. However, complete inactivation of *Pax7*, results in loss of SCs and regeneration deficit.		Not reported.	*Pax7* also plays a role as a chromatin modifier, null results in decondensed chromatin in SCs (Gunther et al., 2013).

(Continued)

Table 1.1 Summary of some genetic mouse mutants and their corresponding phenotypes—cont'd

Transcription factor	Family	Expression/Role	Mutant	Phenotype: Trunk muscle	Phenotype: Limb muscle	Phenotype: Head muscle	Comments
Six1, Six4	Homeo-domain	Expressed in dermomyotome, myotome, limb bud mesenchyme and in myogenic migrating precursors. Also expressed in the head mesenchyme, dorsal root ganglia and a number of developing organs.	**Six1:Six4 null** (Grifone et al., 2005)	Embryo: Increased apoptosis in somitic compartment. Impaired migration of hypaxial myogenic precursors due to a decrease of *Lbx1*, *Met* and *Pax3* expression. Cells that lose *Pax3* expression, lose their identity, fail to activate the *Mrfs* and some of them are misrouted dorsally and detected beneath the ectoderm, others are misrouted medially.		Embryo: Increased apoptosis in BA. Transient muscle deficits. Most of the head muscles, including the masseter, temporalis, pterygoid and EOM are present at E18.5.	Non viable. No rib attachment to the sternum, craniofacial defects.
				Epaxial myogenesis relatively preserved. Hypaxial muscles diminished and most ventral muscles missing. More severe in rostral than caudal levels. Tongue muscles reduced/missing. Diaphragm muscle missing.	No limb muscles.		
Eya 1, Eya2	Eya protein family/ Six protein cofactors	Expressed in ventrolateral extensions of the dermomyotomes of somites, also dorsal root ganglia and a number of developping organs.	**Eya1$^{-/-}$:Eya2$^{+/-}$** (Grifone et al., 2007)	Embryo: Impaired migration of hypaxial myogenic precursors due to a decrease of *Pax3* expression. Cells that lose *Pax3* expression, lose their identity, fail to migrate correctly and to activate the myogenic program similarly to *Six1:Six4* null.		Not reported.	Non viable. *Eya1*$^{-/-}$ fetuses die at birth due to the absence of rib cage closure. *Eya2*$^{-/-}$ mice are viable and fertile.
				Trunk phenotype similar to *Six1:Six4* null.	No limb muscle up to E13.5. Fetus: selective muscle hypoplasia (similar to *Six1*$^{-/-}$), with remaining hypoplasic dorsal muscles in the distal hindlimb, absence of most ventral muscles and a quasi-absence of muscle in the forelimb.		
Tbx1	T-box	Expressed in cranial mesoderm, pharyngeal endoderm, limb buds. Required for robust bilateral transcriptional activation of the Mrfs in the BA; effect is cell autonomous.	**Tbx1 null** (Dastjerdi et al., 2007; Grifone et al., 2008; Jerome and Papaioannou, 2001; Kelly et al., 2004; Kong et al., 2014; Theis et al., 2010; Zhang et al., 2006)	Trunk myogenesis normal. Absence of the trapezius and sternocleidomastoid muscles (head myogenic programme). Tongue muscle is normal (trunk myogenic programme).	Limb myogenesis normal.	E10.5: lack of 2nd and 3rd BA. Early *Musculin* and *Tcf21* expression maintained in 1st BA. *Myf5/Myod* expression lost or very reduced/sporadic, asymmetric. Fetus: Random unilateral loss of 1st BA muscles (masseter, temporalis, pterigoid). Loss of 2nd BA muscles (muscles of the pharynx, larynx and part of the developing trapezius muscle). No apparent EOM phenotype.	Non viable, heart and skeletal defects. Cardiac outflow tract abnormalities, abnormal facial structures, abnormal vertebrae and cleft palate, hypoplasia of the thymus and parathyroid glands.
			Tbx1:Myf5 null (Sambasivan et al., 2009)	Additive phenotype of *Myf5* and *Tbx1* null.	Limb myogenesis apparently normal.	Similar to *Tbx1* null but more severe loss of 1st BA derived muscles.	Non viable.

Transcription factor	Family	Expression/Role	Mutant	Phenotype			Comments
				Trunk muscle	Limb muscle	Head muscle	
Musculin (MyoR), Tcf21 (Capsulin)	**bHLH**	Expressed in paraxial mesodermal cells in the BA and in a number of developping organs. Musculin also expressed in all developing skeletal muscle. Considered as transcriptional repressors that prevent the entry into differentiation programmes.	**Musculin:Tcf21 null** (Lu et al., 2002; von Scheven et al., 2006)	Trunk myogenesis normal.	Limb myogenesis normal.	Embryo: A subset of MPCs of the 1st BA fail to activate the myogenic program and die by apoptosis. Fetus: Complete absence of the major muscles of mastication (1st BA-derived) including the masseter, pterygoid, and temporalis. Other 1st BA–derived muscles, which do not function in mastication (anterior digastric and mylohyoid) present.	Non viable. Proposed role also as activators of the myogenic programme (Moncaut et al., 2012).
Pitx2	**Bicoid-related homeo-domain**	Expressed in cranial MPCs of EOM and BA, periocular mesenchyme, ectoderm of 1st BA, oral ectoderm, cardiac muscle, trunk skeletal muscle, and also in many tissues during development. 3 isoforms (Pitx2a,b,c) Participate in L-R asymetrical pattern formation of the body plan. Required for the development of mulitple organs in a dosage sensitive manner.	**Pitx2 null** (Diehl et al., 2006; Dong et al., 2006; Kitamura et al., 1999; L'Honore et al., 2007; L'Honore et al., 2010; L'Honore et al., 2014; Zacharias et al., 2011)	Pitx2 null mutants form nearly all muscle anlagen even though several of these muscle anlagen are distorted, coupled with malformation of the part of the body to which they attach.	Embryo: delay in Myod expression, recovered by E13.0.	MPCs of EOM and BA lacking *Pitx2* undergo apoptosis. Cell autonomous effect (mesoderm specific KO). No EOM development (EOM formation is dependent on the *Pitx2* dose). MPCs mislocalized around the eye. BA: deficit in 1st BA development. No *Musculin* expression. Reduced *Tbx1* , *Tcf21* expression.	Embryonic lethal (<E13.5). Failure of body wall closure, lung and cardiac defects. *Pitx2:Pitx3* double conditional KO: excessive upregulation of ROS levels leading to DNA damage and apoptosis of differentiating cells in the fetus.

Abreviations: MPCs: muscle progenitor cells; SC: satellite cell; BA: branchial arches; EOM: extraocular muscles

Notes:
- If adult phenotype is not mentioned either the mice are perinatal lethal or phenotypes have not been investigated
- Tongue included as 'trunk derived/somitic' muscle

ACKNOWLEDGMENTS

We acknowledge support from the Institut Pasteur, Fondation pour la Recherche Medicale, Agence Nationale de la Recherche (Laboratoire d'Excellence Revive, Investissement d'Avenir; ANR-10-LABX-73 and ANR-06-BLAN-0039), the Association pour la Recherche sur le Cancer, and the European Research Council (Advanced Research Grant 332893).

REFERENCES

Addis, R. C., & Epstein, J. A. (2013). Induced regeneration—The progress and promise of direct reprogramming for heart repair. *Nature Medicine*, *19*, 829–836.

Aggarwal, V. S., Carpenter, C., Freyer, L., Liao, J., Petti, M., & Morrow, B. E. (2010). Mesodermal Tbx1 is required for patterning the proximal mandible in mice. *Developmental Biology*, *344*, 669–681.

Albini, S., Coutinho, P., Malecova, B., Giordani, L., Savchenko, A., Forcales, S. V., et al. (2013). Epigenetic reprogramming of human embryonic stem cells into skeletal muscle cells and generation of contractile myospheres. *Cell Reports*, *3*, 661–670.

Albini, S., & Puri, P. L. (2010). SWI/SNF complexes, chromatin remodeling and skeletal myogenesis: It's time to exchange! *Experimental Cell Research*, *316*, 3073–3080.

Armour, C., Garson, K., & McBurney, M. W. (1999). Cell-cell interaction modulates myoD-induced skeletal myogenesis of pluripotent P19 cells in vitro. *Experimental Cell Research*, *251*, 79–91.

Asakura, A., Hirai, H., Kablar, B., Morita, S., Ishibashi, J., Piras, B. A., et al. (2007). Increased survival of muscle stem cells lacking the MyoD gene after transplantation into regenerating skeletal muscle. *Proceedings of the National Academy of Sciences of the United States of America*, *104*, 16552–16557.

Asakura, A., Lyons, G. E., & Tapscott, S. J. (1995). The regulation of MyoD gene expression: Conserved elements mediate expression in embryonic axial muscle. *Developmental Biology*, *171*, 386–398.

Asp, P., Blum, R., Vethantham, V., Parisi, F., Micsinai, M., Cheng, J., et al. (2011). Genome-wide remodeling of the epigenetic landscape during myogenic differentiation. *Proceedings of the National Academy of Sciences of the United States of America*, *108*, E149–E158.

Atchley, W. R., Fitch, W. M., & Bronner-Fraser, M. (1994). Molecular evolution of the MyoD family of transcription factors. *Proceedings of the National Academy of Sciences of the United States of America*, *91*, 11522–11526.

Atit, R., Sgaier, S. K., Mohamed, O. A., Taketo, M. M., Dufort, D., Joyner, A. L., et al. (2006). Beta-catenin activation is necessary and sufficient to specify the dorsal dermal fate in the mouse. *Developmental Biology*, *296*, 164–176.

Auerbach, R. (1954). Analysis of the developmental effects of a lethal mutation in the house mouse. *Journal of Experimental Zoology*, *127*, 305–329.

Awaya, T., Kato, T., Mizuno, Y., Chang, H., Niwa, A., Umeda, K., et al. (2012). Selective development of myogenic mesenchymal cells from human embryonic and induced pluripotent stem cells. *PLoS One*, *7*, e51638.

Bajard, L., Relaix, F., Lagha, M., Rocancourt, D., Daubas, P., & Buckingham, M. E. (2006). A novel genetic hierarchy functions during hypaxial myogenesis: Pax3 directly activates Myf5 in muscle progenitor cells in the limb. *Genes & Development*, *20*, 2450–2464.

Ben-Yair, R., & Kalcheim, C. (2008). Notch and bone morphogenetic protein differentially act on dermomyotome cells to generate endothelium, smooth, and striated muscle. *The Journal of Cell Biology*, *180*, 607–618.

Benezra, R., Davis, R. L., Lockshon, D., Turner, D. L., & Weintraub, H. (1990). The protein Id: A negative regulator of helix-loop-helix DNA binding proteins. *Cell, 61*, 49–59.
Berger, S. L. (2007). The complex language of chromatin regulation during transcription. *Nature, 447*, 407–412.
Bergstrom, D. A., Penn, B. H., Strand, A., Perry, R. L., Rudnicki, M. A., & Tapscott, S. J. (2002). Promoter-specific regulation of MyoD binding and signal transduction cooperate to pattern gene expression. *Molecular Cell, 9*, 587–600.
Bergstrom, D. A., & Tapscott, S. J. (2001). Molecular distinction between specification and differentiation in the myogenic basic helix-loop-helix transcription factor family. *Molecular and Cellular Biology, 21*, 2404–2412.
Berkes, C. A., Bergstrom, D. A., Penn, B. H., Seaver, K. J., Knoepfler, P. S., & Tapscott, S. J. (2004). Pbx marks genes for activation by MyoD indicating a role for a homeodomain protein in establishing myogenic potential. *Molecular Cell, 14*, 465–477.
Biesiada, E., Hamamori, Y., Kedes, L., & Sartorelli, V. (1999). Myogenic basic helix-loop-helix proteins and Sp1 interact as components of a multiprotein transcriptional complex required for activity of the human cardiac alpha-actin promoter. *Molecular and Cellular Biology, 19*, 2577–2584.
Biressi, S., Molinaro, M., & Cossu, G. (2007). Cellular heterogeneity during vertebrate skeletal muscle development. *Developmental Biology, 308*, 281–293.
Biressi, S., Tagliafico, E., Lamorte, G., Monteverde, S., Tenedini, E., Roncaglia, E., et al. (2007). Intrinsic phenotypic diversity of embryonic and fetal myoblasts is revealed by genome-wide gene expression analysis on purified cells. *Developmental Biology, 304*(2), 633–651.
Black, B. L., Molkentin, J. D., & Olson, E. N. (1998). Multiple roles for the MyoD basic region in transmission of transcriptional activation signals and interaction with MEF2. *Molecular and Cellular Biology, 18*, 69–77.
Blackwell, T. K., & Weintraub, H. (1990). Differences and similarities in DNA-binding preferences of MyoD and E2A protein complexes revealed by binding site selection. *Science, 250*, 1104–1110.
Blais, A., Tsikitis, M., Acosta-Alvear, D., Sharan, R., Kluger, Y., & Dynlacht, B. D. (2005). An initial blueprint for myogenic differentiation. *Genes & Development, 19*, 553–569.
Blum, B., & Benvenisty, N. (2008). The tumorigenicity of human embryonic stem cells. *Advances in Cancer Research, 100*, 133–158.
Blum, R., Vethantham, V., Bowman, C., Rudnicki, M., & Dynlacht, B. D. (2012). Genome-wide identification of enhancers in skeletal muscle: The role of MyoD1. *Genes & Development, 26*, 2763–2779.
Bober, E., Franz, T., Arnold, H. H., Gruss, P., & Tremblay, P. (1994). Pax-3 is required for the development of limb muscles: A possible role for the migration of dermomyotomal muscle progenitor cells. *Development, 120*, 603–612.
Bolos, V., Peinado, H., Perez-Moreno, M. A., Fraga, M. F., Esteller, M., & Cano, A. (2003). The transcription factor Slug represses E-cadherin expression and induces epithelial to mesenchymal transitions: A comparison with Snail and E47 repressors. *Journal of Cell Science, 116*, 499–511.
Bothe, I., & Dietrich, S. (2006). The molecular setup of the avian head mesoderm and its implication for craniofacial myogenesis. *Developmental Dynamics, 235*, 2845–2860.
Bothe, I., Tenin, G., Oseni, A., & Dietrich, S. (2011). Dynamic control of head mesoderm patterning. *Development, 138*, 2807–2821.
Boukhatmi, H., Frendo, J. L., Enriquez, J., Crozatier, M., Dubois, L., & Vincent, A. (2012). Tup/Islet1 integrates time and position to specify muscle identity in Drosophila. *Development, 139*, 3572–3582.
Braun, T., & Arnold, H. H. (1995). Inactivation of Myf-6 and Myf-5 genes in mice leads to alterations in skeletal muscle development. *The EMBO Journal, 14*, 1176–1186.

Braun, T., Bober, E., Winter, B., Rosenthal, N., & Arnold, H.-H. (1990). Myf-6 a new member of the human gene family of myogenic determination factors: Evidence for a gene cluster on chromosome 12. *The EMBO Journal*, *9*, 821–831.

Braun, T., Buschhausen-Denker, G., Bober, E., Tannich, E., & Arnold, H.-H. (1989). A novel human muscle factor related to but distinct from MyoD1 induces myogenic conversion in 10T1/2 fibroblasts. *The EMBO Journal*, *8*, 701–709.

Braun, T., Rudnicki, M. A., Arnold, H. H., & Jaenisch, R. (1992). Targeted inactivation of the muscle regulatory gene Myf-5 results in abnormal rib development and perinatal death. *Cell*, *71*, 369–382.

Brennan, T. J., Chakraborty, T., & Olson, E. N. (1991). Mutagenesis of the myogenin basic region identifies an ancient protein motif critical for activation of myogenesis. *Proceedings of the National Academy of Sciences of the United States of America*, *88*, 5675–5679.

Brown, C. B., Engleka, K. A., Wenning, J., Min Lu, M., & Epstein, J. A. (2005). Identification of a hypaxial somite enhancer element regulating Pax3 expression in migrating myoblasts and characterization of hypaxial muscle Cre transgenic mice. *Genesis*, *41*, 202–209.

Buckingham, M., & Relaix, F. (2007). The role of Pax genes in the development of tissues and organs: Pax3 and Pax7 regulate muscle progenitor cell functions. *Annual Review of Cell and Developmental Biology*, *23*, 645–673.

Budry, L., Balsalobre, A., Gauthier, Y., Khetchoumian, K., L'Honore, A., Vallette, S., et al. (2012). The selector gene Pax7 dictates alternate pituitary cell fates through its pioneer action on chromatin remodeling. *Genes & Development*, *26*, 2299–2310.

Bulut-Karslioglu, A., Perrera, V., Scaranaro, M., de la Rosa-Velazquez, I. A., van de Nobelen, S., Shukeir, N., et al. (2012). A transcription factor-based mechanism for mouse heterochromatin formation. *Nature Structural & Molecular Biology*, *19*, 1023–1030.

Cai, C. L., Liang, X., Shi, Y., Chu, P. H., Pfaff, S. L., Chen, J., et al. (2003). Isl1 identifies a cardiac progenitor population that proliferates prior to differentiation and contributes a majority of cells to the heart. *Developmental Cell*, *5*, 877–889.

Cao, L., Yu, Y., Bilke, S., Walker, R. L., Mayeenuddin, L. H., Azorsa, D. O., et al. (2010). Genome-wide identification of PAX3-FKHR binding sites in rhabdomyosarcoma reveals candidate target genes important for development and cancer. *Cancer Research*, *70*, 6497–6508.

Cao, Y., Kumar, R. M., Penn, B. H., Berkes, C. A., Kooperberg, C., Boyer, L. A., et al. (2006). Global and gene-specific analyses show distinct roles for Myod and Myog at a common set of promoters. *The EMBO Journal*, *25*, 502–511.

Cao, Y., Yao, Z., Sarkar, D., Lawrence, M., Sanchez, G. J., Parker, M. H., et al. (2010). Genome-wide MyoD binding in skeletal muscle cells: A potential for broad cellular reprogramming. *Developmental Cell*, *18*, 662–674.

Caretti, G., Di Padova, M., Micales, B., Lyons, G. E., & Sartorelli, V. (2004). The Polycomb Ezh2 methyltransferase regulates muscle gene expression and skeletal muscle differentiation. *Genes & Development*, *18*, 2627–2638.

Carvajal, J. J., Keith, A., & Rigby, P. W. (2008). Global transcriptional regulation of the locus encoding the skeletal muscle determination genes Mrf4 and Myf5. *Genes & Development*, *22*, 265–276.

Carvajal, J. J., & Rigby, P. W. (2010). Regulation of gene expression in vertebrate skeletal muscle. *Experimental Cell Research*, *316*, 3014–3018.

Chalepakis, G., & Gruss, P. (1995). Identification of DNA recognition sequences for the Pax3 paired domain. *Gene*, *162*, 267–270.

Chalepakis, G., Wijnholds, J., & Gruss, P. (1994). Pax-3-DNA interaction: Flexibility in the DNA binding and induction of DNA conformational changes by paired domains. *Nucleic Acids Research*, *22*, 3131–3137.

Chen, B., Han, B. H., Sun, X. H., & Lim, R. W. (1997). Inhibition of muscle-specific gene expression by Id3: Requirement of the C-terminal region of the protein for stable expression and function. *Nucleic Acids Research*, *25*, 423–430.

Chen, J. C., & Goldhamer, D. J. (2004). The core enhancer is essential for proper timing of MyoD activation in limb buds and branchial arches. *Developmental Biology*, *265*, 502–512.

Choi, J., Costa, M., Mermelstein, C., Chagas, C., Holtzer, S., & Holtzer, H. (1990). MyoD converts primary dermal fibroblasts, chondroblasts, smooth muscle, and retinal pigmented epithelial cells into striated mononucleated myoblasts and multinucleated myotubes. *Proceedings of the National Academy of Sciences of the United States of America*, *87*, 7988–8022.

Cirillo, L. A., Lin, F. R., Cuesta, I., Friedman, D., Jarnik, M., & Zaret, K. S. (2002). Opening of compacted chromatin by early developmental transcription factors HNF3 (FoxA) and GATA-4. *Molecular Cell*, *9*, 279–289.

Conway, K., Pin, C., Kiernan, J. A., & Merrifield, P. (2004). The E protein HEB is preferentially expressed in developing muscle. *Differentiation*, *72*, 327–340.

Comai, G., Sambasivan, R., Gopalakrishnan, S., & Tajbakhsh, S. (2014). Variations in the efficiency of lineage marking and ablation confound distinctions between myogenic cell populations (*Developmental Cell* in press).

Darabi, R., Arpke, R. W., Irion, S., Dimos, J. T., Grskovic, M., Kyba, M., et al. (2012). Human ES- and iPS-derived myogenic progenitors restore DYSTROPHIN and improve contractility upon transplantation in dystrophic mice. *Cell Stem Cell*, *10*, 610–619.

Darabi, R., Gehlbach, K., Bachoo, R., Kamath, S., Osawa, M., Kamm, K., et al. (2008). Functional skeletal muscle regeneration from differentiating embryonic stem cells. *Nature Medicine*, *14*, 134–143.

Darabi, R., Pan, W., Bosnakovski, D., Baik, J., Kyba, M., & Perlingeiro, R. C. (2011). Functional myogenic engraftment from mouse iPS cells. *Stem Cell Reviews*, *7*, 948–957.

Darabi, R., Santos, F. N., Filareto, A., Pan, W., Koene, R., Rudnicki, M. A., et al. (2011). Assessment of the myogenic stem cell compartment following transplantation of Pax3/Pax7-induced embryonic stem cell-derived progenitors. *Stem Cells*, *29*, 777–790.

Dastjerdi, A., Robson, L., Walker, R., Hadley, J., Zhang, Z., Rodriguez-Niedenfuhr, M., et al. (2007). Tbx1 regulation of myogenic differentiation in the limb and cranial mesoderm. *Developmental Dynamics*, *236*, 353–363.

Davie, J. K., Cho, J. H., Meadows, E., Flynn, J. M., Knapp, J. R., & Klein, W. H. (2007). Target gene selectivity of the myogenic basic helix-loop-helix transcription factor myogenin in embryonic muscle. *Developmental Biology*, *311*, 650–664.

Davis, R. L., Cheng, P. F., Lassar, A. B., & Weintraub, H. (1990). The MyoD DNA binding domain contains a recognition code for muscle-specific gene activation. *Cell*, *60*, 733–746.

Davis, R. L., & Weintraub, H. (1992). Acquisition of myogenic specificity by replacement of three amino acid residues from MyoD into E12. *Science*, *256*, 1027–1030.

Davis, R. L., Weintraub, H., & Lassar, A. B. (1987). Expression of a single transfected cDNA converts fibroblasts to myoblasts. *Cell*, *51*, 987–1000.

de la Serna, I. L., Carlson, K. A., & Imbalzano, A. N. (2001). Mammalian SWI/SNF complexes promote MyoD-mediated muscle differentiation. *Nature Genetics*, *27*, 187–190.

de la Serna, I. L., Ohkawa, Y., Berkes, C. A., Bergstrom, D. A., Dacwag, C. S., Tapscott, S. J., et al. (2005). MyoD targets chromatin remodeling complexes to the myogenin locus prior to forming a stable DNA-bound complex. *Molecular and Cellular Biology*, *25*, 3997–4009.

Degenhardt, K. R., Milewski, R. C., Padmanabhan, A., Miller, M., Singh, M. K., Lang, D., et al. (2010). Distinct enhancers at the Pax3 locus can function redundantly to regulate neural tube and neural crest expressions. *Developmental Biology*, *339*, 519–527.

Dekel, I., Magal, Y., Pearson-White, S., Emerson, C. P., & Shani, M. (1992). Conditional conversion of ES cells to skeletal muscle by an exogenous MyoD1 gene. *The New Biologist*, *4*, 217–224.

Delfini, M. C., & Duprez, D. (2004). Ectopic Myf5 or MyoD prevents the neuronal differentiation program in addition to inducing skeletal muscle differentiation, in the chick neural tube. *Development*, *131*, 713–723.

Delfini, M. C., Hirsinger, E., Pourquie, O., & Duprez, D. (2000). Delta 1-activated notch inhibits muscle differentiation without affecting Myf5 and Pax3 expression in chick limb myogenesis. *Development, 127*, 5213–5224.

Di Padova, M., Caretti, G., Zhao, P., Hoffman, E. P., & Sartorelli, V. (2007). MyoD acetylation influences temporal patterns of skeletal muscle gene expression. *The Journal of Biological Chemistry, 282*, 37650–37659.

Diehl, A. G., Zareparsi, S., Qian, M., Khanna, R., Angeles, R., & Gage, P. J. (2006). Extraocular muscle morphogenesis and gene expression are regulated by Pitx2 gene dose. *Investigative Ophthalmology & Visual Science, 47*, 1785–1793.

Dilworth, F. J., Seaver, K. J., Fishburn, A. L., Htet, S. L., & Tapscott, S. J. (2004). In vitro transcription system delineates the distinct roles of the coactivators pCAF and p300 during MyoD/E47-dependent transactivation. *Proceedings of the National Academy of Sciences of the United States of America, 101*, 11593–11598.

Dong, F., Sun, X., Liu, W., Ai, D., Klysik, E., Lu, M. F., et al. (2006). Pitx2 promotes development of splanchnic mesoderm-derived branchiomeric muscle. *Development, 133*, 4891–4899.

Dong, Z., Yang, N., Yeo, S. Y., Chitnis, A., & Guo, S. (2012). Intralineage directional Notch signaling regulates self-renewal and differentiation of asymmetrically dividing radial glia. *Neuron, 74*, 65–78.

Edmondson, D. G., & Olson, E. N. (1989). A gene with homology to the myc similarity region of MyoD1 is expressed during myogenesis and is sufficient to activate the muscle differentiation program. *Genes & Development, 3*, 628–640.

Epstein, J. A., Shapiro, D. N., Cheng, J., Lam, P. Y., & Maas, R. L. (1996). Pax3 modulates expression of the c-Met receptor during limb muscle development. *Proceedings of the National Academy of Sciences of the United States of America, 93*, 4213–4218.

Etzioni, S., Yafe, A., Khateb, S., Weisman-Shomer, P., Bengal, E., & Fry, M. (2005). Homodimeric MyoD preferentially binds tetraplex structures of regulatory sequences of muscle-specific genes. *The Journal of Biological Chemistry, 280*, 26805–26812.

Faerman, A., Goldhamer, D. J., Puzis, R., Emerson, C. P., Jr., & Shani, M. (1995). The distal human myoD enhancer sequences direct unique muscle-specific patterns of lacZ expression during mouse development. *Developmental Biology, 171*, 27–38.

Faerman, A., Pearson-White, S., Emerson, C., & Shani, M. (1993). Ectopic expression of MyoD1 in mice causes prenatal lethalities. *Developmental Dynamics, 196*, 165–173.

Farah, M. H., Olson, J. M., Sucic, H. B., Hume, R. I., Tapscott, S. J., & Turner, D. L. (2000). Generation of neurons by transient expression of neural bHLH proteins in mammalian cells. *Development, 127*, 693–702.

Ferrari, G., Cusella-De Angelis, G., Coletta, M., Paolucci, E., Stornaiuolo, A., Cossu, G., et al. (1998). Muscle regeneration by bone marrow-derived myogenic progenitors. *Science, 279*, 1528–1530 (see comments, published erratum appears in *Science, 281*(5379), 923 (1998)).

Filareto, A., Parker, S., Darabi, R., Borges, L., Iacovino, M., Schaaf, T., et al. (2013). An ex vivo gene therapy approach to treat muscular dystrophy using inducible pluripotent stem cells. *Nature Communications, 4*, 1549.

Fong, A. P., & Tapscott, S. J. (2013). Skeletal muscle programming and re-programming. *Current Opinion in Genetics & Development, 23*, 568–573.

Fong, A. P., Yao, Z., Zhong, J. W., Cao, Y., Ruzzo, W. L., Gentleman, R. C., et al. (2012). Genetic and epigenetic determinants of neurogenesis and myogenesis. *Developmental Cell, 22*, 721–735.

Forcales, S. V. (2012). The BAF60c-MyoD complex poises chromatin for rapid transcription. *BioArchitecture, 2*, 104–109.

Forcales, S. V., Albini, S., Giordani, L., Malecova, B., Cignolo, L., Chernov, A., et al. (2012). Signal-dependent incorporation of MyoD-BAF60c into Brg1-based SWI/SNF chromatin-remodelling complex. *The EMBO Journal, 31*, 301–316.

Franz, T., Kothary, R., Surani, M. A., Halata, Z., & Grim, M. (1993). The Splotch mutation interferes with muscle development in the limbs. *Anatomy and Embryology*, *187*, 153–160.

Fukushige, T., & Krause, M. (2005). The myogenic potency of HLH-1 reveals wide-spread developmental plasticity in early C. elegans embryos. *Development*, *132*, 1795–1805.

Gage, P. J., Suh, H., & Camper, S. A. (1999). Dosage requirement of Pitx2 for development of multiple organs. *Development*, *126*, 4643–4651.

Gayraud-Morel, B., Chretien, F., Flamant, P., Gomes, D., Zammit, P. S., & Tajbakhsh, S. (2007). A role for the myogenic determination gene Myf5 in adult regenerative myogenesis. *Developmental Biology*, *312*, 13–28.

Gayraud-Morel, B., Chretien, F., Jory, A., Sambasivan, R., Negroni, E., Flamant, P., et al. (2012). Myf5 haploinsufficiency reveals distinct cell fate potentials for adult skeletal muscle stem cells. *Journal of Cell Science*, *125*(Pt. 7), 1738–1749.

Gensch, N., Borchardt, T., Schneider, A., Riethmacher, D., & Braun, T. (2008). Different autonomous myogenic cell populations revealed by ablation of Myf5-expressing cells during mouse embryogenesis. *Development*, *135*, 1597–1604.

Gerber, A. N., Klesert, T. R., Bergstrom, D. A., & Tapscott, S. J. (1997). Two domains of MyoD mediate transcriptional activation of genes in repressive chromatin—A mechanism for lineage determination in myogenesis. *Genes & Development*, *11*, 436–450.

Gilbert, P. M., Havenstrite, K. L., Magnusson, K. E., Sacco, A., Leonardi, N. A., Kraft, P., et al. (2010). Substrate elasticity regulates skeletal muscle stem cell self-renewal in culture. *Science*, *329*, 1078–1081.

Giordani, J., Bajard, L., Demignon, J., Daubas, P., Buckingham, M., & Maire, P. (2007). Six proteins regulate the activation of Myf5 expression in embryonic mouse limbs. *Proceedings of the National Academy of Sciences of the United States of America*, *104*, 11310–11315.

Giordani, L., & Puri, P. L. (2013). Epigenetic control of skeletal muscle regeneration: Integrating genetic determinants and environmental changes. *The FEBS Journal*, *280*, 4014–4025.

Goldhamer, D. J., Brunk, B. P., Faerman, A., King, A., Shani, M., & Emerson, C. P., Jr. (1995). Embryonic activation of the myoD gene is regulated by a highly conserved distal control element. *Development*, *121*, 637–649.

Goldhamer, D. J., Faerman, A., Shani, M., & Emerson, C., Jr. (1992). Regulatory elements that control the lineage-specific expression of myoD. *Science*, *256*, 538–542.

Goudenege, S., Lebel, C., Huot, N. B., Dufour, C., Fujii, I., Gekas, J., et al. (2012). Myoblasts derived from normal hESCs and dystrophic hiPSCs efficiently fuse with existing muscle fibers following transplantation. *Molecular Therapy*, *20*, 2153–2167.

Goulding, M., Lumsden, A., & Paquette, A. J. (1994). Regulation of Pax-3 expression in the dermomyotome and its role in muscle development. *Development*, *120*, 957–971.

Goulding, M., Sterrer, S., Fleming, J., Balling, R., Nadeau, J., Moore, K. J., et al. (1993). Analysis of the Pax-3 gene in the mouse mutant splotch. *Genomics*, *17*, 355–363.

Grifone, R., Demignon, J., Giordani, J., Niro, C., Souil, E., Bertin, F., et al. (2007). Eya1 and Eya2 proteins are required for hypaxial somitic myogenesis in the mouse embryo. *Developmental Biology*, *302*, 602–616.

Grifone, R., Demignon, J., Houbron, C., Souil, E., Niro, C., Seller, M. J., et al. (2005). Six1 and Six4 homeoproteins are required for Pax3 and Mrf expression during myogenesis in the mouse embryo. *Development*, *132*, 2235–2249.

Grifone, R., Jarry, T., Dandonneau, M., Grenier, J., Duprez, D., & Kelly, R. G. (2008). Properties of branchiomeric and somite-derived muscle development in Tbx1 mutant embryos. *Developmental Dynamics*, *237*, 3071–3078.

Guasconi, V., & Puri, P. L. (2009). Chromatin: The interface between extrinsic cues and the epigenetic regulation of muscle regeneration. *Trends in Cell Biology*, *19*, 286–294.

Gunther, S., Kim, J., Kostin, S., Lepper, C., Fan, C. M., & Braun, T. (2013). Myf5-positive satellite cells contribute to Pax7-dependent long-term maintenance of adult muscle stem cells. *Cell Stem Cell*, *13*, 590–601.

Guo, K., Wang, J., Andres, V., Smith, R. C., & Walsh, K. (1995). MyoD-induced expression of p21 inhibits cyclin-dependent kinase activity upon myocyte terminal differentiation. *Molecular and Cellular Biology, 15*, 3823–3829.

Hacker, A., & Guthrie, S. (1998). A distinct developmental programme for the cranial paraxial mesoderm in the chick embryo. *Development, 125*, 3461–3472.

Hajra, K. M., Chen, D. Y., & Fearon, E. R. (2002). The SLUG zinc-finger protein represses E-cadherin in breast cancer. *Cancer Research, 62*, 1613–1618.

Haldar, M., Karan, G., Tvrdik, P., & Capecchi, M. R. (2008). Two cell lineages, myf5 and myf5-independent, participate in mouse skeletal myogenesis. *Developmental Cell, 14*, 437–445.

Hamamori, Y., Wu, H. Y., Sartorelli, V., & Kedes, L. (1997). The basic domain of myogenic basic helix-loop-helix (bHLH) proteins is the novel target for direct inhibition by another bHLH protein, Twist. *Molecular and Cellular Biology, 17*, 6563–6573.

Han, D. W., Tapia, N., Hermann, A., Hemmer, K., Hoing, S., Arauzo-Bravo, M. J., et al. (2012). Direct reprogramming of fibroblasts into neural stem cells by defined factors. *Cell Stem Cell, 10*, 465–472.

Harel, I., Maezawa, Y., Avraham, R., Rinon, A., Ma, H. Y., Cross, J. W., et al. (2012). Pharyngeal mesoderm regulatory network controls cardiac and head muscle morphogenesis. *Proceedings of the National Academy of Sciences of the United States of America, 109*, 18839–18844.

Harel, I., Nathan, E., Tirosh-Finkel, L., Zigdon, H., Guimaraes-Camboa, N., Evans, S. M., et al. (2009). Distinct origins and genetic programs of head muscle satellite cells. *Developmental Cell, 16*, 822–832.

Hasty, P., Bradley, A., Morris, J. H., Edmondson, D. G., Venuti, J. M., Olson, E. N., et al. (1993). Muscle deficiency and neonatal death in mice with a targeted mutation in the myogenin gene. *Nature, 364*, 501–506.

Hayashi, S., & McMahon, A. P. (2002). Efficient recombination in diverse tissues by a tamoxifen-inducible form of Cre: A tool for temporally regulated gene activation/inactivation in the mouse. *Developmental Biology, 244*, 305–318.

Hebert, S. L., Daniel, M. L., & McLoon, L. K. (2013). The role of Pitx2 in maintaining the phenotype of myogenic precursor cells in the extraocular muscles. *PLoS One, 8*, e58405.

Heintzman, N. D., Hon, G. C., Hawkins, R. D., Kheradpour, P., Stark, A., Harp, L. F., et al. (2009). Histone modifications at human enhancers reflect global cell-type-specific gene expression. *Nature, 459*, 108–112.

Henrique, D., Hirsinger, E., Adam, J., Le Roux, I., Pourquie, O., Ish-Horowicz, D., et al. (1997). Maintenance of neuroepithelial progenitor cells by Delta-Notch signalling in the embryonic chick retina. *Current Biology, 7*, 661–670.

Henthorn, P., Kiledjian, M., & Kadesch, T. (1990). Two distinct transcription factors that bind the immunoglobulin enhancer microE5/kappa 2 motif. *Science, 247*, 467–470.

Hinterberger, T. J., Sassoon, D. A., Rhodes, S. J., & Konieczny, S. F. (1991). Expression of the muscle regulatory factor MRF4 during somite and skeletal myofiber development. *Developmental Biology, 147*, 144–156.

Hirsinger, E., Malapert, P., Dubrulle, J., Delfini, M. C., Duprez, D., Henrique, D., et al. (2001). Notch signalling acts in postmitotic avian myogenic cells to control MyoD activation. *Development, 128*, 107–116.

Ho, L., & Crabtree, G. R. (2010). Chromatin remodelling during development. *Nature, 463*, 474–484.

Hopwood, N. D., & Gurdon, J. B. (1990). Activation of muscle genes without myogenesis by ectopic expression of MyoD in frog embryo cells. *Nature, 347*, 197–200.

Hopwood, N. D., Pluck, A., & Gurdon, J. B. (1991). Xenopus Myf-5 marks early muscle cells and can activate muscle genes ectopically in early embryos. *Development, 111*, 551–560.

Horst, D., Ustanina, S., Sergi, C., Mikuz, G., Juergens, H., Braun, T., et al. (2006). Comparative expression analysis of Pax3 and Pax7 during mouse myogenesis. *The International journal of developmental biology*, *50*, 47–54.

Hu, E., Tontonoz, P., & Spiegelman, B. M. (1995). Transdifferentiation of myoblasts by the adipogenic transcription factors PPAR gamma and C/EBP alpha. *Proceedings of the National Academy of Sciences of the United States of America*, *92*, 9856–9860.

Hu, P., Geles, K. G., Paik, J. H., DePinho, R. A., & Tjian, R. (2008). Codependent activators direct myoblast-specific MyoD transcription. *Developmental Cell*, *15*, 534–546.

Huang, P., He, Z., Ji, S., Sun, H., Xiang, D., Liu, C., et al. (2011). Induction of functional hepatocyte-like cells from mouse fibroblasts by defined factors. *Nature*, *475*, 386–389.

Hutcheson, D. A., & Kardon, G. (2009). Genetic manipulations reveal dynamic cell and gene functions: Creating a new view of myogenesis. *Cell Cycle*, *8*, 3675–3678.

Hutcheson, D. A., Zhao, J., Merrell, A., Haldar, M., & Kardon, G. (2009). Embryonic and fetal limb myogenic cells are derived from developmentally distinct progenitors and have different requirements for beta-catenin. *Genes & Development*, *23*, 997–1013.

Hutson, M. R., & Kirby, M. L. (2003). Neural crest and cardiovascular development: A 20-year perspective. *Birth Defects Research. Part C, Embryo Today Reviews*, *69*, 2–13.

Hwang, Y., Suk, S., Lin, S., Tierney, M., Du, B., Seo, T., et al. (2013). Directed in vitro myogenesis of human embryonic stem cells and their in vivo engraftment. *PLoS One*, *8*, e72023.

Iacovino, M., Bosnakovski, D., Fey, H., Rux, D., Bajwa, G., Mahen, E., et al. (2011). Inducible cassette exchange: A rapid and efficient system enabling conditional gene expression in embryonic stem and primary cells. *Stem Cells*, *29*, 1580–1588.

Ieda, M., Fu, J. D., Delgado-Olguin, P., Vedantham, V., Hayashi, Y., Bruneau, B. G., et al. (2010). Direct reprogramming of fibroblasts into functional cardiomyocytes by defined factors. *Cell*, *142*, 375–386.

Ishibashi, J., Perry, R. L., Asakura, A., & Rudnicki, M. A. (2005). MyoD induces myogenic differentiation through cooperation of its NH2- and COOH-terminal regions. *The Journal of Cell Biology*, *171*, 471–482.

Jarad, G., & Miner, J. H. (2009). The Pax3-Cre transgene exhibits a rostrocaudal gradient of expression in the skeletal muscle lineage. *Genesis*, *47*, 1–6.

Jerome, L. A., & Papaioannou, V. E. (2001). DiGeorge syndrome phenotype in mice mutant for the T-box gene, Tbx1. *Nature Genetics*, *27*, 286–291.

Kablar, B., Krastel, K., Ying, C., Asakura, A., Tapscott, S. J., & Rudnicki, M. A. (1997). MyoD and Myf-5 differentially regulate the development of limb versus trunk skeletal muscle. *Development*, *124*, 4729–4738.

Kaji, K., Norrby, K., Paca, A., Mileikovsky, M., Mohseni, P., & Woltjen, K. (2009). Virus-free induction of pluripotency and subsequent excision of reprogramming factors. *Nature*, *458*, 771–775.

Kassar-Duchossoy, L., Gayraud-Morel, B., Gomès, D., Rocancourt, D., Buckingham, M., Shinin, V., et al. (2004). Mrf4 determines skeletal muscle identity in Myf5:Myod double-mutant mice. *Nature*, *431*, 466–471.

Kassar-Duchossoy, L., Giacone, E., Gayraud-Morel, B., Jory, A., Gomes, D., & Tajbakhsh, S. (2005). Pax3/Pax7 mark a novel population of primitive myogenic cells during development. *Genes & Development*, *19*, 1426–1431.

Kaul, A., Koster, M., Neuhaus, H., & Braun, T. (2000). Myf-5 revisited: Loss of early myotome formation does not lead to a rib phenotype in homozygous Myf-5 mutant mice. *Cell*, *102*, 17–19.

Kawakami, K., Sato, S., Ozaki, H., & Ikeda, K. (2000). Six family genes–structure and function as transcription factors and their roles in development. *Bioessays*, *22*, 616–626.

Kazuki, Y., Hiratsuka, M., Takiguchi, M., Osaki, M., Kajitani, N., Hoshiya, H., et al. (2010). Complete genetic correction of iPS cells from Duchenne muscular dystrophy. *Molecular Therapy, 18*, 386–393.

Kelly, R. G. (2013). Craniofacial muscle development. In George T. J. Huang, & I. Thesleff (Eds.), *Stem cells in craniofacial development and regeneration* (pp. 87–107). Hoboken, NJ: John Wiley & Sons, Inc.

Kelly, R. G., Jerome-Majewska, L. A., & Papaioannou, V. E. (2004). The del22q11.2 candidate gene Tbx1 regulates branchiomeric myogenesis. *Human Molecular Genetics, 13*, 2829–2840.

Kim, D., Kim, C. H., Moon, J. I., Chung, Y. G., Chang, M. Y., Han, B. S., et al. (2009). Generation of human induced pluripotent stem cells by direct delivery of reprogramming proteins. *Cell Stem Cell, 4*, 472–476.

Kitamura, K., Miura, H., Miyagawa-Tomita, S., Yanazawa, M., Katoh-Fukui, Y., Suzuki, R., et al. (1999). Mouse Pitx2 deficiency leads to anomalies of the ventral body wall, heart, extra- and periocular mesoderm and right pulmonary isomerism. *Development, 126*, 5749–5758.

Knapp, J. R., Davie, J. K., Myer, A., Meadows, E., Olson, E. N., & Klein, W. H. (2006). Loss of myogenin in postnatal life leads to normal skeletal muscle but reduced body size. *Development, 133*, 601–610.

Knoepfler, P. S., Bergstrom, D. A., Uetsuki, T., Dac-Korytko, I., Sun, Y. H., Wright, W. E., et al. (1999). A conserved motif N-terminal to the DNA-binding domains of myogenic bHLH transcription factors mediates cooperative DNA binding with pbx-Meis1/Prep1. *Nucleic Acids Research, 27*, 3752–3761.

Kok, F. O., Oster, E., Mentzer, L., Hsieh, J. C., Henry, C. A., & Sirotkin, H. I. (2007). The role of the SPT6 chromatin remodeling factor in zebrafish embryogenesis. *Developmental Biology, 307*, 214–226.

Kong, P., Racedo, S. E., Macchiarulo, S., Hu, Z., Carpenter, C., Guo, T., et al. (2014). Tbx1 is required autonomously for cell survival and fate in the pharyngeal core mesoderm to form the muscles of mastication. *Human Molecular Genetics, 23*(16), 4215–4231.

Krause, M., Fire, A., White-Harrison, S., Weintraub, H., & Tapscott, S. (1992). Functional conservation of nematode and vertebrate myogenic regulatory factors. *Journal of Cell Science Supplement, 16*, 111–115.

Krause, M., & Weintraub, H. (1992). CeMyoD expression and myogenesis in *C. elegans*. *Seminars in Developmental Biology, 3*, 277–285.

Kuang, S., Charge, S. B., Seale, P., Huh, M., & Rudnicki, M. A. (2006). Distinct roles for Pax7 and Pax3 in adult regenerative myogenesis. *The Journal of Cell Biology, 172*, 103–113.

Kumar, D., Shadrach, J. L., Wagers, A. J., & Lassar, A. B. (2009). Id3 is a direct transcriptional target of Pax7 in quiescent satellite cells. *Molecular Biology of the Cell, 20*, 3170–3177.

L'Honore, A., Commere, P. H., Ouimette, J. F., Montarras, D., Drouin, J., & Buckingham, M. (2014). Redox regulation by pitx2 and pitx3 is critical for fetal myogenesis. *Developmental Cell, 29*, 392–405.

L'Honore, A., Coulon, V., Marcil, A., Lebel, M., Lafrance-Vanasse, J., Gage, P., et al. (2007). Sequential expression and redundancy of Pitx2 and Pitx3 genes during muscle development. *Developmental Biology, 307*, 421–433.

L'Honore, A., Ouimette, J. F., Lavertu-Jolin, M., & Drouin, J. (2010). Pitx2 defines alternate pathways acting through MyoD during limb and somitic myogenesis. *Development, 137*, 3847–3856.

Lagha, M., Kormish, J. D., Rocancourt, D., Manceau, M., Epstein, J. A., Zaret, K. S., et al. (2008). Pax3 regulation of FGF signaling affects the progression of embryonic progenitor cells into the myogenic program. *Genes & Development, 22*, 1828–1837.

Lagha, M., Sato, T., Bajard, L., Daubas, P., Esner, M., Montarras, D., et al. (2008). Regulation of skeletal muscle stem cell behavior by Pax3 and Pax7. *Cold Spring Harbor Symposia on Quantitative Biology*, *73*, 307–315.

Lagha, M., Sato, T., Regnault, B., Cumano, A., Zuniga, A., Licht, J., et al. (2010). Transcriptome analyses based on genetic screens for Pax3 myogenic targets in the mouse embryo. *BMC Genomics*, *11*, 696.

Lang, D., Brown, C. B., Milewski, R., Jiang, Y. Q., Lu, M. M., & Epstein, J. A. (2003). Distinct enhancers regulate neural expression of Pax7. *Genomics*, *82*, 553–560.

Lapidos, K. A., Chen, Y. E., Earley, J. U., Heydemann, A., Huber, J. M., Chien, M., et al. (2004). Transplanted hematopoietic stem cells demonstrate impaired sarcoglycan expression after engraftment into cardiac and skeletal muscle. *The Journal of Clinical Investigation*, *114*, 1577–1585.

Lassar, A. B., Buskin, J. N., Lockshon, D., Davis, R. L., Apone, S., Hauschka, S. D., et al. (1989). MyoD is a sequence-specific DNA binding protein requiring a region of myc homology to bind to the muscle creatine kinase enhancer. *Cell*, *58*, 823–831.

Lassar, A. B., Davis, R. L., Wright, W. E., Kadesch, T., Murre, C., Voronova, A., et al. (1991). Functional activity of myogenic HLH proteins requires hetero-oligomerization with E12/E47-like proteins in vivo. *Cell*, *66*, 305–315.

Lee, K. S., Kim, H. J., Li, Q. L., Chi, X. Z., Ueta, C., Komori, T., et al. (2000). Runx2 is a common target of transforming growth factor beta1 and bone morphogenetic protein 2, and cooperation between Runx2 and Smad5 induces osteoblast-specific gene expression in the pluripotent mesenchymal precursor cell line C2C12. *Molecular and Cellular Biology*, *20*, 8783–8792.

Lei, H., Fukushige, T., Niu, W., Sarov, M., Reinke, V., & Krause, M. (2010). A widespread distribution of genomic CeMyoD binding sites revealed and cross validated by ChIP-Chip and ChIP-Seq techniques. *PLoS One*, *5*, e15898.

Lemercier, C., To, R. Q., Carrasco, R. A., & Konieczny, S. F. (1998). The basic helix-loop-helix transcription factor Mist1 functions as a transcriptional repressor of myoD. *The EMBO Journal*, *17*, 1412–1422.

Lepper, C., Conway, S. J., & Fan, C. M. (2009). Adult satellite cells and embryonic muscle progenitors have distinct genetic requirements. *Nature*, *460*(7255), 627–631.

Lepper, C., & Fan, C. M. (2010). Inducible lineage tracing of Pax7-descendant cells reveals embryonic origin of adult satellite cells. *Genesis*, *48*, 424–436.

Lepper, C., Partridge, T. A., & Fan, C. M. (2011). An absolute requirement for Pax7-positive satellite cells in acute injury-induced skeletal muscle regeneration. *Development*, *138*, 3639–3646.

Lescroart, F., Kelly, R. G., Le Garrec, J. F., Nicolas, J. F., Meilhac, S. M., & Buckingham, M. (2010). Clonal analysis reveals common lineage relationships between head muscles and second heart field derivatives in the mouse embryo. *Development*, *137*, 3269–3279.

Lessard, J., Wu, J. I., Ranish, J. A., Wan, M., Winslow, M. M., Staahl, B. T., et al. (2007). An essential switch in subunit composition of a chromatin remodeling complex during neural development. *Neuron*, *55*, 201–215.

Lessard, J. A., & Crabtree, G. R. (2010). Chromatin regulatory mechanisms in pluripotency. *Annual Review of Cell and Developmental Biology*, *26*, 503–532.

Li, J., Chen, F., & Epstein, J. A. (2000). Neural crest expression of Cre recombinase directed by the proximal Pax3 promoter in transgenic mice. *Genesis*, *26*, 162–164.

Li, S., Czubryt, M. P., McAnally, J., Bassel-Duby, R., Richardson, J. A., Wiebel, F. F., et al. (2005). Requirement for serum response factor for skeletal muscle growth and maturation revealed by tissue-specific gene deletion in mice. *Proceedings of the National Academy of Sciences of the United States of America*, *102*, 1082–1087.

Lin, Z., Lu, M. H., Schultheiss, T., Choi, J., Holtzer, S., DiLullo, C., et al. (1994). Sequential appearance of muscle-specific proteins in myoblasts as a function of time after cell division: Evidence for a conserved myoblast differentiation program in skeletal muscle. *Cell Motility and the Cytoskeleton, 29*, 1–19.

Liu, Y., Chu, A., Chakroun, I., Islam, U., & Blais, A. (2010). Cooperation between myogenic regulatory factors and SIX family transcription factors is important for myoblast differentiation. *Nucleic Acids Research, 38*, 6857–6871.

Liu, J., Willet, S. G., Bankaitis, E. D., Xu, Y., Wright, C. V., & Gu, G. (2013). Non-parallel recombination limits Cre-LoxP-based reporters as precise indicators of conditional genetic manipulation. *Genesis, 51*, 436–442.

Lu, J., Webb, R., Richardson, J. A., & Olson, E. N. (1999). MyoR: A muscle-restricted basic helix-loop-helix transcription factor that antagonizes the actions of MyoD. *Proceedings of the National Academy of Sciences of the United States of America, 96*, 552–557.

Lu, J. R., Bassel-Duby, R., Hawkins, A., Chang, P., Valdez, R., Wu, H., et al. (2002). Control of facial muscle development by MyoR and capsulin. *Science, 298*, 2378–2381.

Ludolph, D. C., Neff, A. W., Mescher, A. L., Malacinski, G. M., Parker, M. A., & Smith, R. C. (1994). Overexpression of XMyoD or XMyf5 in Xenopus embryos induces the formation of enlarged myotomes through recruitment of cells of nonsomitic lineage. *Developmental Biology, 166*, 18–33.

Ma, Q., Zhou, B., & Pu, W. T. (2008). Reassessment of Isl1 and Nkx2-5 cardiac fate maps using a Gata4-based reporter of Cre activity. *Developmental Biology, 323*, 98–104.

Macharia, R., Otto, A., Valasek, P., & Patel, K. (2010). Neuromuscular junction morphology, fiber-type proportions, and satellite-cell proliferation rates are altered in MyoD(-/-) mice. *Muscle & Nerve, 42*, 38–52.

MacQuarrie, K. L., Yao, Z., Fong, A. P., Diede, S. J., Rudzinski, E. R., Hawkins, D. S., et al. (2013). Comparison of genome-wide binding of MyoD in normal human myogenic cells and rhabdomyosarcomas identifies regional and local suppression of promyogenic transcription factors. *Molecular and Cellular Biology, 33*, 773–784.

Mansouri, A., Stoykova, A., Torres, M., & Gruss, P. (1996). Dysgenesis of cephalic neural crest derivatives in Pax7-/- mutant mice. *Development, 122*, 831–838.

Maroto, M., Reshef, R., Munsterberg, A. E., Koester, S., Goulding, M., & Lassar, A. B. (1997). Ectopic Pax-3 activates MyoD and Myf-5 expression in embryonic mesoderm and neural tissue. *Cell, 89*, 139–148.

Marro, S., Pang, Z. P., Yang, N., Tsai, M. C., Qu, K., Chang, H. Y., et al. (2011). Direct lineage conversion of terminally differentiated hepatocytes to functional neurons. *Cell Stem Cell, 9*, 374–382.

Martin, J. F., Miano, J. M., Hustad, C. M., Copeland, N. G., Jenkins, N. A., & Olson, E. N. (1994). A Mef2 gene that generates a muscle-specific isoform via alternative mRNA splicing. *Molecular and Cellular Biology, 14*, 1647–1656.

Mascarenhas, J. B., Littlejohn, E. L., Wolsky, R. J., Young, K. P., Nelson, M., Salgia, R., et al. (2010). PAX3 and SOX10 activate MET receptor expression in melanoma. *Pigment Cell & Melanoma Research, 23*, 225–237.

Maves, L., Waskiewicz, A. J., Paul, B., Cao, Y., Tyler, A., Moens, C. B., et al. (2007). Pbx homeodomain proteins direct Myod activity to promote fast-muscle differentiation. *Development, 134*, 3371–3382.

McKinnell, I. W., Ishibashi, J., Le Grand, F., Punch, V. G., Addicks, G. C., Greenblatt, J. F., et al. (2008). Pax7 activates myogenic genes by recruitment of a histone methyltransferase complex. *Nature Cell Biology, 10*, 77–84.

Meadows, E., Cho, J. H., Flynn, J. M., & Klein, W. H. (2008). Myogenin regulates a distinct genetic program in adult muscle stem cells. *Developmental Biology, 322*, 406–414.

Meedel, T. H., Farmer, S. C., & Lee, J. J. (1997). The single MyoD family gene of Ciona intestinalis encodes two differentially expressed proteins: Implications for the evolution of chordate muscle gene regulation. *Development, 124*, 1711–1721.

Megeney, L. A., Kablar, B., Garrett, K., Anderson, J. E., & Rudnicki, M. A. (1996). MyoD is required for myogenic stem cell function in adult skeletal muscle. *Genes & Development, 10*, 1173–1183.
Merlino, G., & Helman, L. J. (1999). Rhabdomyosarcoma—Working out the pathways. *Oncogene, 18*, 5340–5348.
Merrell, A. J., & Kardon, G. (2013). Development of the diaphragm—A skeletal muscle essential for mammalian respiration. *The FEBS Journal, 280*, 4026–4035.
Messina, G., Biressi, S., Monteverde, S., Magli, A., Cassano, M., Perani, L., et al. (2010). Nfix regulates fetal-specific transcription in developing skeletal muscle. *Cell, 140*, 554–566.
Michelson, A. M., Abmayr, S. M., Bate, M., Arias, A. M., & Maniatis, T. (1990). Expression of a MyoD family member prefigures muscle pattern in Drosophila embryos. *Genes & Development, 4*, 2086–2097.
Milewski, R. C., Chi, N. C., Li, J., Brown, C., Lu, M. M., & Epstein, J. A. (2004). Identification of minimal enhancer elements sufficient for Pax3 expression in neural crest and implication of Tead2 as a regulator of Pax3. *Development, 131*, 829–837.
Miller, P. J., Dietz, K. N., & Hollenbach, A. D. (2008). Identification of serine 205 as a site of phosphorylation on Pax3 in proliferating but not differentiating primary myoblasts. *Protein Science, 17*, 1979–1986.
Miner, J. H., Miller, J. B., & Wold, B. J. (1992). Skeletal muscle phenotypes initiated by ectopic MyoD in transgenic mouse heart. *Development, 114*, 853–860.
Miner, J. H., & Wold, B. J. (1990). Herculin a fourth member of the MyoD family of myogenic regulatory genes. *Proceedings of the National Academy of Sciences of the United States of America, 87*, 1089–1093.
Molkentin, J. D., Black, B. L., Martin, J. F., & Olson, E. N. (1995). Cooperative activation of muscle gene expression by MEF2 and myogenic bHLH proteins. *Cell, 83*, 1125–1136.
Moncaut, N., Cross, J. W., Siligan, C., Keith, A., Taylor, K., Rigby, P. W., et al. (2012). Musculin and TCF21 coordinate the maintenance of myogenic regulatory factor expression levels during mouse craniofacial development. *Development, 139*, 958–967.
Moore, S., Ribes, V., Terriente, J., Wilkinson, D., Relaix, F., & Briscoe, J. (2013). Distinct regulatory mechanisms act to establish and maintain Pax3 expression in the developing neural tube. *PLoS Genetics, 9*, e1003811.
Mootoosamy, R. C., Dietrich, S., Gustafsson, M. K., Pan, H., Pinney, D. F., Liu, Y., et al. (2002). Distinct regulatory cascades for head and trunk myogenesis. *Development, 129*, 573–583.
Mourikis, P., Gopalakrishnan, S., Sambasivan, R., & Tajbakhsh, S. (2012). Cell-autonomous Notch activity maintains the temporal specification potential of skeletal muscle stem cells. *Development, 139*, 4536–4548.
Mourikis, P., Sambasivan, R., Castel, D., Rocheteau, P., Bizzarro, V., & Tajbakhsh, S. (2012). A critical requirement for Notch signaling in maintenance of the quiescent skeletal muscle stem cell state. *Stem Cells, 30*(2), 243–252.
Mousavi, K., Zare, H., Dell'orso, S., Grontved, L., Gutierrez-Cruz, G., Derfoul, A., et al. (2013). eRNAs promote transcription by establishing chromatin accessibility at defined genomic loci. *Molecular Cell, 51*, 606–617.
Murphy, M., & Kardon, G. (2011). Origin of vertebrate limb muscle: The role of progenitor and myoblast populations. *Current Topics in Developmental Biology, 96*, 1–32.
Murphy, M. M., Lawson, J. A., Mathew, S. J., Hutcheson, D. A., & Kardon, G. (2011). Satellite cells, connective tissue fibroblasts and their interactions are crucial for muscle regeneration. *Development, 138*, 3625–3637.
Murre, C., Bain, G., van Dijk, M. A., Engel, I., Furnari, B. A., Massari, M. E., et al. (1994). Structure and function of helix-loop-helix proteins. *Biochimica et Biophysica Acta, 1218*, 129–135.

Murre, C., McCaw, P. S., & Baltimore, D. (1989). A new DNA binding and dimerization motif in immunoglobulin enhancer binding, daughterless, MyoD, and myc proteins. *Cell, 56*, 777–783.

Murre, C., McCaw, P. S., Vaessin, H., Caudy, M., Jan, L. Y., Jan, Y. N., et al. (1989). Interactions between heterologous helix-loop-helix proteins generate complexes that bind specifically to a common DNA sequence. *Cell, 58*, 537–544.

Nabeshima, Y., Hanaoka, K., Hayasaka, M., Esumi, E., Li, S., Nonaka, I., et al. (1993). Myogenin gene disruption results in perinatal lethality because of severe muscle defect. *Nature, 364*, 532–535.

Nathan, E., Monovich, A., Tirosh-Finkel, L., Harrelson, Z., Rousso, T., Rinon, A., et al. (2008). The contribution of Islet1-expressing splanchnic mesoderm cells to distinct branchiomeric muscles reveals significant heterogeneity in head muscle development. *Development, 135*, 647–657.

Neuhold, L. A., & Wold, B. (1993). HLH forced dimers: Tethering MyoD to E47 generates a dominant positive myogenic factor insulated from negative regulation by Id. *Cell, 74*, 1033–1042.

Niro, C., Demignon, J., Vincent, S., Liu, Y., Giordani, J., Sgarioto, N., et al. (2010). Six1 and Six4 gene expression is necessary to activate the fast-type muscle gene program in the mouse primary myotome. *Developmental Biology, 338*, 168–182.

Nitzan, E., & Kalcheim, C. (2013). Neural crest and somitic mesoderm as paradigms to investigate cell fate decisions during development. *Development, Growth & Differentiation, 55*, 60–78.

Noden, D. M., & Francis-West, P. (2006). The differentiation and morphogenesis of craniofacial muscles. *Developmental Dynamics, 235*, 1194–1218.

Ohkawa, Y., Marfella, C. G., & Imbalzano, A. N. (2006). Skeletal muscle specification by myogenin and Mef2D via the SWI/SNF ATPase Brg1. *The EMBO Journal, 25*, 490–501.

Okita, K., Nakagawa, M., Hyenjong, H., Ichisaka, T., & Yamanaka, S. (2008). Generation of mouse induced pluripotent stem cells without viral vectors. *Science, 322*, 949–953.

Olson, E. N., Arnold, H.-H., Rigby, P. W. J., & Wold, B. J. (1996). Know your neighbors: Three phenotypes in null mutants of the myogenic bHLH gene *MRF4*. *Cell, 85*, 1–4.

Ono, Y., Boldrin, L., Knopp, P., Morgan, J. E., & Zammit, P. S. (2010). Muscle satellite cells are a functionally heterogeneous population in both somite-derived and branchiomeric muscles. *Developmental Biology, 337*, 29–41.

Oustanina, S., Hause, G., & Braun, T. (2004). Pax7 directs postnatal renewal and propagation of myogenic satellite cells but not their specification. *The EMBO Journal, 23*, 3430–3439.

Ozasa, S., Kimura, S., Ito, K., Ueno, H., Ikezawa, M., Matsukura, M., et al. (2007). Efficient conversion of ES cells into myogenic lineage using the gene-inducible system. *Biochemical and Biophysical Research Communications, 357*, 957–963.

Palacios, D., Summerbell, D., Rigby, P. W., & Boyes, J. (2010). Interplay between DNA methylation and transcription factor availability: Implications for developmental activation of the mouse Myogenin gene. *Molecular and Cellular Biology, 30*, 3805–3815.

Pang, Z. P., Yang, N., Vierbuchen, T., Ostermeier, A., Fuentes, D. R., Yang, T. Q., et al. (2011). Induction of human neuronal cells by defined transcription factors. *Nature, 476*, 220–223.

Pari, G., Jardine, K., & McBurney, M. W. (1991). Multiple CArG boxes in the human cardiac actin gene promoter required for expression in embryonic cardiac muscle cells developing in vitro from embryonal carcinoma cells. *Molecular and Cellular Biology, 11*, 4796–4803.

Parker, M. H., Perry, R. L., Fauteux, M. C., Berkes, C. A., & Rudnicki, M. A. (2006). MyoD synergizes with the E-protein HEB beta to induce myogenic differentiation. *26*, 5771–5783.

Pasini, D., Bracken, A. P., Hansen, J. B., Capillo, M., & Helin, K. (2007). The polycomb group protein Suz12 is required for embryonic stem cell differentiation. *Molecular and Cellular Biology*, *27*, 3769–3779.

Patapoutian, A., Yoon, J. K., Miner, J. H., Wang, S., Stark, K., & Wold, B. (1995). Disruption of the mouse MRF4 gene identifies multiple waves of myogenesis in the myotome. *Development*, *121*, 3347–3358.

Patel, T., Tursun, B., Rahe, D. P., & Hobert, O. (2012). Removal of Polycomb repressive complex 2 makes C. elegans germ cells susceptible to direct conversion into specific somatic cell types. *Cell Reports*, *2*, 1178–1186.

Peinado, H., Ballestar, E., Esteller, M., & Cano, A. (2004). Snail mediates E-cadherin repression by the recruitment of the Sin3A/histone deacetylase 1 (HDAC1)/HDAC2 complex. *Molecular and Cellular Biology*, *24*, 306–319.

Penn, B. H., Bergstrom, D. A., Dilworth, F. J., Bengal, E., & Tapscott, S. J. (2004). A MyoD-generated feed-forward circuit temporally patterns gene expression during skeletal muscle differentiation. *Genes & Development*, *18*, 2348–2353.

Pownall, M. E., Gustafsson, M. K., & Emerson, C. P., Jr. (2002). Myogenic regulatory factors and the specification of muscle progenitors in vertebrate embryos. *Annual Review of Cell and Developmental Biology*, *18*, 747–783.

Puri, P., Sartorelli, V., Yang, X. J., Hamamori, Y., Ogryzko, V. V., Howard, B. H., et al. (1997). Differential roles of p300 and PCAF acetyltransferases in muscle differentiation. *Molecular Cell*, *1*, 35–45.

Puri, P. L., & Mercola, M. (2012). BAF60 A, B, and Cs of muscle determination and renewal. *Genes & Development*, *26*, 2673–2683.

Puri, P. L., & Sartorelli, V. (2000). Regulation of muscle regulatory factors by DNA-binding, interacting proteins, and post-transcriptional modifications. *Journal of Cellular Physiology*, *185*, 155–173.

Rada-Iglesias, A., Bajpai, R., Swigut, T., Brugmann, S. A., Flynn, R. A., & Wysocka, J. (2011). A unique chromatin signature uncovers early developmental enhancers in humans. *Nature*, *470*, 279–283.

Rajapakse, I., Perlman, M. D., Scalzo, D., Kooperberg, C., Groudine, M., & Kosak, S. T. (2009). The emergence of lineage-specific chromosomal topologies from coordinate gene regulation. *Proceedings of the National Academy of Sciences of the United States of America*, *106*, 6679–6684.

Rao, L., Tang, W., Wei, Y., Bao, L., Chen, J., Chen, H., et al. (2012). Highly efficient derivation of skeletal myotubes from human embryonic stem cells. *Stem Cell Reviews*, *8*, 1109–1119.

Rawls, A., Morris, J. H., Rudnicki, M., Braun, T., Arnold, H. H., Klein, W. H., et al. (1995). Myogenin's functions do not overlap with those of MyoD or Myf-5 during mouse embryogenesis. *Developmental Biology*, *172*, 37–50.

Rawls, A., Valdez, M. R., Zhang, W., Richardson, J., Klein, W. H., & Olson, E. N. (1998). Overlapping functions of the myogenic bHLH genes MRF4 and MyoD revealed in double mutant mice. *Development*, *125*, 2349–2358.

Razy-Krajka, F., Lam, K., Wang, W., Stolfi, A., Joly, M., Bonneau, R., et al. (2014). Collier/OLF/EBF-dependent transcriptional dynamics control pharyngeal muscle specification from primed cardiopharyngeal progenitors. *Developmental Cell*, *29*, 263–276.

Relaix, F., Demignon, J., Laclef, C., Pujol, J., Santolini, M., Niro, C., et al. (2013). Six homeoproteins directly activate Myod expression in the gene regulatory networks that control early myogenesis. *PLoS Genetics*, *9*, e1003425.

Relaix, F., & Marcelle, C. (2009). Muscle stem cells. *Current Opinion in Cell Biology*, *21*, 748–753.

Relaix, F., Montarras, D., Zaffran, S., Gayraud-Morel, B., Rocancourt, D., Tajbakhsh, S., et al. (2006). Pax3 and Pax7 have distinct and overlapping functions in adult muscle progenitor cells. *The Journal of Cell Biology*, *172*, 91–102.

Relaix, F., Polimeni, M., Rocancourt, D., Ponzetto, C., Schafer, B. W., & Buckingham, M. (2003). The transcriptional activator PAX3-FKHR rescues the defects of Pax3 mutant mice but induces a myogenic gain-of-function phenotype with ligand-independent activation of Met signaling in vivo. *Genes & Development*, *17*, 2950–2965.

Relaix, F., Rocancourt, D., Mansouri, A., & Buckingham, M. (2004). Divergent functions of murine Pax3 and Pax7 in limb muscle development. *Genes & Development*, *18*, 1088–1105.

Relaix, F., Rocancourt, D., Mansouri, A., & Buckingham, M. (2005). A Pax3/Pax7-dependent population of skeletal muscle progenitor cells. *Nature*, *435*, 948–953.

Relaix, F., & Zammit, P. S. (2012). Satellite cells are essential for skeletal muscle regeneration: The cell on the edge returns centre stage. *Development*, *139*, 2845–2856.

Rhodes, S. J., & Konieczny, S. (1989). Identification of MRF4: A new member of the muscle regulatory factor gene family. *Genes & Development*, *3*, 2050–2061.

Ribas, R., Moncaut, N., Siligan, C., Taylor, K., Cross, J. W., Rigby, P. W., et al. (2011). Members of the TEAD family of transcription factors regulate the expression of Myf5 in ventral somitic compartments. *Developmental Biology*, *355*, 372–380.

Richard, A. F., Demignon, J., Sakakibara, I., Pujol, J., Favier, M., Strochlic, L., et al. (2011). Genesis of muscle fiber-type diversity during mouse embryogenesis relies on Six1 and Six4 gene expression. *Developmental Biology*, *359*, 303–320.

Robson, E. J., He, S. J., & Eccles, M. R. (2006). A PANorama of PAX genes in cancer and development. *Nature Reviews. Cancer*, *6*, 52–62.

Roy, K., de la Serna, I. L., & Imbalzano, A. N. (2002). The myogenic basic helix-loop-helix family of transcription factors shows similar requirements for SWI/SNF chromatin remodeling enzymes during muscle differentiation in culture. *The Journal of Biological Chemistry*, *277*, 33818–33824.

Rudnicki, M. A., Braun, T., Hinuma, S., & Jaenisch, R. (1992). Inactivation of *MyoD* in mice leads to up-regulation of the myogenic HLH gene *Myf*-5 and results in apparently normal muscle development. *Cell*, *71*, 383–390.

Rudnicki, M. A., Schneglesberg, P. N. J., Stead, R. H., Braun, T., Arnold, H.-H., & Jaenisch, R. (1993). MyoD or myf-5 is required for the formation of skeletal muscle. *Cell*, *75*, 1351–1359.

Ruggieri, M., Riboldi, G., Brajkovic, S., Bucchia, M., Bresolin, N., Comi, G. P., et al. (2013). Induced neural stem cells: Methods of reprogramming and potential therapeutic applications. *Progress in Neurobiology*, *114*, 15–24.

Sabourin, L. A., Girgis-Gabardo, A., Seale, P., Asakura, A., & Rudnicki, M. A. (1999). Reduced differentiation potential of primary MyoD$^{-/-}$ myogenic cells derived from adult skeletal muscle. *The Journal of Cell Biology*, *144*, 631–643.

Sakurai, H., Okawa, Y., Inami, Y., Nishio, N., & Isobe, K. (2008). Paraxial mesodermal progenitors derived from mouse embryonic stem cells contribute to muscle regeneration via differentiation into muscle satellite cells. *Stem Cells*, *26*, 1865–1873.

Salani, S., Donadoni, C., Rizzo, F., Bresolin, N., Comi, G. P., & Corti, S. (2012). Generation of skeletal muscle cells from embryonic and induced pluripotent stem cells as an in vitro model and for therapy of muscular dystrophies. *Journal of Cellular and Molecular Medicine*, *16*, 1353–1364.

Sambasivan, R., Comai, G., Le Roux, I., Gomes, D., Konge, J., Dumas, G., et al. (2013). Embryonic founders of adult muscle stem cells are primed by the determination gene Mrf4. *Developmental Biology*, *381*, 241–255.

Sambasivan, R., Gayraud-Morel, B., Dumas, G., Cimper, C., Paisant, S., Kelly, R., et al. (2009). Distinct regulatory cascades govern extraocular and pharyngeal arch muscle progenitor cell fates. *Developmental Cell*, *16*, 810–821.

Sambasivan, R., Kuratani, S., & Tajbakhsh, S. (2011). An eye on the head: The development and evolution of craniofacial muscles. *Development, 138*, 2401–2415.

Sambasivan, R., Yao, R., Kissenpfennig, A., Van Wittenberghe, L., Paldi, A., Gayraud-Morel, B., et al. (2011). Pax7-expressing satellite cells are indispensable for adult skeletal muscle regeneration. *Development, 138*, 3647–3656.

Sambasivan, R., & Tajbakhsh, S. (2007). Skeletal muscle stem cell birth and properties. *Seminars in Cell & Developmental Biology, 18*, 870–882.

Santerre, R. F., Bales, K. R., Janney, M. J., Hannon, K., Fisher, L. F., Bailey, C. S., et al. (1993). Expression of bovine myf5 induces ectopic skeletal muscle formation in transgenic mice. *Molecular and Cellular Biology, 13*, 6044–6051.

Sartorelli, V., Webster, K. A., & Kedes, L. (1990). Muscle-specific expression of the cardiac alpha-actin gene requires MyoD1, CArG-box binding factor, and Sp1. *Genes & Development, 4*, 1811–1822.

Schafer, B. W., Blakely, B. T., Darlington, G. J., & Blau, H. M. (1990). Effect of cell history on response to helix-loop-helix family of myogenic regulators. *Nature, 344*, 454–458.

Schuster-Gossler, K., Cordes, R., & Gossler, A. (2007). Premature myogenic differentiation and depletion of progenitor cells cause severe muscle hypotrophy in Delta1 mutants. *Proceedings of the National Academy of Sciences of the United States of America, 104*, 537–542.

Seale, P., Bjork, B., Yang, W., Kajimura, S., Chin, S., Kuang, S., et al. (2008). PRDM16 controls a brown fat/skeletal muscle switch. *Nature, 454*, 961–967.

Seale, P., Sabourin, L. A., Girgis-Gabardo, A., Mansouri, A., Gruss, P., & Rudnicki, M. A. (2000). Pax7 is required for the specification of myogenic satellite cells. *Cell, 102*, 777–786.

Sebastian, S., Faralli, H., Yao, Z., Rakopoulos, P., Palii, C., Cao, Y., et al. (2013). Tissue-specific splicing of a ubiquitously expressed transcription factor is essential for muscle differentiation. *Genes & Development, 27*, 1247–1259.

Seenundun, S., Rampalli, S., Liu, Q. C., Aziz, A., Palii, C., Hong, S., et al. (2010). UTX mediates demethylation of H3K27me3 at muscle-specific genes during myogenesis. *The EMBO Journal, 29*, 1401–1411.

Sekiya, S., & Suzuki, A. (2011). Direct conversion of mouse fibroblasts to hepatocyte-like cells by defined factors. *Nature, 475*, 390–393.

Shih, H. P., Gross, M. K., & Kioussi, C. (2007). Cranial muscle defects of Pitx2 mutants result from specification defects in the first branchial arch. *Proceedings of the National Academy of Sciences of the United States of America, 104*, 5907–5912.

Shklover, J., Weisman-Shomer, P., Yafe, A., & Fry, M. (2010). Quadruplex structures of muscle gene promoter sequences enhance in vivo MyoD-dependent gene expression. *Nucleic Acids Research, 38*, 2369–2377.

Siles, L., Sanchez-Tillo, E., Lim, J. W., Darling, D. S., Kroll, K. L., & Postigo, A. (2013). ZEB1 imposes a temporary stage-dependent inhibition of muscle gene expression and differentiation via CtBP-mediated transcriptional repression. *Molecular and Cellular Biology, 33*, 1368–1382.

Simone, C., Forcales, S. V., Hill, D. A., Imbalzano, A. N., Latella, L., & Puri, P. L. (2004). p38 pathway targets SWI-SNF chromatin-remodeling complex to muscle-specific loci. *Nature Genetics, 36*, 738–743.

Singh, K., & Dilworth, F. J. (2013). Differential modulation of cell cycle progression distinguishes members of the myogenic regulatory factor family of transcription factors. *The FEBS Journal, 280*, 3991–4003.

Skerjanc, I. S., Slack, R. S., & McBurney, M. W. (1994). Cellular aggregation enhances MyoD-directed skeletal myogenesis in embryonal carcinoma cells. *Molecular and Cellular Biology, 14*, 8451–8459.

Soleimani, V. D., Punch, V. G., Kawabe, Y., Jones, A. E., Palidwor, G. A., Porter, C. J., et al. (2012). Transcriptional dominance of Pax7 in adult myogenesis is due to high-affinity recognition of homeodomain motifs. *Developmental Cell*, *22*, 1208–1220.

Soleimani, V. D., Yin, H., Jahani-Asl, A., Ming, H., Kockx, C. E., van Ijcken, W. F., et al. (2012). Snail regulates MyoD binding-site occupancy to direct enhancer switching and differentiation-specific transcription in myogenesis. *Molecular Cell*, *47*, 457–468.

Spicer, D. B., Rhee, J., Cheung, W. L., & Lassar, A. B. (1996). Inhibition of myogenic bHLH and MEF2 transcription factors by the bHLH protein Twist. *Science*, *272*, 1476–1480.

Spitz, F., Demignon, J., Porteu, A., Kahn, A., Concordet, J. P., Daegelen, D., et al. (1998). Expression of myogenin during embryogenesis is controlled by Six/sine oculis homeoproteins through a conserved MEF3 binding site. *Proceedings of the National Academy of Sciences of the United States of America*, *95*, 14220–14225.

Stadtfeld, M., Nagaya, M., Utikal, J., Weir, G., & Hochedlinger, K. (2008). Induced pluripotent stem cells generated without viral integration. *Science*, *322*, 945–949.

Sweetman, D., Goljanek, K., Rathjen, T., Oustanina, S., Braun, T., Dalmay, T., et al. (2008). Specific requirements of MRFs for the expression of muscle specific microRNAs, miR-1, miR-206 and miR-133. *Developmental Biology*, *321*, 491–499.

Szabo, E., Rampalli, S., Risueno, R. M., Schnerch, A., Mitchell, R., Fiebig-Comyn, A., et al. (2010). Direct conversion of human fibroblasts to multilineage blood progenitors. *Nature*, *468*, 521–526.

Tajbakhsh, S. (2009). Skeletal muscle stem cells in developmental versus regenerative myogenesis. *Journal of Internal Medicine*, *266*, 372–389.

Tajbakhsh, S., & Buckingham, M. (2000). The birth of muscle progenitor cells in the mouse: Spatiotemporal considerations. *Current Topics in Developmental Biology*, *48*, 225–268.

Tajbakhsh, S., & Cossu, G. (1997). Establishing myogenic identity during somitogenesis. *Current Opinion in Genetics & Development*, 7, 634–641.

Tajbakhsh, S., Rocancourt, D., Cossu, G., & Buckingham, M. (1997). Redefining the genetic hierarchies controlling skeletal myogenesis: *Pax-3* and *Myf-5* act upstream of *MyoD*. *Cell*, *89*, 127–138.

Tajbakhsh, S., Rocancourt, D., & Buckingham, M. (1996). Muscle progenitor cells failing to respond to positional cues adopt non-myogenic fates in myf-5 null mice. *Nature*, *384*, 266–270.

Takahashi, K., Tanabe, K., Ohnuki, M., Narita, M., Ichisaka, T., Tomoda, K., et al. (2007). Induction of pluripotent stem cells from adult human fibroblasts by defined factors. *Cell*, *131*, 861–872.

Takahashi, K., & Yamanaka, S. (2006). Induction of pluripotent stem cells from mouse embryonic and adult fibroblast cultures by defined factors. *Cell*, *126*, 663–676.

Takeuchi, J. K., & Bruneau, B. G. (2009). Directed transdifferentiation of mouse mesoderm to heart tissue by defined factors. *Nature*, *459*, 708–711.

Tanaka, A., Woltjen, K., Miyake, K., Hotta, A., Ikeya, M., Yamamoto, T., et al. (2013). Efficient and reproducible myogenic differentiation from human iPS cells: Prospects for modeling Miyoshi Myopathy in vitro. *PLoS One*, *8*, e61540.

Tapscott, S. J. (2005). The circuitry of a master switch: Myod and the regulation of skeletal muscle gene transcription. *Development*, *132*, 2685–2695.

Tapscott, S. J., Lassar, A. B., & Weintraub, B. (1992). A novel myoblast enhancer element mediates MyoD transcription. *Molecular and Cellular Biology*, *12*, 4994–5003.

Tedesco, F. S., Dellavalle, A., Diaz-Manera, J., Messina, G., & Cossu, G. (2010). Repairing skeletal muscle: Regenerative potential of skeletal muscle stem cells. *The Journal of Clinical Investigation*, *120*, 11–19.

Theis, S., Patel, K., Valasek, P., Otto, A., Pu, Q., Harel, I., et al. (2010). The occipital lateral plate mesoderm is a novel source for vertebrate neck musculature. *Development*, *137*, 2961–2971.

Thoma, E. C., Maurus, K., Wagner, T. U., & Schartl, M. (2012). Parallel differentiation of embryonic stem cells into different cell types by a single gene-based differentiation system. *Cellular Reprogramming*, *14*, 106–111.

Tian, H., Biehs, B., Warming, S., Leong, K. G., Rangell, L., Klein, O. D., et al. (2011). A reserve stem cell population in small intestine renders Lgr5-positive cells dispensable. *Nature*, *478*, 255–259.

Trapnell, C., Williams, B. A., Pertea, G., Mortazavi, A., Kwan, G., van Baren, M. J., et al. (2010). Transcript assembly and quantification by RNA-Seq reveals unannotated transcripts and isoform switching during cell differentiation. *Nature Biotechnology*, *28*, 511–515.

Tremblay, P., Dietrich, S., Mericskay, M., Schubert, F. R., Li, Z., & Paulin, D. (1998). A crucial role for Pax3 in the development of the hypaxial musculature and the long-range migration of muscle precursors. *Developmental Biology*, *203*, 49–61.

Tzahor, E. (2009). Heart and craniofacial muscle development: A new developmental theme of distinct myogenic fields. *Developmental Biology*, *327*, 273–279.

Tzahor, E., & Evans, S. M. (2011). Pharyngeal mesoderm development during embryogenesis: Implications for both heart and head myogenesis. *Cardiovascular Research*, *91*, 196–202.

Tzahor, E., Kempf, H., Mootoosamy, R. C., Poon, A. C., Abzhanov, A., Tabin, C. J., et al. (2003). Antagonists of Wnt and BMP signaling promote the formation of vertebrate head muscle. *Genes & Development*, *17*, 3087–3099.

Ustanina, S., Carvajal, J., Rigby, P., & Braun, T. (2007). The myogenic factor Myf5 supports efficient skeletal muscle regeneration by enabling transient myoblast amplification. *Stem Cells*, *25*, 2006–2016.

Valdez, M. R., Richardson, J. A., Klein, W. H., & Olson, E. N. (2000). Failure of Myf5 to support myogenic differentiation without myogenin, MyoD, and MRF4. *Developmental Biology*, *219*, 287–298.

Vasyutina, E., & Birchmeier, C. (2006). The development of migrating muscle precursor cells. *Anatomy and Embryology*, *211*(Suppl. 1), 37–41.

Venuti, J. M., Goldberg, L., Chakraborty, T., Olson, E. N., & Klein, W. H. (1991). A myogenic factor from sea urchin embryos capable of programming muscle differentiation in mammalian cells. *Proceedings of the National Academy of Sciences of the United States of America*, *88*, 6219–6223.

Venuti, J. M., Morris, J. H., Vivian, J. L., Olson, E. N., & Klein, W. H. (1995). Myogenin is required for late but not early aspects of myogenesis during mouse development. *The Journal of Cell Biology*, *128*, 563–576.

Vierbuchen, T., Ostermeier, A., Pang, Z. P., Kokubu, Y., Sudhof, T. C., & Wernig, M. (2010). Direct conversion of fibroblasts to functional neurons by defined factors. *Nature*, *463*, 1035–1041.

von Maltzahn, J., Jones, A. E., Parks, R. J., & Rudnicki, M. A. (2013). Pax7 is critical for the normal function of satellite cells in adult skeletal muscle. *Proceedings of the National Academy of Sciences of the United States of America*, *110*, 16474–16479.

von Scheven, G., Bothe, I., Ahmed, M. U., Alvares, L. E., & Dietrich, S. (2006). Protein and genomic organisation of vertebrate MyoR and Capsulin genes and their expression during avian development. *Gene Expression Patterns*, *6*, 383–393.

Walther, C., Guenet, J. L., Simon, D., Deutsch, U., Jostes, B., Goulding, M. D., et al. (1991). Pax: a murine multigene family of paired box-containing genes. *Genomics*, *11*, 424–434.

Wang, A. H., Zare, H., Mousavi, K., Wang, C., Moravec, C. E., Sirotkin, H. I., et al. (2013). The histone chaperone Spt6 coordinates histone H3K27 demethylation and myogenesis. *The EMBO Journal*, *32*, 1075–1086.

Wang, Y., & Jaenisch, R. (1997). Myogenin can substitute for Myf5 in promoting myogenesis but less efficiently. *Development*, *124*, 2507–2513.

Warren, L., Manos, P. D., Ahfeldt, T., Loh, Y. H., Li, H., Lau, F., et al. (2010). Highly efficient reprogramming to pluripotency and directed differentiation of human cells with synthetic modified mRNA. *Cell Stem Cell, 7*, 618–630.

Wasserman, W. W., & Fickett, J. W. (1998). Identification of regulatory regions which confer muscle-specific gene expression. *Journal of Molecular Biology, 278*, 167–181.

Watanabe, S., Hirai, H., Asakura, Y., Tastad, C., Verma, M., Keller, C., et al. (2011). MyoD gene suppression by Oct4 is required for reprogramming in myoblasts to produce induced pluripotent stem cells. *Stem Cells, 29*, 505–516.

Watt, F. M., & Huck, W. T. (2013). Role of the extracellular matrix in regulating stem cell fate. *Nature Reviews. Molecular Cell Biology, 14*, 467–473.

Weintraub, H., Davis, R., Lockshon, D., & Lassar, A. (1990). MyoD binds cooperatively to two sites in a target enhancer sequence: Occupancy of two sites is required for activation. *Proceedings of the National Academy of Sciences of the United States of America, 87*, 5623–5627.

Weintraub, H., Tapscott, S. J., Davis, R. L., Thayer, M. J., Adam, M. A., Lassar, A. B., et al. (1989). Activation of muscle-specific genes in pigment, nerve, fat, liver, and fibroblast cell lines by forced expression of MyoD. *Proceedings of the National Academy of Sciences of the United States of America, 86*, 5434–5438.

Wood, W. M., Etemad, S., Yamamoto, M., & Goldhamer, D. J. (2013). MyoD-expressing progenitors are essential for skeletal myogenesis and satellite cell development. *Developmental Biology, 384*, 114–127.

Wright, W. E., Sassoon, D. A., & Lin, V. K. (1989). Myogenin, a factor regulating myogenesis, has a domain homologous to MyoD. *Cell, 56*, 607–617.

Wu, J., & Lim, R. W. (2005). Regulation of inhibitor of differentiation gene 3 (Id3) expression by Sp2-motif binding factor in myogenic C2C12 cells: Downregulation of DNA binding activity following skeletal muscle differentiation. *Biochimica et Biophysica Acta, 1731*, 13–22.

Wu, J. I., Lessard, J., & Crabtree, G. R. (2009). Understanding the words of chromatin regulation. *Cell, 136*, 200–206.

Wu, J. I., Lessard, J., Olave, I. A., Qiu, Z., Ghosh, A., Graef, I. A., et al. (2007). Regulation of dendritic development by neuron-specific chromatin remodeling complexes. *Neuron, 56*, 94–108.

Wyzykowski, J. C., Winata, T. I., Mitin, N., Taparowsky, E. J., & Konieczny, S. F. (2002). Identification of novel MyoD gene targets in proliferating myogenic stem cells. *Molecular and Cellular Biology, 22*, 6199–6208.

Yablonka-Reuveni, Z., Rudnicki, M. A., Rivera, A. J., Primig, M., Anderson, J. E., & Natanson, P. (1999). The transition from proliferation to differentiation is delayed in satellite cells from mice lacking MyoD. *Developmental Biology, 210*, 440–455.

Yao, Z., Fong, A. P., Cao, Y., Ruzzo, W. L., Gentleman, R. C., & Tapscott, S. J. (2013). Comparison of endogenous and overexpressed MyoD shows enhanced binding of physiologically bound sites. *Skeletal Muscle, 3*, 8.

Yokoyama, S., & Asahara, H. (2011). The myogenic transcriptional network. *Cellular and Molecular Life Sciences, 68*, 1843–1849.

Yokoyama, S., Ito, Y., Ueno-Kudoh, H., Shimizu, H., Uchibe, K., Albini, S., et al. (2009). A systems approach reveals that the myogenesis genome network is regulated by the transcriptional repressor RP58. *Developmental Cell, 17*, 836–848.

Yoo, A. S., Staahl, B. T., Chen, L., & Crabtree, G. R. (2009). MicroRNA-mediated switching of chromatin-remodelling complexes in neural development. *Nature, 460*, 642–646.

Yoon, J. K., Olson, E. N., Arnold, H. H., & Wold, B. J. (1997). Different MRF4 knockout alleles differentially disrupt Myf-5 expression: cis-regulatory interactions at the MRF4/Myf-5 locus. *Developmental Biology, 188*, 349–362.

Yu, J., Hu, K., Smuga-Otto, K., Tian, S., Stewart, R., Slukvin, I. I., et al. (2009). Human induced pluripotent stem cells free of vector and transgene sequences. *Science, 324*, 797–801.

Yu, J., Vodyanik, M. A., Smuga-Otto, K., Antosiewicz-Bourget, J., Frane, J. L., Tian, S., et al. (2007). Induced pluripotent stem cell lines derived from human somatic cells. *Science, 318*, 1917–1920.

Yuan, J., Zhang, S., Liu, Z., Luan, Z., & Hu, G. (2003). Cloning and phylogenetic analysis of an amphioxus myogenic bHLH gene AmphiMDF. *Biochemical and Biophysical Research Communications, 301*, 960–967.

Zacharias, A. L., Lewandoski, M., Rudnicki, M. A., & Gage, P. J. (2011). Pitx2 is an upstream activator of extraocular myogenesis and survival. *Developmental Biology, 349*, 395–405.

Zammit, P. S. (2008). All muscle satellite cells are equal, but are some more equal than others? *Journal of Cell Science, 121*, 2975–2982.

Zammit, P. S., Golding, J. P., Nagata, Y., Hudon, V., Partridge, T. A., & Beauchamp, J. R. (2004). Muscle satellite cells adopt divergent fates: A mechanism for self-renewal? *The Journal of Cell Biology, 166*, 347–357.

Zammit, P. S., Partridge, T. A., & Yablonka-Reuveni, Z. (2006). The skeletal muscle satellite cell: the stem cell that came in from the cold. *The Journal of Histochemistry and Cytochemistry: Official Journal of the Histochemistry Society, 54*, 1177–1191.

Zaret, K. S., & Carroll, J. S. (2011). Pioneer transcription factors: Establishing competence for gene expression. *Genes & Development, 25*, 2227–2241.

Zhang, W., Behringer, R. R., & Olson, E. N. (1995). Inactivation of the myogenic bHLH *MRF4* gene results in up-regulation of myogenin and rib anomalies. *Genes and Development, 9*, 1388–1399.

Zhang, Z., Huynh, T., & Baldini, A. (2006). Mesodermal expression of Tbx1 is necessary and sufficient for pharyngeal arch and cardiac outflow tract development. *Development, 133*, 3587–3595.

Zhao, X., Yu, Q., Huang, L., & Liu, Q. X. (2014). Patterns of positive selection of the myogenic regulatory factor gene family in vertebrates. *PLoS One, 9*, e92873.

Zhou, H., Wu, S., Joo, J. Y., Zhu, S., Han, D. W., Lin, T., et al. (2009). Generation of induced pluripotent stem cells using recombinant proteins. *Cell Stem Cell, 4*, 381–384.

Zhou, Q., Brown, J., Kanarek, A., Rajagopal, J., & Melton, D. A. (2008). In vivo reprogramming of adult pancreatic exocrine cells to beta-cells. *Nature, 455*, 627–632.

Zhou, Q., & Melton, D. A. (2008). Extreme makeover: Converting one cell into another. *Cell Stem Cell, 3*, 382–388.

Zhu, J., Adli, M., Zou, J. Y., Verstappen, G., Coyne, M., Zhang, X., et al. (2013). Genome-wide chromatin state transitions associated with developmental and environmental cues. *Cell, 152*, 642–654.

CHAPTER TWO

Proneural bHLH Genes in Development and Disease

Carol Huang*, Jennifer A. Chan†,1, Carol Schuurmans‡,1

*Department of Pediatrics, Alberta Children's Hospital Research Institute, University of Calgary, Calgary, Alberta, Canada

†Department of Pathology & Laboratory Medicine, Southern Alberta Cancer Research Institute, University of Calgary, Calgary, Alberta, Canada

‡Department of Biochemistry and Molecular Biology, Alberta Children's Hospital Research Institute, Hotchkiss Brain Institute, University of Calgary, Calgary, Alberta, Canada

[1]Corresponding authors: e-mail address: jawchan@ucalgary.ca; cschuurm@ucalgary.ca

Contents

Current Topics in Developmental Biology, Volume 110
ISSN 0070-2153
http://dx.doi.org/10.1016/B978-0-12-405943-6.00002-6

© 2014 Elsevier Inc.
All rights reserved.

Abstract

Proneural genes encode evolutionarily conserved basic-helix–loop–helix transcription factors. In *Drosophila*, proneural genes are required and sufficient to confer a neural identity onto naïve ectodermal cells, inducing delamination and subsequent neuronal differentiation. In vertebrates, proneural genes are expressed in cells that already have a neural identity, but they are still required and sufficient to initiate neurogenesis. In all organisms, proneural genes control neurogenesis by regulating Notch-mediated lateral inhibition and initiating the expression of downstream differentiation genes. The general mode of proneural gene function has thus been elucidated. However, the regulatory mechanisms that spatially and temporally control proneural gene function are only beginning to be deciphered. Understanding how proneural gene function is regulated is essential, as aberrant proneural gene expression has recently been linked to a variety of human diseases—ranging from cancer to neuropsychiatric illnesses and diabetes. Recent insights into proneural gene function in development and disease are highlighted herein.

1. INTRODUCTION TO PRONEURAL GENES

1.1. Identification and functional characterization of proneural genes in *Drosophila*

Proneural genes encode basic-helix–loop–helix (bHLH) transcription factors that were initially identified in *Drosophila melanogaster* based on their ability to confer a neural (rather than epidermal) identity onto naïve ectodermal cells (Ghysen & Dambly-Chaudiere, 1988; Skeath & Carroll, 1994). The original proneural genes identified in *Drosophila* included members of the *achaete-scute* complex (*AS-C*), which is comprised of the four genes *achaete* (*ac*), *scute* (*sc*), *lethal of scute* (*lsc*), and *asense* (*as*) (all but *as* have proneural activity; Garcia-Bellido, 1979; Ghysen & Dambly-Chaudiere, 1988; Villares & Cabrera, 1987). Additional proneural genes were subsequently identified in *Drosophila*, including *atonal* (*ato*) (Jarman, Grau, Jan, & Jan, 1993; Jarman, Grell, Ackerman, Jan, & Jan, 1994), *amos* (Goulding, zur Lage, & Jarman, 2000; Huang, Hsu, & Chien, 2000), and *cato* (Goulding, White, & Jarman, 2000).

Functionally, proneural activity was ascribed based on gain- and loss-of-function studies, which result in the production of more and fewer neuronal structures, respectively. Within the unspecified ectoderm, proneural genes are initially expressed in equivalence groups known as proneural clusters (Skeath & Carroll, 1994; Fig. 2.1). Within each cell in the equivalence group, proneural bHLH transcription factors initiate Notch-mediated lateral

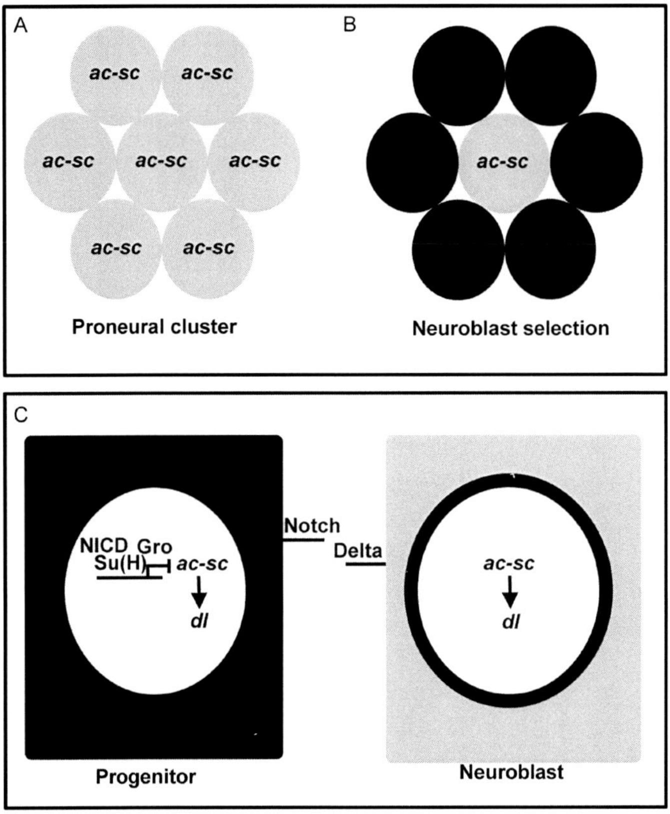

Figure 2.1 Model of lateral inhibition. (A) Within the unspecified ectoderm, cells within each proneural cluster initially express similar levels of the proneural genes (e.g., *ac-sc*). (B) One cell begins expressing *ac-sc* at a higher level, becoming the presumptive neuroblast, while lateral inhibition shuts down the proneural program in neighboring cells, which retain a progenitor phenotype. (C) Higher delta expression on the surface of the presumptive neuroblast signals to neighboring cells through their Notch receptors. Notch signals through the Notch intracellular domain, *gro*, and *Su*(*H*) to suppress *ac-sc* genes, which then cannot transactivate *dl*, keeping neighboring cells in a progenitor state. (See the color plate.)

inhibition in neighboring cells by transactivating the Notch ligands *Delta* (*Dll*) and *Serrate* (*Se*). Notch signaling represses proneural gene expression in the following manner: upon ligand binding, the Notch intracellular domain (NICD) is cleaved and translocates to the nucleus, where it binds suppressor-of-hairless (Su(H)), converting Su(H) to a transactivator of *hairy* and *Enhancer-of-split* (*E*(*spl*)) complex genes (Bray & Furriols, 2001; Davis & Turner, 2001). *E*(*spl*) and *hairy* are neurogenic genes that also encode bHLH transcription factors, but they block rather than promote neurogenesis and bind recognition sites (i.e., N-boxes; CACNAG) that are distinct from those bound by proneural bHLH proteins (i.e., E-boxes; CANNTG; Davis & Turner, 2001; Murre et al., 1989). N-boxes are located in the upstream regulatory regions of proneural genes, and upon E(spl) and hairy binding to these sites, Groucho corepressors are recruited to repress proneural gene expression (Davis & Turner, 2001). Through this process, proneural gene expression is restricted to one (or a few) cells in a proneural cluster—the presumptive neuroblasts (Artavanis-Tsakonas, Rand, & Lake, 1999). Selected neuroblasts "outcompete" their neighbors by initially expressing slightly higher levels of the Notch ligands, a difference that is further amplified through lateral inhibition. Upon lateral inhibition and neuroblast selection, neuroblasts delaminate from the ectodermal epithelium and differentiate under the control of the downstream transcriptional targets of the proneural bHLH factors.

Proneural genes not only promote generic neuronal differentiation but also participate in subtype specification. Indeed, proneural genes are class II bHLH transcription factors, which are characterized by their regionalized patterns of expression. Each proneural gene is expressed in specific populations of neuroblasts in the central nervous system (CNS) and peripheral nervous system (PNS). In the CNS, where neuroblasts form in ordered rows, *ac*, *sc*, and *lsc* are expressed in distinct subsets of these neuroblasts. In the PNS, each sensory structure arises from sensory organ precursors (SOPs) located at precise locations in the embryo, each of which expresses a specific proneural gene (Skeath & Carroll, 1994). For example, *ato* is required and sufficient to specify a chordotonal organ, R8 photoreceptor, or olfactory sense organ identity, depending on its site of expression (Jarman et al., 1993, 1994; Jhaveri, Sen, Reddy, & Rodrigues, 2000), whereas *ac-sc* genes are expressed in extrasensory organ SOPs (Campuzano & Modolell, 1992). The ability of proneural genes to specify more than one neuronal subtype highlights the context specificity of their functions.

Part of the specificity of proneural gene function relates to the structure of the proteins. The bHLH domain is comprised of a basic domain that confers DNA binding and a HLH domain that mediates hetero- or homodimerization (Bertrand, Castro, & Guillemot, 2002; Powell & Jarman, 2008). The common heterodimerization partner of proneural proteins in *Drosophila* is *daughterless* (*da*), which encodes a ubiquitously expressed class I bHLH transcription factor (Cabrera & Alonso, 1991). Interestingly, domain swapping experiments revealed that it is the basic domain that confers subtype specification properties to bHLH proteins. Inserting the basic domain of *ato* into *ac* changes the type of sense organ produced from external sensory to chordotonal organ, highlighting the importance of the basic domain (Chien, Hsiao, Jan, & Jan, 1996). Proneural gene function is also modulated by interactions with other transcriptional regulators and by posttranslational modifications, which are further described for the vertebrate proneural genes below.

1.2. Introduction to vertebrate proneural genes

Not only have proneural genes been conserved throughout evolution, but also members of this gene family have greatly expanded in vertebrates. Vertebrate *ato* orthologues fall into three distinct gene families: *Neurogenin* genes (*Neurog1*, *Neurog2*, *Neurog3*), *Neurogenic differentiation* genes (*NeuroD1*, *NeuroD2*, *Neurod4/Math3/Atoh3*, *Neurod6/Math2/Atoh2*, *Atoh1/Math1*, *Atoh7/Math5*), and *Olig* genes (*Olig1*, *Olig2*, *Olig3*). In contrast, there are only two *AS-C*-related genes in mouse—*Ascl1/Mash1*, which is expressed in the nervous system, and *Ascl2/Mash2*, which is not (Bertrand et al., 2002). To qualify as a proneural gene, also known as a neural determination gene in vertebrates, in loss-of-function mutants, Notch ligand expression and signaling are reduced in the lineages where the proneural gene would normally be expressed (Bertrand et al., 2002). Moreover, the disruption of proneural gene activity results in a loss of specific neural progenitor pools and derivative neurons (Bertrand et al., 2002). Conversely, in gain-of-function assays, vertebrate proneural genes promote cell cycle exit and the generation of ectopic neurons at the expense of glia. Based on this strict definition, only *Neurog1/2*, *Ascl1*, and possibly *Atoh1/Atoh7* are true proneural genes, each functioning in multiple neural (and nonneural) lineages in the PNS and CNS (Fig. 2.2). In contrast, the remainder of the vertebrate bHLH genes functions downstream to regulate neural cell differentiation. It should also be noted that *Ascl1* only functions as a traditional proneural gene in certain cellular contexts; in other contexts, *Ascl1* promotes (rather than inhibits) progenitor cell proliferation and also induces a glial fate (see Section 4.2).

While proneural genes are expressed prior to the epidermal versus neural cell fate choice in *Drosophila*, in vertebrates, proneural genes are primarily expressed in neuroepithelial progenitors that have already acquired a neural identity. Despite this difference, vertebrate and invertebrate proneural genes share several functional characteristics, including their ability to transactivate Notch ligands and initiate Notch signaling, as well as transactivation of

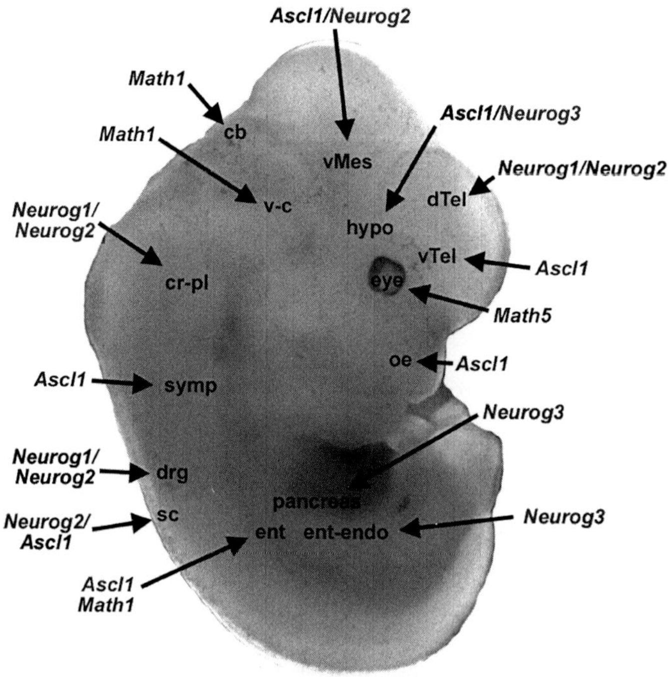

Figure 2.2 Major anatomic sites of proneural gene function in the developing embryo. *Neurog1* and *Neurog2* are active in the dorsal telencephalon (dTel), ventral mesencephalon (vMes), cranial placodes (cr-pl), dorsal root ganglia (drg), and spinal cord (sc). *Neurog3* is active in the developing hypothalamus (hypo), pancreas, and enteroendocrine (ent-endo) system. *Math1/Atoh1* is critical for cerebellar (cb) and enteric (ent) secretory cell development, whereas *Math5/Atoh7* predominantly functions in the eye. *Ascl1* is active in the ventral telencephalon (vTel), ventral mesencephalon (vMes), olfactory epithelium (OE), autonomic system (symp), spinal cord (sc), and enteric autonomic system (ent). (See the color plate.)

neuronal differentiation genes. Notch signaling is also conserved in vertebrates, with Rbpj, a Su(H) ortholog, and NICD-interacting partner, initiating expression of the vertebrate orthologues of the *hairy* and *E(spl)* genes, known as *Hes1–7* (Davis & Turner, 2001; Kageyama, Ohtsuka, & Kobayashi, 2007). Notably, vertebrate Hes proteins function similarly to recruit Groucho-related transducin-like E(spl) (TLE) family corepressors through a conserved WRPW (Trp-Arg-Pro-Trp) domain to repress proneural gene expression (Grbavec & Stifani, 1996).

As in *Drosophila*, vertebrate bHLH proteins form heterodimers with E-proteins that are encoded by three genes—*Tcf2a*, which gives rise to two proteins, E12 and E47 (Ellenberger, Fass, Arnaud, & Harrison, 1994;

Murre et al., 1994, 1989), *Tcf4/E2-2/Itf2/Me2* (Soosaar, Chiaramello, Zuber, & Neuman, 1994), and *Tcf12/HEB* (Hu, Olson, & Kingston, 1992). These E-proteins are not functionally equivalent: mutation of *Tcf4* in Pitt-Hopkins syndrome, for example, is associated with severe motor and mental retardation (Brockschmidt et al., 2007; Forrest, Waite, Martin-Rendon, & Blake, 2013), demonstrating that other E-proteins cannot functionally compensate. Moreover, in development, Math1/Tcf4 (Flora, Garcia, Thaller, & Zoghbi, 2007) and Neurog2/Tcf2a (Li et al., 2012) heterodimers have functions distinct from heterodimers with other E-proteins in the developing cerebellum and neocortex, respectively. Further, each bHLH protein may preferentially bind a slightly different E-box, showing preference for the middle NN residues in the canonical CANNTG E-box sequence (Seo, Lim, Yellajoshyula, Chang, & Kroll, 2007), which may account for functional differences in these genes, and their ability to transactivate distinct target genes. Finally, cofactor interactions can modulate target gene selection by each proneural protein, as described below.

2. PRONEURAL FUNCTIONS OF THE *NEUROGENIN* GENES IN DEVELOPMENT

2.1. *Neurog1* and *Neurog2* identification and functions in the developing PNS

2.1.1 Neurog1 *and* Neurog2 *specify distinct neuronal phenotypes in dorsal root ganglia*

In 1996, two groups—Ma and Anderson and Gradwohl and colleagues—independently identified *Neurog1* and *Neurog2* as novel bHLH genes with potential proneural functions (Gradwohl, Fode, & Guillemot, 1996; Ma, Kintner, & Anderson, 1996). Although they are highly related *ato* orthologues, *Neurog1* and *Neurog2* have distinct expression profiles and control the differentiation of different neuronal populations. Within neural crest cells (NCCs) that are precursors for the developing dorsal root ganglia (DRG), *Neurog2* is expressed earlier, with transcripts detected in newly delaminated and migrating NCCs as well as in the DRG primordium (Ma, Fode, Guillemot, & Anderson, 1999; Perez, Rebelo, & Anderson, 1999). *Neurog1* expression is initiated later, after DRG precursors have started to coalesce into ganglia (Ma et al., 1999). Consistent with temporal differences in their expression, defects in the generation of earlier-born, large diameter trkC^{+} and trkB^{+} sensory neurons are observed in *Neurog2* mutant DRGs, whereas later-born, small diameter trkA^{+} sensory neurons are reduced in number in *Neurog1* mutant DRGs (Ma et al., 1999; Sommer, Ma, & Anderson, 1996). *Neurog1* and *Neurog2* thus share the ability to promote a sensory neuronal identity in the PNS, but the neuronal subtypes generated by each of these proneural genes are distinct.

In *Drosophila*, proneural genes not only select neural precursors but also instruct specific neuronal identities. However, *ato's* ability to specify subtype identities is less stringent and more influenced by cellular context than the *ac-sc* genes (Jarman, Sun, Jan, & Jan, 1995). While *ato* promotes a chordotonal organ identity when misexpressed in the PNS (which is the structure it is normally required to specify), it also induces the formation of some ectopic extra sensory organs, which are not normally associated with *ato* function (Jarman et al., 1995). The context specificity of *ato* function has also been conserved by vertebrate *Neurog* orthologues (Jarman et al., 1995). When *Neurog2* is overexpressed in NCCs *in vitro*, it promotes cell cycle exit and the differentiation of neurons with a sensory identity (Lo, Dormand, Greenwood, & Anderson, 2002). However, when levels of BMP signaling are artificially elevated in these NCC cultures, *Neurog2* instead promotes the ectopic differentiation of autonomic neurons—a neuronal identity that *Neurog2* does not normally promote (Lo et al., 2002). This phenomenon is also seen *in vivo* in *Neurog2* knock-in alleles into the *Ascl1* locus ($Ascl1^{Neurog2KI/KI}$), where *Neurog2* can rescue defects in the differentiation of autonomic neurons, which are normally missing in *Ascl1* mutants (Parras et al., 2002). These results imply that while *Neurog2* can promote generic neuronal differentiation when expressed ectopically, it does not have the capacity to specify its normal neuronal fates—instead, responding to contextual cues to specify neuronal identities (Parras et al., 2002).

2.1.2 Neurog1 *and* Neurog2 *function similar to* Drosophila *proneural genes in the cranial placodes*

Throughout most of the developing PNS and CNS in vertebrates, proneural genes are expressed in progenitors that have already acquired a neural identity. One exception is the cranial placodes, which are specialized regions of the cranial ectoderm that undergo a transitory thickening before acquiring a neural fate and giving rise to neurons that form distal parts of several cranial ganglia (Lassiter, Stark, Zhao, & Zhou, 2013). Cell fate specification in the cranial placodes is closely related to neuroblast selection in *Drosophila*, as vertebrate proneural genes are expressed in uncommitted ectodermal cells in these placodes and confer a neural identity onto unspecified ectoderm (Lassiter et al., 2013). Moreover, upon proneural gene expression and neural cell fate specification, neural precursors delaminate from the placodal epithelium before migrating and differentiating into neurons of the cranial ganglia at defined sites.

Both *Neurog1* and *Neurog2* are expressed in distinct placodal structures in the cranial ectoderm. *Neurog2* is expressed in the geniculate, petrosal, and nodose placodes, which give rise to distal cranial ganglia VII, IX, and X, respectively (Fode et al., 1998), while *Neurog1* is expressed in the otic and trigeminal placodes, which give rise to distal parts of the VIIIth and Vth cranial ganglia, respectively (Ma, Chen, del Barco Barrantes, de la Pompa, & Anderson, 1998). Consistent with *Neurog1* and *Neurog2* functioning as proneural genes in these structures, *Dll1* expression is not initiated in *Neurog1* or *Neurog2* mutant placodal cells. Consequently, neural precursors fail to delaminate and differentiate from these structures, and the distal Vth and VIIIth cranial ganglia do not form in *Neurog1* mutants (Ma et al., 1998), while the distal VIIth, IXth, and Xth cranial ganglia fail to form in *Neurog2* mutants (Fode et al., 1998). *Neurog1* and *Neurog2* thus confer a neural identity onto unspecified ectodermal cells, inducing neural precursor delamination and neuronal differentiation in a manner highly analogous to *Drosophila* proneural genes.

2.2. *Neurog1* and *Neurog2* function in the developing CNS

2.2.1 Neurog2 *specifies distinct neuronal phenotypes in different regions of the neural tube*

Vertebrate proneural genes are expressed in a regionalized manner throughout the neural tube, specifying distinct neuronal identities depending on their site of expression. In the dorsal telencephalon (anlage of the cerebral cortex), *Neurog2* is necessary and sufficient to specify a glutamatergic, pyramidal neuron identity (Fig. 2.3), which go on to populate several brain regions, including the neocortex (Fode et al., 2000; Schuurmans et al., 2004), hippocampus (Galichet, Guillemot, & Parras, 2008; Ozen et al., 2007), piriform cortex (Dixit et al., 2014), and olfactory bulb (Brill et al., 2009; Shaker, Dennis, Kurrasch, & Schuurmans, 2012; Winpenny et al., 2011). In the ventral mesencephalon, *Neurog2* is required for the differentiation of a subset of dopaminergic neurons (Andersson, Jensen, Parmar, Guillemot, & Bjorklund, 2006; Kele et al., 2006). In the ventral spinal cord, *Neurog2* is required for the differentiation of cholinergic motor neurons and different ventral interneuron populations (Mizuguchi et al., 2001; Novitch, Chen, & Jessell, 2001; Scardigli, Schuurmans,

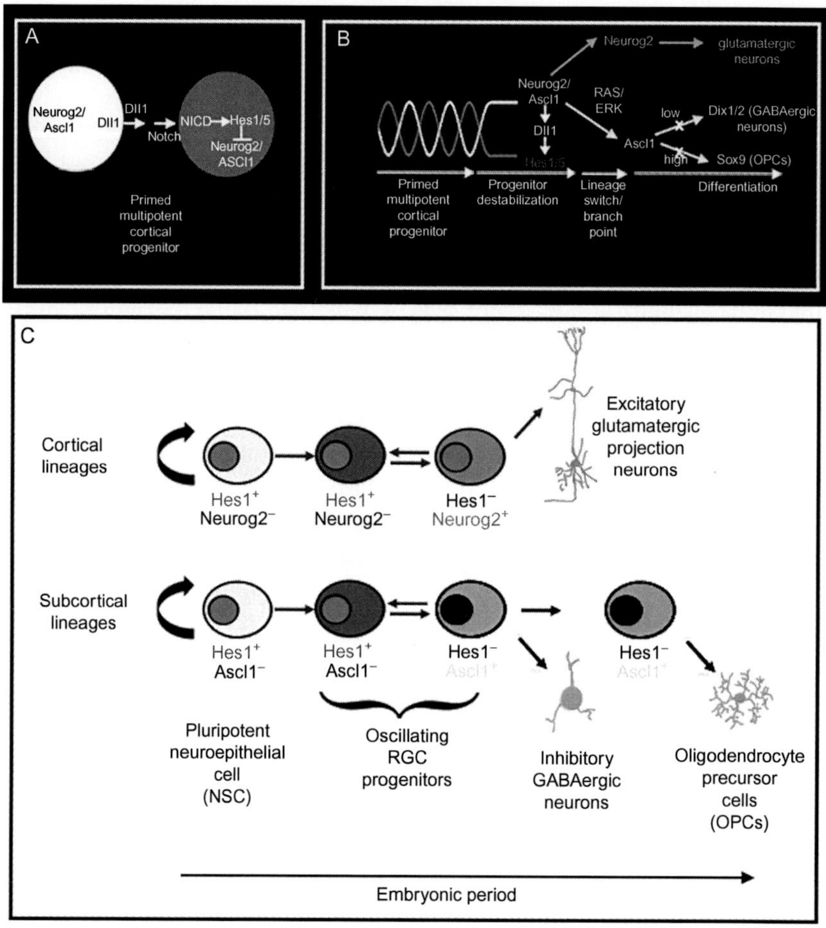

Figure 2.3 Oscillatory cycles of proneural and neurogenic bHLH proteins in cortical progenitors. (A) Both Neurog2 and Ascl1 initiate expression of the Dll1 ligand, which binds to a Notch receptor on a neighboring cell, leading to the translocation of the Notch intracellular domain (NICD) into the nucleus. NICD induces the transcription of *Hes1/5*. Hes1/5 in turn represses *Neurog2* and *Ascl1* transcription, setting up oscillatory cycles of Neurog2 and Hes1/5 proteins. (B) Schematic illustration of predicted Neurog2$^+$Ascl1$^+$/Hes1/5$^+$ oscillatory cycles in cortical progenitors. These cycles must be destabilized for lineage commitment to occur. RAS/ERK signaling promotes stable Ascl1 expression (while blocking Neurog2 expression). (C) Lineage selection in cortical and subcortical progenitors. After Neurog2 is stably expressed in cortical lineages, it promotes a glutamatergic neuronal phenotype in cortical cells. After Ascl1 is stably expressed in subcortical lineages, it promotes a GABAergic neuronal or OPC cell fate. (See the color plate.)

Gradwohl, & Guillemot, 2001). These studies highlight the context specificity of *Neurog2*'s proneural activities in the neural tube.

Different mechanisms underlie *Neurog2*'s ability to specify distinct neuronal identities, including the combinatorial actions of other transcription factors. For example, in the ventral spinal cord, misexpression of *Neurog2* alone promotes cell cycle exit and generic neuronal differentiation but *Neurog2* cannot instruct a motor neuron identity unless coexpressed with other transcription factors, such as *Olig2* (Mizuguchi et al., 2001; Novitch et al., 2001; Scardigli et al., 2001). Moreover, the ratio of *Olig2*/*Neurog2* is important, with higher *Olig2* levels maintaining the motor neuron progenitor pool, while higher levels of *Neurog2* promote motor neuron differentiation (Lee, Lee, Ruiz, & Pfaff, 2005). Neurog2 also forms a complex with retinoic acid receptors, integrating retinoic acid signaling with *Neurog2* proneural activity to drive motor neuron differentiation (Lee, Lee, Lee, & Lee, 2009). Finally, *Neurog2* also functions in a combinatorial fashion with the homeodomain transcription factors encoded by *Isl1* and *Lhx3* to specify a motor neuron identity (Lee et al., 2012; Lee & Pfaff, 2003). This ability of Neurog2 to interact with Isl1 and Lhx3 is regulated by GSK3, which mediates Neurog2's interactions by modulating its phosphorylation status (Ma et al., 2008). Specifically, phosphorylation of S231/S234 in Neurog2 by GSK3 allows Neurog2 to recruit Ldb1 (a linker protein) and the associated transcription factors, Lhx3 and Isl1 (Ma et al., 2008). The ability of *Neurog2* to function combinatorially with other transcription factors to specify distinct neuronal fates is also evident in the ventral mesencephalon, where *Neurog2* alone induces cell cycle exit and generic neuronal differentiation (Roybon, Hjalt, Christophersen, Li, & Brundin, 2008), but when overexpressed together with the nuclear receptor *Nurr1*, it induces a significant number of neurons to acquire a dopaminergic phenotype (Andersson, Irvin, Ahlsio, & Parmar, 2007).

To specify neuronal identities, *Neurog2* initiates the transcription of distinct sets of downstream genes in different neuronal lineages. Downstream transcriptional targets of *Neurog2* have been best analyzed for cortical progenitors (Britz et al., 2006; Fode et al., 2000; Kovach et al., 2013; Mattar et al., 2004, 2008; Parras et al., 2002). A number of *Neurog2*-regulated genes have been identified by genome-wide and subtractive hybridization screens of wild-type and mutant cortical tissue, a handful of which have been validated as direct transcriptional targets via chromatin immunoprecipitation and transcriptional reporter assays (e.g., *Neurod1*, *Neurod4*, *Tbr2*, *Rnd2*, *Dll1*, and *Dll3*; Castro et al., 2006; Gohlke et al., 2008; Henke,

Meredith, Borromeo, Savage, & Johnson, 2009; Mattar et al., 2004; Ochiai et al., 2009; Schuurmans et al., 2004; Seo et al., 2007). However, the binding of bHLH transcription factors to target promoter or enhancer elements is not necessarily sufficient to initiate transcription, as exemplified by the selective transcriptional initiation of a small subset of bound targets by the bHLH protein MyoD in myogenic lineages (Aziz, Liu, & Dilworth, 2010). Moreover, even for those genes that are transactivated by MyoD, the duration of transactivation is important in initiating transcription of early, intermediate, and late target genes, with later onset genes requiring MyoD interactions with other transcription factors and posttranslational modifications (Bergstrom et al., 2002; Blais et al., 2005; Cao et al., 2006; Lluis, Ballestar, Suelves, Esteller, & Munoz-Canoves, 2005). Similarly, in *Neurog2* misexpression studies in an ectopic site in the ventral telencephalon, *Neurog2* acts instructively to induce the ectopic expression of a subset of dorsal telencephalic genes that can be broken down into early, intermediate, and late target genes, based on the time required for ectopic transcripts to be detected, revealing superimposed spatial and temporal constraints (Mattar et al., 2008).

2.2.2 Neurog2 *proneural functions are tightly regulated in the developing neocortex via multiple mechanisms*

The temporally distinct functions of *Neurog2* are illustrated by studies in the dorsal telencephalon, where *Neurog2* only specifies a glutamatergic projection neuron identity before embryonic day (E) 14.5, despite persistent expression beyond this point (Fode et al., 2000; Li et al., 2012; Schuurmans et al., 2004). Several mechanisms underlie such temporal control of *Neurog2* activity. First, *Neurog2* is not stably expressed in dividing cortical progenitors, but rather oscillates in 2–3 h cycles (Shimojo, Ohtsuka, & Kageyama, 2008; Fig. 2.3). These *Neurog2* oscillatory cycles are triggered by Neurog2's ability to initiate the expression of *Dll1*, which activates Notch signaling and *Hes1/5* expression in neighboring cells (Shimojo et al., 2008). Hes1/5 in turn represses *Neurog2* expression, setting up a lateral inhibitory circuit, such that *Neurog2* and *Hes1/5* expression alternates out of phase (Shimojo et al., 2008). In order to induce neuronal differentiation, it has been proposed that these oscillatory cycles must desist and stable *Neurog2* expression should be maintained (as has been shown for *Ascl1*; Imayoshi et al., 2013), but how this switch to stable *Neurog2* expression occurs is unknown (Shimojo et al., 2008). By prolonging the period of *Neurog2* expression, progenitors may have more time to respond to its proneural activity, such that they undergo neuronal differentiation. Indeed, there is a progressive lengthening of G1 phase of the cell cycle at later stages of corticogenesis that correlates with an increase in the number of progenitors undergoing neuronal differentiation (Calegari, Haubensak, Haffner, & Huttner, 2005; Calegari & Huttner, 2003; Caviness, Takahashi, & Nowakowski, 1995; Dehay & Kennedy, 2007; Lukaszewicz, Savatier, Cortay, Kennedy, & Dehay, 2002; Salomoni & Calegari, 2010).

After *Neurog2* is stably expressed, its ability to initiate neuronal differentiation is further regulated (Li et al., 2012; Mattar et al., 2008). One way that *Neurog2* proneural activity is temporally regulated is by GSK3, which is increasingly activated during later stages of cortical development (Li et al., 2012). GSK3 phosphorylates Neurog2 in later stage cortical progenitors on up to nine SP sites, reducing Neurog2's ability to promote neurogenesis (Li et al., 2012). A similar finding was reported in Xenopus, where phosphorylation of Neurog2 SP sites by Cdks, which are also proline-directed serine–threonine kinases, similarly progressively blocks neurogenesis (Ali et al., 2011; Hindley et al., 2012). In Xenopus, a rheostat model has been proposed, which suggests that it is the number of SP sites phosphorylated, rather than the specific sites modified, that impedes Neurog2's proneural activity (Ali et al., 2011; Hindley et al., 2012). We propose that a similar model exists in the murine cortex, where the progressive phosphorylation of Neurog2 by GSK3 favors the formation of Neurog2–E47 heterodimers, which bind DNA and promote neurogenesis less efficiently than Neurog2–Neurog2 homodimers (Li et al., 2012). However, it is important to note that *Neurog2* function is not completely lost in the late cortex, but rather modified, controlling events other than neuronal differentiation, such as cortical progenitor transitions (e.g., *Neurog2* induces *Tbr2* expression to generate intermediate neuronal progenitors), neuronal migration (e.g., *Neurog2* induces *Rnd2* expression), and neuronal morphogenesis (regulated by Neurog2 phosphorylation on Y241; Britz et al., 2006; Ge et al., 2006; Hand & Polleux, 2011; Hand et al., 2005; Heng et al., 2008).

Cell fate specification and differentiation also require extensive genomic reprogramming, which alters the accessibility of specific promoters/enhancers. Recently, it was demonstrated that there is a progressive global compaction of cortical chromatin, during the period when *Neurog2* proneural activity gradually declines, a transition that is under the control of high mobility group A proteins (Kishi, Fujii, Hirabayashi, & Gotoh, 2012). We have also found that cortical chromatin surrounding *Neurog2* and its target genes is increasingly compacted between E12.5 and E15.5

(C. S., unpublished). One possibility is that Neurog2 may modify chromatin around its own targets by recruiting epigenetic modifiers, as shown for other bHLH proteins (Ge et al., 2006; Ju et al., 2004; Penn, Bergstrom, Dilworth, Bengal, & Tapscott, 2004; Seo, Richardson, & Kroll, 2005; Sun et al., 2001; Tapscott, 2005). The related protein Neurog1 binds Brg1, a subunit of the SWI/SNF chromatin-remodeling complex (Seo et al., 2005). *Neurog1* expression is also regulated by epigenetic modifications controlled by the polycomb group proteins (Onoguchi, Hirabayashi, Koseki, & Gotoh, 2012), which confer long-term transcriptional repression (Golbabapour et al., 2013). Future studies will be required to determine the extent to which epigenetic modifications contribute to temporal differences in *Neurog2*'s proneural activity in the developing neocortex.

2.2.3 Neurog1 *and* Neurog2 *have distinct and overlapping functions in the neocortex*

Similar to the PNS, *Neurog1* and *Neurog2* are often expressed in the same CNS lineages, but with distinct temporal expression profiles. In neocortical progenitors, *Neurog2* expression is initiated first, as early as E10, and persists into the postnatal period, whereas *Neurog1* expression is initiated one day later (E11.5) and only persists until ~E15.5 (Fode et al., 2000; Gradwohl et al., 1996; Shaker et al., 2012). During the period when Neurog1 and Neurog2 are both expressed, they are coexpressed in >95% of cortical progenitors (Britz et al., 2006), yet *Neurog1* and *Neurog2* mutant embryos have distinct cortical phenotypes, as described below.

The earliest cortical neurons to differentiate are Cajal–Retzius (CR) cells, which are derived from three sites in the pallial margins (cortical hem, ventral pallium, pallial septum), migrating tangentially to populate the marginal zone (Bielle et al., 2005). Fewer CR neurons differentiate in *Neurog2* mutant cortices, whereas conversely, misexpression of *Neurog2* induces ectopic CR cell genesis, consistent with a proneural role for *Neurog2* in this lineage (Dixit et al., 2014; Imayoshi, Shimogori, Ohtsuka, & Kageyama, 2008). In contrast, more early-born CR neurons are generated in *Neurog1* mutants (Dixit et al., 2014; Mattar et al., 2004), indicating that *Neurog1* is required to limit CR cell genesis and does not have a typical proneural activity in this lineage.

After CR cell genesis is complete, *Neurog1* and *Neurog2* function in a redundant manner to specify a glutamatergic projection neuron identity in early-born, deep-layer cortical neurons (Fode et al., 2000; Schuurmans et al., 2004). Supporting a redundant role for these transcription factors at this later developmental stage, the misspecification of *Neurog2*$^{-/-}$ cortical neurons is only evident in dorsomedial domains, where *Neurog1* expression is lost, and in *Neurog1*$^{-/-}$;*Neurog2*$^{-/-}$ cortices, misspecification defects extend into lateral domains (Fode et al., 2000; Schuurmans et al., 2004). Notably, there is not a loss of neuronal populations in *Neurog2* and *Neurog1*$^{-/-}$;*Neurog2*$^{-/-}$ mutant cortices as *Ascl1* expression is upregulated, compensating for the "proneural" activities of *Neurog1/2*, but specifying an aberrant GABAergic rather than glutamatergic identity (Fode et al., 2000; Schuurmans et al., 2004). However, consistent with *Neurog2* having a genuine proneural activity within the developing cortex, in *Neurog2*$^{-/-}$; *Ascl1*$^{-/-}$ cortices, where *Ascl1* cannot compensate for the loss of *Neurog1* and *Neurog2* in dorsomedial domains, precocious gliogenesis occurs (Nieto, Schuurmans, Britz, & Guillemot, 2001). This is consistent with the idea that proneural genes generally inhibit a glial cell fate, as has been shown for *Neurog1*, which promotes neurogenesis and inhibits astrocyte differentiation *in vitro* (Sun et al., 2001).

2.3. *Neurog3* identification and function in the developing CNS

Neurog3 was identified in a follow-up screen for new bHLH genes, but in contrast to *Neurog1* and *Neurog2*, *Neurog3* is not expressed in NCCs or their PNS derivatives, but is expressed in a regionalized fashion in the developing CNS (Pelling et al., 2011), as well as in the endocrine pancreas and enteroendocrine cells (Sommer et al., 1996). In the CNS, *Neurog3* has been best studied in the hypothalamus, which is a neuroendocrine system (Anthwal et al., 2013; Pelling et al., 2011). Specifically, *Neurog3* is expressed in progenitor cells for the ventromedial hypothalamic (VMH) nucleus and the arcuate (ARC) nucleus (Pelling et al., 2011). However, it remains to be determined whether *Neurog3* functions as a true proneural gene in the developing hypothalamus, since the loss of *Neurog3* does not perturb the expression of *Dll1* or *Hes5*, a Notch ligand and downstream effector, respectively. Nor are the total number of neurons produced in the *Neurog3* mutant hypothalamus altered. Instead, fewer SF1^{+} VMH neurons differentiate in *Neurog3* mutants, while TH^{+} and NPY^{+} neurons are increased in number. Notably, *Neurog3* conditional mutants also display a loss of Pomc expression in the ARC that is associated with postnatal obesity and a loss of leptin sensitivity (Anthwal et al., 2013). Taken together, these results suggest that *Neurog3* may be functioning as a neuronal differentiation (rather than proneural) gene in the hypothalamus. Alternatively, the proneural activity of *Neurog3* may be masked by compensation from *Ascl1*, which is also expressed and required in neuroendocrine lineages in the hypothalamus (see Section 4.2; McNay, Pelling, Claxton, Guillemot, & Ang, 2006).

2.4. *Neurog3* function in the developing pancreas

Although the *Neurogenins* are critical for nervous system development, their "proneural-like" functions are not restricted to the nervous system (Sommer et al., 1996). *Neurog3* is also highly expressed in the endocrine pancreas (Sommer et al., 1996). The pancreas includes both exocrine and endocrine tissues which secrete digestive enzymes and produce metabolic hormones, respectively. Within the exocrine pancreas, acinar cells secrete lipases, proteases, and nucleases, which empty into the pancreatic ductal network. The endocrine pancreas is organized into small clusters of cells called islets of Langerhans, which are composed of glucagon-producing α-cells, insulin-producing β-cells, somatostatin-producing δ-cells, ghrelin-producing ε-cells, and pancreatic polypeptide (PP) cells that regulate diverse processes in nutrient sensing and homeostasis. Glucagon and insulin are the main metabolic hormones that regulate glucose homeostasis, while ghrelin regulates appetite. Somatostatin has a wide range of activities, including regulation of insulin and glucagon secretion, while no known biological function has been ascribed to PP. *Neurog3* is required for the development of all endocrine cell lineages (Fig. 2.4; Gu, Dubauskaite, & Melton, 2002; Jensen et al., 2000).

In embryonic development, the pancreas is derived from the definitive endoderm, formed by fusion of a dorsal and a ventral epithelial bud generated by evagination of the foregut. In rodents, expression of

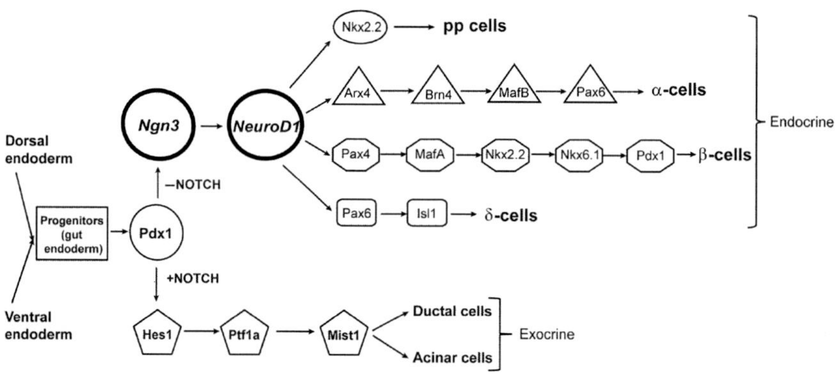

Figure 2.4 Proneural genes dictate lineage selection in pancreatic development. A simplified scheme for the role of transcription factors in the development and differentiation of pancreas based on the temporal expression and the predominant function of each factor. This scheme does not represent direct signaling pathways, but a sequence of gene expression required for differentiation of the endocrine cells. The proneural genes are in bold italic (Avolio et al., 2013; Van Hoof, D'Amour, & German, 2009).

pancreatic-duodenal homeobox 1 (*Pdx1*) in the dorsal domain at E8.5 marks the beginning of pancreatic specification, and the first differentiated glucagon-producing α cells form by E9.5 (Ben-Othman et al., 2013). At E10.5, insulin-producing cells begin to appear, followed by the appearance of somatostatin-producing cells at E15.5 and PP-producing cells at birth (Habener, Kemp, & Thomas, 2005). Cells expressing *Neurog3* are detectable as early as E9.5 in the primordial pancreas; the number of *Neurog3*-positive cells increases until E15.5 and decreases thereafter (Ben-Othman et al., 2013). After birth, *Neurog3* expression is nearly undetectable, at a level several hundred folds below that observed in E15.5 (Gradwohl, Dierich, LeMeur, & Guillemot, 2000). *Neurog3*-null mice develop diabetes and die 1–3 days postnatally, with histological examination showing an absence of the islets of Langerhans, which are present in the wild-type littermates (Gradwohl et al., 2000). At birth, *Neurog3*-mutant mice have no detectable hormone-containing cells in the pancreas, and expression of regulatory genes that participate in cell fate specification in the endocrine pancreas (e.g., *Pax4*, *Pax6*, and *NeuroD1*) is absent, suggesting that endocrine precursors are absent in the *Neurog3* mutant pancreas (Gradwohl et al., 2000). Heterozygous animals, however, are viable and fertile, with normal pancreatic endocrine cell numbers and function (Gradwohl et al., 2000).

In addition to its essential role in endocrine cell lineage determination in the pancreas, recent studies found that *Neurog3* plays a key role in β-cell regeneration in the adult mouse pancreas. In an experimental model of β-cell regeneration, expression of *Neurog3* is strongly induced and is required for regeneration of endocrine cells after injury, as knockdown of *Neurog3* prevents β-cell regeneration (Xu et al., 2008). Furthermore, *Neurog3* expression in adult islets contributes to endocrine maintenance, as inactivation of *Neurog3* in β cells of 1-month-old male mice caused glucose intolerance (Wang et al., 2009). In contrast, in a pancreatectomy model where 60% of the pancreas was surgically removed, β-cell mass regenerated without apparent activation of *Neurog3* (Lee, De Leon, Kaestner, & Stoffers, 2006). Therefore, while the role of *Neurog3* in pancreatic endocrine development is well established, its precise role in the adult pancreas is still debated.

2.5. *Neurog3* function in the enteroendocrine system

The enteroendocrine system consists of endocrine cells of the digestive tract, including insulin- and glucagon-secreting cells in the islets of Langerhans in the pancreas as well as endocrine cells scattered through the digestive epithelium from the stomach to the colon. Enteroendocrine cells represent about 1% of gut epithelial cell population, and they produce more than 15 families of peptide hormones that control glucose homeostasis, exocrine pancreatic secretion, gut motility, and growth and renewal of the gut epithelium. Although they share biochemical similarities with neurons, enteroendocrine cells are derived from endoderm, and they are not modified neurons (Skipper & Lewis, 2000). Similar to endocrine cells of the pancreas, *Neurog3* is also required for cell fate determination of enteroendocrine cells (Lee, Perreault, Brestelli, & Kaestner, 2002). *Neurog3* is transiently expressed in the embryonic gut endoderm, appearing slightly later than in embryonic pancreas, first detectable at E12.5 (Jenny et al., 2002). The number of Neurog3$^+$ cells increases until E15.5, with the highest number found in the duodenum and decreasing in a rostro-caudal manner with fewer found in the colon (Jenny et al., 2002). In adult intestinal epithelium, Neurog3$^+$ cells are present in the villi (Lee et al., 2002). In the stomach, Neurog3$^+$ cells are located in the glandular stomach but not in the squamous epithelium of the forestomach (Lee et al., 2002).

In *Neurog3*-null transgenic mice, however, endocrine cells are not detected in their intestine at birth, and more specifically, cells producing intestinal hormones, i.e., cholecystokinin, secretin, gastrin, serotonin, peptide YY, glucagon-like protein, gastric inhibitory protein, and somatostatin, are not present (Jenny et al., 2002). Notably, expression of *Atoh1*, another bHLH transcription factor that regulates intestinal cell differentiation, is unaffected by *Neurog3* deletion, although *NeuroD* expression is not detectable in the absence of *Neurog3*, placing these factors in *Atoh1–Neurog3–NeuroD* sequence in terms of controlling the development of the enteroendocrine cells (Jenny et al., 2002). In contrast, endocrine cells expressing serotonin and ghrelin are present in the gastric epithelium of *Neurog3*-null mice, while gastrin- and somatostatin-expressing cells are absent (Jenny et al., 2002). Furthermore, there is a significant increase of goblet cells and misexpression of intestine-specific markers in the stomach of *Neurog3* null mice, suggesting that *Neurog3* is important in maintaining the gastric mucosal identity and preventing intestinal metaplasia of the gastric epithelium (Lee et al., 2002). Taken together, it appears that *Neurog3* has different roles in the development of enteroendocrine cells in the stomach and intestine, with endocrine cells in the intestine displaying an absolute requirement for *Neurog3*, while both *Neurog3*-dependent and -independent pathways regulate the development of endocrine cells in the stomach (Jenny et al., 2002). Of note, development of enterocytes is not dependent on *Neurog3* expression (Jenny et al., 2002).

3. *Atoh1/Atoh7* PRONEURAL FUNCTIONS IN DEVELOPMENT

3.1. Introduction to *Atoh* family proneural genes

The mouse *ato* orthologues *Math1/Atoh1* and *Math5/Atoh7* are the two bHLH transcription factors mostly closely related to *Drosophila ato*. *Atoh1* was initially identified and cloned by Akazawa, Ishibashi, Shimizu, Nakanishi, and Kageyama (1995) as a candidate proneural factor based on sequence and structural features. By *in situ* hybridization, *Atoh1* transcripts are detected beginning at E9.5 in the cranial nerve ganglia and in the dorsal neural tube (Akazawa et al., 1995). *Atoh7* was identified in a cDNA library from the embryonic retina using degenerate PCR to isolate *ato*-related genes (Brown et al., 1998). *Atoh7* has a very restricted expression pattern, with transcripts limited to the embryonic retina and the Xth cranial ganglion (Brown et al., 1998).

3.2. *Atoh1* function in rhombic lip derivatives

During development, *Atoh1* is expressed in the region of the dorsal hindbrain neuroepithelium called the rhombic lip (Akazawa et al., 1995; Rose et al., 2009; Wang, Rose, & Zoghbi, 2005). *Atoh1*-dependent neurons are generated in the rhombic lip and migrate to populate several hindbrain and cerebellar nuclei (Rose et al., 2009; Wang et al., 2005). These include neurons in the ventral cochlear nucleus, cochlear granule neurons, precerebellar nuclei, parabrachial nuclei, lateral lemniscal nuclei, parafacial respiratory group/retrotrapezoid nuclei (pFRG/RTN), and deep cerebellar nuclei (Rose et al., 2009; Wang et al., 2005). *Atoh1*-null mice die in the perinatal period from severe central apnea, consistent with the requirement of *Atoh1* for the generation of brainstem respiratory nuclei and normal respiratory rhythmogenesis (Rose et al., 2009).

Atoh1 is also expressed in the cerebellar granule cell precursors that populate the transient cerebellar external granule cell layer (EGL), another derivative of the rhombic lip (Ben-Arie et al., 1997). As an early marker of the granule cell lineage, *Atoh1* is expressed in precursors of the EGL, while they are still in the rhombic lip as early E13 and continues to be expressed after the cells have migrated to form the highly proliferative EGL (Ben-Arie et al., 1997; Machold & Fishell, 2005; Wang et al., 2005). However, the expression of *Atoh1* is temporally restricted, as it is no longer expressed in granule cells once they have exited the cell cycle and migrated to the internal

EGL (Ben-Arie et al., 1997). *Atoh1*-null mice do not produce granule cells and lack an EGL (Ben-Arie et al., 1997). In granule cell precursors, *Atoh1* regulates responsiveness to Sonic Hedgehog (Flora, Klisch, Schuster, & Zoghbi, 2009), a morphogen driving proliferation of granule cell precursors by regulating the transcription factor Gli2 (Wallace, 1999; Wechsler-Reya & Scott, 1999). Conversely, overexpression of *Atoh1* in progenitors perturbs differentiation of granule cell precursors, promoting the expression of early differentiation markers but not resulting in a mature granule cell phenotype (Helms, Gowan, Abney, Savage, & Johnson, 2001). Aside from granule cells, other cells in the cerebellar cortex depend on *Atoh1*, including unipolar brush cells, which are also derivatives of the rhombic lip (Englund et al., 2006).

3.3. *Atoh1* function in inner ear development

In the inner ear, cochlear and vestibular hair cells function as mechanoreceptors to transduce sound and head motion, respectively, into receptor potentials. *Atoh1*, along with *Neurog1* (see Section 2.1.2) and the neuronal differentiation gene *NeuroD1*, regulates the development of these cochlear and vestibular hair cells. *Atoh1*-null mice lack hair cells (Bermingham et al., 1999). Furthermore, during development, the balance of *Atoh1* versus Notch signaling in multipotent progenitor cells in the otic placode is an important determinant of a sensory versus neuronal phenotype (Jeon, Fujioka, Kim, & Edge, 2011). One of the most exciting developments from these studies is that adenoviral vector delivery of *Atoh1* to nonsensory cells in a guinea pig model of sensorineural hearing loss can induce regeneration of hair cells and improve hearing in the mature deaf ear (Izumikawa et al., 2005).

3.4. *Atoh1* function in the intestinal epithelium

The intestinal epithelium is composed of crypts (which contain rapidly cycling crypt base columnar cells, Paneth cells, and transit-amplifying cells that progressively migrate up the crypt-villus axis) and villi (which contain differentiated absorptive enterocytes, goblet cells, and enteroendocrine cells). Within the intestinal epithelium, *Atoh1* is expressed in the crypts and villi of the intestinal epithelium, where it is essential for specification of the secretory cell lineage as well as for regulation of proliferation (Shroyer et al., 2007; Yang, Bermingham, Finegold, & Zoghbi, 2001). *Atoh1*-null mice fail to develop all types of secretory cells (goblet, enteroendocrine, and Paneth cells), while development of absorptive enterocytes is unaffected, consistent with *Atoh1* playing a role in specifying an intestinal progenitor identity (Shroyer et al., 2007; Yang et al., 2001). In contrast, overexpression/ectopic expression of *Atoh1* expands the secretory lineage at the expense of the absorptive enterocyte lineage (VanDussen & Samuelson, 2010). The secretory lineage determination effects of *Atoh1* are mediated through the transcriptional repressor Gfi1 (Shroyer, Wallis, Venken, Bellen, & Zoghbi, 2005). Further development of enteroendocrine cells from the secretory lineage also requires *Neurog3* as discussed above (Jenny et al., 2002). In addition to cell-type specification, *Atoh1* expression is important in cell cycle arrest. The negative regulation of intestinal epithelial cell proliferation may be mediated by repression of Rbpj, which is a downstream effector of Notch signaling, or may be via regulation of p57 and p27 independent of Notch (Kazanjian, Noah, Brown, Burkart, & Shroyer, 2010; Kim & Shivdasani, 2011; Peignon et al., 2011).

3.5. *Atoh7* function in the retina

The retina emerges as an outpocketing of the embryonic diencephalon at ~E9.5 of development (Wawersik, Purcell, & Maas, 2000) and is comprised of multipotent retinal progenitor cells that give rise to seven neural cell types in two developmental waves (Alexiades & Cepko, 1997; Turner & Cepko, 1987; Wetts, Serbedzija, & Fraser, 1989): in the embryonic wave in mouse, retinal ganglion cells (RGCs), cone photoreceptors, and horizontal and amacrine cell interneurons are generated. In the postnatal period, the differentiation of rod photoreceptors, bipolar cell interneurons, and Muller glia cells reaches completion by postnatal day (P) 12.

Atoh7 is expressed very early in retinal progenitors, preceding other proneural genes (Brown et al., 1998; Dixit et al., 2013), and in lineage trace experiments, *Atoh7* expression occurs more frequently in early-born retinal cell lineages (Brzezinski, Prasov, & Glaser, 2012). Consistent with the early onset of its expression, *Atoh7* is required for development of RGCs, which are among the earliest retinal cells generated (Brown, Patel, Brzezinski, & Glaser, 2001; Liu, Mo, & Xiang, 2001; Wang et al., 2001). Notably, the loss of RGCs in *Atoh7* mutants means that RGC axons fail to form the optic nerve, which is even reduced in size in heterozygous animals (Prasov, Nagy, Rudolph, & Glaser, 2012). The optic nerve normally innervates several brain regions, including the suprachiasmatic nucleus, which innervates circadian rhythms (Wee, Castrucci, Provencio, Gan, & Van Gelder, 2002). *Atoh7* mutants thus display defects in photoentrainment and the free-running period (Brzezinski et al., 2005; Wee et al., 2002). Interestingly, in knock-in studies, the neuronal differentiation gene *Neurod1* can replace *Atoh7* function, rescuing defects in RGC differentiation (Mao, Wang, Pan, & Klein, 2008), suggesting that the timing of *Atoh7* expression is critical to its function.

Instead of generating RGCs, *Atoh7* mutants precociously generate later-born retinal cell types, such as cone photoreceptors, amacrine cells, and even Muller glia, the last cell type to differentiate (Feng et al., 2010; Le, Wroblewski, Patel, Riesenberg, & Brown, 2006). A contributing factor to the altered timing of cell differentiation in *Atoh7* mutants is that retinal progenitors proliferate excessively and do not exit the cell cycle at their correct times (Le et al., 2006). Conversely, in gain-of-function studies, *Atoh7* induces the expression of several RGC genes (Yao, Sun, Wang, Xu, & Qian, 2007), suggesting that *Atoh7* is both necessary and sufficient to promote an RGC fate, which are hallmark features of proneural genes. However, the sufficiency of *Atoh7* to promote an RGC fate remains under debate, since heterochronic overexpression of this gene is not sufficient to induce the ectopic formation of RGCs (Prasov & Glaser, 2012). It thus appears that *Atoh7* function is highly context dependent in the retina.

Multiple *Atoh7*-regulated genes have been identified in the retina, including other proneural genes, such as *Neurog2*, which is upregulated in the absence of *Atoh7* (Mu et al., 2005). Cross-regulatory interactions among *Atoh7*, *Neurog2*, and *Hes1* regulate the retinal progenitor cell status and the timing of differentiation (Matter-Sadzinski, Matter, Ong, Hernandez, & Ballivet, 2001; Matter-Sadzinski, Puzianowska-Kuznicka, Hernandez, Ballivet, & Matter, 2005). *Atoh7* function is also regulated in other ways, such as by alternative splicing to generate multiple noncoding transcripts that limit the number of retinal progenitors expressing functional *Atoh7* (Kanadia & Cepko, 2010), although this finding remains under debate (Prasov, Brown, & Glaser, 2010). *Atoh7* expression is also determined by chromatin status (Skowronska-Krawczyk, Ballivet, Dynlacht, & Matter, 2004; Skowronska-Krawczyk et al., 2009).

4. *Ascl1* PRONEURAL FUNCTIONS IN DEVELOPMENT

4.1. *Ascl1* identification and function in the developing PNS

The autonomic nervous system has three subdivisions, including sympathetic, parasympathetic, and enteric. *Ascl1*, a vertebrate *as-sc* orthologue, was initially identified in a sympathoadrenal-derived cell line (Johnson, Birren, & Anderson, 1990) and is expressed in sympathetic (Lo, Johnson, Wuenschell, Saito, & Anderson, 1991), as well as parasympathetic and enteric lineages in the PNS (Guillemot & Joyner, 1993; Guillemot et al., 1993). Consistent with *Ascl1* functioning as a proneural gene in these autonomic lineages, defects in the differentiation of sympathetic chain ganglia, parasympathetic ganglia, and enteric lineages are observed in *Ascl1* mutant embryos (Guillemot et al., 1993; Hirsch, Tiveron, Guillemot, Brunet, & Goridis, 1998). Conversely, misexpression of murine *Ascl1* in NCCs exclusively promotes the formation of noradrenergic, autonomic neurons (Lo et al., 2002). Consistent with the more stringent fate specification properties for *Ascl1* (similar to Drosophila *ac-sc* genes; Jarman et al., 1993, 1995), in $Neurog2^{Ascl1KI/KI}$ mice, in which *Ascl1* has replaced the *Neurog2* coding sequence, *Ascl1* cannot rescue defects in the generation of sensory DRG neurons (Parras et al., 2002). *Ascl1* is also less effective at promoting cell cycle exit in NCCs and is more sensitive to Notch-mediated inhibition of neurogenesis, which makes *Ascl1* an unusual proneural gene (Lo et al., 2002). Notably, the ability of *Ascl1*-overexpressing NCCs to continue proliferating is thought to reflect the normal ability of sympathetic neuroblasts to remain cycling after their fate has been specified (Lo et al., 2002).

The olfactory epithelium (OE) is another region of the PNS where *Ascl1* functions as a proneural gene. Similar to the distal cranial ganglia, the OE arises from ectodermal placodes, which begin to thicken and invaginate to form the olfactory pits between E9.5 and E10.5 (Cuschieri & Bannister, 1975). *Ascl1* is expressed in progenitor cells in the olfactory placode as well as in basal progenitors in the OE (Cau, Gradwohl, Fode, & Guillemot, 1997). In the absence of *Ascl1*, olfactory receptor neurons fail to differentiate, consistent with a proneural function for this gene (Cau et al., 1997; Guillemot et al., 1993).

4.2. *Ascl1* proneural functions in the developing CNS

Ascl1 specifies distinct neuronal phenotypes in different regions of the neural tube. Similar to *Neurog2*, *Ascl1* expression also oscillates in 2–3 h cycles in undifferentiated progenitors due to its ability to transactivate the Notch ligand *Dll1* and initiate Notch-mediated lateral inhibition (Castro et al., 2006, 2011; Imayoshi et al., 2013; Shimojo et al., 2008). Moreover, these *Ascl1* oscillatory cycles retain progenitors in a proliferative state, while sustained *Ascl1* expression promotes differentiation (Imayoshi et al., 2013).

Ascl1 is also expressed in a regionalized fashion in the developing neural tube (Kim, Battiste, Nakagawa, & Johnson, 2008), specifying distinct neuronal fates in each region. In the telencephalon, *Ascl1* expression is highest in ventral progenitors, where it is required and sufficient to specify a GABAergic neuronal phenotype (Berninger, Guillemot, & Gotz, 2007; Britz et al., 2006; Casarosa, Fode, & Guillemot, 1999; Geoffroy et al., 2009; Horton, Meredith, Richardson, & Johnson, 1999; Parras et al., 2002; Poitras, Ghanem, Hatch, & Ekker, 2007; Schuurmans et al., 2004). *Ascl1* also is required for the formation of GABAergic neurons in the cerebellum (Grimaldi, Parras, Guillemot, Rossi, & Wassef, 2009). In the hypothalamus, *Ascl1* is required to promote neurogenesis in the ARC and VMH and also controls *Neurog3* expression (McNay et al., 2006). In the hindbrain, *Ascl1* is required for the differentiation of noradrenergic neurons in the locus coeruleus (Parras et al., 2002), while in the spinal cord, *Ascl1* is required for the generation of V1 interneurons in ventral domains (Battiste et al., 2007; Parras et al., 2002), as well as subsets of dorsal interneurons (Mizuguchi et al., 2006; Parras et al., 2002; Wildner et al., 2006, 2013).

Interestingly, *Ascl1* is also expressed in dorsal telencephalic progenitors, where its functions are less well defined, as GABAergic neurons do not arise from dorsal progenitors but rather differentiate in the ventral telencephalon prior to cortical entry by tangential migration (Anderson, Eisenstat, Shi, & Rubenstein, 1997; Anderson, Kaznowski, Horn, Rubenstein, & McConnell, 2002; Anderson, Marin, Horn, Jennings, & Rubenstein, 2001; Tamamaki, Fujimori, & Takauji, 1997). Recently, it has been reported that in early dorsal telencephalic progenitors, *Ascl1* is required for the differentiation of CR neurons, which are glutamatergic, indicating that *Ascl1* can change its fate specification properties in this region of the neural tube (i.e., *Ascl1* specifies GABAergic identities in ventral domains and glutamatergic fates dorsally; Dixit et al., 2011). Moreover, *Ascl1* is excluded from astrocyte lineages in the spinal cord (Battiste et al., 2007) and is required to inhibit astrocyte differentiation in forebrain progenitors that do not express *Neurog2* or *Math3* (Nieto et al., 2001; Tomita, Moriyoshi, Nakanishi, Guillemot, & Kageyama, 2000) and may promote the early postnatal wave of dorsal oligodendrocyte precursor cell (OPC) differentiation (Parras et al., 2004, 2007). Indeed, *Ascl1* is required for OPC differentiation in the spinal cord (Battiste et al., 2007; Sugimori et al., 2008). Thus, even within a single progenitor domain, *Ascl1* has multiple functions. Moreover, unlike "canonical" proneural genes, *Ascl1* can promote rather than inhibit glial identities (i.e., OPC fate), which are characterized by their proliferative potential.

4.2.1 Ascl1's ability to specify more than one cell fate is regulated by target gene selection and posttranslational modifications

Ascl1 has three known functions in the telencephalon—to promote proliferation or induce the differentiation of GABAergic interneurons or OPCs (Castro et al., 2011; Wilkinson, Dennis, & Schuurmans, 2013), raising the question of how these distinct fate choices are made. The ability of *Ascl1* to induce proliferation is context specific, similar to the *ac-sc* genes in *Drosophila*. For example, misexpression of *Ascl1* in E12.5 cortical progenitors induces robust proliferation that is associated with the acquisition of a glioblast as opposed to a neuronal fate (Li et al., 2014). The most likely cause of these differences is that *Ascl1* selects distinct target genes to control different cell fate choices. Many *Ascl1*-regulated genes have been identified in genome-wide screens (Castro et al., 2011; Gohlke et al., 2008; Long, Cobos, Potter, & Rubenstein, 2009; Wildner et al., 2013). Notably, *Ascl1*'s ability to induce proliferation is due to its ability to transactivate several genes involved in cell cycle progression, including *E2f1*, *Cdk1*, *Cdk2*, and *Skp2* (Castro et al., 2011). Interestingly, this cassette of proliferative genes contains not only E-boxes but also binding sites for Rbpj, suggesting that *Ascl1* promotes proliferation in the context of active Notch signaling (Castro et al., 2011). Moreover, the transactivation of other *Ascl1* target genes is also controlled by cooperative binding with other transcription factors. E-boxes in the *Dll1* promoter, for example, are flanked by binding sites for Brn1/2 and Ascl1, which together initiate *Dll1* transcription (Castro et al., 2006).

The ability of *Ascl1* to select either proliferative (e.g., *E2f1*, *Cdk1*, *Cdk2*, and *Skp2*; Castro et al., 2011), GABAergic neuronal (e.g., *Dlx1/2*; Poitras et al., 2007), or OPC (e.g., *Sox9* and *Hes5*; Castro et al., 2011; Ueno et al., 2012) transcriptional targets is also dependent on cellular context, which is highly influenced by growth factor signaling. In the neural tube, one important signaling pathway is RAS/ERK signaling, which is initiated by growth factors such as EGF and FGF, which bind to their cognate receptor tyrosine kinases to promote neural progenitor cell proliferation (e.g., Imamura, Satoh, Endo, & Takishima, 2008; Lukaszewicz et al., 2002). Regulatory interactions exist between Fgf signaling and the proneural genes (Gonzalez-Quevedo, Lee, Poss, & Wilkinson, 2010; Kataoka & Shimogori, 2008; Ribes et al., 2008). In dissociated cerebral cortical progenitors, Fgfs respecify dissociated cortical progenitors to acquire an aberrant subcortical fate by influencing proneural gene expression (Abematsu et al., 2006; Gabay, Lowell, Rubin, & Anderson, 2003; Hack, Sugimori, Lundberg, Nakafuku, & Gotz, 2004). Specifically, RAS/ERK signaling functions in telencephalic progenitors to turn *Neurog2* expression off and *Ascl1* expression on (Li et al., 2014). Subsequently, levels of RAS/ERK signaling control *Ascl1*'s ability to select distinct cell fates. At high RAS/ERK levels, $Ascl1^{+}$ cortical progenitors preferentially proliferate, acquiring a glioblast identity and ultimately differentiating into OPCs (Li et al., 2014). In contrast, at lower levels of RAS/ERK signaling, *Ascl1* promotes the differentiation of $GABA^{+}$ interneurons (Li et al., 2014).

The ability of RAS/ERK to alter *Ascl1's* fate specification properties *in vivo* is mirrored by changes in Ascl1 target gene selection in transcriptional reporter assays *in vitro* (Li et al., 2014). In P19 cells, at low levels of RAS/ERK signaling, *Ascl1* has an enhanced ability to transactivate a *Dlx1/2*-intergenic enhancer (Li et al., 2014), which is a direct *Ascl1* transcriptional target normally transcribed in $GABA^{+}$ neuronal lineages (Poitras et al., 2007). In contrast, at high levels of RAS/ERK signaling, enhanced transactivation of *Sox9* is observed (Li et al., 2014), which is an *Ascl1* target gene (Castro et al., 2011) expressed at high levels in glioblast progenitors (Kang et al., 2012; Stolt et al., 2003). The effects of RAS/ERK signaling on Ascl1 function occur through the ability of ERK to directly phosphorylate Ascl1 (Li et al., 2014). Current data suggest that phosphorylation effects are additive for Ascl1 (Li et al., 2014), similar to the rheostat model of Neurog2 control (Ali et al., 2011; Hindley et al., 2012), but further work will be required to substantiate this idea.

4.3. *Ascl1* and neuronal reprogramming

Ascl1 is one of the key genes used in the emerging field of neuronal reprogramming, wherein somatic cells are directly converted into neurons by the forced expression of transcription factors (Caiazzo et al., 2011; Pang et al., 2011; Pfisterer et al., 2011; Vierbuchen et al., 2010; Yoo et al., 2011). While neuronal reprogramming of fibroblasts typically involves the misexpression of *Ascl1* along with other transcription factors, such as *Brn2* and *Myt1l* (BAM factors), *Ascl1* is the only one of these three genes that can induce some reprogramming on its own (Marro et al., 2011; Vierbuchen et al., 2010). Mechanistically, *Ascl1* acts as a pioneer factor in fibroblasts by binding its target sites in genomic regions with a specific trivalent signature (H3K4me1, H3K27ac, H3K9me3; Wapinski et al., 2013).

The common use of *Ascl1* in neuronal reprogramming is surprising, given that *Ascl1* not only promotes neuronal differentiation (see Section 4.2) but also promotes both proliferation and an OPC fate (Parras et al., 2004, 2007), both of which may be unwanted from a therapeutic perspective. However, the key effects of *Ascl1* in neuronal reprogramming are not necessarily to respecify cell fates, but to act as a "pioneer" factor (Wapinski et al., 2013). In contrast, *Neurog2* more effectively promotes cell cycle exit and differentiation and could potentially be used as a reprogramming factor. Indeed, *Neurog2* can promote the differentiation of spinal motor neurons when misexpressed with *Isl1* and *Lhx3*, a combination known as NIL factors (Hester et al., 2011; Mazzoni et al., 2013). Moreover, *Neurog2* can also enhance the efficiency of motor neuron recovery in animal models of spinal cord injury, presumably by increasing the differentiation of embryonic neural progenitors injected into the injury site (Perrin et al., 2010). Future studies will be required to determine which proneural gene is the most efficient at neuronal reprogramming.

5. PRONEURAL GENES IN HUMAN DISEASE (TABLE 2.1)

5.1. Human developmental disorders

Few constitutional genetic diseases or developmental malformations have their roots in the abnormal expression of proneural bHLH transcription factors. This may be because mutations in these genes are rare, or there may be sufficient redundancy that mutations do not affect development (or only affect development in the homozygous state), or because the function of the proneural genes is so critical in development that mutation or other alteration is catastrophic and incompatible with viable human embryonic or fetal development.

Congenital central hypoventilation syndrome (CCHS, a.k.a. Ondine's curse) is one developmental disorder that has been tied to mutations in *ASCL1* (de Pontual et al., 2003). A rare disorder of abnormal autonomic respiratory control, CCHS, typically presents at birth as cyanosis during sleep wherein patients breathe normally while awake but hypoventilate during sleep (Axelrod, Chelimsky, & Weese-Mayer, 2006). More severely affected patients hypoventilate in both sleep and awake states. Although anatomic abnormalities have been difficult to discern in CCHS patients, they often manifest a range of symptoms of autonomic nervous system dysfunction including abnormal pupillary control, body temperature control, sweating, and gastrointestinal motility consistent with more global autonomic dysfunction (Axelrod et al., 2006). If the gastrointestinal motility is due to congenital absence of ganglion cells in the gut, this is termed Hirschsprung disease. Because of these diverse associations, CCHS is one of several

Table 2.1 Human diseases associated with mutation or aberrant expression of proneural genes

Gene	Alteration	Human disease	System/tissue affected	References
Atoh1	Upregulation	Medulloblastoma (SHH subgroup)	Cerebellum, granule cell precursor-like cells	Salsano, Pollo, Eoli, Giordana, and Finocchiaro (2004), Ayrault et al. (2010), Flora et al. (2009), Schuller et al. (2008), Yang et al. (2008)
Atoh1	Downregulation	Colonic adenocarcinoma	Intestinal epithelium	Leow et al. (2004), Shroyer et al. (2007), Yang et al. (2001), Bossuyt, De Geest, et al. (2009), Bossuyt, Kazanjian, et al. (2009), Peignon et al. (2011)
Atoh7	Mutation	Persistent hyperplasia of the primary vitreous (PHPV)	Eye	Prasov, Masud, et al. (2012), Prasov, Nagy, et al. (2012)
Ascl1	Mutation	Congenital central hypoventilation syndrome (CCHS)	Autonomic system	de Pontual et al. (2003)
Ascl1	Mutation	Haddad syndrome (CCHS + Hirschsprung disease)	Autonomic system including enteric autonomic	de Pontual et al. (2003)
Ascl1	Upregulation	Diffuse gliomas, some glioblastomas	Brain, glial	Rheinbay et al. (2013), Phillips et al. (2006), Rousseau et al. (2006), Somasundaram et al. (2005), Li et al. (2014)
Ascl1	Upregulation	Neuroendocrine cancers (various)	Lung, thyroid, gastrointestinal, prostate, skin	La Rosa et al. (2013), Jiang et al. (2004), Borges et al. (1997), Linnoila et al. (2000), Rapa et al. (2013), Wang et al. (2007)

Ascl1	Possible polyQ length association	Parkinson disease risk	Midbrain dopaminergic neurons	Ide et al. (2005)
Neurog1	Deletion	Moebius-like syndrome	Brainstem cranial nerve nuclei	Schroder et al. (2013)
Neurog1	Possible SNP association	Schizophrenia risk, cognitive defects	Brain	Ho et al. (2008), Kendler et al. (2000), Schwab et al. (1997), Schwab and Wildenauer (2013), Skol et al. (2003), Zaharieva et al. (2008)
Neurog3	Possible association	Type 2 diabetes risk	Pancreatic endocrine system	del Bosque-Plata et al. (2001), Kim, Warram, Krolewski, and Doria (2001), Chavali et al. (2011), Gupta et al. (2013), Yokoi et al. (2006)
Neurog3	Mutation	Permanent neonatal diabetes (PNDM)	Pancreatic endocrine system	Pinney et al. (2011), Rubio-Cabezas et al. (2011)
Neurog3	Mutation	Congenital malabsorptive diarrhea	Intestinal enteroendocrine cells	Cortina et al. (2007)

disorders known as neurocristopathies (due to the commonality of an underlying defect of cells of neural crest origin) that also includes Hirschsprung disease (Axelrod et al., 2006; Bower & Adkins, 1980). Although most cases of CCHS have underlying mutations in PHOX2B (Amiel et al., 2003; Weese-Mayer et al., 2003), a few cases of CCHS and Haddad syndrome (in which CCHS is associated with Hirschsprung disease) have been reported to harbor heterozygous mutations in *ASCL1* (de Pontual et al., 2003), which is a component of the PHOX-RET developmental pathway (Hirsch et al., 1998; Pattyn, Goridis, & Brunet, 2000). *In vitro*, all of the *ASCL1* mutant alleles identified in CCHS/Haddad syndrome result in defective noradrenergic differentiation of neural progenitors, underscoring the importance of *ASCL1* in the developing autonomic nervous system (de Pontual et al., 2003; Guillemot et al., 1993; Lo et al., 1991).

Other developmental disease associations are even less frequent. Another disease for which loss of a functional proneural gene has been implicated in the causation of a developmental disorder is a Moebius-like syndrome (Schroder et al., 2013). In a single case report, a child with sensorineural deafness, inability to chew and swallow, and a balance disorder due to cranial dysinnervation was found to harbor homozygous microdeletion of chromosome 5q31.1 that spans the *NEUROG1* locus (Schroder et al., 2013). In addition, a homozygous mutation in *ATOH7* that interferes with DNA binding was found in a family with autosomal recessive persistent hyperplasia of the primary vitreous (Prasov, Masud, et al., 2012).

5.2. Neurologic and neuropsychiatric disease

Mutations or polymorphisms in the *NEUROG* genes have similarly been linked to neuropsychiatric and developmental disorders (Ho et al., 2008; Kendler et al., 2000; Schwab & Wildenauer, 2013; Schwab et al., 1997; Skol et al., 2003; Zaharieva et al., 2008). As with data from other large genome-wide association studies, however, the evidence is not strong enough to directly implicate *NEUROG* genes as playing a causative role—yet the possibility is intriguing. *NEUROG1* is located on chromosome 5q31, a region previously implicated in multiple schizophrenia linkage studies (Kendler et al., 2000; Schwab & Wildenauer, 2013; Schwab et al., 1997; Skol et al., 2003; Zaharieva et al., 2008). Patients with single nucleotide polymorphisms lying within *cis*-regulatory regions of *NEUROG1* exhibit cognitive defects in verbal memory, language, and visual-spatial abilities (Ho et al., 2008).

Interestingly, polymorphisms in *ASCL1* polyglutamine length have been associated with increased risk of developing Parkinson's disease (PD), a progressive neurological movement disorder that is caused by the degeneration of midbrain dopaminergic neurons (Ide et al., 2005). Regression analyses also suggest that *ASCL1* and *PHOX2B* are epistatic in increasing PD risk (Ide et al., 2005). Although the genetic and environmental factors that have been implicated in the development of PD are diverse (among which *ASCL1* polymorphisms may represent only a weakly associated factor), the importance of *ASCL1* in PD may be more significant when considered from a therapeutic perspective. Recently, it has been shown that *Ascl1* is an instructive factor (along with other factors including *Nurr1*, *Lmx1a*, *Brn2*, *Myt1*, and *FoxA2*) that can be harnessed for the directed differentiation of dopaminergic neurons (Caiazzo et al., 2011; Liu et al., 2012; Park et al., 2006; Pfisterer et al., 2011; Theka et al., 2013). This discovery has significant implications for potential cell replacement therapies in PD, regardless of the disease etiology.

5.3. Proneural genes in cancer

Sequencing efforts to date do not implicate proneural bHLH genes as classic oncogenes or tumor suppressors that drive cancer growth when genetically altered, as mutations in genes such as *ASCL1*, *NEUROG2*, and *ATOH1* are rare in tumors—having only been reported a few dozen times among more than 8000 samples sequenced and reported in the COSMIC database (cancer.sanger.ac.uk). Nevertheless, the expression of proneural genes at the mRNA and/or protein levels has been found to be dysregulated in a variety of cancers including those that occur in the brain, lung, colon, thyroid, prostate, and other sites (Leonard et al., 2002). Evidence from several of these cancer sites supports the notion that expression of specific proneural genes can affect cancer initiation and maintenance by a variety of mechanisms.

5.3.1 Atoh1 *in nervous system and gastrointestinal cancer*

As *Atoh1* (a.k.a. *Math1*) is intimately involved in cerebellar development and the regulation of cerebellar granule neuron precursor, it is not surprising that *ATOH1* has been found to be central to the biology of the malignant cerebellar tumor medulloblastoma. Medulloblastoma, which histologically resembles primitive neuroectodermal tissue, can be molecularly subclassified into four subgroups—SHH, WNT, Group C/3, and Group D/4—based on integrative genomic analyses (Kool et al., 2008; Northcott et al., 2011;

Taylor et al., 2012). Among them, the SHH group has particularly close ties to granule cell progenitor biology and the expression of *ATOH1* (Gibson et al., 2010; Gilbertson & Ellison, 2008; Salsano et al., 2004; Schuller et al., 2008; Yang et al., 2008). *ATOH1* upregulation is a feature of the SHH group but not other medulloblastoma subtypes (Salsano et al., 2004). Furthermore, in mouse models of medulloblastoma driven by loss of *Patched* or *Smoothened* activation, *Atoh1* is not only highly expressed in tumors, but also the lineage restriction of progenitors to a granule neuron precursor phenotype (as evidence by GNP markers including *Atoh1*) appears to be an essential step in oncogenesis (Schuller et al., 2008; Yang et al., 2008). A functional role for *ATOH1* in SHH-group medulloblastomas is additionally supported by evidence that *Atoh1* regulates the proliferative capacity of granule cell precursors via directly regulating elements of the SHH signaling pathway (Ayrault et al., 2010; Flora et al., 2009). Conversely, *Atoh1* ablation inhibits medulloblastoma tumorigenesis in response to constitutively active *Smoothened* (Flora et al., 2009). Thus, in some contexts such as in the cerebellum, *ATOH1* may function as a competency factor in the transformation process and may also directly regulate components of other bona fide oncogenic pathways.

Interestingly, however, in some nonneural contexts such as the gastrointestinal tract, *ATOH1* may function as a tumor suppressor. In the developing gut, *Atoh1* represses proliferation and alternatively promotes differentiation along secretory lineages or stimulates apoptosis (Bossuyt, Kazanjian, et al., 2009; VanDussen & Samuelson, 2010). *ATOH1* expression is frequently downregulated in human colonic adenocarcinomas (Leow, Romero, Ross, Polakis, & Gao, 2004). In addition, in human cell lines and xenografts, forced *ATOH1* expression inhibits proliferation, and in mice, an intestine-specific *Atoh1* deletion enhances tumorigenesis induced by either chemical carcinogenesis or *APC* mutation/loss (Bossuyt, De Geest, et al., 2009; Peignon et al., 2011).

As discussed previously, *Atoh1* activity in the cerebellar EGL maintains granule cell precursors in a proliferate state (Ben-Arie et al., 1997; Flora et al., 2009). In contrast, *Atoh1* activity in the intestinal epithelium causes cell cycle exit and differentiation (Shroyer et al., 2007; Yang et al., 2001). It is thus not surprising that *ATOH1* has widely differing effects depending on the cancer context.

5.3.2 **ASCL1** *in nervous system cancers*

Like *ATOH1*, *ASCL1* has been similarly implicated as a transcription factor that contributes to oncogenesis in the CNS—but *ASCL1*'s role appears to be more important in the development and/or maintenance of glial tumors rather than neuronal lineage tumors. *ASCL1* transcripts and protein are frequently expressed in the malignant glial brain tumor, glioblastoma (GBM), and in lower grade diffuse gliomas (Phillips et al., 2006; Rheinbay et al., 2013; Rousseau et al., 2006; Somasundaram et al., 2005). Although *ASCL1* expression is highest in oligodendrogliomas (Rousseau et al., 2006), diffuse astrocytomas, and GBMs of the so-called proneural subtype (Rheinbay et al., 2013), *ASCL1* $^+$ cells have been detected within primary GBMs as well. Furthermore, in the cancer stem cell (CSC) population of GBMs, *ASCL1* has been found to maintain or promote key malignant characteristics of the cells (Rheinbay et al., 2013). *ASCL1* knockdown in GBM CSCs decreased the proliferation and self-renewal of the GBM CSCs *in vitro* and increased survival of mice xenotransplanted with GBM CSCs *in vivo* (Rheinbay et al., 2013). Mechanistically, *ASCL1* may do so by controlling Wnt signaling (Rheinbay et al., 2013). In contrast to *ASCL1*, expression of *NEUROG2* can promote the differentiation of glioma cells to neurons and decrease their stem-like properties (Zhao et al., 2012), a finding that supports a potential *NEUROG2–ASCL1* genetic switch that is active both in the developing cortex (see Section 4.2.1) and in gliomas (Li et al., 2014). Aside from playing a role in glioma maintenance once tumors are established, expression of *ASCL1* may be important for the initiation of gliomas as well (Li et al., 2014). In a mouse model of gliomagenesis driven by constitutively active forms of RAS or BRAF in neural progenitors, *Ascl1* expression was increased and *Ascl1* functions were also modulated by ERK-mediated phosphorylation to bias cells toward aberrant proliferative glial fates (Li et al., 2014).

5.3.3 ASCL1 *in neuroendocrine cancers*

As a group, malignant tumors with neuroendocrine differentiation in a variety of disease sites (including small cell lung cancer, medullary thyroid cancer, gastrointestinal and biliary neuroendocrine cancer, small cell prostate cancer, and Merkel cell carcinoma in the skin) show high levels of *ASCL1* expression at the mRNA and protein levels (La Rosa et al., 2013). Interestingly, *ASCL1* levels appear to distinguish these malignancies from more benign well- or moderately differentiated neuroendocrine tumors (La Rosa et al., 2013). Some have observed that *ASCL1* expression correlates with poor prognosis, at least in the setting of neuroendocrine cancers in the lung (Jiang et al., 2004), although others have not found the same association (Hiroshima et al., 2006; La Rosa et al., 2013). The potential functional roles of *ASCL1* in these cancers are diverse. Evidence in the lung and prostate indicate that *ASCL1* is essential for neuroendocrine cell differentiation (Borges et al., 1997; Linnoila et al., 2000; Rapa et al., 2013; Wang et al., 2007). *In vitro* and *in vivo*, forced expression of *Ascl1* is sufficient to promote neuroendocrine differentiation and increase cell viability, proliferation, and self-renewal (Jiang et al., 2009; Linnoila et al., 2000; Rapa et al., 2013). Conversely, silencing of *Ascl1* inhibits neuroendocrine differentiation (Borges et al., 1997; Osada, Tatematsu, Yatabe, Horio, & Takahashi, 2005; Rapa et al., 2013). *Ascl1* may also have roles in regulation of migration (Demelash et al., 2012) and DNA damage and repair responses (Wang, Jensen-Taubman, Keefe, Yang, & Linnoila, 2012). Interestingly, in the setting of prostatic neuroendocrine carcinoma, WNT-11 pathway inhibition showed similar effects to *Ascl1* knockdown (Rapa et al., 2013), echoing the findings in gliomas where *ASCL1* and WNT signaling appear to be interlinked (Rheinbay et al., 2013).

The above lines of evidence solidly place proneural bHLH transcription factors as positive conspirators in oncogenesis across a range of tumors featuring glial, glioneuronal, and neuroendocrine phenotypes, with a major function centering on cell fate selection (both cell type specification and proliferation) in addition to specific functions that impart a selective advantage to cancer cells. These studies also underscore the tissue and context specificity of proneural gene function not only in development but also in cancer. Determining whether and how proneural genes could be manipulated in a therapeutic setting for treatment of cancers, however, is an area for further investigation.

5.4. Proneural genes in diabetes and enteroendocrine disorders

Recent studies have found *NEUROG3* mutations in patients with permanent neonatal diabetes mellitus (PNDM), and allelic variants have been linked to development of type 2 diabetes (Pinney et al., 2011; Rubio-Cabezas et al., 2011).

Initial attempts to understand the role of *NEUROG3* in human disease focused on maturity-onset diabetes of the young (MODY), a rare form of diabetes characterized by an early onset and an autosomal dominant mode of inheritance, since several transcription factors, including *NEUROD1*, a downstream target of *NEUROG3*, have been shown to cause MODY (Malecki et al., 1999). However, no mutations or polymorphisms in the *NEUROG3* gene have been found in MODY patients, although some of these studies have found that certain alleles are associated with common type 2 diabetes (del Bosque-Plata et al., 2001; Kim et al., 2001).

More recently, mutations in *NEUROG3* have been found in two patients with PNDM (Pinney et al., 2011; Rubio-Cabezas et al., 2011). PNDM is a condition characterized by onset of diabetes within the first 6 months of life, a rare condition that affects 1 in 260,000 live births. Both of these are female infants born to nonconsanguineous parents, and they required insulin treatment by day 20 or 5 months of life (Rubio-Cabezas et al., 2011). Of note, both infants were born small for gestational age, underscoring the importance of insulin for intrauterine growth and development. This is a rare cause of PNDM, as other studies failed to find any link between PNDM and *NEUROG3* (Edghill et al., 2010; Nocerino, Colombo, Bonfanti, Iafusco, & Barbetti, 2009; Russo et al., 2011).

Interestingly, these two PNDM patients also have severe malabsorptive diarrhea, the first condition known to be caused by *NEUROG3* mutation in humans. In 2006, Wang et al. reported three male patients with severe congenital malabsorptive diarrhea and a lack of intestinal enteroendocrine cells caused by loss-of-function mutations in *NEUROG3* (Wang et al., 2006). In these three patients, pancreatic exocrine function as well as gut epithelial architecture was normal (Cortina et al., 2007). The only pathological finding was a significant reduction, although not absence, of intestinal enteroendocrine cells. The mechanism linking the absence of enteroendocrine cells to diarrhea has not been determined, as these patients had normal intestinal transit time. It was postulated, however, that *NEUROG3* may affect exocrine function (thereby causing diarrhea) through its downstream target, *NEUROD1*, which is known to regulate cholecystokinin and secretin-expressing enteroendocrine cells (Naya et al., 1997). Two of the three patients developed hyperglycemia at age 8–9 years, while the other patient's glycemic status is unknown as he died unexpectedly at age 35 months from infection. These five patients with *NEUROG3* mutations seem to have a more consistent and severe GI phenotype but variable endocrine phenotype, i.e., two patients developed PNDM, while the other two did not have diabetes until age 8–9. This may be explained by the difference in the role of *NEUROG3* in formation of various endocrine cell types as well as the amount of residual *NEUROG3* function associated with the specific mutation. Since α-cells and other endocrine cells are formed during a narrower window in embryonic development than β-cells (Johansson et al., 2007), any residual *NEUROG3* function may just be enough to allow some β-cell formation and delay the onset of diabetes, but insufficient to support α-cell formation. This may also explain the link of *NEUROG3* gene variants to development of type 2 diabetes.

Recent genome-wide association studies have found many obesity and type 2 diabetes associated genes, and detailed subanalysis of some of these genes found that *NEUROG3* gene variants are involved in pancreatic β-cell function and the risk of developing type 2 diabetes in some but not other populations (Chavali et al., 2011; Gupta et al., 2013; Yokoi et al., 2006). We may speculate that the range of pancreatic endocrine function observed in human conditions mirrors the degree of residual function of *NEUROG3*—while null mutation of *NEUROG3* may cause PNDM, a gene variant that causes a small reduction in *NEUROG3* function will only cause type 2 diabetes when the patient's physiology is being challenged by higher insulin demand associated with obesity or the functional deterioration associated with natural aging.

REFERENCES

Abematsu, M., Kagawa, T., Fukuda, S., Inoue, T., Takebayashi, H., Komiya, S., et al. (2006). Basic fibroblast growth factor endows dorsal telencephalic neural progenitors with the ability to differentiate into oligodendrocytes but not gamma-aminobutyric acidergic neurons. *Journal of Neuroscience Research*, *83*(5), 731–743. http://dx.doi.org/10.1002/jnr.20762.

Akazawa, C., Ishibashi, M., Shimizu, C., Nakanishi, S., & Kageyama, R. (1995). A mammalian helix-loop-helix factor structurally related to the product of Drosophila proneural gene atonal is a positive transcriptional regulator expressed in the developing nervous system. *The Journal of Biological Chemistry*, *270*(15), 8730–8738.

Alexiades, M. R., & Cepko, C. L. (1997). Subsets of retinal progenitors display temporally regulated and distinct biases in the fates of their progeny. *Development*, *124*(6), 1119–1131.

Ali, F., Hindley, C., McDowell, G., Deibler, R., Jones, A., Kirschner, M., et al. (2011). Cell cycle-regulated multi-site phosphorylation of Neurogenin 2 coordinates cell cycling with differentiation during neurogenesis. *Development*, *138*(19), 4267–4277. http://dx.doi.org/10.1242/dev.067900.

Amiel, J., Laudier, B., Attie-Bitach, T., Trang, H., de Pontual, L., Gener, B., et al. (2003). Polyalanine expansion and frameshift mutations of the paired-like homeobox gene PHOX2B in congenital central hypoventilation syndrome. *Nature Genetics*, *33*(4), 459–461. http://dx.doi.org/10.1038/ng1130.

Anderson, S. A., Eisenstat, D. D., Shi, L., & Rubenstein, J. L. (1997). Interneuron migration from basal forebrain to neocortex: Dependence on Dlx genes. *Science*, *278*(5337), 474–476.

Anderson, S. A., Kaznowski, C. E., Horn, C., Rubenstein, J. L., & McConnell, S. K. (2002). Distinct origins of neocortical projection neurons and interneurons in vivo. *Cerebral Cortex*, *12*(7), 702–709.

Anderson, S. A., Marin, O., Horn, C., Jennings, K., & Rubenstein, J. L. (2001). Distinct cortical migrations from the medial and lateral ganglionic eminences. *Development*, *128*(3), 353–363.

Andersson, E. K., Irvin, D. K., Ahlsio, J., & Parmar, M. (2007). Ngn2 and Nurr1 act in synergy to induce midbrain dopaminergic neurons from expanded neural stem and progenitor cells. *Experimental Cell Research, 313*(6), 1172–1180. http://dx.doi.org/10.1016/j.yexcr.2006.12.014.

Andersson, E., Jensen, J. B., Parmar, M., Guillemot, F., & Bjorklund, A. (2006). Development of the mesencephalic dopaminergic neuron system is compromised in the absence of neurogenin 2. *Development, 133*(3), 507–516. http://dx.doi.org/10.1242/dev.02224.

Anthwal, N., Pelling, M., Claxton, S., Mellitzer, G., Collin, C., Kessaris, N., et al. (2013). Conditional deletion of neurogenin-3 using Nkx2.1iCre results in a mouse model for the central control of feeding, activity and obesity. *Disease Models & Mechanisms, 6*(5), 1133–1145. http://dx.doi.org/10.1242/dmm.011916.

Artavanis-Tsakonas, S., Rand, M. D., & Lake, R. J. (1999). Notch signaling: Cell fate control and signal integration in development. *Science, 284*(5415), 770–776.

Avolio, F., Pfeifer, A., Courtney, M., Gjernes, E., Ben-Othman, N., Vieira, A., et al. (2013). From pancreas morphogenesis to β-cell regeneration. *Current Topics in Developmental Biology, 106*, 217–238.

Axelrod, F. B., Chelimsky, G. G., & Weese-Mayer, D. E. (2006). Pediatric autonomic disorders. *Pediatrics, 118*(1), 309–321. http://dx.doi.org/10.1542/peds.2005-3032.

Ayrault, O., Zhao, H., Zindy, F., Qu, C., Sherr, C. J., & Roussel, M. F. (2010). Atoh1 inhibits neuronal differentiation and collaborates with Gli1 to generate medulloblastoma-initiating cells. *Cancer Research, 70*(13), 5618–5627. http://dx.doi.org/10.1158/0008-5472.CAN-09-3740.

Aziz, A., Liu, Q. C., & Dilworth, F. J. (2010). Regulating a master regulator: Establishing tissue-specific gene expression in skeletal muscle. *Epigenetics, 5*(8), 691–695. http://dx.doi.org/10.4161/epi.5.8.13045.

Battiste, J., Helms, A. W., Kim, E. J., Savage, T. K., Lagace, D. C., Mandyam, C. D., et al. (2007). Ascl1 defines sequentially generated lineage-restricted neuronal and oligodendrocyte precursor cells in the spinal cord. *Development, 134*(2), 285–293. http://dx.doi.org/10.1242/dev.02727.

Ben-Arie, N., Bellen, H. J., Armstrong, D. L., McCall, A. E., Gordadze, P. R., Guo, Q., et al. (1997). Math1 is essential for genesis of cerebellar granule neurons. *Nature, 390*(6656), 169–172. http://dx.doi.org/10.1038/36579.

Ben-Othman, N., Courtney, M., Vieira, A., Pfeifer, A., Druelle, N., Gjernes, E., et al. (2013 Feb 1). From pancreatic islet formation to beta-cell regeneration. *Diabetes Res Clin Pract.* pii: S0168-8227(13)00014-4. http://dx.doi.org/10.1016/j.diabres.2013.01.013. [Epub ahead of print].

Bergstrom, D. A., Penn, B. H., Strand, A., Perry, R. L., Rudnicki, M. A., & Tapscott, S. J. (2002). Promoter-specific regulation of MyoD binding and signal transduction cooperate to pattern gene expression. *Molecular Cell, 9*(3), 587–600.

Bermingham, N. A., Hassan, B. A., Price, S. D., Vollrath, M. A., Ben-Arie, N., Eatock, R. A., et al. (1999). Math1: An essential gene for the generation of inner ear hair cells. *Science, 284*(5421), 1837–1841.

Berninger, B., Guillemot, F., & Gotz, M. (2007). Directing neurotransmitter identity of neurones derived from expanded adult neural stem cells. *The European Journal of Neuroscience, 25*(9), 2581–2590. http://dx.doi.org/10.1111/j.1460-9568.2007.05509.x, EJN5509 [pii].

Bertrand, N., Castro, D. S., & Guillemot, F. (2002). Proneural genes and the specification of neural cell types. *Nature Reviews Neuroscience, 3*(7), 517–530. http://dx.doi.org/10.1038/nrn874.

Bielle, F., Griveau, A., Narboux-Neme, N., Vigneau, S., Sigrist, M., Arber, S., et al. (2005). Multiple origins of Cajal-Retzius cells at the borders of the developing pallium. *Nature Neuroscience, 8*(8), 1002–1012. http://dx.doi.org/10.1038/nn1511, nn1511 [pii].

Blais, A., Tsikitis, M., Acosta-Alvear, D., Sharan, R., Kluger, Y., & Dynlacht, B. D. (2005). An initial blueprint for myogenic differentiation. *Genes & Development, 19*(5), 553–569.

Borges, M., Linnoila, R. I., van de Velde, H. J., Chen, H., Nelkin, B. D., Mabry, M., et al. (1997). An achaete-scute homologue essential for neuroendocrine differentiation in the lung. *Nature, 386*(6627), 852–855. http://dx.doi.org/10.1038/386852a0.

Bossuyt, W., De Geest, N., Aerts, S., Leenaerts, I., Marynen, P., & Hassan, B. A. (2009). The atonal proneural transcription factor links differentiation and tumor formation in Drosophila. *PLoS Biology*, 7(2), e40. http://dx.doi.org/10.1371/journal.pbio.1000040.

Bossuyt, W., Kazanjian, A., De Geest, N., Van Kelst, S., De Hertogh, G., Geboes, K., et al. (2009). Atonal homolog 1 is a tumor suppressor gene. *PLoS Biology*, 7(2), e39. http://dx.doi.org/10.1371/journal.pbio.1000039.

Bower, R. J., & Adkins, J. C. (1980). Ondine's curse and neurocristopathy. *Clinical Pediatrics (Phila), 19*(10), 665–668.

Bray, S., & Furriols, M. (2001). Notch pathway: Making sense of suppressor of hairless. *Current Biology, 11*(6), R217–R221.

Brill, M. S., Ninkovic, J., Winpenny, E., Hodge, R. D., Ozen, I., Yang, R., et al. (2009). Adult generation of glutamatergic olfactory bulb interneurons. *Nature Neuroscience, 12*(12), 1524–1533. http://dx.doi.org/10.1038/nn.2416.

Britz, O., Mattar, P., Nguyen, L., Langevin, L. M., Zimmer, C., Alam, S., et al. (2006). A role for proneural genes in the maturation of cortical progenitor cells. *Cerebral Cortex, 16*(Suppl. 1), i138–i151. http://dx.doi.org/10.1093/cercor/bhj168, 16/suppl_1/i138 [pii].

Brockschmidt, A., Todt, U., Ryu, S., Hoischen, A., Landwehr, C., Birnbaum, S., et al. (2007). Severe mental retardation with breathing abnormalities (Pitt-Hopkins syndrome) is caused by haploinsufficiency of the neuronal bHLH transcription factor TCF4. *Human Molecular Genetics, 16*(12), 1488–1494. http://dx.doi.org/10.1093/hmg/ddm099.

Brown, N. L., Kanekar, S., Vetter, M. L., Tucker, P. K., Gemza, D. L., & Glaser, T. (1998). Math5 encodes a murine basic helix-loop-helix transcription factor expressed during early stages of retinal neurogenesis. *Development, 125*(23), 4821–4833.

Brown, N. L., Patel, S., Brzezinski, J., & Glaser, T. (2001). Math5 is required for retinal ganglion cell and optic nerve formation. *Development, 128*(13), 2497–2508.

Brzezinski, J. A., 4th., Brown, N. L., Tanikawa, A., Bush, R. A., Sieving, P. A., Vitaterna, M. H., et al. (2005). Loss of circadian photoentrainment and abnormal retinal electrophysiology in Math5 mutant mice. *Investigative Ophthalmology & Visual Science, 46*(7), 2540–2551. http://dx.doi.org/10.1167/iovs.04-1123.

Brzezinski, J. A., 4th., Prasov, L., & Glaser, T. (2012). Math5 defines the ganglion cell competence state in a subpopulation of retinal progenitor cells exiting the cell cycle. *Developmental Biology, 365*(2), 395–413. http://dx.doi.org/10.1016/j.ydbio.2012.03.006.

Cabrera, C. V., & Alonso, M. C. (1991). Transcriptional activation by heterodimers of the achaete-scute and daughterless gene products of Drosophila. *The EMBO Journal, 10*(10), 2965–2973.

Caiazzo, M., Dell'Anno, M. T., Dvoretskova, E., Lazarevic, D., Taverna, S., Leo, D., et al. (2011). Direct generation of functional dopaminergic neurons from mouse and human fibroblasts. *Nature, 476*(7359), 224–227. http://dx.doi.org/10.1038/nature10284.

Calegari, F., Haubensak, W., Haffner, C., & Huttner, W. B. (2005). Selective lengthening of the cell cycle in the neurogenic subpopulation of neural progenitor cells during mouse brain development. *The Journal of Neuroscience, 25*(28), 6533–6538. http://dx.doi.org/10.1523/JNEUROSCI.0778-05.2005.

Calegari, F., & Huttner, W. B. (2003). An inhibition of cyclin-dependent kinases that lengthens, but does not arrest, neuroepithelial cell cycle induces premature neurogenesis. *Journal of Cell Science, 116*(Pt 24), 4947–4955. http://dx.doi.org/10.1242/jcs.00825.

Campuzano, S., & Modolell, J. (1992). Patterning of the Drosophila nervous system: The achaete-scute gene complex. *Trends in Genetics*, *8*(6), 202–208.

Cao, Y., Kumar, R. M., Penn, B. H., Berkes, C. A., Kooperberg, C., Boyer, L. A., et al. (2006). Global and gene-specific analyses show distinct roles for Myod and Myog at a common set of promoters. *The EMBO Journal*, *25*(3), 502–511.

Casarosa, S., Fode, C., & Guillemot, F. (1999). Mash1 regulates neurogenesis in the ventral telencephalon. *Development*, *126*(3), 525–534.

Castro, D. S., Martynoga, B., Parras, C., Ramesh, V., Pacary, E., Johnston, C., et al. (2011). A novel function of the proneural factor Ascl1 in progenitor proliferation identified by genome-wide characterization of its targets. *Genes & Development*, *25*(9), 930–945. http://dx.doi.org/10.1101/gad.627811, 25/9/930 [pii].

Castro, D. S., Skowronska-Krawczyk, D., Armant, O., Donaldson, I. J., Parras, C., Hunt, C., et al. (2006). Proneural bHLH and Brn proteins coregulate a neurogenic program through cooperative binding to a conserved DNA motif. *Developmental Cell*, *11*(6), 831–844. http://dx.doi.org/10.1016/j.devcel.2006.10.006, S1534-5807(06)00458-8 [pii].

Cau, E., Gradwohl, G., Fode, C., & Guillemot, F. (1997). Mash1 activates a cascade of bHLH regulators in olfactory neuron progenitors. *Development*, *124*(8), 1611–1621.

Caviness, V. S., Jr., Takahashi, T., & Nowakowski, R. S. (1995). Numbers, time and neocortical neuronogenesis: A general developmental and evolutionary model. *Trends in Neurosciences*, *18*(9), 379–383.

Chavali, S., Mahajan, A., Tabassum, R., Dwivedi, O. P., Chauhan, G., Ghosh, S., et al. (2011). Association of variants in genes involved in pancreatic beta-cell development and function with type 2 diabetes in North Indians. *Journal of Human Genetics*, *56*(10), 695–700.

Chien, C. T., Hsiao, C. D., Jan, L. Y., & Jan, Y. N. (1996). Neuronal type information encoded in the basic-helix-loop-helix domain of proneural genes. *Proceedings of the National Academy of Sciences of the United States of America*, *93*(23), 13239–13244.

Cortina, G., Smart, C. N., Farmer, D. G., Bhuta, S., Treem, W. R., Hill, I. D., et al. (2007). Enteroendocrine cell dysgenesis and malabsorption, a histopathologic and immunohistochemical characterization. *Human Pathology*, *38*(4), 570–580.

Cuschieri, A., & Bannister, L. H. (1975). The development of the olfactory mucosa in the mouse: Light microscopy. *Journal of Anatomy*, *119*(Pt 2), 277–286.

Davis, R. L., & Turner, D. L. (2001). Vertebrate hairy and enhancer of split related proteins: Transcriptional repressors regulating cellular differentiation and embryonic patterning. *Oncogene*, *20*(58), 8342–8357. http://dx.doi.org/10.1038/sj.onc.1205094.

Dehay, C., & Kennedy, H. (2007). Cell-cycle control and cortical development. *Nature Reviews. Neuroscience*, *8*(6), 438–450. http://dx.doi.org/10.1038/nrn2097.

del Bosque-Plata, L., Lin, J., Horikawa, Y., Schwarz, P. E., Cox, N. J., Iwasaki, N., et al. (2001). Mutations in the coding region of the neurogenin 3 gene (NEUROG3) are not a common cause of maturity-onset diabetes of the young in Japanese subjects. *Diabetes*, *50*(3), 694–696.

Demelash, A., Rudrabhatla, P., Pant, H. C., Wang, X., Amin, N. D., McWhite, C. D., et al. (2012). Achaete-scute homologue-1 (ASH1) stimulates migration of lung cancer cells through Cdk5/p35 pathway. *Molecular Biology of the Cell*, *23*(15), 2856–2866. http://dx.doi.org/10.1091/mbc.E10-12-1010.

de Pontual, L., Nepote, V., Attie-Bitach, T., Al Halabiah, H., Trang, H., Elghouzzi, V., et al. (2003). Noradrenergic neuronal development is impaired by mutation of the proneural HASH-1 gene in congenital central hypoventilation syndrome (Ondine's curse). *Human Molecular Genetics*, *12*(23), 3173–3180. http://dx.doi.org/10.1093/hmg/ddg339.

Dixit, R., Tachibana, N., Touahri, Y., Zinyk, D., Logan, C., & Schuurmans, C. (2013). Gene expression is dynamically regulated in retinal progenitor cells prior to and during overt cellular differentiation. *Gene Expression Patterns*, *14*(1), 42–54.

Dixit, R., Wilkinson, G., Li, S., Cancino, G., Shaker, T., Adnani, L., et al. (2014). Neurog1 and Neurog2 control two waves of neuronal differentiation in the piriform cortex. *The Journal of Neuroscience*, *34*(2), 539–553.

Dixit, R., Zimmer, C., Waclaw, R. R., Mattar, P., Shaker, T., Kovach, C., et al. (2011). Ascl1 participates in Cajal-Retzius cell development in the neocortex. *Cerebral Cortex*, *21*(11), 2599–2611. http://dx.doi.org/10.1093/cercor/bhr046.

Edghill, E. L., Minton, J. A., Groves, C. J., Flanagan, S. E., Patch, A. M., Rubio-Cabezas, O., et al. (2010). Sequencing of candidate genes selected by beta cell experts in monogenic diabetes of unknown aetiology. *Journal of the Pancreas*, *11*(1), 14–17.

Ellenberger, T., Fass, D., Arnaud, M., & Harrison, S. C. (1994). Crystal structure of transcription factor E47: E-box recognition by a basic region helix-loop-helix dimer. *Genes & Development*, *8*(8), 970–980.

Englund, C., Kowalczyk, T., Daza, R. A., Dagan, A., Lau, C., Rose, M. F., et al. (2006). Unipolar brush cells of the cerebellum are produced in the rhombic lip and migrate through developing white matter. *The Journal of Neuroscience*, *26*(36), 9184–9195. http://dx.doi.org/10.1523/JNEUROSCI.1610-06.2006.

Feng, L., Xie, Z. H., Ding, Q., Xie, X., Libby, R. T., & Gan, L. (2010). MATH5 controls the acquisition of multiple retinal cell fates. *Molecular Brain*, *3*, 36. http://dx.doi.org/10.1186/1756-6606-3-36.

Flora, A., Garcia, J. J., Thaller, C., & Zoghbi, H. Y. (2007). The E-protein Tcf4 interacts with Math1 to regulate differentiation of a specific subset of neuronal progenitors. *Proceedings of the National Academy of Sciences of the United States of America*, *104*(39), 15382–15387. http://dx.doi.org/10.1073/pnas.0707456104, 0707456104 [pii].

Flora, A., Klisch, T. J., Schuster, G., & Zoghbi, H. Y. (2009). Deletion of Atoh1 disrupts Sonic Hedgehog signaling in the developing cerebellum and prevents medulloblastoma. *Science*, *326*(5958), 1424–1427. http://dx.doi.org/10.1126/science.1181453.

Fode, C., Gradwohl, G., Morin, X., Dierich, A., LeMeur, M., Goridis, C., et al. (1998). The bHLH protein NEUROGENIN 2 is a determination factor for epibranchial placode-derived sensory neurons. *Neuron*, *20*(3), 483–494.

Fode, C., Ma, Q., Casarosa, S., Ang, S. L., Anderson, D. J., & Guillemot, F. (2000). A role for neural determination genes in specifying the dorsoventral identity of telencephalic neurons. *Genes & Development*, *14*(1), 67–80.

Forrest, M. P., Waite, A. J., Martin-Rendon, E., & Blake, D. J. (2013). Knockdown of human TCF4 affects multiple signaling pathways involved in cell survival, epithelial to mesenchymal transition and neuronal differentiation. *PLoS One*, *8*(8), e73169. http://dx.doi.org/10.1371/journal.pone.0073169.

Gabay, L., Lowell, S., Rubin, L. L., & Anderson, D. J. (2003). Deregulation of dorsoventral patterning by FGF confers trilineage differentiation capacity on CNS stem cells in vitro. *Neuron*, *40*(3), 485–499, S0896627303006378 [pii].

Galichet, C., Guillemot, F., & Parras, C. M. (2008). Neurogenin 2 has an essential role in development of the dentate gyrus. *Development*, *135*(11), 2031–2041. http://dx.doi.org/10.1242/dev.015115.

Garcia-Bellido, A. (1979). Genetic analysis of the achaete-scute system of Drosophila melanogaster. *Genetics*, *91*(3), 491–520.

Ge, W., He, F., Kim, K. J., Blanchi, B., Coskun, V., Nguyen, L., et al. (2006). Coupling of cell migration with neurogenesis by proneural bHLH factors. *Proceedings of the National Academy of Sciences of the United States of America*, *103*(5), 1319–1324.

Geoffroy, C. G., Critchley, J. A., Castro, D. S., Ramelli, S., Barraclough, C., Descombes, P., et al. (2009). Engineering of dominant active basic helix-loop-helix proteins that are resistant to negative regulation by postnatal central nervous system antineurogenic cues. *Stem Cells*, *27*(4), 847–856. http://dx.doi.org/10.1002/stem.17.

Ghysen, A., & Dambly-Chaudiere, C. (1988). From DNA to form: The achaete-scute complex. *Genes & Development*, *2*(5), 495–501.

Gibson, P., Tong, Y., Robinson, G., Thompson, M. C., Currle, D. S., Eden, C., et al. (2010). Subtypes of medulloblastoma have distinct developmental origins. *Nature*, *468*(7327), 1095–1099. http://dx.doi.org/10.1038/nature09587.

Gilbertson, R. J., & Ellison, D. W. (2008). The origins of medulloblastoma subtypes. *Annual Review of Pathology*, *3*, 341–365. http://dx.doi.org/10.1146/annurev.pathmechdis.3.121806.151518.

Gohlke, J. M., Armant, O., Parham, F. M., Smith, M. V., Zimmer, C., Castro, D. S., et al. (2008). Characterization of the proneural gene regulatory network during mouse telencephalon development. *BMC Biology*, *6*, 15. http://dx.doi.org/10.1186/1741-7007-6-15, 1741-7007-6-15 [pii].

Golbabapour, S., Majid, N. A., Hassandarvish, P., Hajrezaie, M., Abdulla, M. A., & Hadi, A. H. (2013). Gene silencing and polycomb group proteins: An overview of their structure, mechanisms and phylogenetics. *OMICS*, *17*(6), 283–296. http://dx.doi.org/10.1089/omi.2012.0105.

Gonzalez-Quevedo, R., Lee, Y., Poss, K. D., & Wilkinson, D. G. (2010). Neuronal regulation of the spatial patterning of neurogenesis. *Developmental Cell*, *18*(1), 136–147. http://dx.doi.org/10.1016/j.devcel.2009.11.010, S1534-5807(09)00490-0 [pii].

Goulding, S. E., White, N. M., & Jarman, A. P. (2000). cato encodes a basic helix-loop-helix transcription factor implicated in the correct differentiation of Drosophila sense organs. *Developmental Biology*, *221*(1), 120–131. http://dx.doi.org/10.1006/dbio.2000.9677.

Goulding, S. E., zur Lage, P., & Jarman, A. P. (2000). amos, a proneural gene for Drosophila olfactory sense organs that is regulated by lozenge. *Neuron*, *25*(1), 69–78.

Gradwohl, G., Dierich, A., LeMeur, M., & Guillemot, F. (2000). neurogenin3 is required for the development of the four endocrine cell lineages of the pancreas. *Proceedings of the National Academy of Sciences of the United States of America*, *97*(4), 1607–1611.

Gradwohl, G., Fode, C., & Guillemot, F. (1996). Restricted expression of a novel murine atonal-related bHLH protein in undifferentiated neural precursors. *Developmental Biology*, *180*(1), 227–241.

Grbavec, D., & Stifani, S. (1996). Molecular interaction between TLE1 and the carboxyl-terminal domain of HES-1 containing the WRPW motif. *Biochemical and Biophysical Research Communications*, *223*(3), 701–705. http://dx.doi.org/10.1006/bbrc.1996.0959.

Grimaldi, P., Parras, C., Guillemot, F., Rossi, F., & Wassef, M. (2009). Origins and control of the differentiation of inhibitory interneurons and glia in the cerebellum. *Developmental Biology*, *328*(2), 422–433. http://dx.doi.org/10.1016/j.ydbio.2009.02.008.

Gu, G., Dubauskaite, J., & Melton, D. A. (2002). Direct evidence for the pancreatic lineage: NGN3+ cells are islet progenitors and are distinct from duct progenitors. *Development*, *129*(10), 2447–2457.

Guillemot, F., & Joyner, A. L. (1993). Dynamic expression of the murine Achaete-Scute homologue Mash-1 in the developing nervous system. *Mechanisms of Development*, *42*(3), 171–185.

Guillemot, F., Lo, L. C., Johnson, J. E., Auerbach, A., Anderson, D. J., & Joyner, A. L. (1993). Mammalian achaete-scute homolog 1 is required for the early development of olfactory and autonomic neurons. *Cell*, *75*(3), 463–476.

Gupta, V., Vinay, D. G., Sovio, U., Rafiq, S., Kranthi Kumar, M. V., Janipalli, C. S., et al. (2013). Association study of 25 type 2 diabetes related Loci with measures of obesity in Indian sib pairs. *PLoS One*, *8*(1), e53944.

Habener, J. F., Kemp, D. M., & Thomas, M. K. (2005). Minireview: Transcriptional regulation in pancreatic development. *Endocrinology*, *146*(3), 1025–1034. http://dx.doi.org/10.1210/en.2004-1576.

Hack, M. A., Sugimori, M., Lundberg, C., Nakafuku, M., & Gotz, M. (2004). Regionalization and fate specification in neurospheres: The role of Olig2 and Pax6. *Molecular and Cellular Neurosciences*, *25*(4), 664–678. http://dx.doi.org/10.1016/j.mcn.2003.12.012, S1044743103003956 [pii].

Hand, R., Bortone, D., Mattar, P., Nguyen, L., Heng, J. I., Guerrier, S., et al. (2005). Phosphorylation of Neurogenin2 specifies the migration properties and the dendritic morphology of pyramidal neurons in the neocortex. *Neuron*, *48*(1), 45–62.

Hand, R., & Polleux, F. (2011). Neurogenin2 regulates the initial axon guidance of cortical pyramidal neurons projecting medially to the corpus callosum. *Neural Development*, *6*, 30. http://dx.doi.org/10.1186/1749-8104-6-30.

Helms, A. W., Gowan, K., Abney, A., Savage, T., & Johnson, J. E. (2001). Overexpression of MATH1 disrupts the coordination of neural differentiation in cerebellum development. *Molecular and Cellular Neurosciences*, *17*(4), 671–682. http://dx.doi.org/10.1006/mcne.2000.0969.

Heng, J. I., Nguyen, L., Castro, D. S., Zimmer, C., Wildner, H., Armant, O., et al. (2008). Neurogenin 2 controls cortical neuron migration through regulation of Rnd2. *Nature*, *455*(7209), 114–118. http://dx.doi.org/10.1038/nature07198, nature07198 [pii].

Henke, R. M., Meredith, D. M., Borromeo, M. D., Savage, T. K., & Johnson, J. E. (2009). Ascl1 and Neurog2 form novel complexes and regulate Delta-like3 (Dll3) expression in the neural tube. *Developmental Biology*, *328*(2), 529–540. http://dx.doi.org/10.1016/j.ydbio.2009.01.007.

Hester, M. E., Murtha, M. J., Song, S., Rao, M., Miranda, C. J., Meyer, K., et al. (2011). Rapid and efficient generation of functional motor neurons from human pluripotent stem cells using gene delivered transcription factor codes. *Molecular Therapy*, *19*(10), 1905–1912. http://dx.doi.org/10.1038/mt.2011.135.

Hindley, C., Ali, F., McDowell, G., Cheng, K., Jones, A., Guillemot, F., et al. (2012). Post-translational modification of Ngn2 differentially affects transcription of distinct targets to regulate the balance between progenitor maintenance and differentiation. *Development*, *139*(10), 1718–1723. http://dx.doi.org/10.1242/dev.077552.

Hiroshima, K., Iyoda, A., Shida, T., Shibuya, K., Iizasa, T., Kishi, H., et al. (2006). Distinction of pulmonary large cell neuroendocrine carcinoma from small cell lung carcinoma: A morphological, immunohistochemical, and molecular analysis. *Modern Pathology*, *19*(10), 1358–1368. http://dx.doi.org/10.1038/modpathol.3800659.

Hirsch, M. R., Tiveron, M. C., Guillemot, F., Brunet, J. F., & Goridis, C. (1998). Control of noradrenergic differentiation and Phox2a expression by MASH1 in the central and peripheral nervous system. *Development*, *125*(4), 599–608.

Ho, B. C., Epping, E., Wang, K., Andreasen, N. C., Librant, A., & Wassink, T. H. (2008). Basic helix-loop-helix transcription factor NEUROG1 and schizophrenia: Effects on illness susceptibility, MRI brain morphometry and cognitive abilities. *Schizophrenia Research*, *106*(2–3), 192–199. http://dx.doi.org/10.1016/j.schres.2008.08.009.

Horton, S., Meredith, A., Richardson, J. A., & Johnson, J. E. (1999). Correct coordination of neuronal differentiation events in ventral forebrain requires the bHLH factor MASH1. *Molecular and Cellular Neurosciences*, *14*(4–5), 355–369. http://dx.doi.org/10.1006/mcne.1999.0791, S1044-7431(99)90791-1 [pii].

Hu, J. S., Olson, E. N., & Kingston, R. E. (1992). HEB, a helix-loop-helix protein related to E2A and ITF2 that can modulate the DNA-binding ability of myogenic regulatory factors. *Molecular and Cellular Biology*, *12*(3), 1031–1042.

Huang, M. L., Hsu, C. H., & Chien, C. T. (2000). The proneural gene amos promotes multiple dendritic neuron formation in the Drosophila peripheral nervous system. *Neuron*, *25*(1), 57–67.

Ide, M., Yamada, K., Toyota, T., Iwayama, Y., Ishitsuka, Y., Minabe, Y., et al. (2005). Genetic association analyses of PHOX2B and ASCL1 in neuropsychiatric disorders:

Evidence for association of ASCL1 with Parkinson's disease. *Human Genetics*, *117*(6), 520–527. http://dx.doi.org/10.1007/s00439-005-1342-8.

Imamura, O., Satoh, Y., Endo, S., & Takishima, K. (2008). Analysis of extracellular signal-regulated kinase 2 function in neural stem/progenitor cells via nervous system-specific gene disruption. *Stem Cells*, *26*(12), 3247–3256. http://dx.doi.org/10.1634/stemcells.2008-0578, 2008–0578 [pii].

Imayoshi, I., Isomura, A., Harima, Y., Kawaguchi, K., Kori, H., Miyachi, H., et al. (2013). Oscillatory control of factors determining multipotency and fate in mouse neural progenitors. *Science*, *342*(6163), 1203–1208. http://dx.doi.org/10.1126/science.1242366.

Imayoshi, I., Shimogori, T., Ohtsuka, T., & Kageyama, R. (2008). Hes genes and neurogenin regulate non-neural versus neural fate specification in the dorsal telencephalic midline. *Development*, *135*(15), 2531–2541. http://dx.doi.org/10.1242/dev.021535.

Izumikawa, M., Minoda, R., Kawamoto, K., Abrashkin, K. A., Swiderski, D. L., Dolan, D. F., et al. (2005). Auditory hair cell replacement and hearing improvement by Atoh1 gene therapy in deaf mammals. *Nature Medicine*, *11*(3), 271–276. http://dx.doi.org/10.1038/nm1193.

Jarman, A. P., Grau, Y., Jan, L. Y., & Jan, Y. N. (1993). atonal is a proneural gene that directs chordotonal organ formation in the Drosophila peripheral nervous system. *Cell*, *73*(7), 1307–1321.

Jarman, A. P., Grell, E. H., Ackerman, L., Jan, L. Y., & Jan, Y. N. (1994). Atonal is the proneural gene for Drosophila photoreceptors. *Nature*, *369*(6479), 398–400. http://dx.doi.org/10.1038/369398a0.

Jarman, A. P., Sun, Y., Jan, L. Y., & Jan, Y. N. (1995). Role of the proneural gene, atonal, in formation of Drosophila chordotonal organs and photoreceptors. *Development*, *121*(7), 2019–2030.

Jenny, M., Uhl, C., Roche, C., Duluc, I., Guillermin, V., Guillemot, F., et al. (2002). Neurogenin3 is differentially required for endocrine cell fate specification in the intestinal and gastric epithelium. *The EMBO Journal*, *21*(23), 6338–6347.

Jensen, J., Heller, R. S., Funder-Nielsen, T., Pedersen, E. E., Lindsell, C., Weinmaster, G., et al. (2000). Independent development of pancreatic alpha- and beta-cells from neurogenin3-expressing precursors: A role for the notch pathway in repression of premature differentiation. *Diabetes*, *49*(2), 163–176.

Jeon, S. J., Fujioka, M., Kim, S. C., & Edge, A. S. (2011). Notch signaling alters sensory or neuronal cell fate specification of inner ear stem cells. *The Journal of Neuroscience*, *31*(23), 8351–8358. http://dx.doi.org/10.1523/JNEUROSCI.6366-10.2011.

Jhaveri, D., Sen, A., Reddy, G. V., & Rodrigues, V. (2000). Sense organ identity in the Drosophila antenna is specified by the expression of the proneural gene atonal. *Mechanisms of Development*, *99*(1–2), 101–111.

Jiang, T., Collins, B. J., Jin, N., Watkins, D. N., Brock, M. V., Matsui, W., et al. (2009). Achaete-scute complex homologue 1 regulates tumor-initiating capacity in human small cell lung cancer. *Cancer Research*, *69*(3), 845–854. http://dx.doi.org/10.1158/0008-5472.CAN-08-2762.

Jiang, S. X., Kameya, T., Asamura, H., Umezawa, A., Sato, Y., Shinada, J., et al. (2004). hASH1 expression is closely correlated with endocrine phenotype and differentiation extent in pulmonary neuroendocrine tumors. *Modern Pathology*, *17*(2), 222–229. http://dx.doi.org/10.1038/modpathol.3800038.

Johansson, K. A., Dursun, U., Jordan, N., Gu, G., Beermann, F., Gradwohl, G., et al. (2007). Temporal control of neurogenin3 activity in pancreas progenitors reveals competence windows for the generation of different endocrine cell types. *Developmental Cell*, *12*(3), 457–465.

Johnson, J. E., Birren, S. J., & Anderson, D. J. (1990). Two rat homologues of Drosophila achaete-scute specifically expressed in neuronal precursors. *Nature*, *346*(6287), 858–861. http://dx.doi.org/10.1038/346858a0.

Ju, B. G., Solum, D., Song, E. J., Lee, K. J., Rose, D. W., Glass, C. K., et al. (2004). Activating the PARP-1 sensor component of the groucho/TLE1 corepressor complex mediates a CaMKinase IIdelta-dependent neurogenic gene activation pathway. *Cell*, *119*(6), 815–829. http://dx.doi.org/10.1016/j.cell.2004.11.017, S0092867404010931 [pii].

Kageyama, R., Ohtsuka, T., & Kobayashi, T. (2007). The Hes gene family: Repressors and oscillators that orchestrate embryogenesis. *Development*, *134*(7), 1243–1251. http://dx.doi.org/10.1242/dev.000786, dev.000786 [pii].

Kanadia, R. N., & Cepko, C. L. (2010). Alternative splicing produces high levels of noncoding isoforms of bHLH transcription factors during development. *Genes & Development*, *24*(3), 229–234. http://dx.doi.org/10.1101/gad.1847110.

Kang, P., Lee, H. K., Glasgow, S. M., Finley, M., Donti, T., Gaber, Z. B., et al. (2012). Sox9 and NFIA coordinate a transcriptional regulatory cascade during the initiation of gliogenesis. *Neuron*, *74*(1), 79–94. http://dx.doi.org/10.1016/j.neuron.2012.01.024.

Kataoka, A., & Shimogori, T. (2008). Fgf8 controls regional identity in the developing thalamus. *Development*, *135*(17), 2873–2881. http://dx.doi.org/10.1242/dev.021618.

Kazanjian, A., Noah, T., Brown, D., Burkart, J., & Shroyer, N. F. (2010). Atonal homolog 1 is required for growth and differentiation effects of notch/gamma-secretase inhibitors on normal and cancerous intestinal epithelial cells. *Gastroenterology*, *139*(3), 918–928. http://dx.doi.org/10.1053/j.gastro.2010.05.081, 928 e911–916.

Kele, J., Simplicio, N., Ferri, A. L., Mira, H., Guillemot, F., Arenas, E., et al. (2006). Neurogenin 2 is required for the development of ventral midbrain dopaminergic neurons. *Development*, *133*(3), 495–505.

Kendler, K. S., Myers, J. M., O'Neill, F. A., Martin, R., Murphy, B., MacLean, C. J., et al. (2000). Clinical features of schizophrenia and linkage to chromosomes 5q, 6p, 8p, and 10p in the Irish Study of High-Density Schizophrenia Families. *The American Journal of Psychiatry*, *157*(3), 402–408.

Kim, E. J., Battiste, J., Nakagawa, Y., & Johnson, J. E. (2008). Ascl1 (Mash1) lineage cells contribute to discrete cell populations in CNS architecture. *Molecular and Cellular Neurosciences*, *38*(4), 595–606. http://dx.doi.org/10.1016/j.mcn.2008.05.008.

Kim, T. H., & Shivdasani, R. A. (2011). Genetic evidence that intestinal Notch functions vary regionally and operate through a common mechanism of Math1 repression. *The Journal of Biological Chemistry*, *286*(13), 11427–11433. http://dx.doi.org/10.1074/jbc.M110.188797.

Kim, S. H., Warram, J. H., Krolewski, A. S., & Doria, A. (2001). Mutation screening of the neurogenin-3 gene in autosomal dominant diabetes. *The Journal of Clinical Endocrinology and Metabolism*, *86*(5), 2320–2322.

Kishi, Y., Fujii, Y., Hirabayashi, Y., & Gotoh, Y. (2012). HMGA regulates the global chromatin state and neurogenic potential in neocortical precursor cells. *Nature Neuroscience*. http://dx.doi.org/10.1038/nn.3165.

Kool, M., Koster, J., Bunt, J., Hasselt, N. E., Lakeman, A., van Sluis, P., et al. (2008). Integrated genomics identifies five medulloblastoma subtypes with distinct genetic profiles, pathway signatures and clinicopathological features. *PLoS One*, *3*(8), e3088. http://dx.doi.org/10.1371/journal.pone.0003088.

Kovach, C., Dixit, R., Li, S., Mattar, P., Wilkinson, G., Elsen, G. E., et al. (2013). Neurog2 simultaneously activates and represses alternative gene expression programs in the developing neocortex. *Cerebral Cortex*, *23*(8), 1884–1900. http://dx.doi.org/10.1093/cercor/bhs176.

La Rosa, S., Marando, A., Gatti, G., Rapa, I., Volante, M., Papotti, M., et al. (2013). Achaete-scute homolog 1 as a marker of poorly differentiated neuroendocrine carcinomas of different sites: A validation study using immunohistochemistry and quantitative real-time polymerase chain reaction on 335 cases. *Human Pathology*, *44*(7), 1391–1399. http://dx.doi.org/10.1016/j.humpath.2012.11.013.

Lassiter, R. N., Stark, M. R., Zhao, T., & Zhou, C. (2013). Signaling mechanisms controlling cranial placode neurogenesis and delamination. *Developmental Biology*, *389*(1), 39–49. http://dx.doi.org/10.1016/j.ydbio.2013.11.025.

Le, T. T., Wroblewski, E., Patel, S., Riesenberg, A. N., & Brown, N. L. (2006). Math5 is required for both early retinal neuron differentiation and cell cycle progression. *Developmental Biology*, *295*(2), 764–778. http://dx.doi.org/10.1016/j.ydbio.2006.03.055.

Lee, S., Cuvillier, J. M., Lee, B., Shen, R., Lee, J. W., & Lee, S. K. (2012). Fusion protein Isl1-Lhx3 specifies motor neuron fate by inducing motor neuron genes and concomitantly suppressing the interneuron programs. *Proceedings of the National Academy of Sciences of the United States of America*, *109*(9), 3383–3388. http://dx.doi.org/10.1073/pnas.1114515109.

Lee, C. S., De Leon, D. D., Kaestner, K. H., & Stoffers, D. A. (2006). Regeneration of pancreatic islets after partial pancreatectomy in mice does not involve the reactivation of neurogenin-3. *Diabetes*, *55*(2), 269–272.

Lee, S., Lee, B., Lee, J. W., & Lee, S. K. (2009). Retinoid signaling and neurogenin2 function are coupled for the specification of spinal motor neurons through a chromatin modifier CBP. *Neuron*, *62*(5), 641–654. http://dx.doi.org/10.1016/j.neuron.2009.04.025.

Lee, S. K., Lee, B., Ruiz, E. C., & Pfaff, S. L. (2005). Olig2 and Ngn2 function in opposition to modulate gene expression in motor neuron progenitor cells. *Genes & Development*, *19*(2), 282–294. http://dx.doi.org/10.1101/gad.1257105.

Lee, C. S., Perreault, N., Brestelli, J. E., & Kaestner, K. H. (2002). Neurogenin 3 is essential for the proper specification of gastric enteroendocrine cells and the maintenance of gastric epithelial cell identity. *Genes & Development*, *16*(12), 1488–1497.

Lee, S. K., & Pfaff, S. L. (2003). Synchronization of neurogenesis and motor neuron specification by direct coupling of bHLH and homeodomain transcription factors. *Neuron*, *38*(5), 731–745.

Leonard, J. H., Cook, A. L., Van Gele, M., Boyle, G. M., Inglis, K. J., Speleman, F., et al. (2002). Proneural and proneuroendocrine transcription factor expression in cutaneous mechanoreceptor (Merkel) cells and Merkel cell carcinoma. *International Journal Cancer. Journal International du Cancer*, *101*(2), 103–110. http://dx.doi.org/10.1002/ijc.10554.

Leow, C. C., Romero, M. S., Ross, S., Polakis, P., & Gao, W. Q. (2004). Hath1, down-regulated in colon adenocarcinomas, inhibits proliferation and tumorigenesis of colon cancer cells. *Cancer Research*, *64*(17), 6050–6057. http://dx.doi.org/10.1158/0008-5472.CAN-04-0290.

Li, S., Mattar, P., Dixit, R., Lawn, S. O., Wilkinson, G., Kinch, C., et al. (2014). RAS/ERK signaling functions as a proneural lineage switch in cortical development and gliomagenesis. *Journal of Neuroscience*, *34*(6), 2169–2190. http://dx.doi.org/10.1523/JNEUROSCI.4077-13.2014.

Li, S., Mattar, P., Zinyk, D., Singh, K., Chaturvedi, C. P., Kovach, C., et al. (2012). GSK3 temporally regulates neurogenin 2 proneural activity in the neocortex. *The Journal of Neuroscience*, *32*(23), 7791–7805. http://dx.doi.org/10.1523/JNEUROSCI.1309-12.2012.

Linnoila, R. I., Zhao, B., DeMayo, J. L., Nelkin, B. D., Baylin, S. B., DeMayo, F. J., et al. (2000). Constitutive achaete-scute homologue-1 promotes airway dysplasia and lung neuroendocrine tumors in transgenic mice. *Cancer Research*, *60*(15), 4005–4009.

Liu, X., Li, F., Stubblefield, E. A., Blanchard, B., Richards, T. L., Larson, G. A., et al. (2012). Direct reprogramming of human fibroblasts into dopaminergic neuron-like cells. *Cell Research*, *22*(2), 321–332. http://dx.doi.org/10.1038/cr.2011.181.

Liu, W., Mo, Z., & Xiang, M. (2001). The Ath5 proneural genes function upstream of Brn3 POU domain transcription factor genes to promote retinal ganglion cell development.

Proceedings of the National Academy of Sciences of the United States of America, 98(4), 1649–1654. http://dx.doi.org/10.1073/pnas.98.4.1649.

Lluis, F., Ballestar, E., Suelves, M., Esteller, M., & Munoz-Canoves, P. (2005). E47 phosphorylation by p38 MAPK promotes MyoD/E47 association and muscle-specific gene transcription. *The EMBO Journal, 24*(5), 974–984.

Lo, L., Dormand, E., Greenwood, A., & Anderson, D. J. (2002). Comparison of the generic neuronal differentiation and neuron subtype specification functions of mammalian achaete-scute and atonal homologs in cultured neural progenitor cells. *Development, 129*(7), 1553–1567.

Lo, L. C., Johnson, J. E., Wuenschell, C. W., Saito, T., & Anderson, D. J. (1991). Mammalian achaete-scute homolog 1 is transiently expressed by spatially restricted subsets of early neuroepithelial and neural crest cells. *Genes & Development, 5*(9), 1524–1537.

Long, J. E., Cobos, I., Potter, G. B., & Rubenstein, J. L. (2009). Dlx1&2 and Mash1 transcription factors control MGE and CGE patterning and differentiation through parallel and overlapping pathways. *Cerebral Cortex, 19*(Suppl 1), i96–i106. http://dx.doi.org/10.1093/cercor/bhp045.

Lukaszewicz, A., Savatier, P., Cortay, V., Kennedy, H., & Dehay, C. (2002). Contrasting effects of basic fibroblast growth factor and neurotrophin 3 on cell cycle kinetics of mouse cortical stem cells. *The Journal of Neuroscience, 22*(15), 6610–6622, 20026666 22/15/6610 [pii].

Ma, Q., Chen, Z., del Barco Barrantes, I., de la Pompa, J. L., & Anderson, D. J. (1998). Neurogenin1 is essential for the determination of neuronal precursors for proximal cranial sensory ganglia. *Neuron, 20*(3), 469–482.

Ma, Q., Fode, C., Guillemot, F., & Anderson, D. J. (1999). Neurogenin1 and neurogenin2 control two distinct waves of neurogenesis in developing dorsal root ganglia. *Genes & Development, 13*(13), 1717–1728.

Ma, Q., Kintner, C., & Anderson, D. J. (1996). Identification of neurogenin, a vertebrate neuronal determination gene. *Cell, 87*(1), 43–52.

Ma, Y. C., Song, M. R., Park, J. P., Henry Ho, H. Y., Hu, L., Kurtev, M. V., et al. (2008). Regulation of motor neuron specification by phosphorylation of neurogenin 2. *Neuron, 58*(1), 65–77. http://dx.doi.org/10.1016/j.neuron.2008.01.037, S0896-6273(08)00132-3 [pii].

Machold, R., & Fishell, G. (2005). Math1 is expressed in temporally discrete pools of cerebellar rhombic-lip neural progenitors. *Neuron, 48*(1), 17–24. http://dx.doi.org/10.1016/j.neuron.2005.08.028.

Malecki, M. T., Jhala, U. S., Antonellis, A., Fields, L., Doria, A., Orban, T., et al. (1999). Mutations in NEUROD1 are associated with the development of type 2 diabetes mellitus. *Nature Genetics, 23*(3), 323–328.

Mao, C. A., Wang, S. W., Pan, P., & Klein, W. H. (2008). Rewiring the retinal ganglion cell gene regulatory network: Neurod1 promotes retinal ganglion cell fate in the absence of Math5. *Development, 135*(20), 3379–3388. http://dx.doi.org/10.1242/dev.024612.

Marro, S., Pang, Z. P., Yang, N., Tsai, M. C., Qu, K., Chang, H. Y., et al. (2011). Direct lineage conversion of terminally differentiated hepatocytes to functional neurons. *Cell Stem Cell, 9*(4), 374–382. http://dx.doi.org/10.1016/j.stem.2011.09.002.

Mattar, P., Britz, O., Johannes, C., Nieto, M., Ma, L., Rebeyka, A., et al. (2004). A screen for downstream effectors of Neurogenin2 in the embryonic neocortex. *Developmental Biology, 273*(2), 373–389. http://dx.doi.org/10.1016/j.ydbio.2004.06.013, S0012-1606(04)00429-4 [pii].

Mattar, P., Langevin, L. M., Markham, K., Klenin, N., Shivji, S., Zinyk, D., et al. (2008). Basic helix-loop-helix transcription factors cooperate to specify a cortical projection neuron identity. *Molecular and Cellular Biology, 28*(5), 1456–1469. http://dx.doi.org/10.1128/MCB.01510-07, MCB.01510-07 [pii].

Matter-Sadzinski, L., Matter, J. M., Ong, M. T., Hernandez, J., & Ballivet, M. (2001). Specification of neurotransmitter receptor identity in developing retina: The chick ATH5 promoter integrates the positive and negative effects of several bHLH proteins. *Development, 128*(2), 217–231.

Matter-Sadzinski, L., Puzianowska-Kuznicka, M., Hernandez, J., Ballivet, M., & Matter, J. M. (2005). A bHLH transcriptional network regulating the specification of retinal ganglion cells. *Development, 132*(17), 3907–3921. http://dx.doi.org/10.1242/dev.01960.

Mazzoni, E. O., Mahony, S., Closser, M., Morrison, C. A., Nedelec, S., Williams, D. J., et al. (2013). Synergistic binding of transcription factors to cell-specific enhancers programs motor neuron identity. *Nature Neuroscience, 16*(9), 1219–1227. http://dx.doi.org/10.1038/nn.3467.

McNay, D. E., Pelling, M., Claxton, S., Guillemot, F., & Ang, S. L. (2006). Mash1 is required for generic and subtype differentiation of hypothalamic neuroendocrine cells. *Molecular Endocrinology, 20*(7), 1623–1632. http://dx.doi.org/10.1210/me.2005-0518.

Mizuguchi, R., Kriks, S., Cordes, R., Gossler, A., Ma, Q., & Goulding, M. (2006). Ascl1 and Gsh1/2 control inhibitory and excitatory cell fate in spinal sensory interneurons. *Nature Neuroscience, 9*(6), 770–778. http://dx.doi.org/10.1038/nn1706.

Mizuguchi, R., Sugimori, M., Takebayashi, H., Kosako, H., Nagao, M., Yoshida, S., et al. (2001). Combinatorial roles of olig2 and neurogenin2 in the coordinated induction of pan-neuronal and subtype-specific properties of motoneurons. *Neuron, 31*(5), 757–771.

Mu, X., Fu, X., Sun, H., Beremand, P. D., Thomas, T. L., & Klein, W. H. (2005). A gene network downstream of transcription factor Math5 regulates retinal progenitor cell competence and ganglion cell fate. *Developmental Biology, 280*(2), 467–481. http://dx.doi.org/10.1016/j.ydbio.2005.01.028.

Murre, C., Bain, G., van Dijk, M. A., Engel, I., Furnari, B. A., Massari, M. E., et al. (1994). Structure and function of helix-loop-helix proteins. *Biochimica et Biophysica Acta, 1218*(2), 129–135.

Murre, C., McCaw, P. S., Vaessin, H., Caudy, M., Jan, L. Y., Jan, Y. N., et al. (1989). Interactions between heterologous helix-loop-helix proteins generate complexes that bind specifically to a common DNA sequence. *Cell, 58*(3), 537–544, 0092-8674(89)90434-0 [pii].

Naya, F. J., Huang, H. P., Qiu, Y., Mutoh, H., DeMayo, F. J., Leiter, A. B., et al. (1997). Diabetes, defective pancreatic morphogenesis, and abnormal enteroendocrine differentiation in BETA2/neuroD-deficient mice. *Genes & Development, 11*(18), 2323–2334.

Nieto, M., Schuurmans, C., Britz, O., & Guillemot, F. (2001). Neural bHLH genes control the neuronal versus glial fate decision in cortical progenitors. *Neuron, 29*(2), 401–413, S0896-6273(01)00214-8 [pii].

Nocerino, V., Colombo, C., Bonfanti, R., Iafusco, D., & Barbetti, F. (2009). Mutations in IAPP and NEUROG3 genes are not a common cause of permanent neonatal/infancy/childhood-onset diabetes. *Diabetic Medicine, 26*(6), 660–661.

Northcott, P. A., Korshunov, A., Witt, H., Hielscher, T., Eberhart, C. G., Mack, S., et al. (2011). Medulloblastoma comprises four distinct molecular variants. *Journal of Clinical Oncology, 29*(11), 1408–1414. http://dx.doi.org/10.1200/JCO.2009.27.4324.

Novitch, B. G., Chen, A. I., & Jessell, T. M. (2001). Coordinate regulation of motor neuron subtype identity and pan-neuronal properties by the bHLH repressor Olig2. *Neuron, 31*(5), 773–789.

Ochiai, W., Nakatani, S., Takahara, T., Kainuma, M., Masaoka, M., Minobe, S., et al. (2009). Periventricular notch activation and asymmetric Ngn2 and Tbr2 expression in pair-generated neocortical daughter cells. *Molecular and Cellular Neurosciences, 40*(2), 225–233. http://dx.doi.org/10.1016/j.mcn.2008.10.007, S1044-7431(08)00275-3 [pii].

Onoguchi, M., Hirabayashi, Y., Koseki, H., & Gotoh, Y. (2012). A noncoding RNA regulates the neurogenin1 gene locus during mouse neocortical development. *Proceedings of the National Academy of Sciences of the United States of America, 109*(42), 16939–16944. http://dx.doi.org/10.1073/pnas.1202956109.

Osada, H., Tatematsu, Y., Yatabe, Y., Horio, Y., & Takahashi, T. (2005). ASH1 gene is a specific therapeutic target for lung cancers with neuroendocrine features. *Cancer Research, 65*(23), 10680–10685. http://dx.doi.org/10.1158/0008-5472.CAN-05-1404.

Ozen, I., Galichet, C., Watts, C., Parras, C., Guillemot, F., & Raineteau, O. (2007). Proliferating neuronal progenitors in the postnatal hippocampus transiently express the proneural gene Ngn2. *The European Journal of Neuroscience, 25*(9), 2591–2603. http://dx.doi.org/10.1111/j.1460-9568.2007.05541.x.

Pang, Z. P., Yang, N., Vierbuchen, T., Ostermeier, A., Fuentes, D. R., Yang, T. Q., et al. (2011). Induction of human neuronal cells by defined transcription factors. *Nature, 476*(7359), 220–223. http://dx.doi.org/10.1038/nature10202.

Park, C. H., Kang, J. S., Shin, Y. H., Chang, M. Y., Chung, S., Koh, H. C., et al. (2006). Acquisition of in vitro and in vivo functionality of Nurr1-induced dopamine neurons. *The FASEB Journal, 20*(14), 2553–2555. http://dx.doi.org/10.1096/fj.06-6159fje.

Parras, C. M., Galli, R., Britz, O., Soares, S., Galichet, C., Battiste, J., et al. (2004). Mash1 specifies neurons and oligodendrocytes in the postnatal brain. *The EMBO Journal, 23*(22), 4495–4505. http://dx.doi.org/10.1038/sj.emboj.7600447, 7600447 [pii].

Parras, C. M., Hunt, C., Sugimori, M., Nakafuku, M., Rowitch, D., & Guillemot, F. (2007). The proneural gene Mash1 specifies an early population of telencephalic oligodendrocytes. *The Journal of Neuroscience, 27*(16), 4233–4242. http://dx.doi.org/10.1523/JNEUROSCI.0126-07.2007, 27/16/4233 [pii].

Parras, C. M., Schuurmans, C., Scardigli, R., Kim, J., Anderson, D. J., & Guillemot, F. (2002). Divergent functions of the proneural genes Mash1 and Ngn2 in the specification of neuronal subtype identity. *Genes & Development, 16*(3), 324–338. http://dx.doi.org/10.1101/gad.940902.

Pattyn, A., Goridis, C., & Brunet, J. F. (2000). Specification of the central noradrenergic phenotype by the homeobox gene Phox2b. *Molecular and Cellular Neurosciences, 15*(3), 235–243. http://dx.doi.org/10.1006/mcne.1999.0826.

Peignon, G., Durand, A., Cacheux, W., Ayrault, O., Terris, B., Laurent-Puig, P., et al. (2011). Complex interplay between beta-catenin signalling and Notch effectors in intestinal tumorigenesis. *Gut, 60*(2), 166–176. http://dx.doi.org/10.1136/gut.2009.204719.

Pelling, M., Anthwal, N., McNay, D., Gradwohl, G., Leiter, A. B., Guillemot, F., et al. (2011). Differential requirements for neurogenin 3 in the development of POMC and NPY neurons in the hypothalamus. *Developmental Biology, 349*(2), 406–416. http://dx.doi.org/10.1016/j.ydbio.2010.11.007.

Penn, B. H., Bergstrom, D. A., Dilworth, F. J., Bengal, E., & Tapscott, S. J. (2004). A MyoD-generated feed-forward circuit temporally patterns gene expression during skeletal muscle differentiation. *Genes & Development, 18*(19), 2348–2353.

Perez, S. E., Rebelo, S., & Anderson, D. J. (1999). Early specification of sensory neuron fate revealed by expression and function of neurogenins in the chick embryo. *Development, 126*(8), 1715–1728.

Perrin, F. E., Boniface, G., Serguera, C., Lonjon, N., Serre, A., Prieto, M., et al. (2010). Grafted human embryonic progenitors expressing neurogenin-2 stimulate axonal sprouting and improve motor recovery after severe spinal cord injury. *PLoS One, 5*(12), e15914. http://dx.doi.org/10.1371/journal.pone.0015914.

Pfisterer, U., Kirkeby, A., Torper, O., Wood, J., Nelander, J., Dufour, A., et al. (2011). Direct conversion of human fibroblasts to dopaminergic neurons. *Proceedings of the National Academy of Sciences of the United States of America, 108*(25), 10343–10348. http://dx.doi.org/10.1073/pnas.1105135108.

Phillips, H. S., Kharbanda, S., Chen, R., Forrest, W. F., Soriano, R. H., Wu, T. D., et al. (2006). Molecular subclasses of high-grade glioma predict prognosis, delineate a pattern of disease progression, and resemble stages in neurogenesis. *Cancer Cell*, *9*(3), 157–173. http://dx.doi.org/10.1016/j.ccr.2006.02.019, S1535-6108(06)00056-0 [pii].

Pinney, S. E., Oliver-Krasinski, J., Ernst, L., Hughes, N., Patel, P., Stoffers, D. A., et al. (2011). Neonatal diabetes and congenital malabsorptive diarrhea attributable to a novel mutation in the human neurogenin-3 gene coding sequence. *The Journal of Clinical Endocrinology and Metabolism*, *96*(7), 1960–1965.

Poitras, L., Ghanem, N., Hatch, G., & Ekker, M. (2007). The proneural determinant MASH1 regulates forebrain Dlx1/2 expression through the I12b intergenic enhancer. *Development*, *134*(9), 1755–1765. http://dx.doi.org/10.1242/dev.02845, dev.02845 [pii].

Powell, L. M., & Jarman, A. P. (2008). Context dependence of proneural bHLH proteins. *Current Opinion in Genetics & Development*, *18*(5), 411–417. http://dx.doi.org/10.1016/j.gde.2008.07.012, S0959-437X(08)00091-9 [pii].

Prasov, L., Brown, N. L., & Glaser, T. (2010). A critical analysis of Atoh7 (Math5) mRNA splicing in the developing mouse retina. *PLoS One*, *5*(8), e12315. http://dx.doi.org/10.1371/journal.pone.0012315.

Prasov, L., & Glaser, T. (2012). Pushing the envelope of retinal ganglion cell genesis: Context dependent function of Math5 (Atoh7). *Developmental Biology*, *368*(2), 214–230. http://dx.doi.org/10.1016/j.ydbio.2012.05.005.

Prasov, L., Masud, T., Khaliq, S., Mehdi, S. Q., Abid, A., Oliver, E. R., et al. (2012). ATOH7 mutations cause autosomal recessive persistent hyperplasia of the primary vitreous. *Human Molecular Genetics*, *21*(16), 3681–3694. http://dx.doi.org/10.1093/hmg/dds197.

Prasov, L., Nagy, M., Rudolph, D. D., & Glaser, T. (2012). Math5 (Atoh7) gene dosage limits retinal ganglion cell genesis. *Neuroreport*, *23*(10), 631–634. http://dx.doi.org/10.1097/WNR.0b013e328355f260.

Rapa, I., Volante, M., Migliore, C., Farsetti, A., Berruti, A., Vittorio Scagliotti, G., et al. (2013). Human ASH-1 promotes neuroendocrine differentiation in androgen deprivation conditions and interferes with androgen responsiveness in prostate cancer cells. *Prostate*, *73*(11), 1241–1249. http://dx.doi.org/10.1002/pros.22679.

Rheinbay, E., Suva, M. L., Gillespie, S. M., Wakimoto, H., Patel, A. P., Shahid, M., et al. (2013). An aberrant transcription factor network essential for Wnt signaling and stem cell maintenance in glioblastoma. *Cell Reports*, *3*(5), 1567–1579. http://dx.doi.org/10.1016/j.celrep.2013.04.021.

Ribes, V., Stutzmann, F., Bianchetti, L., Guillemot, F., Dolle, P., & Le Roux, I. (2008). Combinatorial signalling controls Neurogenin2 expression at the onset of spinal neurogenesis. *Developmental Biology*, *321*(2), 470–481. http://dx.doi.org/10.1016/j.ydbio.2008.06.003.

Rose, M. F., Ren, J., Ahmad, K. A., Chao, H. T., Klisch, T. J., Flora, A., et al. (2009). Math1 is essential for the development of hindbrain neurons critical for perinatal breathing. *Neuron*, *64*(3), 341–354. http://dx.doi.org/10.1016/j.neuron.2009.10.023.

Rousseau, A., Nutt, C. L., Betensky, R. A., Iafrate, A. J., Han, M., Ligon, K. L., et al. (2006). Expression of oligodendroglial and astrocytic lineage markers in diffuse gliomas: Use of YKL-40, ApoE, ASCL1, and NKX2-2. *Journal of Neuropathology and Experimental Neurology*, *65*(12), 1149–1156. http://dx.doi.org/10.1097/01.jnen.0000248543.90304.2b.

Roybon, L., Hjalt, T., Christophersen, N. S., Li, J. Y., & Brundin, P. (2008). Effects on differentiation of embryonic ventral midbrain progenitors by Lmx1a, Msx1, Ngn2, and Pitx3. *The Journal of Neuroscience*, *28*(14), 3644–3656. http://dx.doi.org/10.1523/JNEUROSCI.0311-08.2008.

Rubio-Cabezas, O., Jensen, J. N., Hodgson, M. I., Codner, E., Ellard, S., Serup, P., et al. (2011). Permanent neonatal diabetes and enteric anendocrinosis associated with biallelic mutations in NEUROG3. *Diabetes*, *60*(4), 1349–1353.

Russo, L., Iafusco, D., Brescianini, S., Nocerino, V., Bizzarri, C., Toni, S., et al. (2011). Permanent diabetes during the first year of life: Multiple gene screening in 54 patients. *Diabetologia, 54*(7), 1693–1701.

Salomoni, P., & Calegari, F. (2010). Cell cycle control of mammalian neural stem cells: Putting a speed limit on G1. *Trends in Cell Biology, 20*(5), 233–243. http://dx.doi.org/10.1016/j.tcb.2010.01.006.

Salsano, E., Pollo, B., Eoli, M., Giordana, M. T., & Finocchiaro, G. (2004). Expression of MATH1, a marker of cerebellar granule cell progenitors, identifies different medulloblastoma sub-types. *Neuroscience Letters, 370*(2–3), 180–185. http://dx.doi.org/10.1016/j.neulet.2004.08.053.

Scardigli, R., Schuurmans, C., Gradwohl, G., & Guillemot, F. (2001). Crossregulation between Neurogenin2 and pathways specifying neuronal identity in the spinal cord. *Neuron, 31*(2), 203–217, S0896-6273(01)00358-0 [pii].

Schroder, J. C., Lassig, A. K., Galetzka, D., Peters, A., Castle, J. C., Diederich, S., et al. (2013). A boy with homozygous microdeletion of NEUROG1 presents with a congenital cranial dysinnervation disorder [Moebius syndrome variant]. *Behavioral and Brain Functions, 9*, 7. http://dx.doi.org/10.1186/1744-9081-9-7.

Schuller, U., Heine, V. M., Mao, J., Kho, A. T., Dillon, A. K., Han, Y. G., et al. (2008). Acquisition of granule neuron precursor identity is a critical determinant of progenitor cell competence to form Shh-induced medulloblastoma. *Cancer Cell, 14*(2), 123–134. http://dx.doi.org/10.1016/j.ccr.2008.07.005.

Schuurmans, C., Armant, O., Nieto, M., Stenman, J. M., Britz, O., Klenin, N., et al. (2004). Sequential phases of cortical specification involve Neurogenin-dependent and -independent pathways. *The EMBO Journal, 23*(14), 2892–2902.

Schwab, S. G., Eckstein, G. N., Hallmayer, J., Lerer, B., Albus, M., Borrmann, M., et al. (1997). Evidence suggestive of a locus on chromosome 5q31 contributing to susceptibility for schizophrenia in German and Israeli families by multipoint affected sib-pair linkage analysis. *Molecular Psychiatry, 2*(2), 156–160.

Schwab, S. G., & Wildenauer, D. B. (2013). Genetics of psychiatric disorders in the GWAS era: An update on schizophrenia. *European Archives of Psychiatry and Clinical Neuroscience, 263*(Suppl. 2), S147–S154. http://dx.doi.org/10.1007/s00406-013-0450-z.

Seo, S., Lim, J. W., Yellajoshyula, D., Chang, L. W., & Kroll, K. L. (2007). Neurogenin and NeuroD direct transcriptional targets and their regulatory enhancers. *The EMBO Journal, 26*(24), 5093–5108. http://dx.doi.org/10.1038/sj.emboj, 7601923 [pii].

Seo, S., Richardson, G. A., & Kroll, K. L. (2005). The SWI/SNF chromatin remodeling protein Brg1 is required for vertebrate neurogenesis and mediates transactivation of Ngn and NeuroD. *Development, 132*(1), 105–115.

Shaker, T., Dennis, D., Kurrasch, D. M., & Schuurmans, C. (2012). Neurog1 and Neurog2 coordinately regulate development of the olfactory system. *Neural Development*, 7(1), 28. http://dx.doi.org/10.1186/1749-8104-7-28.

Shimojo, H., Ohtsuka, T., & Kageyama, R. (2008). Oscillations in notch signaling regulate maintenance of neural progenitors. *Neuron, 58*(1), 52–64. http://dx.doi.org/10.1016/j.neuron.2008.02.014, S0896-6273(08)00166-9 [pii].

Shroyer, N. F., Helmrath, M. A., Wang, V. Y., Antalffy, B., Henning, S. J., & Zoghbi, H. Y. (2007). Intestine-specific ablation of mouse atonal homolog 1 (Math1) reveals a role in cellular homeostasis. *Gastroenterology, 132*(7), 2478–2488. http://dx.doi.org/10.1053/j.gastro.2007.03.047.

Shroyer, N. F., Wallis, D., Venken, K. J., Bellen, H. J., & Zoghbi, H. Y. (2005). Gfi1 functions downstream of Math1 to control intestinal secretory cell subtype allocation and differentiation. *Genes & Development, 19*(20), 2412–2417. http://dx.doi.org/10.1101/gad.1353905.

Skeath, J. B., & Carroll, S. B. (1994). The achaete-scute complex: Generation of cellular pattern and fate within the Drosophila nervous system. *The FASEB Journal*, *8*(10), 714–721.

Skipper, M., & Lewis, J. (2000). Getting to the guts of enteroendocrine differentiation. *Nature Genetics*, *24*(1), 3–4.

Skol, A. D., Young, K. A., Tsuang, D. W., Faraone, S. V., Haverstock, S. L., Bingham, S., et al. (2003). Modest evidence for linkage and possible confirmation of association between NOTCH4 and schizophrenia in a large Veterans Affairs Cooperative Study sample. *American Journal of Medical Genetics. Part B, Neuropsychiatric Genetics*, *118B*(1), 8–15. http://dx.doi.org/10.1002/ajmg.b.10055.

Skowronska-Krawczyk, D., Ballivet, M., Dynlacht, B. D., & Matter, J. M. (2004). Highly specific interactions between bHLH transcription factors and chromatin during retina development. *Development*, *131*(18), 4447–4454.

Skowronska-Krawczyk, D., Chiodini, F., Ebeling, M., Alliod, C., Kundzewicz, A., Castro, D., et al. (2009). Conserved regulatory sequences in Atoh7 mediate non-conserved regulatory responses in retina ontogenesis. *Development*, *136*(22), 3767–3777. http://dx.doi.org/10.1242/dev.033449.

Somasundaram, K., Reddy, S. P., Vinnakota, K., Britto, R., Subbarayan, M., Nambiar, S., et al. (2005). Upregulation of ASCL1 and inhibition of Notch signaling pathway characterize progressive astrocytoma. *Oncogene*, *24*(47), 7073–7083. http://dx.doi.org/10.1038/sj.onc.1208865.

Sommer, L., Ma, Q., & Anderson, D. J. (1996). Neurogenins, a novel family of atonal-related bHLH transcription factors, are putative mammalian neuronal determination genes that reveal progenitor cell heterogeneity in the developing CNS and PNS. *Molecular and Cellular Neurosciences*, *8*(4), 221–241.

Soosaar, A., Chiaramello, A., Zuber, M. X., & Neuman, T. (1994). Expression of basic-helix-loop-helix transcription factor ME2 during brain development and in the regions of neuronal plasticity in the adult brain. *Brain Research Molecular Brain Research*, *25*(1–2), 176–180.

Stolt, C. C., Lommes, P., Sock, E., Chaboissier, M. C., Schedl, A., & Wegner, M. (2003). The Sox9 transcription factor determines glial fate choice in the developing spinal cord. *Genes & Development*, *17*(13), 1677–1689. http://dx.doi.org/10.1101/gad.259003.

Sugimori, M., Nagao, M., Parras, C. M., Nakatani, H., Lebel, M., Guillemot, F., et al. (2008). Ascl1 is required for oligodendrocyte development in the spinal cord. *Development*, *135*(7), 1271–1281. http://dx.doi.org/10.1242/dev.015370, dev.015370 [pii].

Sun, Y., Nadal-Vicens, M., Misono, S., Lin, M. Z., Zubiaga, A., Hua, X., et al. (2001). Neurogenin promotes neurogenesis and inhibits glial differentiation by independent mechanisms. *Cell*, *104*(3), 365–376.

Tamamaki, N., Fujimori, K. E., & Takauji, R. (1997). Origin and route of tangentially migrating neurons in the developing neocortical intermediate zone. *The Journal of Neuroscience*, *17*(21), 8313–8323.

Tapscott, S. J. (2005). The circuitry of a master switch: Myod and the regulation of skeletal muscle gene transcription. *Development*, *132*(12), 2685–2695.

Taylor, M. D., Northcott, P. A., Korshunov, A., Remke, M., Cho, Y. J., Clifford, S. C., et al. (2012). Molecular subgroups of medulloblastoma: The current consensus. *Acta Neuropathologica*, *123*(4), 465–472. http://dx.doi.org/10.1007/s00401-011-0922-z.

Theka, I., Caiazzo, M., Dvoretskova, E., Leo, D., Ungaro, F., Curreli, S., et al. (2013). Rapid generation of functional dopaminergic neurons from human induced pluripotent stem cells through a single-step procedure using cell lineage transcription factors. *Stem Cells Translational Medicine*, *2*(6), 473–479. http://dx.doi.org/10.5966/sctm.2012-0133.

Tomita, K., Moriyoshi, K., Nakanishi, S., Guillemot, F., & Kageyama, R. (2000). Mammalian achaete-scute and atonal homologs regulate neuronal versus glial fate determination in the central nervous system. *The EMBO Journal*, *19*(20), 5460–5472.

Turner, D. L., & Cepko, C. L. (1987). A common progenitor for neurons and glia persists in rat retina late in development. *Nature*, *328*(6126), 131–136.

Ueno, T., Ito, J., Hoshikawa, S., Ohori, Y., Fujiwara, S., Yamamoto, S., et al. (2012). The identification of transcriptional targets of Ascl1 in oligodendrocyte development. *Glia*, *60*(10), 1495–1505. http://dx.doi.org/10.1002/glia.22369.

VanDussen, K. L., & Samuelson, L. C. (2010). Mouse atonal homolog 1 directs intestinal progenitors to secretory cell rather than absorptive cell fate. *Developmental Biology*, *346*(2), 215–223. http://dx.doi.org/10.1016/j.ydbio.2010.07.026.

Van Hoof, D., D'Amour, K. A., & German, M. S. (2009). Derivation of insulin-producing cells from human embryonic stem cells. *Stem Cell Research*, *3*, 73–87.

Vierbuchen, T., Ostermeier, A., Pang, Z. P., Kokubu, Y., Sudhof, T. C., & Wernig, M. (2010). Direct conversion of fibroblasts to functional neurons by defined factors. *Nature*, *463*(7284), 1035–1041. http://dx.doi.org/10.1038/nature08797.

Villares, R., & Cabrera, C. V. (1987). The achaete-scute gene complex of D. melanogaster: Conserved domains in a subset of genes required for neurogenesis and their homology to myc. *Cell*, *50*(3), 415–424.

Wallace, V. A. (1999). Purkinje-cell-derived Sonic hedgehog regulates granule neuron precursor cell proliferation in the developing mouse cerebellum. *Current Biology*, *9*(8), 445–448.

Wang, J., Cortina, G., Wu, S. V., Tran, R., Cho, J. H., Tsai, M. J., et al. (2006). Mutant neurogenin-3 in congenital malabsorptive diarrhea. *The New England Journal of Medicine*, *355*(3), 270–280.

Wang, X. Y., Dakir el, H., Naizhen, X., Jensen-Taubman, S. M., DeMayo, F. J., & Linnoila, R. I. (2007). Achaete-scute homolog-1 linked to remodeling and preneoplasia of pulmonary epithelium. *Laboratory Investigation*, *87*(6), 527–539. http://dx.doi.org/10.1038/labinvest.3700552.

Wang, S., Jensen, J. N., Seymour, P. A., Hsu, W., Dor, Y., Sander, M., et al. (2009). Sustained Neurog3 expression in hormone-expressing islet cells is required for endocrine maturation and function. *Proceedings of the National Academy of Sciences of the United States of America*, *106*(24), 9715–9720.

Wang, X. Y., Jensen-Taubman, S. M., Keefe, K. M., Yang, D., & Linnoila, R. I. (2012). Achaete-scute complex homolog-1 promotes DNA repair in the lung carcinogenesis through matrix metalloproteinase-7 and O(6)-methylguanine-DNA methyltransferase. *PLoS One*, 7(12), e52832. http://dx.doi.org/10.1371/journal.pone.0052832.

Wang, S. W., Kim, B. S., Ding, K., Wang, H., Sun, D., Johnson, R. L., et al. (2001). Requirement for math5 in the development of retinal ganglion cells. *Genes & Development*, *15*(1), 24–29.

Wang, V. Y., Rose, M. F., & Zoghbi, H. Y. (2005). Math1 expression redefines the rhombic lip derivatives and reveals novel lineages within the brainstem and cerebellum. *Neuron*, *48*(1), 31–43. http://dx.doi.org/10.1016/j.neuron.2005.08.024.

Wapinski, O. L., Vierbuchen, T., Qu, K., Lee, Q. Y., Chanda, S., Fuentes, D. R., et al. (2013). Hierarchical mechanisms for direct reprogramming of fibroblasts to neurons. *Cell*, *155*(3), 621–635. http://dx.doi.org/10.1016/j.cell.2013.09.028.

Wawersik, S., Purcell, P., & Maas, R. L. (2000). Pax6 and the genetic control of early eye development. *Results and Problems in Cell Differentiation*, *31*, 15–36.

Wechsler-Reya, R. J., & Scott, M. P. (1999). Control of neuronal precursor proliferation in the cerebellum by Sonic Hedgehog. *Neuron*, *22*(1), 103–114.

Wee, R., Castrucci, A. M., Provencio, I., Gan, L., & Van Gelder, R. N. (2002). Loss of photic entrainment and altered free-running circadian rhythms in math5 −/− mice. *The Journal of Neuroscience*, *22*(23), 10427–10433.

Weese-Mayer, D. E., Berry-Kravis, E. M., Zhou, L., Maher, B. S., Silvestri, J. M., Curran, M. E., et al. (2003). Idiopathic congenital central hypoventilation syndrome: Analysis of genes pertinent to early autonomic nervous system embryologic development and identification of mutations in PHOX2b. *American Journal of Medical Genetics. Part A, 123A*(3), 267–278. http://dx.doi.org/10.1002/ajmg.a.20527.

Wetts, R., Serbedzija, G. N., & Fraser, S. E. (1989). Cell lineage analysis reveals multipotent precursors in the ciliary margin of the frog retina. *Developmental Biology, 136*(1), 254–263.

Wildner, H., Das Gupta, R., Brohl, D., Heppenstall, P. A., Zeilhofer, H. U., & Birchmeier, C. (2013). Genome-wide expression analysis of Ptf1a- and Ascl1-deficient mice reveals new markers for distinct dorsal horn interneuron populations contributing to nociceptive reflex plasticity. *The Journal of Neuroscience, 33*(17), 7299–7307. http://dx.doi.org/10.1523/JNEUROSCI.0491-13.2013.

Wildner, H., Muller, T., Cho, S. H., Brohl, D., Cepko, C. L., Guillemot, F., et al. (2006). dILA neurons in the dorsal spinal cord are the product of terminal and non-terminal asymmetric progenitor cell divisions, and require Mash1 for their development. *Development, 133*(11), 2105–2113. http://dx.doi.org/10.1242/dev.02345.

Wilkinson, G., Dennis, D., & Schuurmans, C. (2013). Proneural genes in neocortical development. *Neuroscience, 253*, 256–273. http://dx.doi.org/10.1016/j.neuroscience.2013.08.029.

Winpenny, E., Lebel-Potter, M., Fernandez, M. E., Brill, M. S., Gotz, M., Guillemot, F., et al. (2011). Sequential generation of olfactory bulb glutamatergic neurons by Neurog2-expressing precursor cells. *Neural Development, 6*, 12. http://dx.doi.org/10.1186/1749-8104-6-12.

Xu, X., D'Hoker, J., Stange, G., Bonne, S., De Leu, N., Xiao, X., et al. (2008). Beta cells can be generated from endogenous progenitors in injured adult mouse pancreas. *Cell, 132*(2), 197–207.

Yang, Q., Bermingham, N. A., Finegold, M. J., & Zoghbi, H. Y. (2001). Requirement of Math1 for secretory cell lineage commitment in the mouse intestine. *Science, 294*(5549), 2155–2158. http://dx.doi.org/10.1126/science.1065718.

Yang, Z. J., Ellis, T., Markant, S. L., Read, T. A., Kessler, J. D., Bourboulas, M., et al. (2008). Medulloblastoma can be initiated by deletion of patched in lineage-restricted progenitors or stem cells. *Cancer Cell, 14*(2), 135–145. http://dx.doi.org/10.1016/j.ccr.2008.07.003.

Yao, J., Sun, X., Wang, Y., Xu, G., & Qian, J. (2007). Math5 promotes retinal ganglion cell expression patterns in retinal progenitor cells. *Molecular Vision, 13*, 1066–1072.

Yokoi, N., Kanamori, M., Horikawa, Y., Takeda, J., Sanke, T., Furuta, H., et al. (2006). Association studies of variants in the genes involved in pancreatic beta-cell function in type 2 diabetes in Japanese subjects. *Diabetes, 55*(8), 2379–2386.

Yoo, A. S., Sun, A. X., Li, L., Shcheglovitov, A., Portmann, T., Li, Y., et al. (2011). MicroRNA-mediated conversion of human fibroblasts to neurons. *Nature, 476*(7359), 228–231. http://dx.doi.org/10.1038/nature10323.

Zaharieva, I., Georgieva, L., Nikolov, I., Kirov, G., Owen, M. J., O'Donovan, M. C., et al. (2008). Association study in the 5q31-32 linkage region for schizophrenia using pooled DNA genotyping. *BMC Psychiatry, 8*, 11. http://dx.doi.org/10.1186/1471-244X-8-11.

Zhao, J., He, H., Zhou, K., Ren, Y., Shi, Z., Wu, Z., et al. (2012). Neuronal transcription factors induce conversion of human glioma cells to neurons and inhibit tumorigenesis. *PLoS One*, 7(7), e41506. http://dx.doi.org/10.1371/journal.pone.0041506.

CHAPTER THREE

The *Hand2* Gene Dosage Effect in Developmental Defects and Human Congenital Disorders

Masaru Tamura*[,1], Takanori Amano[†], Toshihiko Shiroishi[†,1]

*Technology and Development Team for Mouse Phenotyping Analysis, RIKEN BioResource Center, Tsukuba, Ibaraki, Japan

[†]Mammalian Genetics Laboratory, Genetic Strains Research Center, National Institute of Genetics, Mishima, Shizuoka, Japan

[1]Corresponding authors: e-mail address: matamura@brc.riken.jp; tshirois@nig.ac.jp

Contents

Abstract

Heart- and neural crest derivatives-expressed (Hand) proteins belong to the Twist family of the basic helix–loop–helix (bHLH) transcription factors, and play crucial roles in the development of several organs. They form heterodimers with Twist1 via their HLH domain. Disruption of the expression balance between Hand2 and Twist1 causes limb malformation, indicating that the expression level of Hand2 relative to Twist1 is essential for limb development. Mutations of the *TWIST1* and *TWIST2* genes are involved in human diseases. Although, the functions of the Hand proteins are indispensable for limb, heart, and craniofacial development, mutations of the *Hand* genes that are causative of human diseases remain elusive. Recently, comparative analyses of a human chromosomal disorder, partial trisomy distal 4q, and its mouse model, which is a spontaneously occurring mutant, clearly demonstrated that over dosage of *Hand2* results in developmental defects of limbs, craniofacial, and lumbar vertebrae, and that trisomy of the *Hand2* gene directly causes a human congenital disorder. In this review, we focus on gene dosage effect of *Hand2* in limb, heart, and craniofacial development, and discuss its implication in human diseases.

Current Topics in Developmental Biology, Volume 110
ISSN 0070-2153
http://dx.doi.org/10.1016/B978-0-12-405943-6.00003-8

© 2014 Elsevier Inc.
All rights reserved.

1. INTRODUCTION

Basic helix–loop–helix (bHLH) transcription factors have an evolutionarily conserved domain that comprises a short stretch of basic amino acids and two alpha helices separated by a loop motif (Murre et al., 1989). Phenotype analyses of mutant flies and mice revealed that the bHLH transcription factors play essential roles to regulate cell-fate determination and the development of several organs (Massari & Murre, 2000). To date, extensive searches for homologs of bHLH genes in higher vertebrate species, including mouse and human, have been carried out, and many homologs and novel bHLH transcription factors have been identified. From a functional aspect, they are divided into several groups, such as neurogenic differentiation (NeuroD), Hairy/enhancer-of-split (Hes), and myogenic differentiation antigen (MyoD) families, which act on neural differentiation, neural inhibition, and myogenic differentiation, respectively. On the other hand, based on their expression patterns, the bHLH transcription factors are categorized into two classes. One is the ubiquitously expressed Class A, which includes E12, E47, and HEB; and the other is tissue-specifically expressed Class B, which includes MyoD, Neurog1, and Mash1. The Class B proteins form homo- and/or hetero dimers with Class A or other Class B proteins through their HLH domains, and bind to genomic DNA through their basic amino acid stretch, which recognizes a consensus sequence, CANNTG, termed an E-box, which resides in *cis*-regulatory elements of downstream target genes (Murre et al., 1989; Wilson-Rawls, Rhee, & Rawls, 2004).

Heart- and neural crest derivatives-expressed 2 (Hand2), also known as dHand, Hed, and Thing2, is a typical bHLH transcription factor. It belongs to the Twist family together with Hand1, which is also knows as eHand, Hxt, and Thing1 (Fig. 3.1). Hand2 forms a heterodimer with Twist1, which also belongs to the Twist family (Firulli et al., 2005). Haploinsufficiency of human TWIST1 resulting from mutation of the *TWIST1* gene causes a congenital disorder, Saethre–Chotzen Syndrome (OMIM 101400), which manifests as craniofacial anomaly, facial asymmetry, premature closing of suture, polydactyly, and clinodactyly (el Ghouzzi et al., 1997; Howard et al., 1997; Jabs, 2001). Indeed, approximately 80% of Saethre–Chotzen Syndrome patients have a point mutation or deletion in *TWIST1* (Gripp, Zackai, & Stolle, 2000; Jabs, 2001; Johnson et al., 1998). Moreover, it was reported that mutations of the class A bHLH transcription factor,

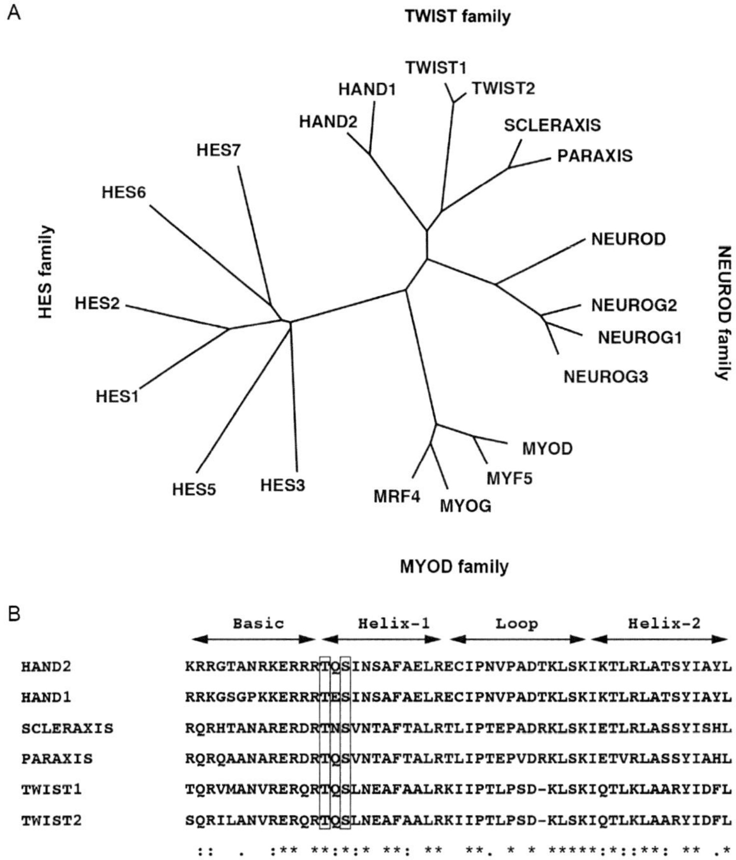

Figure 3.1 Phylogeny of HAND proteins in the bHLH family. (A) The phylogenetic tree of human HAND, MyoD, NeuroD, Hes, and Twist. The tree was generated by comparing the amino acid sequences of the bHLH domains. Hand proteins belong to the TWIST family. (B) Amino acid alignment of the bHLH domains in human TWIST family proteins. Asterisks (*) indicate identical amino acids throughout the TWIST family proteins. Colons and dots indicate conserved sequences and similar substitutions, respectively. Dashed lines indicate gaps. Open boxes indicate the two conserved phosphorylation sites, which correspond to the sites of Thr125 and Ser127 in the Twist1 protein.

TCF12, which forms a dimer with Twist1, as does Hand2, is also responsible for Saethre–Chotzen Syndrome (Sharma et al., 2013).

Although, the Hand proteins form heterodimers with the Twist1 protein and play pivotal roles during the heart and limb development, human diseases caused by mutations in the *Hand* genes are largely unknown. Recently, a genetic study of a mouse mutant revealed that over dosage of *Hand2* caused numerous developmental defects, and is responsible for several symptoms of a human chromosomal disorder named partial trisomy distal 4q (Tamura et al., 2013). In this review, we focus on the gene dosage effect of *Hand2* on the cardiovascular, craniofacial, and limb development and its implication in human diseases.

2. CLONING OF *HAND* GENES AND THEIR EXPRESSION PATTERNS

Several research groups independently cloned the mouse *Hand1* and *Hand2* genes using the Yeast two-hybrid system with the E-protein as the bait, or by screening of homologous sequences in cDNA libraries using known sequences encoding the bHLH domain as probes (Cross et al., 1995; Cserjesi, Brown, Lyons, & Olson, 1995; Hollenberg, Sternglanz, Cheng, & Weintraub, 1995; Srivastava, Cserjesi, & Olson, 1995). These studies revealed that the Hand1 and Hand2 proteins are members of the Hand subfamily, and belong to the Twist family, which include Scleraxis, Paraxis, and Twist1 (Fig. 3.1). Twist1 is a mouse homolog of the Drosophila bHLH transcription factor, Twist, which functions in the formation of Dorso-Ventral patterning or in determining mesodermal cell fate (Leptin, 1991). The human *HAND1* and *HAND2* genes were cloned by *in silico* screening of databases and by screening of a fetal heart cDNA library (Knöfler, Meinhardt, Vasicek, Husslein, & Egarter, 1998; Russell, Kemp, Wang, Brody, & Izumo, 1998). The human HAND1 protein shares 92% amino acid identity with mouse Hand1, and 98% identity with the bHLH domain of mouse Hand1. The Hand2 proteins are evolutionarily more conserved, and human and mouse Hand2 have a completely identical amino acid sequence.

Expression of the mouse *Hand2* gene is first observed in maternally derived deciduum at embryonic day 7.5 (E7.5). Its expression in mouse embryos subsequently starts at E7.75 in the lateral mesoderm, which migrates toward the anterior most end of the embryo, and forms a horseshoe-shaped cardiac crescent, which is the future heart tissue. This *Hand2* expression continues until the linear heart tube stage at E8.5. Until cardiac looping begins, *Hand2* expression is restricted to the right ventricle and outflow tract. At the same stage, *Hand1* is specifically expressed in the left ventricle.

Before limb development starts, *Hand2* is expressed in the lateral plate mesoderm (Charité, McFadden, & Olson, 2000). Subsequently, a gradient expression pattern of *Hand2* is observed in the portion where the forelimb bud will be developed. At E9.5, it is restricted to the posterior portion of mesenchymal cells in the developing limb bud. At E9.75–E11.5, its expression is upregulated and expanded toward the anterior of the limb bud, but is never expressed in the distal-anterior region. *Hand2* expression continues in the portion that forms the second to fifth digits and proximal region of the fore limb bud at E12.5 (Charité et al., 2000).

Hand2 is also expressed in the first and second branchial arch, neural crest cell, sympathetic ganglia, the gastrointestinal tract, and the cardiovascular system in the later developmental stages. The neural crest cells contribute to the craniofacial structure, sympathetic nervous system, and extra-adrenal chromaffin cells. It was reported that *Hand2* is expressed in the adult heart in mice (Thattaliyath, Livi, Steinhelper, Toney, & Firulli, 2002).

Gata4 regulates *Hand2* expression during cardiovascular development (Zeisberg et al., 2005). Cardiogenesis-specific *Gata4* conditional knockout (KO) mice, using the *Nkx2.5-Cre* driver, showed a heart phenotype similar to that of *Hand2* KO mice (see Section 3), including hypoplasia of the right ventricle. Notably, reduced expression of *Hand2* was observed in the *Gata4* conditional KO mice (Zeisberg et al., 2005). All these facts support the view that Gata4 is an upstream regulator of *Hand2*. The bHLH domain of HAND2 physically interacts with the C-terminal Zinc finger domain of GATA4 to activate expression of cardiogenesis-related genes via binding to their promoters (Dai, Cserjesi, Markham, & Molkentin, 2002). The combination of HAND2 and GATA4 with transcription factors MEF2C and TBX5 is able to reprogram fibroblast cells to cardiac myocytes *in vitro* (Song et al., 2012). Therefore, HAND2 is a therapeutic target of human heart injury, such as myocardial infarction.

A 208-bp genomic sequence that is evolutionarily conserved and located 6.6 kb upstream of the *Hand2* transcription start site is necessary for expression of the *Hand2* gene in the branchial arch (Charité et al., 2001). This branchial arch-specific enhancer contains four homeodomain binding sites, and the Distal-less family transcription factor, Dlx6, directly regulates the *Hand2* expression in the branchial arch by binding to these sites. Dlx6 is also an effector of endothelin-1 (Edn1), and Edn1 signaling might be mediated by MADS box-type transcription factor Mef2c (Verzi et al., 2007). Edn1 is necessary for *Hand2* expression in the branchial arch (Barron et al., 2011; Thomas, Yamagishi, Overbeek, Olson, & Srivastava, 1998).

In the mouse embryo, *Hand1* expression is first observed before E7.5 in the extra embryonic tissue, including the extra embryonic mesoderm, ectoplacental corn, york sac, and corion (Cserjesi et al., 1995). In the embryonic tissues, *Hand1* expression starts in the lateral mesoderm at E7.5 (Srivastava et al., 1997). This expression timing in the embryonic tissues is faster than that of *Hand2*. At E7.75, *Hand1* expression is not observed in the cardiac crescent, whereas *Hand2* is expressed in that region. At E8.5, *Hand1* is expressed in the developing heart tube. Subsequently, its expression is restricted to the left ventricle at E9.5 (Cserjesi et al., 1995). This expression pattern continues until E13.5 (Cserjesi et al., 1995; Firulli, McFadden, Lin, Srivastava, & Olson, 1998; Thomas et al., 1998). Unlike *Hand2, Hand1* is not detectable in the early mouse limb bud. After establishment of the zone of polarizing activity (ZPA) (see Section 3), *Hand1* expression starts in a portion of mesenchymal cells of the anterior limb bud at E11.5 (Fernandez-Teran, Piedra, Rodriguez-Rey, Talamillo, & Ros, 2003). The expression is restricted to the digit 1 region at E13.5, and then downregulated in the limb bud by E14.5. The anteriorly localized pattern of *Hand1* expression is evolutionarily conserved between mouse and chicken. However, in the chicken autopod, *Hand1* is also observed in the developing tendon, implying a divergence in function of Hand1 homologs in vertebrates. *Hand1* is also expressed in other tissues, including the umbilical cord, branchial arch, and gut (Cserjesi et al., 1995; Firulli et al., 1998). The typical expression patterns of *Hand1* and *Hand2* are summarized in Table 3.1. Interestingly, the expression patterns of *Hand1* and *Hand2* are complementary to one another in certain organs, such as in the developing heart and limb.

3. DEVELOPMENTAL FUNCTIONS OF HAND2

Developmental functions of the vertebrate *Hand* genes have been investigated by phenotype analyses of mice mutated in these genes. The first conventional KO mouse of *Hand2* was generated soon after cloning of the gene (Srivastava et al., 1997). *Hand2* comprises two exons, and this structure is evolutionally conserved from teleosts to mammals, including humans (Firulli, 2003). In the first KO mouse, a neomycin resistance cassette replaced a segment containing the two exons. This *Hand2* KO mouse showed extensive apoptosis in the region of the prospective right ventricle. As a consequence, the right ventricle was not formed, and the mice only developed a left ventricle, and died between E9.5 and E10.5 (Srivastava et al., 1997; Thomas et al., 1998). Notably, *Hand1* KO mice also died at E9.5 from defects in the extra embryonic tissues and arrest of the heart tube formation (Firulli et al., 1998; Riley, Anson-Cartwright, & Cross, 1998). At this stage, in the *Hand2* KO mouse, expression of several marker genes of

Table 3.1 Expression patterns of *Hand1* and *Hand2*

Hand1	**Hand2**
Extra embryonic tissue (E7.5)	
Extra embryonic mesoderm	Decidual
Alantois	
Trophoblast	
Maternal mesoderm	
Embryo (E7.75 –)	
Lateral plate mesoderm	Lateral plate mesoderm
Cardiovascular system	
Cardiac neural crest cell	Cardiac neural crest cell
Left ventricle	Right ventricle
Outflow tract	Outflow tract
Pericardium	Pericardium
Umbilical vein/artery	
Vitellin vein/artery	
Craniofacial	
Cranial neural crest cell	Cranial neural crest cell
Brachial arch	Brachial arch
Mandible	
Limb	
Anterior side of limb bud	Posterior side of limb bud
Internal organ	
Adrenal gland	Adrenal gland
Sympathetic ganglia	Gut
Gut	

cardiovascular development were reduced, suggesting that *Hand2* plays key roles in the maintenance of right ventricle progenitor cells and morphogenesis of the right ventricle. It was reported that the Nebulette (*Nebl*) gene is the downstream target of Hand2 during the cardiovascular development

(Villanueva et al., 2002). In that study, differential display analysis was carried out using cDNAs of the *Hand2* KO and wild-type mice. *Nebl* expression level in the heart of the *Hand2* KO mouse was fourfold lower than that of wild-type mouse. In addition to *Nebl*, neuropilin-1, the receptor for the 165-amino acid form of vascular endothelial growth factor (VEGF) was reported as a downstream target of Hand2 by subtraction analysis using heart cDNAs of the *Hand2* KO and wild-type mice (Yamagishi, Olson, & Srivastava, 2000). Atrial naturetic peptide (*Anp*) was also reported to be a downstream target of Hand2 (Thattaliyath, Firulli, & Firulli, 2002).

Since development of *Hand2* KO homozygous mouse is arrested at E10.5, several research groups have been generated conditional *Hand2* KO mice to elucidate the function of *Hand2* in various tissues and organs at later developmental stages (Galli et al., 2010; Hendershot et al., 2007; Morikawa, D'Autréaux, Gershon, & Cserjesi, 2007; Tsuchihashi et al., 2011). In addition, a conditional KO mouse with a tissue-specific enhancer, and a knockin (KI) mouse, in which a mutation was introduced in the basic amino acid stretch of the bHLH domain (Liu et al., 2009), were also generated. These genetically engineered mutant mice are summarized in Table 3.2.

To elucidate the function of the *Hand2* gene in the development of the cardiac neural crest linage, a mouse mutant with a floxed *Hand2* allele was crossed with a *Wnt1-Cre* driver, and its phenotype was carefully characterized (Morikawa & Cserjesi, 2008). This mutant mouse exhibited pulmonary stenosis, an interrupted aortic artery, an anomalous origin of right subclavian

Table 3.2 List of *Hand2* conventional and conditional KO mice

Strain name	KO type	Deletion/modified region	Reference
$Hand2^{tm1Dsr}$	Conventional	Exon1 and Exon2	Srivastava et al. (1997)
$Hand2^{tm1Eno}$	Tissue specific	Enhancer (Brachial arch)	Yanagisawa, Clouthier, Richardson, Charité, and Olson (2003)
$Hand2^{tm1Cse}$	Conditional	Exon1	Morikawa et al. (2007)
$Hand2^{tm1Majh}$	Conditional	Exon1	Hendershot et al. (2007)
$Hand2^{tm1.1Zllr}$	Conditional	Exon1 and Exon2	Galli et al. (2010)
$Hand2^{tm2Dsr}$	Conditional	Exon1 and Exon2	Tsuchihashi et al. (2011)
$Hand2^{tm2Eno}$	Knock in	Basic domain	Liu et al. (2009)

artery, a retroesophageal right subclavian artery, and a ventricular septal defect (VSD), indicating that *Hand2* is essential for the development of the cardiovascular system in mouse.

Both homozygotes and heterozygotes of the floxed mutant allele (*Hand2*^{loxP}) showed no phenotype, because the floxed mutant allele produces the normal Hand2 protein. In contrast, the phenotype of the compound heterozygote (*Hand2*$^{loxP/-}$) of the floxed mutant allele and the *Hand2* KO allele showed a cleft palate, and died during the perinatal period (Morikawa et al., 2007; Xiong et al., 2009). Expression analysis revealed that the expression pattern of the *Hand2* gene in *Hand2*$^{loxP/-}$ was the same as that of the wild-type mouse. On the other hand, the *Hand2* expression level in the compound heterozygous mouse (*Hand2*$^{loxP/-}$) was significantly reduced. These results suggested that reduced levels of the Hand2 protein result in the cleft palate phenotype.

To elucidate the function of *Hand2* in craniofacial development, a branchial arch-specific *Hand2* enhancer, which is located 6.6 kb upstream of the *Hand2* in the mouse genome, was eliminated (Yanagisawa et al., 2003). This mutant mouse showed no *Hand2* expression in the ventrolateral region in the first and second branchial arches, and exhibited phenotypes including a cleft palate, mandibular hypoplasia (micrognathia), cartilage malformation, and perinatal death. These results indicated that *Hand2* plays an essential role in patterning and development of skeletal elements derived from the first and second branchial arches. It was also inferred that an additional ventral domain-specific *Hand2* enhancer exists, because expression of *Hand2* remained specifically in the ventral domain of the first and second branchial arches.

Hand2 is involved in osteogenesis and directly interacts with the Runt-related transcription factor 2 (Runx2), also known as Cbfa1, Aml3, Pebp2a, and Osf2. *Runx2* is a master gene of bone formation (Ducy, Zhang, Geoffroy, Ridall, & Karsenty, 1997; Komori et al., 1997; Otto et al., 1997), and disruption of *Runx2* leads to failure of osteogenesis. *Runx2* KO mice never develop bones, whereas formation of cartilage normally occurs, and the mice die from respiratory failure. The Runx2 protein binds to a consensus sequence, CCAACCAC, which termed osteoblast specific element-2 (OSE2) and exists in the promoter regions of bone formation-related genes. Runx2 regulates the expression of osteogenesis-related genes during osteogenesis. The N-terminal domain of the Hand2 protein interacts directly with DNA binding "runt" domain of Runx2, and inhibits transcriptional activity of the bone formation-related genes, thereby negatively controlling differentiation of mesenchymal progenitor cells into osteoblast cells. Indeed, the branchial arch-specific *Hand2* KO mouse shows insufficient mineralization and ectopic bone formation during intramembranous ossification (Funato et al., 2009). Moreover, osteogenesis-specific forced expression of a *Hand2* transgene, driven by the collagen II promoter, caused significant reduction in osteogenesis (Abe et al., 2010).

The vertebrate limb is a three-dimensional structure that develops along anteroposterior (A–P), dorsoventral (D–V), and proximodistal (P–D) axes. It has been an excellent model to study the concept of embryological morphogenesis for many years (Zeller, Lopez-Rios, & Zuniga, 2009). Two signaling centers are responsible for the formation of complicated limb structures. A thickened distal epithelium, referred to as the apical ectodermal ridge (AER), produces several fibroblast growth factors (FGFs), and plays roles in the formation of the proximodistal axis and elongation of the limb bud. A portion of the posterior mesenchyme, referred to as the ZPA, is necessary for specification of limb identity along the A–P axis. A member of *Hh* family, Sonic hedgehog (*Shh*), plays a pivotal role in A–P axis formation during limb development (Fig. 3.2A) (Harfe et al., 2004; McMahon, Ingham, & Tabin, 2003; Riddle, Johnson, Laufer, & Tabin, 1993; Tickle, 1981).

Many lines of evidence have shown that Hand2 positively regulates *Shh* expression as an upstream factor in the limb bud. A long-range enhancer named Mammalian-Fish Conserved Sequence 1 (MFCS1), also known as ZPA regulating sequence (ZRS), regulates *Shh* expression (Lettice et al., 2003; Sagai et al., 2004). Transcription factors that bind to MFCS1 are still largely unknown. Chromatin immunoprecipitation (ChIP) suggested that direct binding of some homeodomain proteins, such as Pbx1 and Hoxd13, to MFCS1 (Capellini et al., 2006). The double KO mouse of the *Hoxa* and *Hoxd* clusters exhibited a loss of posterior expression of *Shh* in the limb buds (Kmita et al., 2005), which is consistent with the result of the ChIP experiment. Members of two different groups of ETS transcription factors have opposing effects on *Shh* expression via direct binding to MFCS1 (Lettice et al., 2012). GABPα and ETS1 positively regulate *Shh* expression in the posterior mesenchyme, whereas Ets-variant 4 and 5 (Etv4/5) localize *Shh* expression posteriorly by suppressing its anterior expression. A balanced occupancy of the two groups of ETS factors on MFCS1 is needed to control the precise expression level of *Shh* (Fig. 3.2B). Recent ChIP-qPCR analysis illustrated that the Hand2 protein is enriched at the MFCS1 enhancer in a limb bud-specific manner. Overexpression of Hand2 was sufficient to induce luciferase activity in culture cells transfected with a reporter construct

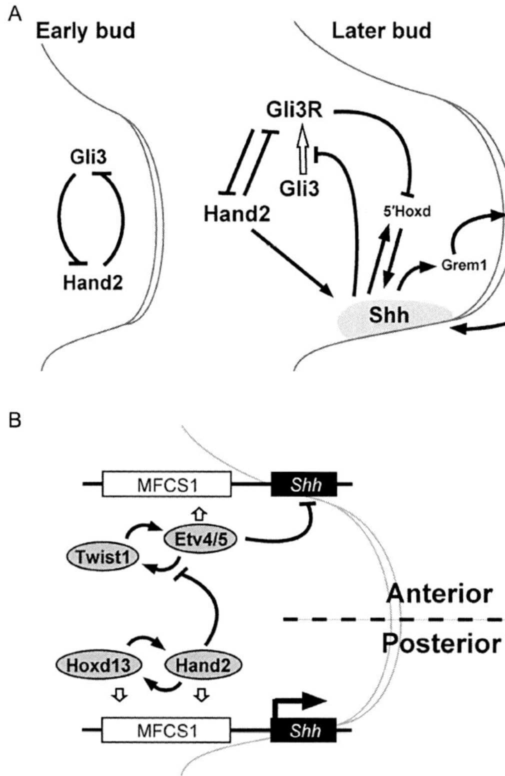

Figure 3.2 Gene regulatory network in the developing limb bud. (A) In the early limb bud, the A–P axis is formed initially by a mutual suppression of *Hand2* and *Gli3*. In the later limb bud, Hand2 and 5′Hoxd induce *Shh* in the zone of polarizing activity (ZPA). The Shh signal inhibits processing of Gli3, which produces a repressor form of Gli3 (Gli3R). The concentration of Gli3R is high in the anterior mesenchyme and is gradually lowered toward the ZPA. (B) Regional differences in regulatory mechanisms for *Shh* expression. Binding of Hand2 and Hoxd13 to MFCS1 (open arrows) induces *Shh* in the posterior limb bud, whereas binding of Etv4/5 to MFCS1 (open arrow) suppresses *Shh* in the anterior limb bud. Hand2 also plays a role in inhibiting the interaction between Twist1 and Etv5.

containing the luciferase gene under the control of MFCS1 (Galli et al., 2010). These results strongly support a hypothesis that Hand2 directly binds to the limb enhancer to upregulate *Shh*. Furthermore, Hand2 forms a protein complex with Hoxd13 in the mesenchymal cells of limb buds (Galli et al., 2010). Hoxd proteins are also essential as upstream factors for *Shh* expression; thus, two independent *trans* activators, Hand2 and Hoxd13, may act synergistically in initiation of *Shh* transcription (Fig. 3.2B). When

Hand2 is specifically depleted in the limb buds, the resulting limb shows loss of the autopod and forms a single zeugopodial bone, namely, a shrunken skeletal pattern along the A–P axis (Galli et al., 2010). The skeletal pattern is similar to that in the *Shh* mutant limb. This result is consistent with the idea that Hand2 directly regulates *Shh* expression, and suggests that Hand2 is a vital transcription factor to build the identity of the A–P axis.

During the initiation of limb budding, a Zn finger protein named Gli3 is expressed anteriorly, and its action restricts posterior expression of genes such as 5′Hoxd and Gremlin1 (Grem1). Many lines of evidence suggest a genetic interaction between anteriorly expressed Gli3 and posteriorly expressed Hand2 (Fig. 3.2A). A classical mouse mutant, *Extra-toes* (*Xt*), whose *Gli3* allele is disrupted, shows polydactylous limbs without the A–P identity. In the *Xt* limb buds, expression domains of *Hand2* and *5′Hoxd* are anteriorly expanded, and cover overall distal mesodermal cells that develop into the future autopod (Litingtung, Dahn, Li, Fallon, & Chiang, 2002; te Welscher et al., 2002). The loss of the A–P polarity in the *Xt* limb is most likely caused by the anterior expansion of posteriorly expressed genes. Conversely, targeted disruption of *Hand2* resulted in posterior expansion of anteriorly expressed genes, such as *Gli3* and *Alx4*, in the limb bud (Galli et al., 2010). Thus, Hand2 plays a key role in posteriorization to maintain *5′Hoxd* and *Grem1* by suppressing anterior factors (Fig. 3.2A). The limbs of the mouse *Gli3* mutant (*Xt*) had mirror-image duplication of the digits, but its proximal skeletal structures (stylopod and zeugopod) showed an asymmetrical pattern. Moreover, the skeletal pattern of the *Xt* limb was unaltered in the genetic background of the *Shh* KO homozygote, suggesting that polydactyly in the limb bud of the *Gli3* KO homozygote is Shh-signal independent (Litingtung et al., 2002; te Welscher et al., 2002). By contrast, the limb of the *Gli3/Hand2* double mutant mouse showed an almost symmetrical pattern in both the proximal skeletal structure and the distal digits, with a severely disrupted A–P axis (Galli et al., 2010). Taken together, these results suggested that the antagonistic interaction between *Gli3* and *Hand2* is required for the initiation of the A–P axis formation in the developing limb bud before *Shh* expression begins (Fig. 3.2A).

4. GENE DOSAGE EFFECT OF *HAND2* IN MOUSE EMBRYOGENESIS

Hand2 forms a heterodimer with Twist1 (Firulli et al., 2005). Heterozygotes of the *Twist1* KO mutant mouse exhibited an anterior ectopic expression of *Shh*, leading to excess digits in the anterior limb bud (Bourgeois et al., 1998; el Ghouzzi et al., 1997; Zhang et al., 2010). This suggests that *Twist1* has an inhibitory effect on the *Shh* expression in the normal limb bud. The opposing actions of Hand2 and Twist1 in A–P patterning are controlled in a dose dependent manner. Polydactyly in heterozygotes of the *Twist1* KO mouse is rescued by intercrossing with the *Hand2* KO mouse, which produces a lower level of Hand2 in the limb buds (Firulli et al., 2005). Thus, Twist1 is a key regulator to maintain an appropriate level of the Hand2 activity by antagonizing the function of Hand2 in the developing limb bud.

The dimer formation of Twist1 with Hand2 is affected by posttranslational phosphorylation at Thr125 and Ser127 in the Twist1 protein (Fig. 3.1B). Overexpression of Twist1 with T125A and S127A double mutations, which make Twist1 constitutively phosphorylated, resulted in polydactyly (Firulli, Redick, Conway, & Firulli, 2007). Overexpression of the phosphorylated form of Twist1 resulted in skeletal structures that were severely reduced in size, though digit numbers were normal. Therefore, in addition to the expression level of Hand2, affinity in the dimerization of Twist1 with Hand2 is another factor that controls limb development. Hand2 and Twist1 also have the potential to bind to Etv5, a negative regulator of *Shh* expression. Like the single *Twist1* mutant, the Etv4/5 double mutant showed polydactyly, which might have been caused by ectopic *Shh* expression in the anterior limb bud (Mao, McGlinn, Huang, Tabin, & McMahon, 2009; Zhang, Verheyden, Hassell, & Sun, 2009). This phenotype was rescued by reduced *Hand2* expression in the genetic background of the *Hand2* KO heterozygote; thus, Etv factors and Hand2 act in the same genetic pathway (Zhang et al., 2010). In the anterior limb bud, the Twist1–Etv5 complex may effectively suppress *Shh* expression, whereas in the posterior limb bud, Hand2 may interfere with dimerization of the Twist1–Etv5 complex, and thereby protect *Shh* expression from transcriptional inhibition by the Twist1–Etv5 complex (Zhang et al., 2010). Taken together, Hand2 has at least dual functions in the formation of the A–P axis. One is suppression of the anteriorizing function of Twist1 and Etv5 via protein–protein interaction. The other is induction of *Shh* expression by binding directly to the MFCS1 together with Hoxd13, which confers posterior identity on the mesenchymal cells (Fig. 3.2B and C).

Expression of the Hand2 protein in the heart is directly regulated by micro-RNAs. micro-RNAs are small non-coding RNA molecules of 21–22 bases in length, which are transcribed from vertebrate genomes. They recognize mRNAs in a sequence specific manner, and cause degradation or translation inhibition of target mRNAs (He & Hannon, 2004; Timmons, 2002). The *miR-1* subfamily consists of *miR-1-1* and *miR-1-2*, which encode the same mature micro-RNA, and are expressed specifically in cardiac and skeletal muscle precursor cells. *miR-1* regulates *Hand2* posttranscriptionally, and forced expression of *miR-1* leads to hypotrophy of the ventricular myocardium (Zhao, Samal, & Srivastava, 2005). Moreover, it was reported that targeted deletion of *miR-1-2* caused a fourfold increase in the Hand2 protein level without any change in the amount of *Hand2* mRNA, which resulted in a VSD phenotype (Zhao et al., 2007), indicating that overexpression of the Hand2 protein might cause the congenital heart defect. These results suggested that abundance of the Hand2 protein is genetically regulated and is essential for heart development. The posttranscriptional regulation of the Hand2 level is observed not only in the embryonic stage, but also after birth. It was recently reported that gain of calcineurin/Nfat signaling and reduction of *miR-25* expression combine to re-initiate production of the Hand2 protein in the mouse myocardium. The overexpression of *Hand2* causes hypertrophy of the myocardium in the adult mouse (Dirkx et al., 2013).

5. DISRUPTION OF *HAND2* DOSAGE CAUSES HUMAN DISEASES

The human *HAND2* gene is located on human chromosome 4, at 4q34.1 (UCSC Human Genome Browser; GRCh37/hg19). Duplication of the distal end of the long arm of human chromosome 4, which includes 4q34, leads to the congenital disorder, "Partial trisomy distal 4q" (denoted 4q+), whose typical phenotypes include growth retardation, psychomotor retardation, microcephaly, epicantic folds, high nasal bridge, short philtrum, micrognathia, low set/malformed ear, preaxial polydactyly, epilepsy, renal hypoplasia, and congenital heart disease (Battaglia et al., 2005; Lundin, Zech, Sjörs, Wadelius, & Annerén, 2002; Lurie, 2005; Rinaldi et al., 2003). Based on comparison of the phenotypes and the chromosome aberration segments, several chromosomal intervals have been suggested as candidate regions for some of the phenotypes. However, causative gene(s) of each phenotype have remained unclear. In general, chromosome aberrations in human diseases are complex. It is very difficult to identify the chromosomal region and causative gene(s) responsible for the phenotype, because chromosome aberration-related diseases are rare, and the genetic background varies across patient families. In some cases of 4q+, duplication and deletion of 4q (Rossi et al., 2009) or inter chromosomal duplication or deletion (Battaglia et al., 2005; Oka, Nakagome, Honda, & Arima, 1978; Rinaldi et al., 2003) occurred in the same patient. Furthermore, patients with the pure 4q duplication, who could provide the key to identifying the causative gene, are very rare (Celle et al., 2000; Goodman, Capone, Hennessey, & Thomas, 1997; Jeziorowska et al., 1993; Mikelsaar, Lurie, & Ilus, 1996; Shashi, Berry, Santos, & Pettenati, 1999; Vogel, Siebers, & Gunkel, 1975; Zollino, Zampino, Torrioli, Pomponi, & Neri, 1995).

In 2005, two patients were reported who provided meaningful information to identify causative genes (Otsuka et al., 2005). These patients are Japanese siblings, and have simple duplication of 4q31.22–q35.2 without any other chromosomal alteration, including translocation or deletion. They showed all of the typical phenotypes of partial trisomy distal 4q, but not congenital heart disease, suggesting that responsible gene(s) of typical 4q+ phenotypes are located on 4q31.22–q35.2. Before this report, we found a semi-dominant mouse mutant, which arose spontaneously and showed preaxial polydactyly (Masuya, Sagai, Wakana, Moriwaki, & Shiroishi, 1995) (Fig. 3.3A and B). This mutation named "Recombination induced mutation 4" (*Rim4*) was mapped to mouse chromosome 6 (Chr6). Homozygotes of *Rim4* show VSD (Fig. 3.3C and D) and tibial hemimelia associated with massive preaxial polydactyly, and the majority of *Rim4* homozygous

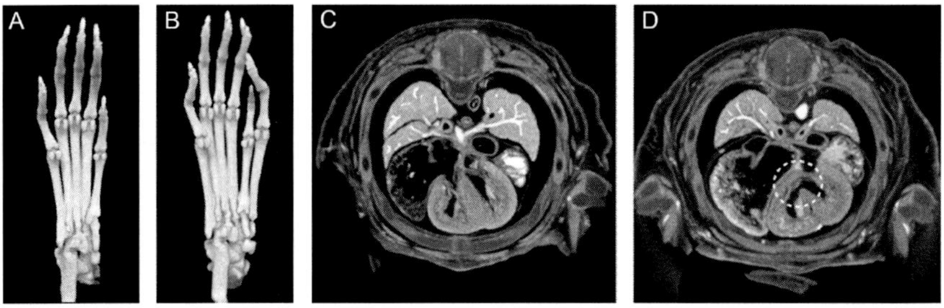

Figure 3.3 Limb and heart phenotypes of the mouse *Rim4* mutant, a model of human partial trisomy distal 4q. (A and B) The right leg of volume rendering micro-CT image of wild-type (A) and a *Rim4* heterozygote (B) at 3 months of age. *Rim4* shows preaxial polydactyly. (C and D) Heart defects in the mouse *Rim4* mutant. Enhanced contrast μ-CT cross heart images of E14.5 embryos of the wild-type (A) and the *Rim4* homozygote (B). A white dotted circle indicates a ventricular septal defect (VSD). The preaxial polydactyly and VSD phenotype are completely rescued by rebalancing the gene dosage of *Hand2* in a cross of the *Rim4* mutant with the *Hand2* KO mouse.

newborns die shortly after birth in the genetic background of inbred strain C57BL/10J. Detailed genetic analysis revealed that Chr6 of *Rim4* has an insertion of a 6.5 Mb genomic fragment derived from chromosome 8 (Chr8), which is syntenic to the distal end of the long arm of human Chr4, 4q32.3–4q34.1 (Tamura et al., 2013). Notably, 4q trisomic region of the patients reported by Otsuka et al. completely includes the region that is syntenic with the mouse genome and is duplicated in *Rim4*. These patients and *Rim4* shared typical phenotypes, including preaxial polydactyly and craniofacial anomaly, suggesting that the gene responsible for the common phenotype between 4q+ patients and *Rim4* is located on the duplicated region of *Rim4* (Table 3.3). We determined the breakpoints of the translocation between Chr6 and the 6.5 Mb insertion from Chr8, and found that the inserted Chr8 fragment contains 17 genes, including *Hand2*.

This finding prompted us to test whether *Hand2* is a causative gene of the *Rim4* phenotype, including the preaxial polydactyly and the craniofacial anomaly. The most plausible possibility was that increased dosage of *Hand2* leading to overexpression of the Hand2 protein is causally related to the developmental defects of *Rim4*. Therefore, we tried to rebalance the gene dosage of *Hand2* in the *Rim4* mutant by a genetic cross of this mutant with the conventional *Hand2* KO mice. Note that, except for *Hand2*, all genes

Table 3.3 Comparison of phenotypes between 4q+ patients reported by Otsuka et al. (2005) and *Rim4*

Phenotype	Patients	*Rim4*
Growth retardation	+	*Rim4/+*
Mental retardation	+	ND
Craniofacial anomaly	+	*Rim4/+*
Epilepsy	+	ND
Preaxial polydactyly	+	*Rim4/+*
Clavicular anomaly	+	*Rim4/Rim4*
Scoliosis	+	*Rim4/+*
Lumbar vertebra defects	+	*Rim4/Rim4*
Renal defects	+	–
Congenital heart defects	–	*Rim4/Rim4*

ND: not determined; *Rim4/+*, observed in *Rim4* heterozygote; *Rim4/Rim4*, observed in *Rim4* homozygote.

located in the duplicated Chr8 fragment in *Rim4* should retain their increased dosage in the resultant compound heterozygotes (*Rim4/+; Hand2+/−*) and double homozygous mutants (*Rim4/Rim4; Hand2−/−*). The result of this mating experiment clearly showed that the double mutant of *Rim4* and *Hand2* KO did not exhibit any first digit deformities, VSD and perinatal death, indicating that increased expression of Hand2 is responsible for these defects. This was further supported by a mouse transgenesis experiment. Increased *Hand2* gene dosage by transgenesis of the BAC-DNA containing *Hand2* recaptured the *Rim4* phenotypes, such as first digit deformities, hemimelia, and decreased number of lumbar vertebrae. After this finding, a decreased number of lumbar vertebrae were observed in a 4q+ patient reported by Otsuka et al., indicating that Hand2 is involved in lumbar vertebrae development (Tamura et al., 2013).

In humans, deletion of the distal end of the long arm of chromosome 4 (4q), which contains the *HAND2* locus, leads to development of the congenital disorder named "4q syndrome" (denoted 4q−). This syndrome shows Robin sequence, growth retardation, mental retardation, congenital heart disease (i.e., VSD and atrial septal defects), digital anomaly (i.e., clinodactyly and oligodactyly), and characteristic craniofacial dismorphology, including a cleft palate. By comparing the chromosomal deletion segments of various patient families and their corresponding phenotypes, a chromosomal region of 4q31–q34 has been reported to be responsible for most of the 4q− phenotypes, including heart and limb defects (Robertson, O'Day, & Bankier, 1998). In addition, by comparing abnormalities in digit morphology and the chromosome aberration segments, 4q32–35 appeared to be the candidate region that harbors a causative gene of split hand/foot malformation (Niedrist, Lurie, & Schinzel, 2009). However, direct proof that a reduced level of Hand2 causes the complex phenotype of 4q− requires further studies.

As mentioned above, the *Nebl* gene is the downstream target of Hand2 during cardiovascular development in mice. Human *NEBL* has been mapped to human chromosome 10, at 10p13–12, by FISH analysis. A congenital disorder named "DiGeorge syndrome" (DGS) or "Velocardiofacial syndrome" (VCFS), which is associated with cardiac and craniofacial abnormalities, also maps to this chromosomal region. It is inferred that haploinsufficiency of a gene(s) residing in a segment of 10p13–12 could cause the DGS/VCFS phenotypes (Daw et al., 1996). A genomic fragment including the *NEBL* gene is deleted heterozygously in cell lines derived from two female DGS patients with the proximal deletion of chromosome 10p14–p13 (Villanueva et al., 2002). One patient showed a cardiac defect, cleft palate, craniofacial anomaly, and developmental retardation. The other patient showed microcephaly, microphthalmia, and hypotelorism. Therefore, it is possible that lower levels of HAND2 are responsible for the DGS-related phenotype via reduced expression level of the *NEBL* gene.

To clarify the correlation between the *HAND2* genotype and congenital heart disease phenotypes, mutations of *HAND2* were explored in sporadic Chinese patients who were diagnosed with congenital heart disease (Shen et al., 2010). In that study, three missense mutations (P11R, S36N, and V83L), one synonymous mutation (H14H) and three mutations in the untranslated region (241A > G, 604C > T, and 3237T > A) were identified in 12 out of 131 patients. These results suggested that *HAND2* might be a candidate gene of congenital defect of the right ventricle and outflow tract (Shen et al., 2010).

6. FUTURE PERSPECTIVES

It is intriguing that only expression level changes of Hand2, but not its ectopic expression, elicits several developmental defects, including preaxial polydactyly in mice and humans. Accumulating data have provided evidence that balanced expression of *Hand2* and *Twist1* is essential for proper limb development, and disruption of the expression balance between these two genes causes developmental defects.

Human 4q syndrome (4q −), which might be caused by lower levels of Hand2, shows several characteristic developmental defects. By contrast, heterozygotes of *Hand2* KO mice exhibit no phenotype. Similar species difference in the *Hand2* dosage effect was also observed in 4q+ patients. The 4q+ patients reported by Otsuka et al. did not show congenital heart defects such as VSD, but mouse *Rim4* mutants show this phenotype. At present, we cannot answer the question of what is responsible for this species difference. It is, however, notable that sensitivity of the gene expression level might be dependent on the genetic background, even in the same species. In fact, overexpression of Hand2 in *Rim4* heterozygotes resulted in preaxial polydactyly in the C57BL/10 genetic background, but no digit phenotype is observed in a genetic background of wild-derived mouse strain MSM/Ms (Masuya et al., personal communication). Compound heterozygotes of the *Hand2* floxed and *Hand2* KO mutant mice, which have lower Hand2 levels than *Hand2* single KO heterozygotes, shows cleft palate, as is often observed in patients with 4q− (Morikawa et al., 2007). Thus, it is will be interesting to determine whether sensitivity to the level of Hand2 in human patients also reflects differences in their genetic backgrounds.

Though genetic analysis of the mouse *Rim4* mutant has opened up the study of the effect the Hand2 level in human diseases, factors interacting with Hand2 in heart development, craniofacial development, and bone formation are still largely unknown. Thus, molecular mechanisms underlying the effect of the expression level of Hand2 in these developmental processes are main issues to be addressed in future study. Further analysis is necessary to clarify whether reduced Hand2 expression is responsible for the 4q− syndrome and DGS/Velocardiofacial syndrome. We sincerely hope that accumulating knowledge of the developmental functions of Hand2 will lead to the development of therapeutic approaches, drug discovery, and improvement in quality of life of the patients with these syndromes.

ACKNOWLEDGMENTS

This work was supported by the Japan Society for the Promotion of Science (12013551 to M. T.), and by the Ministry of Education, Culture, Sports, Science and Technology in Japan (17018033 to T. S.).

REFERENCES

Abe, M., Michikami, I., Fukushi, T., Abe, A., Maeda, Y., Ooshima, T., et al. (2010). Hand2 regulates chondrogenesis in vitro and in vivo. *Bone, 46*, 1359–1368.

Barron, F., Woods, C., Kuhn, K., Bishop, J., Howard, M. J., & Clouthier, D. E. (2011). Downregulation of Dlx5 and Dlx6 expression by Hand2 is essential for initiation of tongue morphogenesis. *Development, 138*, 2249–2259.

Battaglia, A., Chen, Z., Brothman, A. R., Morelli, S., Palumbos, J. C., Carey, J. C., et al. (2005). Karyotype/phenotype correlations in duplication 4q: Evidence for a critical region within 4q27–28 for preaxial defects. *American Journal of Medical Genetics. Part A, 134A*, 334–337.

Bourgeois, P., Bolcato-Bellemin, A. L., Danse, J. M., Bloch-Zupan, A., Yoshiba, K., Stoetzel, C., et al. (1998). The variable expressivity and incomplete penetrance of the twist-null heterozygous mouse phenotype resemble those of human Saethre–Chotzen syndrome. *Human Molecular Genetics*, 7, 945–957.

Capellini, T. D., Di Giacomo, G., Salsi, V., Brendolan, A., Ferretti, E., Srivastava, D., et al. (2006). Pbx1/Pbx2 requirement for distal limb patterning is mediated by the hierarchical control of Hox gene spatial distribution and Shh expression. *Development, 133*, 2263–2273.

Celle, L., Lee, L., Rintoul, N., Savani, R. C., Long, W., Mennuti, M. T., et al. (2000). Duplication of chromosome region 4q28.3–qter in monozygotic twins with discordant phenotypes. *American Journal of Medical Genetics, 94*, 125–140.

Charité, J., McFadden, D. G., Merlo, G., Levi, G., Clouthier, D. E., Yanagisawa, M., et al. (2001). Role of Dlx6 in regulation of an endothelin-1-dependent, dHAND branchial arch enhancer. *Genes & Development, 15*, 3039–3049.

Charité, J., McFadden, D. G., & Olson, E. N. (2000). The bHLH transcription factor dHAND controls Sonic hedgehog expression and establishment of the zone of polarizing activity during limb development. *Development, 127*, 2461–2470.

Cross, J. C., Flannery, M. L., Blanar, M. A., Steingrimsson, E., Jenkins, N. A., Copeland, N. G., et al. (1995). Hxt encodes a basic helix–loop–helix transcription factor that regulates trophoblast cell development. *Development, 121*, 2513–2523.

Cserjesi, P., Brown, D., Lyons, G. E., & Olson, E. N. (1995). Expression of the novel basic helix–loop–helix gene eHAND in neural crest derivatives and extraembryonic membranes during mouse development. *Developmental Biology, 170*, 664–678.

Dai, Y. S., Cserjesi, P., Markham, B. E., & Molkentin, J. D. (2002). The transcription factors GATA4 and dHAND physically interact to synergistically activate cardiac gene expression through a p300-dependent mechanism. *The Journal of Biological Chemistry, 277*, 24390–24398.

Daw, S. C. M., Taylor, C., Kraman, M., Call, K., Mao, J., Schuffenhauer, S., et al. (1996). A common region of 10p deleted in DiGeorge and velocardiofacial syndromes. *Nature Genetics, 13*, 458–461.

Dirkx, E., Gladka, M. M., Philippen, L. E., Armand, A. S., Kinet, V., Leptidis, S., et al. (2013). Nfat and miR-25 cooperate to reactivate the transcription factor Hand2 in heart failure. *Nature Cell Biology, 15*, 1282–1293.

Ducy, P., Zhang, R., Geoffroy, V., Ridall, A. L., & Karsenty, G. (1997). Osf2/Cbfa1: A transcriptional activator of osteoblast differentiation. *Cell, 89*, 747–754.

el Ghouzzi, V., Le Merrer, M., Perrin-Schmitt, F., Lajeunie, E., Benit, P., Renier, D., et al. (1997). Mutations of the TWIST gene in the Saethre–Chotzen syndrome. *Nature Genetics, 15*, 42–46.

Fernandez-Teran, M., Piedra, M. E., Rodriguez-Rey, J. C., Talamillo, A., & Ros, M. A. (2003). Expression and regulation of eHAND during limb development. *Developmental Dynamics, 226*, 690–701.

Firulli, A. B. (2003). A HANDful of questions: The molecular biology of the heart and neural crest derivatives (HAND)-subclass of basic helix–loop–helix transcription factors. *Gene, 312*, 27–40.

Firulli, B. A., Krawchuk, D., Centonze, V. E., Vargesson, N., Virshup, D. M., Conway, S. J., et al. (2005). Altered Twist1 and Hand2 dimerization is associated with Saethre–Chotzen syndrome and limb abnormalities. *Nature Genetics, 37*, 373–381.

Firulli, A. B., McFadden, D. G., Lin, Q., Srivastava, D., & Olson, E. N. (1998). Heart and extra-embryonic mesodermal defects in mouse embryos lacking the bHLH transcription factor Hand1. *Nature Genetics, 18*, 266–270.

Firulli, B. A., Redick, B. A., Conway, S. J., & Firulli, A. B. (2007). Mutations within helix I of Twist1 result in distinct limb defects and variation of DNA binding affinities. *The Journal of Biological Chemistry, 282*, 27536–27546.

Funato, N., Chapman, S. L., McKee, M. D., Funato, H., Morris, J. A., Shelton, J. M., et al. (2009). Hand2 controls osteoblast differentiation in the branchial arch by inhibiting DNA binding of Runx2. *Development, 136*, 615–625.

Galli, A., Robay, D., Osterwalder, M., Bao, X., Bénazet, J. D., Tariq, M., et al. (2010). Distinct roles of Hand2 in initiating polarity and posterior *Shh* expression during the onset of mouse limb bud development. *PLoS Genetics, 6*(4), e1000901.

Goodman, B. K., Capone, G. T., Hennessey, J., & Thomas, G. H. (1997). Familial tandem duplication of bands q31.1 to q32.3 on chromosome 4 with mild phenotypic effect. *American Journal of Medical Genetics, 73*, 119–124.

Gripp, K. W., Zackai, E. H., & Stolle, C. A. (2000). Mutations in the human *TWIST* gene. *Human Mutation, 15*, 150–155.

Harfe, B. D., Scherz, P. J., Nissim, S., Tian, H., McMahon, A. P., & Tabin, C. J. (2004). Evidence for an expansion-based temporal Shh gradient in specifying vertebrate digit identities. *Cell, 118*, 517–528.

He, L., & Hannon, G. J. (2004). MicroRNAs: Small RNAs with a big role in gene regulation. *Nature Reviews. Genetics, 5*, 522–531.

Hendershot, T. J., Liu, H., Sarkar, A. A., Giovannucci, D. R., Clouthier, D. E., Abe, M., et al. (2007). Expression of Hand2 is sufficient for neurogenesis and cell type-specific gene expression in the enteric nervous system. *Developmental Dynamics, 236*, 93–105.

Hollenberg, S. M., Sternglanz, R., Cheng, P. F., & Weintraub, H. (1995). Identification of a new family of tissue-specific basic helix–loop–helix proteins with a two-hybrid system. *Molecular and Cell Biology, 15*, 3813–3822.

Howard, T. D., Paznekas, W. A., Green, E. D., Chiang, L. C., Ma, N., Ortiz De Luna, R. I., et al. (1997). Mutations in TWIST, a basic helix–loop–helix transcription factor, in Saethre–Chotzen syndrome. *Nature Genetics, 15*, 36–41.

Jabs, E. W. (2001). A TWIST in the fate of human osteoblasts identifies signaling molecules involved in skull development. *The Journal of Clinical Investigation, 107*, 1075–1077.

Jeziorowska, A., Ciesla, W., Houck, G. E., Jr., Yao, X. L., Harris, M. S., Truszczak, B., et al. (1993). Cytogenetic and molecular identification of a de novo direct duplication of the long arm of chromosome 4(q21.3–q31.3). *American Journal of Medical Genetics, 46*, 83–87.

Johnson, D., Horsley, S. W., Moloney, D. M., Oldridge, M., Twigg, S. R., Walsh, S., et al. (1998). A comprehensive screen for TWIST mutations in patients with craniosynostosis identifies a new microdeletion syndrome of chromosome band 7p21.1. *American Journal of Human Genetics, 63*, 1282–1293.

Kmita, M., Tarchini, B., Zakany, J., Logan, M., Tabin, C. J., & Duboule, D. (2005). Early developmental arrest of mammalian limbs lacking HoxA/HoxD gene function. *Nature, 435*, 1113–1116.

Knöfler, M., Meinhardt, G., Vasicek, R., Husslein, P., & Egarter, C. (1998). Molecular cloning of the human Hand1 gene/cDNA and its tissue-restricted expression in cytotrophoblastic cells and heart. *Gene, 224*, 77–86.

Komori, T., Yagi, H., Nomura, S., Yamaguchi, A., Sasaki, K., Deguchi, K., et al. (1997). Targeted disruption of Cbfa1 results in a complete lack of bone formation owing to maturational arrest of osteoblasts. *Cell, 89*, 755–764.

Leptin, M. (1991). twist and snail as positive and negative regulators during Drosophila mesoderm development. *Genes & Development, 5*, 1568–1576.

Lettice, L. A., Heaney, S. J., Purdie, L. A., Li, L., de Beer, P., Oostra, B. A., et al. (2003). A long-range Shh enhancer regulates expression in the developing limb and fin and is associated with preaxial polydactyly. *Human Molecular Genetics, 12*, 1725–1735.

Lettice, L. A., Williamson, I., Wiltshire, J. H., Peluso, S., Devenney, P. S., Hill, A. E., et al. (2012). Opposing functions of the ETS factor family define Shh spatial expression in limb buds and underlie polydactyly. *Developmental Cell, 22*, 459–467.

Litingtung, Y., Dahn, R. D., Li, Y., Fallon, J. F., & Chiang, C. (2002). Shh and Gli3 are dispensable for limb skeleton formation but regulate digit number and identity. *Nature, 418*, 979–983.

Liu, N., Barbosa, A. C., Chapman, S. L., Bezprozvannaya, S., Qi, X., Richardson, J. A., et al. (2009). DNA binding-dependent and -independent functions of the Hand2 transcription factor during mouse embryogenesis. *Development, 136*, 933–942.

Lundin, C., Zech, L., Sjörs, K., Wadelius, C., & Annerén, G. (2002). Trisomy 4q syndrome: Presentation of a new case and review of the literature. *Annales de Génétique, 45*, 53–57.

Lurie, I. W. (2005). Partial trisomy 4q and preaxial limb defects. *American Journal of Medical Genetics. Part A, 138A*, 304–305.

Mao, J., McGlinn, E., Huang, P., Tabin, C. J., & McMahon, A. P. (2009). Fgf-dependent Etv4/5 activity is required for posterior restriction of Sonic Hedgehog and promoting outgrowth of the vertebrate limb. *Developmental Cell, 16*, 600–606.

Massari, M. E., & Murre, C. (2000). Helix–loop–helix proteins: Regulators of transcription in eucaryotic organisms. *Molecular and Cell Biology, 20*, 429–440.

Masuya, H., Sagai, T., Wakana, S., Moriwaki, K., & Shiroishi, T. (1995). A duplicated zone of polarizing activity in polydactylous mouse mutants. *Genes & Development, 9*, 1645–1653.

McMahon, A. P., Ingham, P. W., & Tabin, C. J. (2003). Developmental roles and clinical significance of hedgehog signaling. *Current Topics in Developmental Biology, 53*, 1–114.

Mikelsaar, R. V., Lurie, I. W., & Ilus, T. E. (1996). "Pure" partial trisomy 4q25–qter owing to a de novo 4;22 translocation. *Journal of Medical Genetics, 33*, 344–345.

Morikawa, Y., & Cserjesi, P. (2008). Cardiac neural crest expression of Hand2 regulates outflow and second heart field development. *Circulation Research, 103*, 1422–1429.

Morikawa, Y., D'Autréaux, F., Gershon, M. D., & Cserjesi, P. (2007). Hand2 determines the noradrenergic phenotype in the mouse sympathetic nervous system. *Developmental Biology, 307*, 114–126.

Murre, C., McCaw, P. S., Vaessin, H., Caudy, M., Jan, L. Y., Jan, Y. N., et al. (1989). Interactions between heterologous helix–loop–helix proteins generate complexes that bind specifically to a common DNA sequence. *Cell, 58*, 537–544.

Niedrist, D., Lurie, I. W., & Schinzel, A. (2009). 4q32–q35 and 6q16–q22 are valuable candidate regions for split hand/foot malformation. *European Journal of Human Genetics, 17*, 1086–1091.

Oka, S., Nakagome, Y., Honda, T., & Arima, M. (1978). A case of distal 4q trisomy due to familial (4;5)(q31;p15) translocation. *The Japanese Journal of Human Genetics, 23*, 167–172.

Otsuka, T., Fujinaka, H., Imamura, M., Tanaka, Y., Hayakawa, H., & Tomizawa, S. (2005). Duplication of chromosome 4q: Renal pathology of two siblings. *American Journal of Medical Genetics. Part A, 134A*, 330–333.

Otto, F., Thornell, A. P., Crompton, T., Denzel, A., Gilmour, K. C., Rosewell, I. R., et al. (1997). Cbfa1, a candidate gene for cleidocranial dysplasia syndrome, is essential for osteoblast differentiation and bone development. *Cell, 89*, 765–771.

Riddle, R. D., Johnson, R. L., Laufer, E., & Tabin, C. (1993). Sonic hedgehog mediates the polarizing activity of the ZPA. *Cell, 75*, 1401–1416.

Riley, P., Anson-Cartwright, L., & Cross, J. C. (1998). The Hand1 bHLH transcription factor is essential for placentation and cardiac morphogenesis. *Nature Genetics, 18*, 271–275.

Rinaldi, R., De Bernardo, C., Assumma, M., Grammatico, B., Buffone, E., Poscente, M., et al. (2003). Cytogenetic and molecular characterization of a de novo 4q24qter duplication and correlation to the associated phenotype. *American Journal of Medical Genetics. Part A, 118A*, 122–126.

Robertson, S. P., O'Day, K., & Bankier, A. (1998). The 4q-syndrome: Delineation of the minimal critical region to within band 4q31. *Clinical Genetics, 53*, 70–73.

Rossi, M. R., DiMaio, M. S., Xiang, B., Lu, K., Kaymakcalan, H., Seashore, M., et al. (2009). Clinical and genomic characterization of distal duplications and deletions of chromosome 4q: Study of two cases and review of the literature. *American Journal of Medical Genetics, 149A*, 2788–2794.

Russell, M. W., Kemp, P., Wang, L., Brody, L. C., & Izumo, S. (1998). Molecular cloning of the human HAND2 gene. *Biochimica et Biophysica Acta, 1443*, 393–399.

Sagai, T., Masuya, H., Tamura, M., Shimizu, K., Yada, Y., Wakana, S., et al. (2004). Phylogenetic conservation of a limb-specific, cis-acting regulator of Sonic hedgehog (*Shh*). *Mammalian Genome, 15*, 23–34.

Sharma, V. P., Fenwick, A. L., Brockop, M. S., McGowan, S. J., Goos, J. A., Hoogeboom, A. J., et al. (2013). Mutations in TCF12, encoding a basic helix–loop–helix partner of TWIST1, are a frequent cause of coronal craniosynostosis. *Nature Genetics*, *45*, 304–307.

Shashi, V., Berry, M. N., Santos, C., & Pettenati, M. J. (1999). Partial duplication of 4q12q13 leads to a mild phenotype. *American Journal of Medical Genetics*, *86*, 51–53.

Shen, L., Li, X. F., Shen, A. D., Wang, Q., Liu, C. X., Guo, Y. J., et al. (2010). Transcription factor HAND2 mutations in sporadic Chinese patients with congenital heart disease. *Chinese Medical Journal*, *123*, 1623–1627.

Song, K., Nam, Y. J., Luo, X., Qi, X., Tan, W., Huang, G. N., et al. (2012). Heart repair by reprogramming non-myocytes with cardiac transcription factors. *Nature*, *485*, 599–604.

Srivastava, D., Cserjesi, P., & Olson, E. N. (1995). A subclass of bHLH proteins required for cardiac morphogenesis. *Science*, *270*, 1995–1999.

Srivastava, D., Thomas, T., Lin, Q., Kirby, M. L., Brown, D., & Olson, E. N. (1997). Regulation of cardiac mesodermal and neural crest development by the bHLH transcription factor, dHAND. *Nature Genetics*, *16*, 154–160.

Tamura, M., Hosoya, M., Fujita, M., Iida, T., Amano, T., Maeno, A., et al. (2013). Overdosage of Hand2 causes limb and heart defects in the human chromosomal disorder partial trisomy distal 4q. *Human Molecular Genetics*, *22*, 2471–2481.

te Welscher, P., Zuniga, A., Kuijper, S., Drenth, T., Goedemans, H. J., Meijlink, F., et al. (2002). Progression of vertebrate limb development through SHH-mediated counteraction of GLI3. *Science*, *298*, 827–830.

Thattaliyath, B. D., Firulli, B. A., & Firulli, A. B. (2002). The basic-helix–loop–helix transcription factor HAND2 directly regulates transcription of the atrial naturetic peptide gene. *Journal of Molecular and Cellular Cardiology*, *34*, 1335–1344.

Thattaliyath, B. D., Livi, C. B., Steinhelper, M. E., Toney, G. M., & Firulli, A. B. (2002). HAND1 and HAND2 are expressed in the adult-rodent heart and are modulated during cardiac hypertrophy. *Biochemical and Biophysical Research Communications*, *297*, 870–875.

Thomas, T., Yamagishi, H., Overbeek, P. A., Olson, E. N., & Srivastava, D. (1998). The bHLH factors, dHAND and eHAND, specify pulmonary and systemic cardiac ventricles independent of left–right sidedness. *Developmental Biology*, *196*, 228–236.

Tickle, C. (1981). The number of polarizing region cells required to specify additional digits in the developing chick wing. *Nature*, *289*, 295–298.

Timmons, L. (2002). The long and short of siRNAs. *Molecular Cell*, *10*, 435–437.

Tsuchihashi, T., Maeda, J., Shin, C. H., Ivey, K. N., Black, B. L., Olson, E. N., et al. (2011). Hand2 function in second heart field progenitors is essential for cardiogenesis. *Developmental Biology*, *351*, 62–69.

Verzi, M. P., Agarwal, P., Brown, C., McCulley, D. J., Schwarz, J. J., & Black, B. L. (2007). The transcription factor MEF2C is required for craniofacial development. *Developmental Cell*, *12*, 645–652.

Villanueva, M. P., Aiyer, A. R., Muller, S., Pletcher, M. T., Liu, X., Emanuel, B., et al. (2002). Genetic and comparative mapping of genes dysregulated in mouse hearts lacking the Hand2 transcription factor gene. *Genomics*, *80*, 593–600.

Vogel, W., Siebers, J. W., & Gunkel, J. (1975). Phenotypic variation in partial trisomy 4q. *Humangenetik*, *28*, 103–112.

Wilson-Rawls, J., Rhee, J. M., & Rawls, A. (2004). Paraxis is a basic helix–loop–helix protein that positively regulates transcription through binding to specific E-box elements. *The Journal of Biological Chemistry*, *279*, 37685–37692.

Xiong, W., He, F., Morikawa, Y., Yu, X., Zhang, Z., Lan, Y., et al. (2009). Hand2 is required in the epithelium for palatogenesis in mice. *Developmental Biology*, *330*, 131–141.

Yamagishi, H., Olson, E. N., & Srivastava, D. (2000). The basic helix–loop–helix transcription factor, dHAND, is required for vascular development. *The Journal of Clinical Investigation*, *105*, 261–270.

Yanagisawa, H., Clouthier, D. E., Richardson, J. A., Charité, J., & Olson, E. N. (2003). Targeted deletion of a branchial arch-specific enhancer reveals a role of dHAND in craniofacial development. *Development, 130*, 1069–1078.

Zeisberg, E. M., Ma, Q., Juraszek, A. L., Moses, K., Schwartz, R. J., Izumo, S., et al. (2005). Morphogenesis of the right ventricle requires myocardial expression of Gata4. *The Journal of Clinical Investigation, 115*, 1522–1531.

Zeller, R., Lopez-Rios, J., & Zuniga, A. (2009). Vertebrate limb bud development: Moving towards integrative analysis of organogenesis. *Nature Reviews. Genetics, 10*, 845–858.

Zhang, Z., Sui, P., Dong, A., Hassell, J., Cserjesi, P., Chen, Y. T., et al. (2010). Preaxial polydactyly: Interactions among ETV, TWIST1 and HAND2 control anterior–posterior patterning of the limb. *Development, 137*, 3417–3426.

Zhang, Z., Verheyden, J. M., Hassell, J. A., & Sun, X. (2009). FGF-regulated Etv genes are essential for repressing Shh expression in mouse limb buds. *Developmental Cell, 16*, 607–613.

Zhao, Y., Ransom, J. F., Li, A., Vedantham, V., von Drehle, M., Muth, A. N., et al. (2007). Dysregulation of cardiogenesis, cardiac conduction, and cell cycle in mice lacking miRNA-1-2. *Cell, 129*, 303–317.

Zhao, Y., Samal, E., & Srivastava, D. (2005). Serum response factor regulates a muscle-specific microRNA that targets Hand2 during cardiogenesis. *Nature, 436*, 214–220.

Zollino, M., Zampino, G., Torrioli, G., Pomponi, M. G., & Neri, G. (1995). Further contribution to the description of phenotypes associated with partial 4q duplication. *American Journal of Medical Genetics, 57*, 69–73.

CHAPTER FOUR

E Proteins in Lymphocyte Development and Lymphoid Diseases

Ian Belle[1], Yuan Zhuang
Department of Immunology, Duke University Medical Center, Durham North Carolina, USA
[1]Corresponding author: e-mail address: ilb2@duke.edu

Contents

Abstract

As members of the basic helix-loop-helix (bHLH) family of transcription factors, E proteins function in the immune system by directing and maintaining a vast transcriptional network that regulates cell survival, proliferation, differentiation, and function. Proper activity of this network is essential to the functionality of the immune system. Aberrations in E protein expression or function can cause numerous defects, ranging from impaired lymphocyte development and immunodeficiency to aberrant function, cancer, and autoimmunity. Additionally, disruption of inhibitor of DNA-binding (Id) proteins, natural inhibitors of E proteins, can induce additional defects in development and function. Although E proteins have been investigated for several decades, their study continues to yield novel and exciting insights into the workings of the immune system. The goal of this chapter is to discuss the various classical roles of E proteins in lymphocyte development and highlight new and ongoing research into how these roles, if compromised, can lead to disease.

Current Topics in Developmental Biology, Volume 110
ISSN 0070-2153
http://dx.doi.org/10.1016/B978-0-12-405943-6.00004-X
© 2014 Elsevier Inc.
All rights reserved.

1. INTRODUCTION

The immune system maintains the health of the host by identifying, responding to, and subsequently eliminating harmful pathogens. These processes involve a multitude of cell types, which are loosely separated into two main branches of the immune system, the innate immune system and the adaptive immune system. A great deal of research has gone into elucidating the mechanisms and pathways involved in the development of the immune system. Of particular interest have been the developmental pathways of B and T cell development. As members of the adaptive immune system, B and T cells undergo an exceptionally complicated developmental process, including the acquisition of a diverse repertoire of antigen receptor specificities capable of recognizing virtually any antigen, previously encountered by the host or otherwise. Upon recognizing cognate antigen, B and T cells further adapt and evolve to better counter an identified threat. Following elimination of a pathogen, cells of the adaptive immune system form a pool of memory cells, capable of responding to a new challenge by the same pathogen with even greater rapidity and efficiency.

These processes supply the host with an effective, adaptive defense; however, the complex developmental and regulatory pathways that control the adaptive immune system can also be harmful if they are disrupted by genetic mutations. Production of B or T cells capable of responding to host proteins can initiate a destructive autoimmune response against critical tissues and organ systems in the body. Additionally, the high expression of particular lymphocyte-specific genes poses a potential problem as well. Translocation of various oncogenes to the transcriptional control of lymphocyte-specific regulatory elements, notably those of the antigen receptor genes, is a frequent event in tumorigenesis and is very common in leukemias and lymphomas. As such, it is critical that mechanisms exist to ensure that the immune system is kept in balance. These mechanisms have been and continue to be the subject of intense research.

One of the major regulatory mechanisms in directing lymphocyte development and function that has been frequently implicated in disease processes is the E protein transcriptional network. E proteins are members of the larger basic helix-loop-helix (bHLH) family and are widely expressed within the immune system. These proteins have been demonstrated to play critical roles at nearly every step of B and T cell development and function, from acquisition of a functional antigen receptor to cell survival and proliferation to maintaining proper functionality during an immune response. This chapter will focus primarily on the roles of E proteins in the development of B and T cells, their function within the immune system, and how these roles, when compromised, lead to severe consequences for the host.

2. E PROTEINS

E proteins are a family of transcription factors comprising a subgroup of the much larger basic bHLH family (Ephrussi, Church, Tonegawa, & Gilbert, 1985). The bHLH protein family comprises a group of widely expressed transcription factors involved in the development and maintenance of numerous cell types. bHLH proteins have been categorized into several classes. Most notable are the Class I bHLH proteins, which are widely expressed within the immune system and on which the majority of this chapter will be focused (Henthorn, Kiledjian, & Kadesch, 1990). These proteins recognize a canonical CANNTG DNA sequence, termed an E box. As such, Class I bHLH proteins are referred to as E proteins.

The E protein family is defined by the presence of several main protein domains: a C-terminal basic DNA-binding domain (the b in bHLH) and a helix-loop-helix domain (the HLH) comprising a pair of closely spaced alpha helices (Murre, McCaw, & Baltimore, 1989). These HLH domains facilitate the dimerization of bHLH proteins, an event that is required for their transcriptional activity (Murre & Baltimore, 1993). The bHLH domain has also been shown to interact with p300, a major component of the cell's ubiquitous transcriptional machinery (Eckner, Yao, Oldread, & Livingston, 1996). bHLH proteins also contain two transcriptional activation domains, AD1 and AD2 (Aronheim, Shiran, Rosen, & Walker, 1993). AD2 is located within the central portion of the protein and is capable of driving expression of reporter constructs containing bHLH-regulated genes. AD1 is located at the N-terminus and has been shown to recruit the SAGA chromatin-remodeling complex (Massari et al., 1999).

Class I bHLH proteins (E proteins) include the E2A (also referred to as TCF-3), HEB (also referred to as TCF-12), and E2-2 (also referred to as TCF-4). The *E2A* and *HEB* genes encode several proteins by way of alternative splicing. The *E2A* gene encodes the proteins E12 and E47, while the *HEB* gene encodes the canonical HEB protein (HEBcan) as well as a shorter alternative variant (HEBalt) (Wang et al., 2006). While E47 is capable of readily binding DNA as a homodimer, E12 contains a different basic region, allowing it to only function efficiently as a heterodimer with other bHLH proteins (Sun & Baltimore, 1991). Mutations within these dimerization and DNA-binding regions can be disastrous, as shown by the recent discovery of a dominant-negative mutation in the E47 protein, which led to agammaglobulinemia and severe immunodeficiency (Boisson et al., 2013). Within developing B cells, dimers of *E2A* gene products are the predominant E protein transcriptional regulators, whereas developing T cells utilize primarily heterodimers of *E2A* and *HEB* gene products (Sawada & Littman, 1993). E protein dimers function by regulating a large array of genes. Dimerization between particular E proteins subtly alters the complex's preferred DNA-binding sequence, suggesting a similar alteration in the set of genes being regulated (Hsu et al., 1994; Hu, Olson, & Kingston, 1992). E proteins have been shown to function as both transcriptional activators as well as transcriptional repressors, maintaining a vast transcriptional network (Greenbaum, Lazorchak, & Zhuang, 2004).

E protein activity is regulated by the inhibitor of differentiation (Id) gene family (Benezra, Davis, Lockshon, Turner, & Weintraub, 1990). Id proteins are similar to E proteins in that they also contain a conserved bHLH domain capable of dimerization with E proteins. However, Id proteins lack the basic DNA-binding domain. This lack of a DNA-binding domain effectively prevents the protein dimer from binding DNA and directing transcription. As such, Id proteins essentially inhibit E protein activity by out-competing functional E protein–E protein dimer formation in favor of nonfunctional E protein–Id protein dimer formation. This process effectively reverses the E protein transcriptional network, shutting down transcription of genes promoted by E proteins and removing repression of genes kept silent. In this way, E proteins can be thought of as a transcriptional "switch," maintaining a network of gene expression until "switched off" by upregulation of Id proteins.

Within the immune system, the primary active Id protein family members are *Id3* and *Id2* (Rivera, Johns, Quan, Johnson, & Murre, 2000; Sun, Copeland, Jenkins, & Baltimore, 1991). In lymphocytes, Id proteins are upregulated upon receipt of an activating signal. *Id3* is rapidly upregulated following lymphocyte activation, while *Id2* is upregulated more slowly (Bain et al., 2001). This suggests that Id3 and Id2 function in a semiredundant manner after activation to modulate the E protein transcriptional network, although their individual unique roles are not yet well understood.

2.1. E proteins and cell-cycle control

One of the major global transcriptional programs controlled by E proteins is the control of survival and cell-cycle progression. E2A-deficient cells develop an aggressive T cell leukemia (discussed in more detail below) characterized by rapid proliferation of developing thymocytes (Bain et al., 1997). Interestingly, restoration of E2A by ectopic expression in these cells did not arrest proliferation of these tumor cells, but rather resulted in cell death, indicating a role for E2A in cell survival as well (Engel & Murre, 1999). Subsequent experiments have demonstrated that E proteins do indeed regulate cell survival of developing lymphocytes, as conditional deletion of E2A was sufficient to drive cell death in otherwise healthy cells (Lazorchak, Wojciechowski, Dai, & Zhuang, 2006). Additional work has further implicated E proteins in regulating cell-cycle progression. Removal of E2A in some cell types was found to result in reduced levels of cyclin D2 and cyclin D3 and impaired entry into cell cycle (Zhao, Vilardi, Neely, & Choi, 2001). Interestingly, other work indicated that loss of E2A resulted in increased proliferation in lymphocytes (Park, Nolan, & Sun, 1999). Additionally, restoration of E2A in these cells halted growth. Thus, it appears that E proteins are capable of differentially regulating cell-cycle progression in a cell type-specific manner.

3. E PROTEINS IN LYMPHOCYTE DEVELOPMENT

As mentioned above, E proteins play many critical roles in lymphocyte development, particularly in the development of B and T cells. Developing B and T lymphocytes progress through a series of developmental stages in a highly regulated manner. E proteins have been shown to play a critical role in these processes.

3.1. Antigen receptor recombination

One of the primary characteristics of B and T cells is the acquisition of a highly diverse repertoire of antigen receptors (reviewed in detail here; Krangel, 2003). These antigen receptors are used to recognize and respond to antigens either directly, in the case of B cells, or after uptake, processing and subsequent presentation in the context of major histocompatibility complex (MHC) molecules in the case of T cells. These antigen receptor genes are the immunoglobulin heavy chain (IgH) and light chain (IgL) in B cells and the T cell receptor (TCR) genes in T cells. These antigen receptor genes share a unique structure, containing numerous sets of similar gene segments upstream of a conserved domain, termed the constant (C) region. These gene segments are divided into three subsets, termed variable (V), diversity (D), and joining (J) units. Upon successful recombination, the IgH and IgL chains combine to form the B cell receptor (BCR), also termed antibody. T cells contain two separate pairs of antigen receptor genes, giving rise to two distinct subsets of T cells. The TCR alpha and beta genes are capable of dimerization, producing an αβ TCR, while the TCR gamma and delta genes pair to produce a γδ TCR. Cells bearing these TCRs are referred to as αβ and γδ T cells, respectively. Interestingly, the *TCRδ* locus is housed within the *TCRα* locus, such that recombination of the *TCRα* locus removes a large part of the *TCRδ* locus, preventing further development toward the γδ lineage. While the mechanisms of recombination are similar between B cells, αβ T cells, and γδ T cells, the process and timing of recombination varies in each cell type. The regulation of these processes will be discussed in greater, cell type-specific detail later.

Acquisition of a functional antigen receptor occurs by splicing of the antigen receptor genes in a highly regulated manner. This splicing process is dependent on the recombination activating gene (RAG) family, known targets of E proteins (Hsu et al., 2003; Yu et al., 1999). RAG-mediated splicing begins upon transcriptional activation of the antigen receptor genes, a process that has been shown to be regulated in part by E proteins. RAG proteins recognize splice sites located between the various V, D, and J gene segments and facilitate the joining of these various gene segments into a functional antigen receptor gene. These events proceed in a highly regulated manner. Prior to RAG-mediated recombination, transcription of the germline IgH gene (in developing B cells) or the *TCRβ*, *TCRγ*, and *TCRδ* genes (in developing T cells) is initiated, opening up the chromatin environment surrounding the genes and making them more readily accessible to the recombination machinery. Animals lacking the *E2A* gene display an inability to initiate germline transcription of the IgH gene, leading to a block in B cell development, suggesting that E proteins play a role in initiating germline antigen receptor gene transcription (Bain et al., 1994; Zhuang, Soriano, & Weintraub, 1994). Further research showed that E proteins indeed play a direct role in activating germline transcription of antigen receptor loci and that ectopic expression of E47 alone is capable of initiating germline IgH transcripts in non-B cell lines (Choi, Shen, Radomska, Eckhardt, & Kadesch, 1996; Schlissel, Voronova, & Baltimore, 1991). Upon successful recombination of *IgH* (in B cells) the functional protein pairs with a conserved binding partner, the surrogate light chain to form a primordial antigen receptor. Successful recombination of the *TCRβ* gene is analogous to this process in T cells, pairing with the *pre-Tα* gene. This event allows the developing lymphocyte to receive a signal, leading to upregulation of Id proteins and subsequent reversal of E protein activity. This leads to repression of RAG genes as well as promotion of cell-cycle progression, proliferation, metabolic activity and expression of antiapoptotic genes, notably *Bcl-2* (Maraskovsky et al., 1997). Additionally, successful recombination of both the *TCRγ* and *TCRδ* genes will result in the formation of a functional γδ TCR and yields a functional γδ T cell. Unsuccessful recombination resulting in an inability to pair with the surrogate light chain (in the case of B cells) or pre-Tα (in the case of αβT cells) will result in cell death, as failure to reverse E protein transcription will result in a lack of nutrient uptake, proliferation, and poor expression of antiapoptotic genes.

Following this proliferative burst, Id protein expression subsides, allowing E protein-mediated transcriptional control to resume (Bain et al., 2001). Proliferation ceases and RAG expression resumes, allowing recombination of the *IgL* genes in B cells and the *TCRα* gene in T cells (Yu et al., 1999). Successful recombination of the *IgL* or *TCRα* genes results in the development of a complete antigen receptor. The highly variable nature of this process results in the generation of a lymphocyte pool with a wide array of antigen specificities, although not all of these specificities are useful or desirable, notably in the case of autoreactive antigen receptors.

3.2. Lymphocyte selection

Following completion of V(D)J recombination, developing B and T cells must undergo a process of selection. Developing B or T cells capable of recognizing some form of antigen receive a signal through the antigen receptor, resulting in upregulation of Id proteins and a process similar, but not identical, to that described above will take place (these processes will be discussed in more detail below). *RAG* expression ceases, preventing further recombination of antigen receptor genes. Antiapoptotic genes begin to be expressed, promoting the survival of the cell. Any developing B or T cell that is incapable of recognizing some form of antigen will fail to receive this signal, resulting in cell death. While a B cell must simply express a functional antigen receptor, T cells must express a receptor capable of recognizing antigen peptide presented by MHC molecules (Anderson, Jenkinson, Moore, & Owen, 1993). Successful recognition of a peptide-MHC complex transmits a signal to the T cell that allows the cell to fully mature. This process is termed positive selection (Hogquist et al., 1994). Additionally, any B or T cells capable of recognizing host-derived antigens will be deleted via apoptosis in a process termed negative selection (Buch, Rieux-Laucat, Förster, & Rajewsky, 2002). Should a B or T cell develop bearing a TCR capable of recognizing self-antigens (termed an autoreactive TCR), it will receive a signal that is inherently different than one that yields positive selection. Receipt of a negative selection signal will result in cell death. This process ensures that the immune system will not recognize the host as a foreign pathogen and will not initiate immune responses against it. This lymphocyte-host nonaggression pact is referred to as self-tolerance and ensures that no lymphocytes capable of initiating an immune response against the host are permitted to develop. Negative selection, however, occasionally fails due to genetic defects or environmental perturbations. When this occurs, autoreactive B and/or T cells initiate an immune response against the host, resulting in a condition generally referred to as autoimmune disease. As mentioned above, autoimmune diseases destroy critical body tissues, resulting in dramatically decreased quality of life or even death.

Thus, proper recombination of the antigen receptor loci, and by extension, the role of E proteins in these processes, is critical to the survival of the health and survival of the host. An inability to produce an antigen receptor will result in severe immunodeficiency, rendering the host susceptible to numerous opportunistic pathogens. On the other hand, a failure to properly dispose of autoreactive cells can result in inappropriate lymphocyte-mediated destruction of body tissues, leading to reduced quality of life or even death.

4. E PROTEINS IN B CELL DEVELOPMENT

E proteins also play unique roles specific to B or T cells. Development of B cells from hematopoietic stem cells occurs in a highly regulated fashion, with E proteins, most significantly E2A, playing critical roles from a very early stage (Bain et al., 1994). During this process, developing B lymphocyte progenitors gradually lose the properties of hematopoietic stem cells while gaining B cell characteristics, becoming progressively more committed to the B cell lineage (see Fig. 4.1). A progenitor cell begins the journey to becoming a B cell when E protein activity is initiated in a common lymphoid progenitor cell (Dias, Månsson, Gurbuxani, Sigvardsson, & Kee, 2008). Activation of the E protein transcriptional network initiates

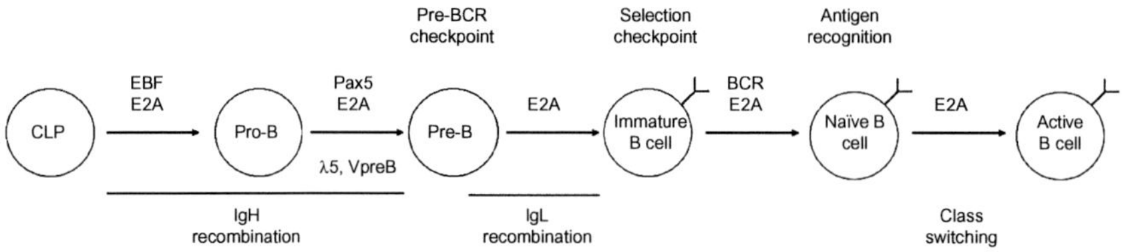

Figure 4.1 Diagram of B cell development. Major developmental events involving E proteins and major E protein targets are indicated.

expression of a series of additional transcription factors that cooperatively direct B cell development. The first of these is the early B cell factor (EBF) gene, which has been shown to be critically important in B cell development, permitting B cell progenitors to develop into pro-B cells (Lin & Grosschedl, 1995). Prior to its activation, the *EBF* gene is located at the transcriptionally repressive nuclear periphery (Lin, Benner, et al., 2012). Upon activation by E2A, the *EBF* gene relocates to a centralized, transcriptionally permissive environment and transcription begins (Lin, Benner, et al., 2012). Together, E2A and EBF begin to cooperatively initiate antigen receptor recombination (Romanow et al., 2000). Similar to the case of the *EBF* gene, the *IgH* locus also repositions itself away from the repressive nuclear periphery and DJ joining proceeds. It is likely that this repositioning facilitates not only germline transcription but also facilitates DNA accessibility for the RAG complex, as mice lacking EBF fail to generate B cells, as development is blocked at the pro-B cell stage prior to DJ rearrangement (Lin & Grosschedl, 1995). EBF has also recently been shown to promote B cell development by repressing the *Id2* and *Id3* genes, effectively ensuring that E2A activity is allowed to continue (Thal et al., 2009).

The coordinate activities of E2A and EBF also function to promote the transcription of the *Pax5* (also known as BSAP) gene (Nutt, Heavey, Rolink, & Busslinger, 1999; Urbánek, Wang, Fetka, Wagner, & Busslinger, 1994). As with mice lacking E2A or EBF, mice lacking Pax5 also show a block in B cell development at the pro-B cell stage (Nutt et al., 1999). Unlike *E2A* or *EBF*-null B cells however, *Pax5*-deficient B cells have been shown to be capable of abandoning the B cell developmental pathway and adopting different cell fates, including T cells and macrophages (Nutt et al., 1999). To this end, Pax5-deficient B cells are capable of expressing several genes normally expressed in other cell types, including *pre-Tα* and macrophage colony-stimulating factor, while these genes are silenced in WT B cells, suggesting that Pax5 plays a role in silencing gene sets utilized by other lymphocyte lineages. Furthermore, experiments using conditional deletion of *E2A* in developing B cells have indicated that Pax5 alone is capable of driving B cell development beyond the pre-B cell stage (Kwon et al., 2008). However, these *E2A*-deficient mature B cells display impaired functionality, indicating that E proteins play additional distinct roles in B cell development and function apart from their roles in conjunction with Pax5 activity (Kwon et al., 2008). As such E2A and Pax5 function to promote commitment to the B cell lineage.

In addition to promoting commitment to the B cell lineage, Pax5 has been shown to play a role in *IgH* recombination. While EBF-deficient pro-B cells contain unrearranged *IgH* loci, Pax5-deficient pro-B cells complete DJ recombination, but fail to complete V-DJ recombination (Fuxa et al., 2004). There are several possible explanations for this phenomenon. First, Pax5-binding sites have been located near the V gene segments of the *IgH* locus. Additionally, Pax5 has been shown to regulate the conformation and location of the *IgH* locus. Prior to *Pax5* expression, the *IgH* locus is located at the nuclear periphery; upon *Pax5* expression, the locus relocates to the interior of the nucleus, suggesting that the location of the locus may play a role in the recombination process (Fuxa et al., 2004). Additional research into gene expression and RAG-mediated recombination suggests that localization at the nuclear periphery results in impaired V-DJ recombination. It is likely that these alterations in gene localization function to modulate the ability of RAG complexes to access the DNA (Chan et al., 2013). It is possible that E proteins or their cooperative activity with transcription factors such as Pax5 and/or EBF mediate this repositioning, as binding sites for these factors exist in several regulatory regions flanking the locus (Henthorn et al., 1990). Furthermore, Pax5 has been shown to promote chromatin interactions between recombined DJ segments and the distant V gene segments, bringing the two into close contact (Fuxa et al., 2004). In-depth analysis into these chromatin interactions has indicated that the distant V gene segments form large-scale looping structures, providing visual confirmation of earlier biochemical data (Sayegh, Jhunjhunwala, Riblet, & Murre, 2005). Thus, in addition to driving commitment to the B cell lineage, E2A and Pax5 also promote further development along the B cell developmental pathway by promoting completion of IgH recombination.

Following completion of *IgH* recombination, the newly rearranged IgH protein pairs with the surrogate light chain (composed of λ5 and V-pre-B, both E2A targets) and Igα and Igβ signaling chains, forming the pre-B cell receptor (Loffert et al., 1994). Developing B cells that fail to produce a functional, in-frame *IgH* locus are destined for apoptosis, while those cells that successfully express a functional pre-B cell receptor become pre-B cells and undergo several subsequent, simultaneous events. First, further *IgH* recombination is blocked in a process referred to as "allelic exclusion," preventing simultaneous expression of multiple BCR specificities in a single B cell (Loffert et al., 1994). This prevents B cells from reacting to multiple, potentially unrelated, antigens. Second, a short burst of proliferation is initiated, allowing amplification of successfully recombined IgH chains (Yankee & Clark, 1999). During this time, E protein expression transiently declines, along with the E2A targets *RAG1* and *RAG2*, preventing further recombination (Yu et al., 1999). Following this period of proliferation, E protein expression resumes and recombination of the *Igκ* and *Igλ* light chain (IgL) loci begins (Hardy & Hayakawa, 2001).

Similar to *IgH* recombination, E proteins also play roles in *IgL* recombination. Ectopic expression of E2A proteins in addition to the RAG genes has been shown to activate *IgL* transcription as well as V-J recombination in transformed cells (Romanow et al., 2000). E2A has been demonstrated to promote *IgL* recombination by inducing chromatin remodeling in conjunction with IRF-4 (Lazorchak, Schlissel, & Zhuang, 2006). Interestingly, these genes are also present during the pro-B cell stage, although IgL recombination does not occur. It is possible that EBF provides a mechanism for preventing *IgL* expression, as coexpression of E2A and EBF has been shown to drive DJ recombination on the *IgH* locus, while recombination of the *IgL* loci does not occur (Goebel et al., 2001).

Following successful recombination of an Igλ or Igκ chain, the recombined IgL protein pairs with the previously recombined IgH protein to produce the BCR, also termed antibody. In the event that the resulting BCR recognizes a host antigen, it is deleted through apoptosis in a process called "negative selection" (Nossal, 1983). Those B cells that do not recognize host antigen are then released from the bone marrow to colonize the body's peripheral lymphoid organs. By preventing B cells capable of responding to host antigens from completing the developmental process, the B cell pool becomes tolerant to its host. Disruption of this process often leads to an autoimmune response against the host, with B cells often producing antibodies against antigens such as DNA and ribosomal proteins as well as other host proteins particular to individual autoimmune diseases.

As discussed above, the E protein network plays critical roles in B cell development, ranging from transcriptional regulation of key developmental genes to antigen receptor recombination, but the mechanisms underpinning this regulatory network have largely remained elusive. However, recent research has begun to bring these mechanisms to light. With the development of large-scale biochemistry and DNA sequencing, it has become possible to examine transcription factor activity on a global scale. These experiments have revealed intriguing roles for E2A beyond simply turning genes on or off. E2A has been shown to regulate gene transcription on several levels. Recent work has demonstrated that global chromatin accessibility is regulated, both positively and negatively, in part by E proteins (Lin, Jhunjhunwala, et al., 2010). E proteins have been shown to be capable of recruiting chromatin-remodeling complexes to the various genes they regulate (Lazorchak, Schlissel, et al., 2006; Sakamoto et al., 2012). E protein-mediated chromatin remodeling has been shown to be increasingly finely regulated by many of the transcription factors described above. In concert with EBF, E proteins have been shown to fine-tune the E protein network by initiating chromatin remodeling at loci jointly regulated by E2A and EBF prior to transcription initiation, making these loci effectively "poised" for transcription (Treiber et al., 2010). In this manner, E protein activity initiates a complex global transcriptional program that develops in concert with the cells themselves, evolving as additional developmental coregulators such as EBF and Pax5 are progressively activated (Lin, Jhunjhunwala, et al., 2010). These events fine-tune the E protein network and facilitate the completion of upcoming developmental processes, while shutting down activities occurring in previously completed developmental stages.

5. E PROTEIN ROLES IN MATURE B CELLS

Even after B cells complete development, the E protein network remains an integral part of B cell activity. B cells that pass the selection checkpoint are released into the periphery where they patrol the body and protect it from pathogens. Upon recognizing a pathogen via the BCR, the B cell is activated and undergoes a number of changes, many of them involving E proteins. In resting B cells, E2A levels are low, while E2A is highly expressed in activated B cells (Quong, Harris, Swain, & Murre, 1999). Shortly after BCR stimulation, the Id3 gene is rapidly upregulated (Deed, Hara, Atherton, Peters, & Norton, 1997). B cells lacking the Id3 gene display impaired proliferation, consistent with the role of E proteins in regulating cell cycle (Pan, Sato, Frederick, Sun, & Zhuang, 1999). Additionally, activated B cells undergo class switching, altering their BCR to better respond to the activating pathogen (Goldfarb, Flores, &

Lewandowska, 1996). Notably, B cells lacking the *E2A* gene fail to initiate class switching, although most other events in B cell activation proceed normally, including expression of the activation markers CD69 and CD44 (Quong et al., 1999). This is largely due to the role of E proteins in regulating the expression of AID, the molecule responsible for class switching (Sayegh, Quong, Agata, & Murre, 2003). Further supporting the role of E proteins in class switching, ectopic expression of the *Id1* gene in B cells yields an impairment similar to the one observed in *E2A*-deficient cells (Goldfarb et al., 1996).

While the roles of E2A in B cell development have been studied in depth, the *HEB* and *E2-2* genes have also been shown to play roles in B cell development. Unlike mice lacking *E2A*, mice lacking either *E2-2* or *HEB* are able to successfully produce mature B cells; they display a dramatic reduction in numbers of pro-B cells, suggesting roles for these genes in promoting survival at the pro-B cell stage (Zhuang, Cheng, & Weintraub, 1996).

6. E PROTEIN ROLES IN T CELL DEVELOPMENT

Similar to developing B cells, E proteins have been shown to play critical roles in T cell development as well. Whereas homodimers of *E2A* gene products predominate in developing B cells, heterodimers of *E2A* and *HEB* gene products predominate in thymocytes (Sawada & Littman, 1993). T cell development begins when early thymic progenitor cells migrate from the bone marrow to the thymus. The thymus is a complex organ, containing two major regions, the outer cortex and the inner medulla (see Fig. 4.2). Similar to developing B cells, thymocytes must undergo recombination of their antigen receptor genes. During this process, they must also migrate to particular regions of the thymus for development to proceed properly. Interestingly, E proteins have been shown to play important roles in the completion of this migratory process, while disruption of the E protein network can produce dramatic impairments in thymocyte development.

Developing thymocytes are broadly classified by their expression of the CD4 and CD8 coreceptor molecules as well as expression of an αβ or γδ TCR. Upon entry into the thymus, thymocytes express neither CD4 nor CD8 and are referred to as DN cells. These cells migrate into the thymic cortex and initiate recombination of the *TCRβ*, *TCRγ*, and *TCRδ* genes. This process is dependent on CXCR4, a chemokine receptor (Plotkin, Prockop, Lepique, & Petrie, 2003). Expression of CXCR4 has been shown

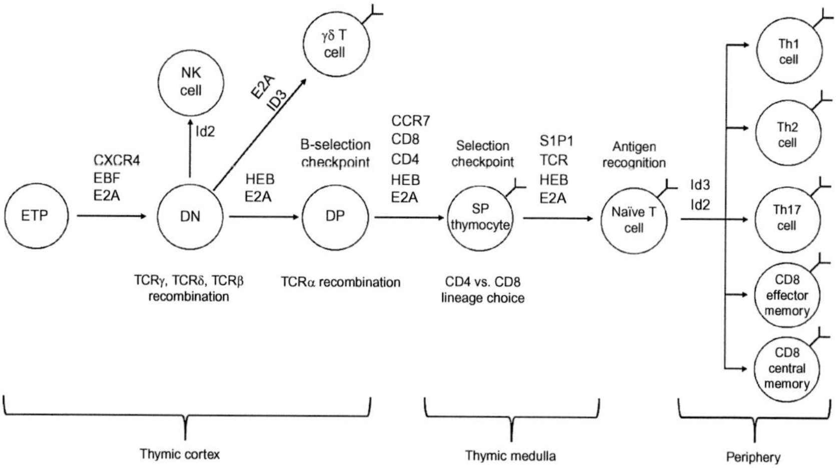

Figure 4.2 Diagram of T cell development. Major developmental events involving E proteins and major E protein targets are indicated.

to be regulated in part by E proteins, as cells lacking E2A and HEB fail to upregulate this marker (Jones & Zhuang, 2007). As in developing B cells, expression of E proteins is required for activation of germline antigen receptor transcription, although full activation of the *TCRβ*, *TCRγ*, and *TCRδ* loci has been shown to require expression of both E2A as well as HEB (Ghosh, Romanow, & Murre, 2001; Jia, Dai, & Zhuang, 2008; Langerak, Wolvers-Tettero, Van Gastel-Mol, Oud, & Van Dongen, 2001). Again, germline transcription of the antigen receptor genes is coupled with subnuclear repositioning of the DNA, paving the way for RAG-mediated recombination (Schlimgen, Reddy, Singh, & Krangel, 2008). Successful recombination of the *TCRγ* and *TCRδ* genes yields a γδ T cell, while successful recombination of the *TCRβ* gene allows the TCRβ protein to pair with the pre-T cell receptor alpha (pre-Tα) chain, analogous to the surrogate light chain of B cells (Kruisbeek et al., 2000). This process, called β-selection, initiates a burst of proliferation as well as initiation of *CD8* and *CD4* expression, becoming double-positive (DP) thymocytes (Barndt, Dai, & Zhuang, 2000). Again, E proteins play key roles in these processes. Upon pre-Tα expression, *Id3* expression is upregulated, inhibiting E protein activity (Engel, Johns, Bain, Rivera, & Murre, 2001). It is likely that E proteins play a role in restricting passage of DN thymocytes to the DP stage, as elimination of E2A is sufficient to permit DN thymocytes to transition to the DP stage without a rearranged *TCRβ* allele (Engel et al., 2001). Additionally, the *HEB* gene has been shown to be required for proper expression of *pre-Tα*, leading to a block in T cell development at the β-selection checkpoint (Herblot, Steff, Hugo, Aplan, & Hoang, 2000). Furthermore, overexpression of Id3 recapitulates the phenotype observed in HEB-deficient thymocytes, generating a block in thymocyte development, partly by blocking *pre-Tα* expression (Blom et al., 1999).

Following β-selection, a developing αβ T cell downregulates *CXCR4* and migrates to the thymic medulla, a process requiring upregulation of the CCR7 chemokine receptor. As with CXCR4 expression, CCR7 expression is also regulated by E proteins, albeit in an opposite (repressive) manner (Jones & Zhuang, 2007). Recombination of the *TCRβ* genes ceases, with the unrearranged allele relocating to the nuclear periphery so as to prevent further recombination (Chan et al., 2013). Removal of E proteins allows for spontaneous upregulation of *CCR7* along with continued development of thymocytes in the absence of β-selection. This defect will be discussed in more detail below. Upon migration into the thymic medulla, recombination of the *TCRβ* genes ceases and *TCRα* recombination begins (Von Boehmer, 1994). Following successful *TCRα* recombination, a functional αβ TCR can be expressed on the cell surface. Cells expressing an αβ TCR then undergo the processes of positive and negative selection. As mentioned above, in order to successfully pass the positive selection checkpoint, a given TCR must be capable of recognizing antigenic peptide presented by either MHC class I or MHC class II (reviewed in detail here; Robey & Fowlkes, 1994). Thymocytes-recognizing antigen presented by MHC class I will lose *CD4* expression, becoming CD8 single-positive (SP) cells, while those recognizing antigen presented by MHC class II will lose *CD8* expression and become CD4 SP cells. Again, E proteins play key roles in positive selection. Upon receiving a signal through the TCR, *Id3* is rapidly upregulated (Bain et al., 2001). *Id3* upregulation is coupled with upregulation of S1P1, a molecule required for exit from the thymus (Matloubian et al., 2004). When E2A and HEB are removed in DP thymocytes, it appears that Id protein upregulation is no longer required for continued development, permitting developing T cells to bypass the need for any signal through the TCR (Jones & Zhuang, 2007), removing the need for an antigen receptor signal results in the development of T cells lacking a functional TCR. Interestingly, removal of E2A and HEB in developing T cells results in a complete block in the development of CD4 T cells, suggesting that the CD8 lineage is a "default" developmental pathway.

Disruption of Id proteins in developing T cells also results in numerous developmental defects. Elimination of the *Id3* gene results in several major phenotypes. The first is the preferential development and/or expansion of a unique subset of γδ T cells (Ueda-Hayakawa, Mahlios, & Zhuang, 2009). These cells all bear a TCR using the Vγ1.1 and Vδ6.3 gene segments and share many characteristics with innate immune cells, including cytokine production, expression of *PLZF* and dependence on the signaling adaptor protein SAP (Verykokakis et al., 2010). This population is developed very early in life and is maintained through self-renewal. Deletion of Id3 in these Vγ1.1/Vδ6.3$^+$ cells was found to cause a dramatic expansion early in life, although Vγ1.1/Vδ6.3$^+$ cells developed later in life did not share this property (Zhang, Dai, Li, & Zhuang, 2013). Interestingly, this phenotype shows a peculiar strain-dependence, as Id3 deficiency on the C57BL/6 background results in a strong phenotype, while the same deletion on the 129/sv background does not (Azuara, Grigoriadou, Lembezat, Nagler Anderson, & Pereira, 2001). This difference in γδ T cell development was traced to a strain-specific mutation in the *Id2* gene resulting in weaker Id2 expression in C57BL/6 mice (Zhang, Lin, Dai, & Zhuang, 2014). *Id3*-deficient mice have also been shown to have defects in thymocyte selection (Rivera et al., 2000). A disproportionately small number of DP thymocytes are capable of making the leap to mature SP cells, suggesting that the early upregulation of *Id3* plays a critical role in the development of a large number of thymocytes. These defects in T cell development are coupled with the initiation of an autoimmune disease reminiscent of human Sjogren's Syndrome (SS) (Li, Dai, & Zhuang, 2004). These data also suggest that Id2 is capable of compensating for the loss of Id3, at least partially. While Id2 can compensate for some roles of Id3 in thymocyte development, it also plays unique roles as well. Disruption of the *Id2* gene has been shown to result in a block in natural killer (NK) and natural killer-T (NKT) cell development, while αβ and γδ development proceeds seemingly normally (Boos, Yokota, Eberl, & Kee, 2007). Further investigation into the roles of E proteins in NK cell development indicated that combined deletion of *Id2* and *Id3* in DN thymocytes was sufficient to drive a dramatic expansion of invariant NKT (iNKT) cells (Li, Wu, Jiang, & Zhuang, 2013). E proteins were also shown to have a role in regulating the proliferation of these iNKT cells (Hu et al., 2013; Li et al., 2013; Verykokakis et al., 2013). Intriguingly, partial restoration of Id proteins resulted in a switch from iNKT development to γδ T cell development, suggesting that fine regulation of E protein activity is required in the development of various T cell subsets (Li et al., 2013).

While the removal of Id3 or Id2 alone permits relatively normal αβ T cell development, removal of both Id2 and Id3 in DP thymocytes results in profound developmental aberrations. While removal of the E2A and HEB genes results in the development of large numbers of CD8 T cells, many of which lack a functional TCR, deletion of *Id2* and *Id3* results in a complete lack of $CD8^+$ T cells (Jones-Mason et al., 2012). Additionally, development of $CD4^+$ T cells is severely restricted. Taken together with the results derived from *E2A/HEB* double-knockout T cells, these data indicate that E proteins are required for the development of $CD4^+$ T cells.

7. ROLES OF E PROTEINS IN MATURE T CELLS

Following maturation within the thymus, naïve T cells migrate to the periphery. There, they protect the host from invading pathogens. Upon recognition of a pathogen via the TCR, a naïve T cell becomes activated and begins to adapt its response to the current pathogenic challenge. During this process, a naïve T cell response will further develop into one of a number of effector responses, each with its own signature array of inflammatory mediators. For example, an intracellular pathogen, such as a virus, induces a T helper type 1 (Th1) response, including production of IFN-γ. This response features significant activity by CD8 cytotoxic cells, which eliminate infected host cells. By contrast, extracellular pathogens give rise to other responses. A Th2 response is characterized by production of IL-4, IL-5, and IL-13 as well as marked recruitment of B cell activity, which aids in removal of blood-borne pathogens.

As ever, E proteins play important roles in regulating these processes. Id3 expression has been shown to be relatively high in naïve T cells (Miyazaki et al., 2011). Elimination of the *Id3* gene results in spontaneous upregulation of surface markers characteristic of differentiated, effector-memory cells (Miyazaki et al., 2011). These results indicate that Id3-mediated suppression of E proteins in naïve T cells is required to prevent spontaneous maturation into effector cells. These results were further supported by experiments investigating the various roles of Id2 and Id3 in naïve T cells (Yang et al., 2011). Id3 expression was found to correlate with long-lived memory cell formation, while Id2 was associated with the development of short-lived, effector-memory cells (Yang et al., 2011). Additionally, deletion of either *Id2* or *Id3* resulted in a failure to generate effector-memory or long-lived memory cells, respectively. The role of Id3 in supporting memory cell formation is likely regulated in part by Blimp-1, as high *Blimp-1* expression downregulated *Id3* and limited the generation of long-lived memory cells (Ji et al., 2011). Id proteins have also been implicated in the adoption of particular T helper responses. A large-scale gene-association study found that *Id2* and *Id3* are differentially regulated in particular T helper populations (Lund et al., 2007). This evidence is further supported by the finding that mice lacking *Id2* display disproportionately Th2-skewed T cell responses (Kusunoki et al., 2003). The role of Id2 in this developmental skewing is in fact profound enough to affect autoimmune disease progression. An analysis of *Id2*-deficient T cells demonstrated that removal of *Id2* in T cells was sufficient to protect mice from experimentally induced autoimmune encephalomyelitis, a Th17-mediated disease (Lin, Jones-Mason, et al., 2012). Although pathogenic cells were developed, they did so at a greatly reduced rate, as proapoptotic proteins were found to be upregulated in *Id2*-deficient T cells, leading to increased cell death (Lin, Jones-Mason, et al., 2012). These results further highlight the role of the E protein system in maintaining and regulating the T cell response to antigen.

8. E PROTEINS IN LYMPHOID DISEASES

As demonstrated above, E proteins play critical roles in the highly complex developmental pathways and subsequent functions of B and T cells. These roles include controlling cell proliferation, ensuring developmental checkpoints remain enforced, maintaining self-tolerance and regulating the effector functions of mature cells. In general, E proteins tend to function as gatekeepers, ensuring a cell does not proceed to another developmental stage before it has completed the steps required to do so. By enforcing these checkpoints, E proteins keep the immune system operating normally. Because E proteins play so many roles in these sensitive processes, it is critical that their function remain undisturbed, as serious consequences could result from defects in these processes. Failures in selection or antigen receptor generation could lead to severe immunodeficiency. Failures in selection could lead to autoimmune reactions. Failures in cell-cycle regulation could lead to tumor formation. The remainder of this chapter will focus on the ways in which these developmental mechanisms can be co-opted to induce diseases, particularly autoimmunity and cancer.

8.1. E proteins in autoimmunity

SS is an autoimmune disease characterized by progressive destruction of the salivary and lachrymal glands, resulting in impaired saliva and tear production (Bloch, Buchanan, Wohl, & Bunim, 1965). This gland destruction is mediated by lymphocytic infiltration into the gland tissue, provoking an inflammatory response within the gland (Greenspan, Daniels, Talal, & Sylvester, 1974). This inflammatory response results in development of fibrotic scar tissue within the gland, leading to increasingly poor secretory function (Tarpley, Anderson, & White, 1974). In humans, SS primarily affects women, typically becoming apparent in the fourth or fifth decade of life. Among available mouse models, the *Id3* knockout mouse has been established as a unique model for human primary SS (Lee, Gauna, Pauley, Park, & Cha, 2012).

The *Id3*-deficient mouse develops many of the same characteristics of human SS (Li et al., 2004). Significantly impaired gland function is readily apparent in *Id3* knockout mice within 2 months of age. By 4 months, pronounced lymphocytic infiltration begins to occur. The initial infiltrating lymphocytes are primarily αβ T cells, with B cells arriving shortly thereafter. Beyond 6 months of age, mice begin to show outward signs of disease: increased water consumption and lesions around the face due to repeated scratching of dry eyes. Autoantibody production, notably anti-Ro and anti-La, typically begins around 1 year of age (Alexander, Hirsch, Arnett, Provost, & Stevens, 1981; Alexander, Arnett, Provost, & Stevens, 1983). Historically, SS was thought of as a Th1-mediated disease, as human patients frequently displayed elevated levels of IFN-γ, although animal models did not typically share this phenotype (Dalavanga, Drosos, & Moutsopoulos, 1985; Mahlios & Zhuang, 2011; Szodoray, Alex, Brun, Centola, & Jonsson, 2004). Additionally, it has been demonstrated that removal of Id3 in T cells alone was sufficient to initiate the disease process (Guo et al., 2011). Adoptive transfer of *Id3*-deficient T cells into sublethally irradiated WT hosts was also sufficient to transfer disease (Li et al., 2004). However, additional cell types have proved to be important disease mediators as well, as depletion of B cells was capable of improving disease symptoms in mice (Hayakawa, Tedder, & Zhuang, 2007). In addition to B cell involvement, mice suffering from SS were found to contain markedly elevated numbers of mast cells within the gland tissue (Mahlios & Zhuang, 2011). Together, the immune cells drive extensive fibrotic remodeling of the gland tissue, leading to degradation of gland function. Interestingly, elevated E protein activity has been implicated in driving fibrotic remodeling in other tissues, suggesting an additional potential nonlymphoid role in gland impairment (Slattery, Mcmorrow, & Ryan, 2006). Furthermore, with later discoveries of additional effector T cell subtypes and the cytokines they produce, notably IL-13, it became clear that the disease process was in fact far more complicated than originally thought (McKenzie et al., 1993).

Identification of IL-13 as a part of the Th2 effector response led to reevaluation of several immune processes, such as allergic processes and epithelial inflammation (De Vries, 1998; Grünig et al., 1998). This paradigm shift was not limited to typical foreign-antigen responses, as elevated levels of IL-13 were soon identified in patients with SS symptoms (Mitsias et al., 2002). This discovery was also observed in the *Id3*-deficient mouse model of SS (Mahlios & Zhuang, 2011). Additional research into the role of IL-13 in SS pathogenesis has been enlightening. Neutralization of circulating IL-13 was sufficient to improve gland function in *Id3*-deficient mice and caused a mild reduction in the numbers of mast cells residing within the gland tissue, although the source of IL-13 remained elusive (Mahlios & Zhuang, 2011). As IL-13 is known to be produced by T cells, elimination of αβ T cells in *Id3*-deficient mice was sufficient to reduce IL-13 concentrations in serum to near-WT levels, though IL-13 was still slightly elevated. Elimination of αβ T cells was also capable of impeding the disease process; disease symptoms still appeared in *Id3/TCRβ* double-deficient mice, albeit much later than in Id3 knockout mice (Mahlios & Zhuang, 2011).

The above results indicate that SS is a highly complex disorder with numerous cell types involved in its pathogenesis. Many of the cell types known to be involved ($CD4^+$ T cells, γδ T cells, B cells, mast cells, etc.) likely contribute to gland impairment in multiple ways, ranging from gland infiltration to cytokine production to tissue remodeling. Recent research in both humans and animal models has highlighted the importance of IL-13, particularly T cell-derived IL-13, in disease progression. These findings raise the possibility of IL-13 as a potential diagnostic tool or therapeutic target in the treatment of patients with SS.

8.2. E proteins in cancer

The prevalent roles of E proteins in regulating gene expression and developmental progression ensure that lymphocytes develop and function in a way that is beneficial to the host. We have already discussed how defects in lymphocyte development and selection can turn the immune system against the host. The role of E proteins in regulating gene expression can also be dysregulated, oftentimes resulting in tumorigenesis.

8.2.1 Burkitt Lymphoma

Many B cell cancers arise from a translocation event between the *IgH* promoter, which is constitutively active in B cells, and the *c-Myc* oncogene, a prosurvival transcription factor. Upon placing *c-Myc* under the control of the

IgH promoter, the resultant B cell is highly resistant to apoptosis. This event can lead to several different varieties of B cell cancers, including both Burkitt's Lymphoma (BL) and Diffuse Large B Cell Lymphoma (DLBCL). Intriguingly, recent work has shown that, upon B cell activation, the c-Myc locus becomes transcriptionally active and frequently repositions itself adjacent to the IgH locus, greatly facilitating this translocation event (Osborne et al., 2007). While both BL and DLBCL often share certain underlying mutations, recent research has shown them to be quite different. For example, low-dose cyclophosphamide, doxorubicin, vincristine, and prednisone (CHOP) therapy has been successfully used in the treatment of DLBCL, however, much larger doses have been shown to be required for effective treatment of BL (Butler & Hainsworth, 1993). Additionally, use of treatments capable of passing the blood-brain barrier have been shown to be required for proper treatment of BL, while these measures have proven unnecessary for DLBCL (Bishop, Rao, & Wilson, 2000). These remarkable differences between cancers sharing such similar mutations have led to a great deal of research into the genetic differences underlying these disparities.

Several groups have recently performed extensive comparative analysis of BL and DLBCL (Love et al., 2012; Richter et al., 2012; Schmitz et al., 2012). These groups discovered that, while BL and DLBCL may share an initial mutation, many secondary mutations are unique to each tumor. Indeed, more than two thirds of BL tumors have been found to contain mutations within the *Id3* gene (Love et al., 2012; Richter et al., 2012). Interestingly, the majority of these mutations were determined to be within the bHLH region of the protein and were frequently nonsense or frameshift mutations (Love et al., 2012; Richter et al., 2012). These results suggest that the secondary mutations within the *Id3* locus are loss-of-function mutations, resulting in dysregulation of the E protein transcriptional network. Intriguingly, these mutations were almost entirely absent from DLBCL tumors. Additional work by Staudt and colleagues corroborated this finding and additionally identified frequent mutations within the *E2A* gene (Schmitz et al., 2012). While mutations within the *Id3* gene were biallelic and predominantly resulted in a loss-of-function, the mutations within the *E2A* gene typically monoallelic and resulted in elevated transcription. Strikingly, in BL tumors bearing E2A mutations, nearly all mutations were restricted to the bHLH or DNA-binding domains of the E47 splice-variant, while the E12 variant remained unchanged. In many *E2A* alleles containing mutations within the DNA-binding domain, these changes resulted in alterations in the canonical "CANNTG" binding sequence preferred by E protein dimers (Schmitz et al., 2012). Additionally, E47 was expressed at a higher level than E12 in these tumors, suggesting a nonredundant role for these proteins in BL pathogenesis.

While BL tumors have been shown to contain secondary mutations in *Id3* and/or *E2A*, the effects of these mutations are not well understood. As mentioned above, mutations in *Id3* and *E2A* are typically found within the bHLH region. In addition to increased expression of E47 (in the case of E2A mutations), one of the primary effects of these mutations has been shown to be an inhibition of dimerization between E proteins and Id proteins (Schmitz et al., 2012). In the case of Id3 and E2A, this inhibition results in an inability of Id3 to reverse the E protein transcriptional network. In BL tumors bearing Id3 mutations, this has been shown to result in increased progression through the cell cycle. Expression of several genes involved in cell-cycle initiation, including *CDK7*, *E2F1*, and *MCM10* were found to be elevated in BL cells bearing *Id3* mutations. Additionally, introduction of individual BL-derived *Id3* mutant alleles into BL lines containing wild-type *Id3* genes resulted in a significantly reduced proportion of cells in G1 as well as an increased proportion of cells entering S phase (Love et al., 2012). Furthermore, reintroduction of wild-type *Id3* into BL cells expressing *Id3* mutants was sufficient to slow the rate at which cells entered the cell cycle (Love et al., 2012).

In addition to altering progression into the cell cycle, mutations in *Id3* and *E2A* have been shown to have notable effects on cell survival. Reintroduction of E2A expression has been shown to result in significantly increased cell death in several tumor lines, while alteration of E2A activity is a major oncogenic factor in primary BL cells (Engel & Murre, 1999; Schmitz et al., 2012). Recent work has shown that E protein regulation of PI3K signaling may be responsible for this increased survival (Schmitz et al., 2012). PI3Ks regulate a wide range of biological processes by generating lipid-based second-messenger molecules (Okkenhaug & Vanhaesebroeck, 2003). In lymphocytes, PI3K plays an important role in antigen receptor and coreceptor signal transduction. Notably, PI3K signaling is known to activate AKT, which in turn regulates cell growth, survival, and metabolism (Stocker et al., 2002; Vanhaesebroeck & Alessi, 2000). PI3K is activated within seconds of tyrosine phosphorylation of antigen receptor molecules, notably Igα and Igβ (also known as CD79a and CD79b, respectively) (Okada, Maeda, Iwamatsu, Gotoh, & Kurosaki, 2000; Tuveson, Carter, Soltoff, & Fearon, 1993). Following PI3K activation, AKT is phosphorylated and activates survival, proliferation, and metabolic pathways, largely through activation of the mTOR pathway (Mendoza, Er, & Blenis, 2011). This pathway is negatively regulated by tyrosine phosphatase proteins, notably SHP-1 (Sathish et al., 2001). SHP-1 dephosphorylates tyrosine residues on antigen receptor and coreceptor molecules, inhibiting activation of AKT. In BL tumors bearing mutations in *E2A* or *Id3*, levels of SHP-1 were found to be lower than in WT cells; additionally, knockdown of E2A was found to increase SHP-1 levels while simultaneously decreasing phospho-AKT levels (Schmitz et al., 2012). These results suggest that, in addition to regulating entry into the cell cycle, disruptions in the E protein system can also alter antigen receptor signaling to further promote growth, survival, and nutrient uptake.

Taken together, the results discussed above suggest that mutations targeting the E protein transcriptional network are capable of supporting the oncogenic nature of *c-Myc* translocation in a manner distinct from other *c-Myc*-driven cancers. It has been suggested that these mutations, as well as their downstream effects, may be used as diagnostic criteria, allowing clinicians to better identify cases of BL and separate them from other, similar cancers, thereby allowing improved, more specialized treatment. It is also possible that a more complete understanding of the mechanisms and cellular pathways involved in and unique to BL may allow for the development of new therapeutic options, yielding improved prognoses for patients.

8.2.2 E2A–PBX1 translocation in B cell acute lymphocytic leukemia

As mentioned previously, E proteins control a large transcriptional network, regulating events such as lymphocyte development, cell survival and proliferation. Loss of E or Id proteins can cause significant defects in these processes. Alterations in the gene networks regulated by E proteins can have similarly deleterious effects. E proteins have also been found to play a role in other types of B cell cancers, notably pre-B cell Acute Lymphocytic Leukemia (pre-B ALL). The t(1;19)(q23;p13.3) translocation is found in approximately 25% of patients with pre-B ALL. In the vast majority of these patients, the translocation event joins the promoter and transcriptional activation domain of the E2A gene with the DNA-binding region of the Pre-B Cell Homeobox-1 (PBX1) gene. While *E2A* is widely expressed throughout the body, *PBX1* is not normally expressed within the immune system. PBX1 mediates its transcriptional activity by forming part of a molecular complex that includes Class 1 Homeobox (HOX) proteins as well as *Meis1* and *pKnox1* (Knoepfler, Calvo, Chen, Antonarakis, & Kamps, 1997). In addition to disrupting one copy of the E2A gene, which has been shown to lead to an increased incidence of lymphoid tumors, this fusion event results in constitutive activation of many of the genes regulated by PBX1 in cells expressing E2A. This transactivation has been shown to require both the DNA-binding region of PBX1 as well as the transcriptional activation domain of E2A, as disruption of either of these sequences eliminates the tumorigenicity of the E2A–PBX1 fusion protein (Monica et al., 1994). Monica and colleagues further characterized the effects of mutations within the E2A portion of E2A–PBX1, noting that disruption of the AD1 domain resulted in a loss of transcriptional activation in both lymphoid and fibroblast lines, while disruption of AD2 only impaired expression in fibroblast lines, indicating potential separate roles for the AD1 and AD2 domains in tumorigenesis (Monica et al., 1994).

While both E2A and E2A–PBX1 are potent transcriptional activators in lymphoid cells, PBX1 alone is not (Lu & Kamps, 1996). In fact, expression of PBX1 has been shown to inhibit some of the activity of other HOX proteins (Lu & Kamps, 1996). Intriguingly, the E2A–PBX1 fusion protein produces further aberrations in transcriptional activity beyond simple abnormal expression of PBX1. While normal activity of PBX1 requires interactions with Meis1 and pKnox1, the E2A–PBX1 fusion protein is incapable of interacting with these proteins (Knoepfler et al., 1997). This is likely due to the fact that the residues required for interaction with Meis1 and pKnox1 are contained in the portion of PBX1 replaced by the E2A gene (Knoepfler et al., 1997). Thus, the transcriptional activity of the E2A–PBX1 protein results in expression of a different set of genes than does native PBX1. Later research into the tumorigenicity of the E2A–PBX1 fusion protein yielded several intriguing gene targets, in particular, *Wnt16* and *EB-1*. *Wnt16* is a member of the Wnt family of proteins, which are involved in promoting cell growth. The Wnt family of proteins is expressed in numerous cell types and has been shown to be frequently mutated in several types of cancers (Polakis, 2000). Interestingly, *Wnt16* is not normally expressed in pre-B cells (McWhirter et al., 1999). However, a role for the Wnt signaling pathway in early hematopoiesis and proliferation of developing B cells has been described (Austin et al., 1997; Reya et al., 2000; Van Den Berg et al., 1998). These results indicate that the E2A–PBX1 fusion protein may exert some of its negative effects by aberrantly activating the Wnt signaling pathway, leading to inappropriate proliferation of pre-B cells.

EB-1 was identified by Kamps and colleagues while investigating genes activated by E2A–PBX1 (Fu et al., 1999). The *EB-1* gene encodes protein containing a phosphotyrosine-binding element and two SAM domains, similar to other genes playing roles in tyrosine kinase signal transduction and cell proliferation. EB-1 expression is normally restricted to the brain and testes, however, upon introduction of E2A–PBX1, EB-1 expression in developing B cells increased nearly 100-fold. Such a result suggests that E2A–PBX1-driven EB1 expression may alter one or more of the signaling pathways responsible for regulating cell growth.

Taken together, the above observations indicate that the E2A–PBX1 translocation leads to pre-B ALL in part by disrupting the sensitive E protein transcriptional network as well as initiating the aberrant expression of several tissue-specific genes via the unique transcriptional activity of E2A–PBX1. Several of these genes have been shown to regulate cell growth and proliferation, further amplifying the oncogenic effects of the initial translocation event. The Wnt pathway has been well studied in the context of neurobiology, particularly in Alzheimer's disease (Caricasole et al., 2003). The association of Wnt pathway activity in E2A–PBX1 tumorigenesis may provide opportunities for the development of new therapeutic strategies (Barker & Clevers, 2006).

8.2.3 E proteins in T cell cancers

E proteins have also been shown to play significant roles in T cell leukemias and lymphomas as well (Yan et al., 1997). While mice deficient in E2A often die perinatally, those that survive their infancy often develop an aggressive T cell lymphoma comprised of immature thymic progenitors. It is likely that the loss of E2A results in enhanced proliferation, as reintroduction of E2A proteins is able to induce cell death in E2A-deficient tumors (Engel & Murre, 1999). Furthermore, as mentioned above, DN thymocytes lacking E proteins are capable of spontaneously bypassing the β-selection checkpoint. Given that one of the consequences of β-selection is rapid proliferation during the period of Id3-mediated E protein suppression, disruption of E protein-mediated cell-cycle control seems a likely culprit in the development of these tumors. In fact, recent work has demonstrated that loss of E2A is indeed a major player in Sézary Syndrome, a subtype of human T cell lymphoma (Steininger et al., 2011). Restoration of E2A in these tumor cells resulted in a reduction in proliferation and increased cell death, corroborating previous data implicating E proteins in cell-cycle control. E2A-deficient tumors are characterized by a greatly enlarged thymus with an abnormal architecture, lymphadenopathy and frequent metastasis to other lymphoid and nonlymphoid organs. These tumorigenic T cells expressed little to no surface TCR, highlighting their immaturity. The lack of surface TCR on these cells is likely due to the removal of the requirement for β-selection. Interestingly, particular tumor incidences were found to be made up of either $CD4^+/CD8^+$ cells, $CD4^{low}/CD8^+$, or $CD4^-/CD8^+$ cells, lending support the role of E proteins in regulating T cell lineage choice.

Last, recent work investigating the *Id3*-deficient mouse model of SS revealed an intriguing finding. Mice lacking the Id3 gene were found to occasionally develop a lymphoma comprised of TCRγδ$^+$ T cells, a condition which is very rare occurrence in humans (Lin, Jhunjhunwala, et al., 2010). These lymphomas presented themselves primarily with splenomegaly, hepatomegaly, and lymphadenopathy. Additionally, tumor cells were found within the bone marrow, kidneys, lungs, and thymus. Indeed, splenic involvement was such that the overall architecture was almost completely destroyed. Interestingly, the majority of these lymphomas were comprised of cells bearing TCRs featuring the Vγ1.1 gene segment, likely due to the preferential development of these cells in *Id3*-deficient mice. Some lymphomas consisted of cells using the Vγ3 gene segment, but none were found to use Vγ2. This suggests that Id3 may play some role in tumor suppression in addition to its role in suppressing the development of Vγ1.1$^+$ cells. It is possible that the *Id3*-deficient mouse model may also be useful in the study of human γδ T cell lymphoma, as no additional model of this disease is known to exist.

9. CONCLUSION

Since their discovery, E proteins and their inhibitors, Id proteins, have proven to be major players in the immune system at virtually every stage. Even though they were first discovered decades ago, ongoing research continues to uncover new intricacies of the E proteins. Recent research has repeatedly demonstrated the roles of E and Id proteins in maintaining the state of a lymphocyte, preventing them from improper maturation, proliferation, and functional activity (Miyazaki et al., 2011). Breakdowns in the E protein system can cause serious defects in the immune system, causing aberrant activity and autoimmunity (Guo et al., 2011). Subtle changes in E protein function, even those that alter dimerization properties can also produce profound immunodeficiencies (Boisson et al., 2013). New research has also shown that E protein functionality can differentiate types of cancer formerly thought to be highly similar (Love et al., 2012; Richter et al., 2012). Because E proteins and Id proteins play such numerous and such varied roles in lymphocyte development and function, sometimes in a cell type-specific manner, and because their expression level must remain finely tuned, their use as potential therapeutic targets has been limited. For example, loss of E protein activity blocks B cell development, but can also lead to aberrant T cell development (Bain et al., 1994, 1997). Furthermore, simply restoring E protein activity (or removal of Id proteins) can also yield deleterious results in these models (Engel & Murre, 1999). As such, more research is required to better understand the mechanisms by which breakdowns in the E protein system contribute to developmental and functional abnormalities in particular cell types and particular stages of development.

Fortunately, recent research into the targets of the E protein network has been promising. The evolution of large-scale sequencing has identified numerous downstream gene targets of E proteins, as well as transcription factors, some of them cell type-specific, involved in fine-tuning the E protein network (Lin, Jhunjhunwala, et al., 2010). However, further research is necessary to more fully understand which E protein functions are ubiquitous and which are unique to particular cell types or cell stages. By better understanding how the E protein network is regulated in particular cell types, E protein activity can be better connected to the pathways that modulate its downstream effects. A more complete understanding of the ways in which E proteins function to regulate downstream events in the many cellular processes involved in development could yield therapeutic targets useful in the treatment of the many diseases that can result from abnormalities in the E protein system.

ACKNOWLEDGMENTS

The authors would like to thank Baojun Zhang, Yen-yu Lin, Elizabeth Chan, Jia Li, and Sumedha Roy for helpful comments during the preparation of this chapter.

REFERENCES

Alexander, E. L., Arnett, F. C., Provost, T. T., & Stevens, M. B. (1983). Sjögren's syndrome: Association of anti-Ro (SS-A) antibodies with vasculitis, hematologic abnormalities, and serologic hyperreactivity. *Annals of Internal Medicine*, *98*(2), 155–159.

Alexander, E., Hirsch, T., Arnett, F., Provost, T., & Stevens, M. (1981). Ro (SSA) and La (SSB) antibodies in the clinical spectrum of Sjogren's syndrome. *The Journal of Rheumatology*, *9*(2), 239–246.

Anderson, G., Jenkinson, E. J., Moore, N. C., & Owen, J. J. (1993). MHC class II-positive epithelium and mesenchyme cells are both required for T-cell development in the thymus. *Nature*, *362*, 70–73.

Aronheim, A., Shiran, R., Rosen, A., & Walker, M. D. (1993). The E2A gene product contains two separable and functionally distinct transcription activation domains. *Proceedings of the National Academy of Sciences*, *90*(17), 8063–8067.

Austin, T. W., Solar, G. P., Ziegler, F. C., Liem, L., & Matthews, W. (1997). A role for the Wnt gene family in hematopoiesis: Expansion of multilineage progenitor cells. *Blood*, *89*(10), 3624–3635.

Azuara, V., Grigoriadou, K., Lembezat, M. P., Nagler Anderson, C., & Pereira, P. (2001). Strain-specific TCR repertoire selection of IL-4-producing Thy-1dull γ δ thymocytes. *European Journal of Immunology*, *31*(1), 205–214.

Bain, G., Cravatt, C. B., Loomans, C., Alberola-Ila, J., Hedrick, S. M., & Murre, C. (2001). Regulation of the helix-loop-helix proteins, E2A and Id3, by the Ras-ERK MAPK cascade. *Nature Immunology*, *2*(2), 165–171.

Bain, G., Engel, I., Maandag, E. R., Te Riele, H., Voland, J. R., Sharp, L. L., et al. (1997). E2A deficiency leads to abnormalities in alphabeta T-cell development and to rapid development of T-cell lymphomas. *Molecular and Cellular Biology*, *17*(8), 4782–4791.

Bain, G., Maandag, E. C. R., Izon, D. J., Amsen, D., Kruisbeek, A. M., Weintraub, B. C., et al. (1994). E2A proteins are required for proper B cell development and initiation of immunoglobulin gene rearrangements. *Cell*, *79*(5), 885–892.

Barker, N., & Clevers, H. (2006). Mining the Wnt pathway for cancer therapeutics. *Nature Reviews Drug Discovery*, *5*(12), 997–1014.

Barndt, R. J., Dai, M., & Zhuang, Y. (2000). Functions of E2A-HEB heterodimers in T-cell development revealed by a dominant negative mutation of HEB. *Molecular and Cellular Biology*, *20*(18), 6677–6685.

Benezra, R., Davis, R. L., Lockshon, D., Turner, D. L., & Weintraub, H. (1990). The protein Id: A negative regulator of helix-loop-helix DNA binding proteins. *Cell*, *61*(1), 49–59.

Bishop, P. C., Rao, V. K., & Wilson, W. H. (2000). Burkitt's lymphoma: Molecular pathogenesis and treatment. *Cancer Investigation*, *18*(6), 574–583.

Bloch, K., Buchanan, W., Wohl, M., & Bunim, J. (1965). Sjogren's syndrome: A clinical, pathological, and serological study of sixty-two cases. *Medicine*, *44*(3), 187–231.

Blom, B., Heemskerk, M. H., Verschuren, M. C., Van Dongen, J. J., Stegmann, A. P., Bakker, A. Q., et al. (1999). Disruption of αβ but not of γδ T cell development by overexpression of the helix-loop-helix protein Id3 in committed T cell progenitors. *The EMBO Journal*, *18*(10), 2793–2802.

Boisson, B., Wang, Y.-D., Bosompem, A., Ma, C. S., Lim, A., Kochetkov, T., et al. (2013). A recurrent dominant negative E47 mutation causes agammaglobulinemia and BCR-B cells. *The Journal of Clinical Investigation*, *123*(11), 4781.

Boos, M. D., Yokota, Y., Eberl, G., & Kee, B. L. (2007). Mature natural killer cell and lymphoid tissue-inducing cell development requires Id2-mediated suppression of E protein activity. *The Journal of Experimental Medicine*, *204*(5), 1119–1130.

Buch, T., Rieux-Laucat, F., Förster, I., & Rajewsky, K. (2002). Failure of HY-specific thymocytes to escape negative selection by receptor editing. *Immunity*, *16*(5), 707–718.

Butler, R. D., & Hainsworth, J. D. (1993). Optimal therapy for small noncleaved cell lymphoma. In W. D. Dana (Ed.), *Malignant lymphomas, including Hodgkin's disease: diagnosis, management, and special problems* (pp. 65–67). Boston, MA: Springer.

Caricasole, A., Copani, A., Caruso, A., Caraci, F., Iacovelli, L., Sortino, M. A., et al. (2003). The Wnt pathway, cell-cycle activation and β-amyloid: Novel therapeutic strategies in Alzheimer's disease? *Trends in Pharmacological Sciences*, *24*(5), 233–238.

Chan, E. A., Teng, G., Corbett, E., Choudhury, K. R., Bassing, C. H., Schatz, D. G., et al. (2013). Peripheral subnuclear positioning suppresses Tcrb recombination and segregates Tcrb alleles from RAG2. *Proceedings of the National Academy of Sciences*, *110*(48), E4628–E4637.

Choi, J. K., Shen, C.-P., Radomska, H., Eckhardt, L., & Kadesch, T. (1996). E47 activates the Ig-heavy chain and TdT loci in non-B cells. *The EMBO Journal*, *15*(18), 5014.

Dalavanga, Y., Drosos, A., & Moutsopoulos, H. (1985). Labial salivary gland immunopathology in Sjogren's syndrome. *Scandinavian Journal of Rheumatology. Supplement, 61*, 67–70.
De Vries, J. E. (1998). The role of IL-13 and its receptor in allergy and inflammatory responses. *Journal of Allergy and Clinical Immunology, 102*(2), 165–169.
Deed, R. W., Hara, E., Atherton, G. T., Peters, G., & Norton, J. D. (1997). Regulation of Id3 cell cycle function by Cdk-2-dependent phosphorylation. *Molecular and Cellular Biology, 17*(12), 6815–6821.
Dias, S., Månsson, R., Gurbuxani, S., Sigvardsson, M., & Kee, B. L. (2008). E2A proteins promote development of lymphoid-primed multipotent progenitors. *Immunity, 29*(2), 217–227.
Eckner, R., Yao, T.-P., Oldread, E., & Livingston, D. M. (1996). Interaction and functional collaboration of p300/CBP and bHLH proteins in muscle and B-cell differentiation. *Genes & Development, 10*(19), 2478–2490.
Engel, I., Johns, C., Bain, G., Rivera, R. R., & Murre, C. (2001). Early thymocyte development is regulated by modulation of E2A protein activity. *The Journal of Experimental Medicine, 194*(6), 733–746.
Engel, I., & Murre, C. (1999). Ectopic expression of E47 or E12 promotes the death of E2A-deficient lymphomas. *Proceedings of the National Academy of Sciences, 96*(3), 996–1001.
Ephrussi, A., Church, G. M., Tonegawa, S., & Gilbert, W. (1985). B lineage-specific interactions of an immunoglobulin enhancer with cellular factors in vivo. *Science, 227*(4683), 134–140.
Fu, X., Roberts, W. G., Nobile, V., Shapiro, R., & Kamps, M. P. (1999). mAngiogenin-3, a target gene of oncoprotein E2a-Pbx1, encodes a new angiogenic member of the angiogenin family. *Growth Factors, 17*(2), 125–137.
Fuxa, M., Skok, J., Souabni, A., Salvagiotto, G., Roldan, E., & Busslinger, M. (2004). Pax5 induces V-to-DJ rearrangements and locus contraction of the immunoglobulin heavy-chain gene. *Genes & Development, 18*(4), 411–422.
Ghosh, J. K., Romanow, W. J., & Murre, C. (2001). Induction of a diverse T cell receptor γ/δ repertoire by the helix-loop-helix proteins E2A and HEB in nonlymphoid cells. *The Journal of Experimental Medicine, 193*(6), 769–776.
Goebel, P., Janney, N., Valenzuela, J. R., Romanow, W. J., Murre, C., & Feeney, A. J. (2001). Localized gene-specific induction of accessibility to V (D) J recombination induced by E2A and early B cell factor in nonlymphoid cells. *The Journal of Experimental Medicine, 194*(5), 645–656.
Goldfarb, A. N., Flores, J. P., & Lewandowska, K. (1996). Involvement of the E2A basic helix-loop-helix protein in immunoglobulin heavy chain class switching. *Molecular Immunology, 33*(11), 947–956.
Greenbaum, S., Lazorchak, A. S., & Zhuang, Y. (2004). Differential functions for the transcription factor E2A in positive and negative gene regulation in pre-B lymphocytes. *Journal of Biological Chemistry, 279*(43), 45028–45035.
Greenspan, J., Daniels, T., Talal, N., & Sylvester, R. (1974). The histopathology of Sjögren's syndrome in labial salivary gland biopsies. *Oral Surgery, Oral Medicine, Oral Pathology, 37*(2), 217–229.
Grünig, G., Warnock, M., Wakil, A. E., Venkayya, R., Brombacher, F., Rennick, D. M., et al. (1998). Requirement for IL-13 independently of IL-4 in experimental asthma. *Science, 282*(5397), 2261–2263.
Guo, Z., Li, H., Han, M., Xu, T., Wu, X., & Zhuang, Y. (2011). Modeling Sjögren's syndrome with Id3 conditional knockout mice. *Immunology Letters, 135*(1), 34–42.
Hardy, R. R., & Hayakawa, K. (2001). B cell development pathways. *Annual Review of Immunology, 19*(1), 595–621.

Hayakawa, I., Tedder, T. F., & Zhuang, Y. (2007). B-lymphocyte depletion ameliorates Sjögren's syndrome in Id3 knockout mice. *Immunology*, *122*(1), 73–79.

Henthorn, P., Kiledjian, M., & Kadesch, T. (1990). Two distinct transcription factors that bind the immunoglobulin enhancer microE5/kappa 2 motif. *Science*, *247*(4941), 467–470.

Herblot, S., Steff, A.-M., Hugo, P., Aplan, P. D., & Hoang, T. (2000). SCL and LMO1 alter thymocyte differentiation: Inhibition of E2A-HEB function and pre-Tα chain expression. *Nature Immunology*, *1*(2), 138–144.

Hogquist, K. A., Jameson, S. C., Heath, W. R., Howard, J. L., Bevan, M. J., & Carbone, F. R. (1994). T cell receptor antagonist peptides induce positive selection. *Cell*, *76*(1), 17–27.

Hsu, H.-L., Huang, L., Tsan, J. T., Funk, W., Wright, W. E., Hu, J.-S., et al. (1994). Preferred sequences for DNA recognition by the TAL1 helix-loop-helix proteins. *Molecular and Cellular Biology*, *14*(2), 1256–1265.

Hsu, L.-Y., Lauring, J., Liang, H.-E., Greenbaum, S., Cado, D., Zhuang, Y., et al. (2003). A conserved transcriptional enhancer regulates RAG gene expression in developing B cells. *Immunity*, *19*(1), 105–117.

Hu, J.-S., Olson, E., & Kingston, R. (1992). HEB, a helix-loop-helix protein related to E2A and ITF2 that can modulate the DNA-binding ability of myogenic regulatory factors. *Molecular and Cellular Biology*, *12*(3), 1031–1042.

Hu, T., Wang, H., Simmons, A., Bajaña, S., Zhao, Y., Kovats, S., et al. (2013). Increased level of e protein activity during invariant NKT development promotes differentiation of invariant NKT2 and invariant NKT17 subsets. *The Journal of Immunology*, *191*(10), 5065–5073.

Ji, Y., Pos, Z., Rao, M., Klebanoff, C. A., Yu, Z., Sukumar, M., et al. (2011). Repression of the DNA-binding inhibitor Id3 by Blimp-1 limits the formation of memory CD8+ T cells. *Nature Immunology*, *12*(12), 1230–1237.

Jia, J., Dai, M., & Zhuang, Y. (2008). E proteins are required to activate germline transcription of the TCR Vβ8. 2 gene. *European Journal of Immunology*, *38*(10), 2806–2820.

Jones, M. E., & Zhuang, Y. (2007). Acquisition of a functional T cell receptor during T lymphocyte development is enforced by HEB and E2A transcription factors. *Immunity*, *27*(6), 860–870.

Jones-Mason, M. E., Zhao, X., Kappes, D., Lasorella, A., Iavarone, A., & Zhuang, Y. (2012). E protein transcription factors are required for the development of CD4+ lineage T cells. *Immunity*, *36*(3), 348–361.

Knoepfler, P. S., Calvo, K. R., Chen, H., Antonarakis, S. E., & Kamps, M. P. (1997). Meis1 and pKnox1 bind DNA cooperatively with Pbx1 utilizing an interaction surface disrupted in oncoprotein E2a-Pbx1. *Proceedings of the National Academy of Sciences*, *94*(26), 14553–14558.

Krangel, M. S. (2003). Gene segment selection in V (D) J recombination: Accessibility and beyond. *Nature Immunology*, *4*(7), 624–630.

Kruisbeek, A. M., Haks, M. C., Carleton, M., Wiest, D. L., Michie, A. M., & Zúñiga-Pflücker, J. C. (2000). Branching out to gain control: How the pre-TCR is linked to multiple functions. *Immunology Today*, *21*(12), 637–644.

Kusunoki, T., Sugai, M., Katakai, T., Omatsu, Y., Iyoda, T., Inaba, K., et al. (2003). TH2 dominance and defective development of a CD8+ dendritic cell subset in Id2-deficient mice. *Journal of Allergy and Clinical Immunology*, *111*(1), 136–142.

Kwon, K., Hutter, C., Sun, Q., Bilic, I., Cobaleda, C., Malin, S., et al. (2008). Instructive role of the transcription factor E2A in early B lymphopoiesis and germinal center B cell development. *Immunity*, *28*(6), 751–762.

Langerak, A. W., Wolvers-Tettero, I. L., Van Gastel-Mol, E. J., Oud, M. E., & Van Dongen, J. J. (2001). Basic helix-loop-helix proteins E2A and HEB induce immature T-cell receptor rearrangements in nonlymphoid cells. *Blood, 98*(8), 2456–2465.

Lazorchak, A. S., Schlissel, M. S., & Zhuang, Y. (2006). E2A and IRF-4/Pip promote chromatin modification and transcription of the immunoglobulin κ locus in pre-B cells. *Molecular and Cellular Biology, 26*(3), 810–821.

Lazorchak, A. S., Wojciechowski, J., Dai, M., & Zhuang, Y. (2006). E2A promotes the survival of precursor and mature B lymphocytes. *The Journal of Immunology, 177*(4), 2495–2504.

Lee, B. H., Gauna, A. E., Pauley, K. M., Park, Y.-J., & Cha, S. (2012). Animal models in autoimmune diseases: Lessons learned from mouse models for Sjögren's syndrome. *Clinical Reviews in Allergy & Immunology, 42*(1), 35–44.

Li, H., Dai, M., & Zhuang, Y. (2004). AT cell intrinsic role of Id3 in a mouse model for primary Sjögren's syndrome. *Immunity, 21*(4), 551–560.

Li, J., Wu, D., Jiang, N., & Zhuang, Y. (2013). Combined deletion of Id2 and Id3 genes reveals multiple roles for E proteins in invariant NKT cell development and expansion. *The Journal of Immunology, 191*(10), 5052–5064.

Lin, Y. C., Benner, C., Mansson, R., Heinz, S., Miyazaki, K., Miyazaki, M., et al. (2012). Global changes in the nuclear positioning of genes and intra-and interdomain genomic interactions that orchestrate B cell fate. *Nature Immunology, 13*(12), 1196–1204.

Lin, H., & Grosschedl, R. (1995). Failure of B-cell differentiation in mice lacking the transcription factor EBF. *Nature, 376*(6537), 263–267.

Lin, Y. C., Jhunjhunwala, S., Benner, C., Heinz, S., Welinder, E., Mansson, R., et al. (2010). A global network of transcription factors, involving E2A, EBF1 and Foxo1, that orchestrates B cell fate. *Nature Immunology, 11*(7), 635–643. http://www.nature.com/ni/journal/v11/n7/suppinfo/ni.1891_S1.html.

Lin, Y. C., Jhunjhunwala, S., Benner, C., Heinz, S., Welinder, E., Mansson, R., et al. (2010). Mutation of inhibitory helix-loop-helix protein Id3 causes γδ T-cell lymphoma in mice. *Blood, 116*(25), 5615–5621.

Lin, Y.-Y., Jones-Mason, M. E., Inoue, M., Lasorella, A., Iavarone, A., Li, Q.-J., et al. (2012). Transcriptional regulator Id2 is required for the CD4 T cell immune response in the development of experimental autoimmune encephalomyelitis. *The Journal of Immunology, 189*(3), 1400–1405.

Loffert, D., Schaal, S., Ehlich, A., Hardy, R. R., Zou, Y. R., Müller, W., et al. (1994). Early B cell development in the mouse: Insights from mutations introduced by gene targeting. *Immunological Reviews, 137*(1), 135–153.

Love, C., Sun, Z., Jima, D., Li, G., Zhang, J., Miles, R., et al. (2012). The genetic landscape of mutations in Burkitt lymphoma. *Nature Genetics, 44*(12), 1321–1325.

Lu, Q., & Kamps, M. P. (1996). Structural determinants within Pbx1 that mediate cooperative DNA binding with pentapeptide-containing Hox proteins: Proposal for a model of a Pbx1-Hox-DNA complex. *Molecular and Cellular Biology, 16*(4), 1632–1640.

Lund, R. J., Löytömäki, M., Naumanen, T., Dixon, C., Chen, Z., Ahlfors, H., et al. (2007). Genome-wide identification of novel genes involved in early Th1 and Th2 cell differentiation. *The Journal of Immunology, 178*(6), 3648–3660.

Mahlios, J., & Zhuang, Y. (2011). Contribution of IL-13 to early exocrinopathy in Id3 −/− mice. *Molecular Immunology, 49*(1), 227–233.

Maraskovsky, E., O'Reilly, L. A., Teepe, M., Corcoran, L. M., Peschon, J. J., & Strasser, A. (1997). Bcl-2 can rescue T lymphocyte development in interleukin-7 receptor-deficient mice but not in mutant rag-1 −/− Mice. *Cell, 89*(7), 1011–1019.

Massari, M. E., Grant, P. A., Pray-Grant, M. G., Berger, S. L., Workman, J. L., & Murre, C. (1999). A conserved motif present in a class of helix-loop-helix proteins activates transcription by direct recruitment of the SAGA complex. *Molecular Cell, 4*(1), 63–73.

Matloubian, M., Lo, C. G., Cinamon, G., Lesneski, M. J., Xu, Y., Brinkmann, V., et al. (2004). Lymphocyte egress from thymus and peripheral lymphoid organs is dependent on S1P receptor 1. *Nature, 427*(6972), 355–360.

McKenzie, A., Culpepper, J., De Waal Malefyt, R., Briere, F., Punnonen, J., Aversa, G., et al. (1993). Interleukin 13, a T-cell-derived cytokine that regulates human monocyte and B-cell function. *Proceedings of the National Academy of Sciences, 90*(8), 3735–3739.

McWhirter, J. R., Neuteboom, S. T. C., Wancewicz, E. V., Monia, B. P., Downing, J. R., & Murre, C. (1999). *Proceedings of the National Academy of Sciences, 96*(20), 11464–11469.

Mendoza, M. C., Er, E. E., & Blenis, J. (2011). The Ras-ERK and PI3K-mTOR pathways: Cross-talk and compensation. *Trends in Biochemical Sciences, 36*(6), 320–328.

Mitsias, D., Tzioufas, A., Veiopoulou, C., Zintzaras, E., Tassios, I., Kogopoulou, O., et al. (2002). The Th1/Th2 cytokine balance changes with the progress of the immunopathological lesion of Sjogren's syndrome. *Clinical & Experimental Immunology, 128*(3), 562–568.

Miyazaki, M., Rivera, R. R., Miyazaki, K., Lin, Y. C., Agata, Y., & Murre, C. (2011). The opposing roles of the transcription factor E2A and its antagonist Id3 that orchestrate and enforce the naive fate of T cells. *Nature Immunology, 12*(10), 992–1001.

Monica, K., LeBrun, D. P., Dedera, D. A., Brown, R., & Cleary, M. L. (1994). Transformation properties of the E2a-Pbx1 chimeric oncoprotein: Fusion with E2a is essential, but the Pbx1 homeodomain is dispensable. *Molecular and Cellular Biology, 14*(12), 8304–8314.

Murre, C., & Baltimore, D. (1993). The helix-loop-helix motif: Structure and function. *Cold Spring Harbor Monograph Series, 22*, 861.

Murre, C., McCaw, P. S., & Baltimore, D. (1989). A new DNA binding and dimerization motif in immunoglobulin enhancer binding, daughterless, MyoD and myc proteins. *Cell, 56*(5), 777–783.

Nossal, G. (1983). Cellular mechanisms of immunologic tolerance. *Annual Review of Immunology, 1*(1), 33–62.

Nutt, S. L., Heavey, B., Rolink, A. G., & Busslinger, M. (1999). Commitment to the B-lymphoid lineage depends on the transcription factor Pax5. *Nature, 401*(6753), 556–562.

Okada, T., Maeda, A., Iwamatsu, A., Gotoh, K., & Kurosaki, T. (2000). BCAP: The tyrosine kinase substrate that connects B cell receptor to phosphoinositide 3-kinase activation. *Immunity, 13*(6), 817–827.

Okkenhaug, K., & Vanhaesebroeck, B. (2003). PI3K in lymphocyte development, differentiation and activation. *Nature Reviews Immunology, 3*(4), 317–330.

Osborne, C. S., Chakalova, L., Mitchell, J. A., Horton, A., Wood, A. L., Bolland, D. J., et al. (2007). Myc dynamically and preferentially relocates to a transcription factory occupied by Igh. *PLoS Biology, 5*(8), e192.

Pan, L., Sato, S., Frederick, J. P., Sun, X.-H., & Zhuang, Y. (1999). Impaired immune responses and B-cell proliferation in mice lacking the Id3 gene. *Molecular and Cellular Biology, 19*(9), 5969–5980.

Park, S. T., Nolan, G. P., & Sun, X.-H. (1999). Growth inhibition and apoptosis due to restoration of E2A activity in T cell acute lymphoblastic leukemia cells. *The Journal of Experimental Medicine, 189*(3), 501–508.

Plotkin, J., Prockop, S. E., Lepique, A., & Petrie, H. T. (2003). Critical role for CXCR4 signaling in progenitor localization and T cell differentiation in the postnatal thymus. *The Journal of Immunology, 171*(9), 4521–4527.

Polakis, P. (2000). Wnt signaling and cancer. *Genes and Development, 14*(15), 1837–1851.

Quong, M. W., Harris, D. P., Swain, S. L., & Murre, C. (1999). E2A activity is induced during B-cell activation to promote immunoglobulin class switch recombination. *The EMBO Journal, 18*(22), 6307–6318.

Reya, T., O'Riordan, M., Okamura, R., Devaney, E., Willert, K., Nusse, R., et al. (2000). Wnt signaling regulates B lymphocyte proliferation through a LEF-1 dependent mechanism. *Immunity, 13*(1), 15–24.
Richter, J., Schlesner, M., Hoffmann, S., Kreuz, M., Leich, E., Burkhardt, B., et al. (2012). Recurrent mutation of the ID3 gene in Burkitt lymphoma identified by integrated genome, exome and transcriptome sequencing. *Nature Genetics, 44*(12), 1316–1320.
Rivera, R. R., Johns, C. P., Quan, J., Johnson, R. S., & Murre, C. (2000). Thymocyte selection is regulated by the helix-loop-helix inhibitor protein, Id3. *Immunity, 12*(1), 17–26.
Robey, E., & Fowlkes, B. (1994). Selective events in T cell development. *Annual Review of Immunology, 12*(1), 675–705.
Romanow, W. J., Langerak, A. W., Goebel, P., Wolvers-Tettero, I. L., Van Dongen, J. J., Feeney, A. J., et al. (2000). E2A and EBF act in synergy with the V (D) J recombinase to generate a diverse immunoglobulin repertoire in nonlymphoid cells. *Molecular Cell, 5*(2), 343–353.
Sakamoto, S., Wakae, K., Anzai, Y., Murai, K., Tamaki, N., Miyazaki, M., et al. (2012). E2A and CBP/p300 act in synergy to promote chromatin accessibility of the immunoglobulin κ locus. *The Journal of Immunology, 188*(11), 5547–5560.
Sathish, J. G., Johnson, K. G., Leroy, F. G., Fuller, K. J., Hallett, M. B., Brennan, P., et al. (2001). Requirement for CD28 costimulation is lower in SHP 1 deficient T cells. *European Journal of Immunology, 31*(12), 3649–3658.
Sawada, S., & Littman, D. (1993). A heterodimer of HEB and an E12-related protein interacts with the CD4 enhancer and regulates its activity in T-cell lines. *Molecular and Cellular Biology, 13*(9), 5620–5628.
Sayegh, C., Jhunjhunwala, S., Riblet, R., & Murre, C. (2005). Visualization of looping involving the immunoglobulin heavy-chain locus in developing B cells. *Genes & Development, 19*(3), 322–327.
Sayegh, C. E., Quong, M. W., Agata, Y., & Murre, C. (2003). E-proteins directly regulate expression of activation-induced deaminase in mature B cells. *Nature Immunology, 4*(6), 586–593.
Schlimgen, R. J., Reddy, K. L., Singh, H., & Krangel, M. S. (2008). Initiation of allelic exclusion by stochastic interaction of Tcrb alleles with repressive nuclear compartments. *Nature Immunology, 9*(7), 802–809.
Schlissel, M., Voronova, A., & Baltimore, D. (1991). Helix-loop-helix transcription factor E47 activates germ-line immunoglobulin heavy-chain gene transcription and rearrangement in a pre-T-cell line. *Genes & Development, 5*(8), 1367–1376.
Schmitz, R., Young, R. M., Ceribelli, M., Jhavar, S., Xiao, W., Zhang, M., et al. (2012). Burkitt lymphoma pathogenesis and therapeutic targets from structural and functional genomics. *Nature, 490*(7418), 116–120.
Slattery, C., Mcmorrow, T., & Ryan, M. P. (2006). Overexpression of E2A proteins induces epithelial-mesenchymal transition in human renal proximal tubular epithelial cells suggesting a potential role in renal fibrosis. *FEBS Letters, 580*(17), 4021–4030.
Steininger, A., Möbs, M., Ullmann, R., Köchert, K., Kreher, S., Lamprecht, B., et al. (2011). Genomic loss of the putative tumor suppressor gene E2A in human lymphoma. *The Journal of Experimental Medicine, 208*(8), 1585–1593.
Stocker, H., Andjelkovic, M., Oldham, S., Laffargue, M., Wymann, M. P., Hemmings, B. A., et al. (2002). Living with lethal PIP3 levels: Viability of flies lacking PTEN restored by a PH domain mutation in Akt/PKB. *Science, 295*(5562), 2088–2091.
Sun, X.-H., & Baltimore, D. (1991). An inhibitory domain of E12 transcription factor prevents DNA binding in E12 homodimers but not in E12 heterodimers. *Cell, 64*(2), 459–470.

Sun, X.-H., Copeland, N. G., Jenkins, N. A., & Baltimore, D. (1991). Id proteins Id1 and Id2 selectively inhibit DNA binding by one class of helix-loop-helix proteins. *Molecular and Cellular Biology*, *11*(11), 5603–5611.

Szodoray, P., Alex, P., Brun, J., Centola, M., & Jonsson, R. (2004). Circulating cytokines in primary Sjögren's syndrome determined by a multiplex cytokine array system. *Scandinavian Journal of Immunology*, *59*(6), 592–599.

Tarpley, T. M., Jr., Anderson, L. G., & White, C. L. (1974). Minor salivary gland involvement in Sjögren's syndrome. *Oral Surgery, Oral Medicine, Oral Pathology*, *37*(1), 64–74.

Thal, M. A., Carvalho, T. L., He, T., Kim, H.-G., Gao, H., Hagman, J., et al. (2009). Ebf1-mediated down-regulation of Id2 and Id3 is essential for specification of the B cell lineage. *Proceedings of the National Academy of Sciences*, *106*(2), 552–557.

Treiber, T., Mandel, E. M., Pott, S., Györy, I., Firner, S., Liu, E. T., et al. (2010). Early B cell factor 1 regulates B cell gene networks by activation, repression, and transcription-independent poising of chromatin. *Immunity*, *32*(5), 714–725.

Tuveson, D. A., Carter, R. H., Soltoff, S. P., & Fearon, D. T. (1993). CD19 of B cells as a surrogate kinase insert region to bind phosphatidylinositol 3-kinase. *Science*, *260*(5110), 986–989.

Ueda-Hayakawa, I., Mahlios, J., & Zhuang, Y. (2009). Id3 restricts the developmental potential of γδ lineage during thymopoiesis. *The Journal of Immunology*, *182*(9), 5306–5316.

Urbánek, P., Wang, Z.-Q., Fetka, I., Wagner, E. F., & Busslinger, M. (1994). Complete block of early B cell differentiation and altered patterning of the posterior midbrain in mice lacking Pax5BSAP. *Cell*, *79*(5), 901–912.

Van Den Berg, D. J., Sharma, A. K., Bruno, E., & Hoffman, R. (1998). Role of members of the Wnt gene family in human hematopoiesis. *Blood*, *92*(9), 3189–3202.

Vanhaesebroeck, B., & Alessi, D. (2000). The PI3K–PDK1 connection: More than just a road to PKB. *Biochemical Journal*, *346*, 561–576.

Verykokakis, M., Boos, M. D., Bendelac, A., Adams, E. J., Pereira, P., & Kee, B. L. (2010). Inhibitor of DNA binding 3 limits development of murine Slam-associated adaptor protein-dependent "innate" γδ T cells. *PLoS ONE*, *5*(2), e9303.

Verykokakis, M., Krishnamoorthy, V., Iavarone, A., Lasorella, A., Sigvardsson, M., & Kee, B. L. (2013). Essential functions for ID proteins at multiple checkpoints in invariant NKT cell development. *The Journal of Immunology*, *191*(12), 5973–5983. http://dx.doi.org/10.4049/jimmunol.1301521.

Von Boehmer, H. (1994). Positive selection of lymphocytes. *Cell*, *76*(2), 219–228.

Wang, D., Claus, C. L., Vaccarelli, G., Braunstein, M., Schmitt, T. M., Zúñiga-Pflücker, J. C., et al. (2006). The basic helix-loop-helix transcription factor HEBAlt is expressed in pro-T cells and enhances the generation of T cell precursors. *The Journal of Immunology*, 177(1), 109–119.

Yan, W., Young, A. Z., Soares, V. C., Kelley, R., Benezra, R., & Zhuang, Y. (1997). *Molecular and Cellular Biology*, *17*(12), 7317–7327.

Yang, C. Y., Best, J. A., Knell, J., Yang, E., Sheridan, A. D., Jesionek, A. K., et al. (2011). The transcriptional regulators Id2 and Id3 control the formation of distinct memory CD8+ T cell subsets. *Nature Immunology*, *12*(12), 1221–1229.

Yankee, T., & Clark, E. (1999). Signaling through the B cell antigen receptor in developing B cells. *Reviews in Immunogenetics*, *2*(2), 185–203.

Yu, W., Misulovin, Z., Suh, H., Hardy, R. R., Jankovic, M., Yannoutsos, N., et al. (1999). Coordinate regulation of RAG1 and RAG2 by cell type-specific DNA elements 5' of RAG2. *Science*, *285*(5430), 1080–1084.

Zhang, B., Dai, M., Li, Q.-J., & Zhuang, Y. (2013). Tracking proliferative history in lymphocyte development with Cre-mediated sister chromatid recombination. *PLoS Genetics*, *9*(10), e1003887.

Zhang, B., Lin, Y.-Y., Dai, M., & Zhuang, Y. (2014). Id3 and Id2 act as a dual safety mechanism in regulating the development and population size of innate-like γδ T cells. *The Journal of Immunology*, *192*, 1055–1063, 1302694.

Zhao, F., Vilardi, A., Neely, R. J., & Choi, J. K. (2001). Promotion of cell cycle progression by basic helix-loop-helix E2A. *Molecular and Cellular Biology*, *21*(18), 6346–6357.

Zhuang, Y., Cheng, P., & Weintraub, H. (1996). B-lymphocyte development is regulated by the combined dosage of three basic helix-loop-helix genes, E2A, E2-2, and HEB. *Molecular and Cellular Biology*, *16*(6), 2898–2905.

Zhuang, Y., Soriano, P., & Weintraub, H. (1994). The helix-loop-helix gene E2A is required for B cell formation. *Cell*, *79*(5), 875–884.

CHAPTER FIVE

Id Proteins: Small Molecules, Mighty Regulators

Flora Ling, Bin Kang, Xiao-Hong Sun[1]
Immunobiology Cancer Research Program, Oklahoma Medical Research Foundation, Oklahoma City, Oklahoma, USA
[1]Corresponding author: e-mail address: sunx@omrf.org

Contents

Abstract

The family of inhibitor of differentiation (Id) proteins is a group of evolutionarily conserved molecules, which play important regulatory roles in organisms ranging from *Drosophila* to humans. Id proteins are small polypeptides harboring a helix-loop-helix (HLH) motif, which are best known to mediate dimerization with other basic HLH proteins, primarily E proteins. Because Id proteins do not possess the basic amino acids adjacent to the HLH motif necessary for DNA binding, Id proteins inhibit the function of E protein homodimers, as well as heterodimers between E proteins and tissue-specific bHLH proteins. However, Id proteins have also been shown to have E protein-independent functions. The Id genes are broadly but differentially expressed in a variety of cell types. Transcription of the Id genes is controlled by transcription factors such as C/EBPβ and Egr as well as by signaling pathways triggered by different stimuli, which include bone morphogenic proteins, cytokines, and ligands of T cell receptors. In general, Id proteins are capable of inhibiting the differentiation of progenitors of different cell types, promoting cell-cycle progression, delaying cellular senescence, and facilitating cell migration. These properties of Id proteins enable them to play significant roles in stem cell maintenance, vasculogenesis, tumorigenesis and metastasis, the development of the immune system, and energy metabolism. In this review, we intend to highlight

Current Topics in Developmental Biology, Volume 110
ISSN 0070-2153
http://dx.doi.org/10.1016/B978-0-12-405943-6.00005-1
© 2014 Elsevier Inc.
All rights reserved.

the current understanding of the function of Id proteins and discuss gaps in our knowledge about the mechanisms whereby Id proteins exert their diverse effects in multiple cellular processes.

1. INTRODUCTION

The inhibitor of differentiation (Id) family of helix-loop-helix (HLH) proteins is a group of evolutionarily conserved molecules, which play important regulatory roles in organisms ranging from *Drosophila* to humans. Because of their dynamic patterns of expression, the Id family has recently attracted much attention. In this chapter, we will review current advances in our knowledge about the general function of Id proteins and regulation of the Id genes. We will then discuss in broad terms the role of Id proteins in several representative cellular processes.

2. THE STRUCTURE AND FUNCTION OF ID PROTEINS

Mammals have four Id proteins, Id1 to Id4 (Benezra, Davis, Lockshon, Turner, & Weintraub, 1990; Christy et al., 1991; Riechmann, van Cruchten, & Sablitzky, 1994; Sun, Copeland, Jenkins, & Baltimore, 1991), that share extensive sequence homology in the HLH motif (Fig. 5.1A), which mediates the dimerization of Id proteins with basic HLH proteins, primarily the group of E proteins encoded by the *E2A* (giving rise to E12 and E47), *HEB*, and *E2-2* genes (Massari & Murre, 2000). Because Id proteins do not have the basic amino acids necessary for DNA binding, they serve as naturally occurring dominant-negative inhibitors of E proteins by forming nonfunctional heterodimers (Fig. 5.1B). Outside of the HLH motif, there is little similarity among the four proteins except for a few patches of homology whose functional significance has not been well defined. Therefore, the four Id proteins have one common function, namely, to inhibit the DNA-binding activity of E proteins. By sequestering E proteins, Id proteins also impair the function of their heterodimers with other tissue-specific bHLH proteins (Massari & Murre, 2000).

Although the affinities of different Id proteins for each of the E proteins have not been biochemically quantified, it is safe to say that the Id proteins have redundant functions in sequestering the four E proteins, which also resemble each other in terms of their structures and functions. Hence, it

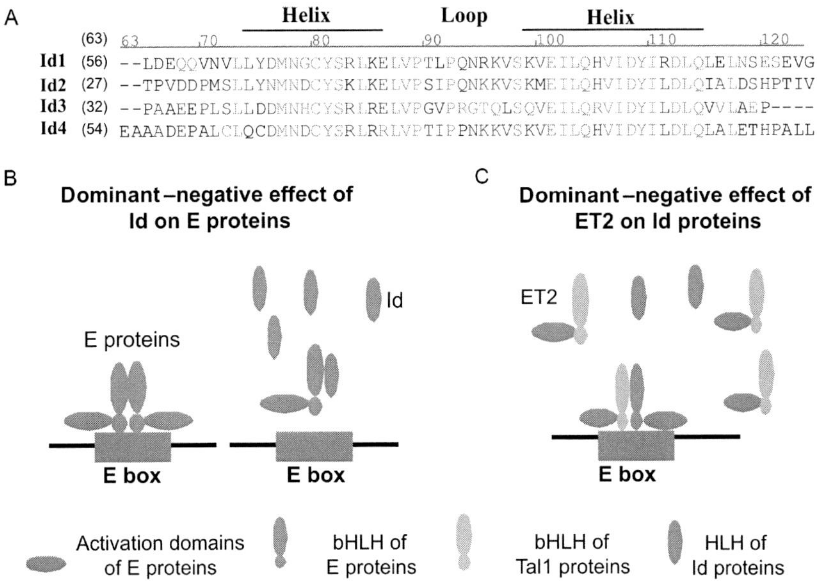

Figure 5.1 Structural similarity of Id proteins and their mode of action. (A) Sequence alignment of the HLH motif of the four human Id proteins. Identical amino acids are shown in red; conserved changes in blue and weak homologies in green. Schematic diagrams of the effect of Id proteins on the DNA binding activity of all E proteins (B) and the ability of ET2 to compete with all Id proteins to dimerize with E proteins and activate transcription of E protein target genes (C). ET2 is an artificial hybrid molecule between E47 and Tal1. The symbols for each functional domain of Id, Tal1 and E proteins are indicated by the legend included in the figure. (See the color plate.)

is important to remember that the net E protein activity in a given cell is determined by the total concentration of all E proteins subtracted by that of all Id proteins, assuming that influences by posttranslational modifications of E and Id proteins are equivalent. Therefore, overexpressing Id proteins has been employed as an effective approach to delineate the collective function of E proteins (Kim, Peng, & Sun, 1999; Morrow, Mayer, Perez, Adlam, & Siu, 1999; Sun, 1994). Conversely, to study the combined role of Id proteins, an artificial molecule, designated ET2, has been created and exploited (Cochrane, Zhao, Welner, & Sun, 2009; Hu et al., 2013; Zhang, Zhao, & Sun, 2013). ET2 contains the N-terminal 573 amino acids of E47 including the two known transcription activation domains and the C-terminal 143 amino acids of SCL/Tal1 consisting of the bHLH domain (Cochrane et al., 2009; Park & Sun, 1998; Park, Nolan, & Sun, 1999). Because the bHLH domain of SCL/Tal1 does not dimerize with itself or Id proteins yet has a high affinity for E proteins, ET2 can interact with

E proteins avidly and bind to E box DNA sequences, and since ET2 contains the transcription activation domains of E47, heterodimers between ET2 and E proteins can activate transcription of E protein target genes. Importantly, ET2 has the ability to compete with Id proteins to associate with E proteins and neutralize the inhibitory effects of Id proteins. The ET2 protein, when overexpressed, can thus be utilized to learn the impacts of ablating multiple Id proteins.

Apart from the ability of Id proteins to inhibit the function of E proteins, they have also been shown to interact with other proteins without the HLH motifs. For example, Id2, but not other Id proteins, has been shown to associate with Rb proteins via its HLH domain, which leads to an antagonistic relationship between these two proteins (Lasorella et al., 2002; Lasorella, Iavarone, & Israel, 1996; Lasorella, Noseda, Beyna, Yokota, & Iavarone, 2000). Additionally, the HLH domain of Id1 has been reported to bind to caveolin 1, a membrane-associated molecule-regulating integrin activity (Zhang et al., 2007). Id3 has been shown to coimmunoprecipitate with the Pax5 protein and inhibit its transcriptional activity (Roberts, Deed, Inoue, Norton, & Sharrocks, 2001). Therefore, it should not be automatically assumed that Id proteins act only by counteracting E protein functions. In this regard, the ET2 protein will be useful to test if an effect of Id proteins in a particular cellular process can be negated by ET2 proteins, which do not sequester Id proteins but restore E protein activity. If Id proteins work independently of E proteins in that situation, ET2 would not affect the function of Id proteins.

3. REGULATION OF ID GENE EXPRESSION

The four Id genes are differentially regulated but often have overlapping patterns of expression. Although it is impossible to delineate the regulation of each Id gene in every cell type, the regulation of Id gene has been elucidated in a number of situations, and the general principles obtained are helpful in considering the control of Id gene expression in a specific cell type or under a specific condition of interest. First, several Id genes have been shown to be activated by bone morphogenic proteins via Smads in a variety of cell types (Hollnagel, Oehlmann, Heymer, Ruther, & Nordheim, 1999; Katagiri et al., 2002; Miyazono & Miyazawa, 2002; Nakahiro, Kurooka, Mori, Sano, & Yokota, 2010; Vrijens et al., 2013). Moreover, the Id1 gene has been shown to have an enhancer element located at the 3′ end of the gene which mediates transcriptional activation by C/EBPβ and STAT5 transcription factors (Saisanit & Sun, 1995, 1997; Xu, Nie, Kim, & Sun, 2003). This mode of regulation may explain why Id1 expression is high in myeloid cells where the C/EBP family of transcription factors plays important roles, and why Id1 is inducible by cytokines such as IL-3 and GM-CSF, which activate STAT5 (Cochrane et al., 2009). Another inflammatory cytokine, IL-6, also stimulates Id1 expression (Cochrane et al., 2009; Maeda et al., 2009). Id2 expression has also been suggested to be activated by C/EBPβ (Karaya et al., 2005). In contrast, the Id3 gene is known to be downstream of the Ras/MAPK pathway and activated by the Egr transcription factors (Bain et al., 2001). This mechanism is well documented in the context of T cell receptor signaling and enables Id3 to participate in several checkpoints during T cell development.

In general, the Id2 mRNA is expressed rather ubiquitously and abundantly, whereas Id4 transcription is limited to a small number of cell types. The Id1 and Id3 genes are often expressed transiently during development and in response to stimulations. If all Id proteins have similar functions, the up and downregulation of Id1 or Id3 production on top of the basal level of Id2 proteins would mean that the threshold of total Id activity needs to be altered under particular circumstances. Alternatively, the function of Id2 proteins is regulated by posttranslational mechanisms. Indeed, Id2 and Id3 have been shown to be modified by cyclin-dependent kinases, which impair its ability to associate with E proteins (Deed, Hara, Atherton, Peters, & Norton, 1997; Hara, Hall, & Peters, 1997). Whether this modification plays a significant role in most cell types is not clear. Furthermore, it is possible that each Id protein preferentially binds to distinct targets, thus having a specific function. However, this idea remains to be vigorously tested.

4. ID PROTEINS IN STEM CELL MAINTENANCE

Stem cells are defined by their ability to self-renew over the lifetime of an organism and to differentiate into various lineages. Since Id proteins are able to block the activity of E protein transcription factors necessary for development, it seems reasonable that Id proteins would have a function in maintaining stem cells by inhibiting differentiation. This is supported by the fact that expression of Id proteins is typically high in embryonic and adult stem/progenitor cells but levels decrease as the cells differentiate. In this section, we will discuss two examples where Id proteins play a significant role in stem cell maintenance.

4.1. Hematopoietic stem cells

Hematopoietic stem cells (HSCs) are responsible for generating all the blood cells for the life of an individual. The extensively studied HSCs are divided into two populations, short-term (ST) and long-term (LT) HSCs. Both populations are capable of repopulating all blood cell lineages and only differ in their ability to self-renewal with ST-HSCs and LT-HSCs having limited and unlimited self-renewal capacity, respectively. HSCs are contained within the LSK population meaning that cells are negative for any lineage markers (Lin−) and positive for c-kit and Sca-1 markers (Ikuta & Weissman, 1992; Spangrude, Heimfeld, & Weissman, 1988). However, the LSK fraction still contains a heterogeneous mixture of cells that include multipotent progenitors and ST- and LT-HSCs. LSK constitutes 0.5% of total bone marrow cells, while the rare LT-HSCs only accounts for 0.003% of the bone marrow (Morrison & Weissman, 1994). It has been shown that a subset of LSK that express Id1 is enriched in LT-HSCs, and loss of Id1 resulted in a decrease in the number of LT-HSCs (Perry et al., 2007). Consistent with the reduced number of LT-HSCs in $Id1^{-/-}$ mice, serial bone marrow transplantation assays to test for self-renewal showed impairment in secondary engraftment but not primary engraftment (Jankovic et al., 2007; Perry et al., 2007). Id1 has also been shown to play a role in the bone marrow niche to support hematopoiesis (Suh et al., 2009). Id3, however, does not appear to play a role in LT-HSC maintenance since transplantation assays with Id3-deficient bone marrow did not show any impairment in LT engraftment (Perry et al., 2007).

The direct role of Id2 in HSC maintenance has not been examined, but it has been shown that expression of Id2 and Id3 in $CD34^{+}$ stem cells can inhibit the development of dendritic cell precursors (Spits, Couwenberg, Bakker, Weijer, & Uittenbogaart, 2000). Additionally, Id2 expression has been found to be regulated by growth factor independent 1 (Gfi-1) which is a zinc-finger protein that functions as a transcriptional repressor (Li, Ji, Klarmann, & Keller, 2010). Gfi-1 has been found to play a role in HSC maintenance and can bind to the Id2 promoter to inhibit Id2 transcription (Li et al., 2010; Zeng, Yucel, Kosan, Klein-Hitpass, & Moroy, 2004). Knocking down Id2 expression in $Gfi\text{-}1^{-/-}$ mice can partially rescue the impaired B cell and myeloid cell development in $Gfi\text{-}1^{-/-}$ mice (Li et al., 2010). Id2 also plays a role in erythroid differentiation as overexpression and knocking down Id2 promoted and hindered the production of erythroid lineage cells, respectively (Ji et al., 2008). These observations suggest that regulation of Id2 expression is important in hematopoiesis, and therefore, Id2 might also have a role in HSC maintenance.

Since HSCs are responsible for repopulating all blood cells for the life of an individual, they can be susceptible to the effects of chronic inflammation caused by infection or autoimmunity. Interestingly, LPS treatment stimulates Id1 expression in HSC populations (Zhao, Ling, Wang, & Sun, 2013). When low-grade chronic inflammation was induced with daily injections of 1 μg/ml of LPS for 30 days, the LT repopulating potential of wild-type HSC was greatly impaired. This effect of LPS treatment was attributed to a transient induction of inflammatory cytokines such as TNFα and IL-10 as well as to the increased turnover of HSCs. However, $Id1^{-/-}$ mice exhibited attenuated responses to the same treatment, namely, sustained LT engraftment compared to untreated cells and unaffected HSC quiescence, which were accompanied by lower levels of induction of TNFα and IL-10 (Zhao et al., 2013). Id1 expression in LSK is also elevated in a mouse model of systemic lupus erythematosis, which is known to impair hematopoiesis (Beyan, Beyan, & Turan, 2007; Maeda et al., 2009). These findings suggest that upregulation of Id1 expression at least in part mediates the HSC damage caused by LPS induced TLR signaling or autoimmunity.

4.2. Neural stem cells

Neural stem cells are self-renewing cells responsible for generating all the major cell types of the nervous system (Sequerra, Costa, Menezes, & Hedin-Pereira, 2013). A rare subset of $GFAP^+$ astrocytes with stem cell characteristics was found to express high levels of Id1 (Nam & Benezra, 2009). This $GFAP^+Id1^{high}$ astrocyte was shown to be able to divide asymmetrically to produce a stem cell with high levels of Id1 and a differentiated cell with lower levels of Id1 (Nam & Benezra, 2009). Therefore, it seems that high levels of Id1 expression are needed to retain self-renewal capacity and maintain the NSC pool (Nam & Benezra, 2009).

Id proteins are thought to maintain NSCs by regulating lineage commitment and preventing NSCs from prematurely differentiating. For example, Id2 and Id4 have been shown to block oligodendrocyte commitment by inhibiting OLIG1 and OLIG2, which are bHLH proteins important in oligodendrocyte development (Samanta & Kessler, 2004). Additionally, Id1 to Id3 have been shown to be instrumental for blocking premature differentiation by sustaining expression of Hes1, which inhibits expression of proneural genes. Low concentrations of Id proteins can prevent Hes1 from negatively regulating itself through a feedback loop, but disrupting Hes1 inhibition of proneural genes requires a higher threshold of Id proteins. Since excess Id proteins are quickly degraded, Id proteins do not reach the necessary concentration to affect Hes1 inhibition of proneural genes and only alleviates Hes1-mediated autorepression (Bai et al., 2007). Id proteins have also been shown to inhibit neuronal differentiation by interfering with the binding of the NeuroD/E47 complex to E boxes (Jung et al., 2010). Taken together, Id proteins appear to preserve NSC self-renewal capacity by preventing differentiation.

Id proteins could also regulate NSC self-renewal by promoting proliferation. It was found that p53 functions as a transcriptional repressor of Id1 and Id2, and $p53^{-/-}$ NSCs have elevated Id expression and proliferation (Liu et al., 2013; Paolella et al., 2011). This would have important implications in the treatment of cancer since p53 is inactivated in a majority of glioblastomas (Paolella et al., 2011).

5. ID PROTEINS IN VASCULOGENESIS

Id1 and Id3 double knockout mice die mid-gestation from multiple cardiac defects and malformation of vasculature, thus highlighting the importance of this group of molecules in vasculogenesis (Lyden et al., 1999; Zhao et al., 2011). The Id1 and Id3 genes share overlapping expression patterns in the cardiovascular system as retaining one of either Id1 or Id3 alleles is sufficient to allow for the survival of $Id1^{-/-}Id3^{+/-}$ or $Id1^{+/-}Id3^{-/-}$ mice (Lyden et al., 1999). Bone morphogenic proteins are known to stimulate the transcription of the Id genes and regulate vascular biology by impacting both endothelial cells and vascular smooth muscle cells through distinct mechanisms. It has been shown that upregulation of Id1 expression mediates BMP-induced activation and migration of endothelial cells (Valdimarsdottir et al., 2002). Ectopic expression of Id1 also delays the senescence of human endothelial cells and promotes the angiogenic activity of fully differentiated endothelial cells, thus providing a potential therapeutic strategy (Nishiyama et al., 2005; Tang, Gordon, Nickoloff, & Foreman, 2002).

In contrast, Id1 and Id3 are coordinately induced in a BMP-dependent manner in hypoxic pulmonary vascular smooth muscle cells (Lowery et al., 2010). Ectopic expression of Id3 in vascular smooth muscle cells leads to cell-cycle inhibition, which may be important for prevention of pulmonary hypertension (Yang et al., 2013). Indeed, expression of BMP receptors and Id1/3 has been found to be reduced in mouse models of pulmonary hypertension. More importantly, BMP type II receptor mutations are responsible for over 70% of the cases of heritable pulmonary arterial hypertension.

With respect to atherosclerosis, Id3 has been shown to play complex roles. Id3 expression is induced by hyperlipidemia and oxidized LDL and is thought to contribute to increased vascular smooth muscle cell proliferation (Taylor et al., 2006). Angiotensin II also promotes regional hyperplasia of ascending aorta through upregulation of Id3 expression (Owens et al., 2010). Paradoxically, a single-nucleotide polymorphism (SNP) in the human Id3 gene was independently associated with carotid intima-media thickness. This SNP results in an Id3 mutant that displays reduced affinities to E proteins (Doran et al., 2010). If Id3 functions in the context of atheroprotection by inhibiting the function of E proteins, the SNP could be considered as a loss of function mutation. However, if Id3 acts independently of E proteins, the reduced affinity of the Id3 mutant to E proteins could mean less sequestration of Id3 by E proteins and thus a gain of Id3 function. It is possible that the atheroprotective function of Id3 may be independent of its effect on vascular smooth muscle cells but instead related to the attraction of B cells to the sites of atherosclerosis and exclusion of macrophages from the sites (Doran et al., 2012; Lipinski et al., 2012).

Finally, Id1 and Id3 have been shown to play crucial roles in angiogenesis during tumor growth. Although $Id1^{-/-}Id3^{+/-}$ or $Id1^{+/-}Id3^{-/-}$ mice can develop to adulthood, they cannot support tumor xenografts (Lyden et al., 1999). This was shown to be due to failures in the recruitment of bone marrow-derived endothelial and hematopoietic precursors (Lyden et al., 2001). A similar mechanism has also been attributed to the impaired angiogenesis of spontaneous tumors resulting from Pten deficiency (Ruzinova et al., 2003). In the absence of Id1, these tumors such as uterine carcinomas display necrosis and hemorrhage (Ruzinova et al., 2003). Ablation of the Id1 gene in tumor endothelial cells leads to downregulation of several proangiogenic genes, such as alpha6 and beta4 integrins, matrix metalloprotease-2, and fibroblast growth factor receptor-1 (Ruzinova et al., 2003). Id1 has also been shown to diminish the expression of an angiogenesis inhibitor, thrombospondin-1 (Volpert et al., 2002). However, angiogenesis in skin tumors developed by chemical treatment is not affected by loss of Id1, thus suggesting tumor-specific dependence on Id1 in angiogenesis (Sikder et al., 2003). Alternatively, expression of other Id proteins such as Id3 could support angiogenesis in the skin. Consistent with the angiogenic activity of Id proteins, Id1 overexpression in endothelial progenitor cells has been linked to increased capacity of angiogenesis in a number of tumors such as ovarian cancer (Su et al., 2013). Moreover, epigenetic silencing of the Id4 gene has been found to be associated with reduced angiogenesis in glioblastomas and a better prognosis (Martini et al., 2013). Therefore, targeting specific Id proteins could potentially be used as a strategy for tumor therapy.

6. ID PROTEINS IN CANCER

Id proteins have been implicated in a variety of tumor types either as a driver for tumorigenesis or as an indicator of tumor progression. It is impractical to discuss the role of different Id proteins in every type of tumor. However, it is helpful to examine several general functions of Id proteins in relationship to fundamental processes in cancer biology.

First, the Id1 and Id3 genes are regulated during cell cycle and are generally transcribed in the G1 phase (Barone, Pepperkok, Peverali, & Philipson, 1994; Deed et al., 1997; Prabhu, Ignatova, Park, & Sun, 1997; Ruzinova & Benezra, 2003). They are also considered immediate-early genes in response to serum stimulation (Christy et al., 1991). Overexpression of Id1 promotes the growth of NIH3T3 fibroblasts by facilitating the G1 to S phase transition (Prabhu et al., 1997). On the other hand, over production of E47, an E protein transcription factor, arrests cell cycle at the same transition in NIH3T3 cells (Peverali et al., 1994). These effects are consistent with the fact that E proteins are involved in the transcriptional activation of the p21 gene, which encodes an inhibitor of cyclin-dependent kinases (Prabhu et al., 1997). However, while Id proteins have consistently been found to promote tumorigenesis, their correlation with the downregulation of p21 expression has been complicated in different types of cancers. For example, knocking down both Id1 and Id3 (Id1/3KD) have been shown to impair the self-renewal of colon cancer-initiating cells and cause reduced tumor growth and increased sensitivity to chemotherapy (O'Brien et al., 2012). Paradoxically, p21 levels were significantly reduced in Id1/3KD cancer-initiating cells, and ectopic expression of p21 in these cells partially restores the tumor initiating activity, suggesting the dependence of p21 on tumorigenesis. Considering that multiple transcription factors are involved in p21 expression, it is reasonable to believe that p21 expression is stimulated by other factors in these cells.

Second, Id proteins, primarily Id1, have been shown to delay the onset of cellular senescence as demonstrated in Id1-deficient mouse embryonic fibroblasts (MEFs) or Id1 knockdown human diploid fibroblasts (Alani, Young, & Shifflett, 2001; Zheng, Wang, Xue, Zhang, & Tong, 2004). This has been shown to be mediated by transcriptional repression of the p16/Ink4a gene, whose promoter contains E boxes to which E protein transcription factors bind (Alani et al., 2001; Zheng et al., 2004). It has also been reported that Id1 and the Ets family of transcription factors have opposing roles in regulating p16/Ink4a expression (Ohtani et al., 2001). In addition to fibroblasts, Id1 has been shown to play similar roles in human fetal cardiomyocyte, keratinocytes and endothelial cells (Ball & Levine, 2005; Nickoloff et al., 2000; Tang et al., 2002). Moreover, Id1 is thought to synergize with oncogenic ras to induce metastatic breast cancer by subverting cellular senescence (Swarbrick, Roy, Allen, & Bishop, 2008). Interestingly, Id4 exhibits an opposite function by promoting cellular senescence in a prostate cancer cell line (Carey, Knowell, Chinaranagari, & Chaudhary, 2013). However, the molecular mechanism whereby Id4 exerts this function is not known.

Third, Id proteins have been found to be instrumental for the metastasis of a variety of cancer cells. For breast cancer, overexpression of Id1 was initially shown to block the differentiation of the murine mammary epithelial cell line, SCp2, which normally arrests growth and functionally differentiates in response to a basement membrane and lactogenic hormones (Desprez et al., 1998). SCp2 cells that constitutively express Id1 secrete a matrix metalloproteinase (MMP) and invade the basement membrane even though they remain anchorage dependent for growth and do not form tumors in nude mice (Desprez et al., 1998). On the other hand, downregulation of Id1 by antisense RNA, siRNA, or small peptide inhibitors diminished the expression of MMP and the metastatic spread of 4T1 breast cancer cells or led to cell-cycle arrest and apoptosis of MCF7 and MDA-MB231 breast cancer cells (Fong et al., 2003; Mern, Hoppe-Seyler, Hoppe-Seyler, Hasskarl, & Burwinkel, 2010; Tobin, Sims, Lundgren, Lehn, & Landberg, 2011). Furthermore, high levels of Id1 expression has been detected in lung metastasis of breast cancer cells negative for estrogen receptor, progesterone receptor and Her2 amplification. Id1 and Id3 have been shown to be critical for tumor initiation in primary sites and in metastatic colonization in the lung, as well as subsequent extravasation into the lung parenchyma (Gupta et al., 2007). During metastatic colonization, Id1 induces a mesenchymal to epithelial transition, which appears to be crucial for lung colonization (Stankic et al., 2013). In contrast, Id2 expression is variable in breast cancer samples, and the cytoplasmic localization of Id2 appeared to be associated with reduced invasive capacity and less aggressive tumor phenotypes (Itahana et al., 2003; Stighall, Manetopoulos, Axelson, & Landberg, 2005). However, Wazir et al. showed significant association of elevated Id1 and Id2 expression with worse clinical outcomes of breast cancer (Wazir, Jiang, Sharma, Newbold, & Mokbel, 2013).

In prostate cancer, Id1 and Id3 expression is linked to higher grades of prostate cancer (Ouyang, Wang, Lee, Tsao, & Wong, 2002; Sharma, Patel, & Chaudhary, 2012). Silencing either Id1 or Id3 attenuates cell cycle by upregulation of cyclin-dependent kinase inhibitors, p21 and p27 (Asirvatham, Schmidt, & Chaudhary, 2006; Sharma et al., 2012). Id1 has been found to promote prostate cancer cell survival by activation of NF-κB and induce androgen-independent growth of prostate cancer cells through upregulation of EGF receptor (Ling et al., 2003, 2004). Id1 also interacts with caveolin-1, an integrin-binding protein, to promote epithelial–mesenchymal transition and increase cell migration, as well as to increase resistance to taxol-induced apoptosis (Zhang et al., 2007). In addition to breast and prostate cancers, Id proteins have been documented to enhance cell proliferation and invasiveness of immortalized esophageal epithelial cells, non-small cell lung cancer cells, hepatocellular carcinoma, thyroid cancer cells, and pancreatic cancer cells (Ao et al., 2012; Cheung, Yip, Tsao, Ching, & Cheung, 2011; Ciarrocchi, Piana, Valcavi, Gardini, & Casali, 2011; Pillai et al., 2011; Shuno et al., 2010).

In myeloid leukemia, an ANOVA analysis of the microarray data from a dataset of NCBI GEO (GSE4170) revealed that levels of Id1 expression in $CD34^{+}$ cells in peripheral blood of patients diagnosed with chronic myelogenous leukemia (CML) were proportional to the progression of the disease ($p < 0.001$), namely, the levels are the highest at the most advanced stage (blast crisis phase) and the lowest at the chronic phase (Fig. 5.2). It remains to be determined if the Id1 are important for the self-renewal of leukemic stem cells. Id1 expression is believed to be induced by BCR-ABL via STAT5, and C/EBPα also drives Id1 expression which in turn helps maintain the myeloid phenotype of the leukemic cells (Cammenga et al., 2003; Nieborowska-Skorska et al., 2006; Wagner et al., 2006). These modes of regulation are similar to the control mechanisms of Id1 transcription found in the IL-3-dependent BaF3 cell line (Mui, Wakao, Kinoshita, Kitamura, & Miyajima, 1996; Saisanit & Sun, 1997; Xu et al., 2003). Furthermore, Id1-mediated production of matrix metalloproteinase 9 has been shown to be

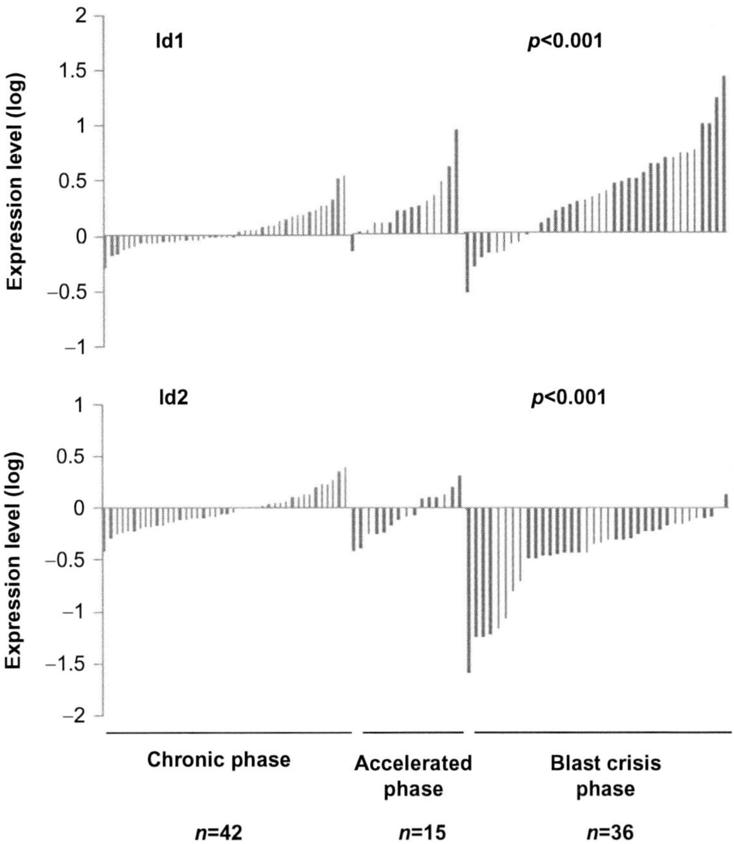

Figure 5.2 Id1 and Id2 expression in chronic myelogenous leukemia. Microarray analysis of Id1 and Id2 expression in $CD34^+$ cells in the peripheral blood of CML patients at the indicated disease stage. The number of samples in each group is as indicated. Statistical significance was determined using multigroup comparison using ANOVA F test. (See the color plate.)

responsible for the invasiveness of leukemic cells (Nieborowska-Skorska et al., 2006). In contrast, Id2 expression is significantly downregulated during the blast crisis phase of CML (Fig. 5.2), thus suggesting that Id1 and Id2 are regulated differently in CML leukemic stem cells and could also have different functions in this context.

Despite the broad roles of Id proteins in tumorigenesis and metastasis, the molecular mechanisms by which Id proteins exert their diverse functions remain incompletely understood. Id proteins are best known as inhibitors of bHLH E proteins. Indeed, they have been shown to antagonize

E protein-mediated transcription of p21 and p16 cell-cycle regulators (Alani et al., 2001; Mern et al., 2010; O'Brien et al., 2012; Prabhu et al., 1997; Swarbrick et al., 2008; Zheng et al., 2004). However, alterations in the expression of these regulators are not sufficient to explain the function of Id proteins in promoting cancer cell invasion and migration. Although both overexpression of Id proteins and ablation of the *E2A* gene cause T cell lymphoma (Bain et al., 1997; Kim et al., 1999; Yan et al., 1997), loss of E proteins have not been associated with many of the tumorigenic and metastatic effects of Id proteins. Furthermore, Asirvatham et al. has shown that Id1 to Id3 are capable of stimulating the expression of genes involved in survival, proliferation, and invasion in prostate cancer cell lines with or without E proteins, thus suggesting that Id proteins can function independently of inhibition of E proteins (Asirvatham, Carey, & Chaudhary, 2007).

Several E protein-independent mechanisms of Id action have been identified. For example, Id1 was found to interact with caveolin-1, which leads to activation of the Akt signaling pathway and augmentation of prostate cancer cell invasion (Zhang et al., 2007). Id1 has also been shown to bind to the p65 subunit of NF-κB and enhance the transcription of NF-κB target genes (Peng et al., 2012). Id1-induced NF-κB activation is also attributed to the elevated expression of survivin and the resistance to apoptosis in head and neck squamous cell carcinoma (Lin et al., 2010). Interestingly, overexpression of Id1 in developing T cells also potentiates NF-κB activation, a phenomenon not observed in E protein-deficient T cells (Kim et al., 2002; Yang, Liou, & Sun, 2006). Therefore, a better understanding of the mechanisms underlying Id functions would greatly improve the design of targeted therapy for cancers in which Id proteins play key roles.

7. ID PROTEINS IN THE IMMUNE SYSTEM

The immune system consists of mainly B and T lymphocytes and myeloid lineage cells including macrophages, granulocytes, NK, and dendritic cells. These cells originate from HSCs and initially differentiate in the bone marrow or thymus. Maturation of some of the lineages occurs in peripheral lymphoid organs such as the spleen, lymph nodes, and lymphoid tissues in the intestine. Like the development of other tissues or organs, the generation of immune system involves a series of cell fate choices and the balance between Id and E proteins play a significant role in these processes.

One of the fundamental lineage decisions is the lymphoid versus myeloid choice. It is well established that E proteins play crucial roles in the commitment and differentiation of multipotent progenitors into B and T cell lineages (Kee, 2009; Murre, 2005). E proteins are present at relatively steady levels in different subsets of hematopoietic progenitors, but Id1 and Id2 levels are very low once the cells are committed to the lymphoid lineages. E proteins have been demonstrated to drive the lymphoid fate while suppressing myeloid differentiation. E2A deficiency results in reduction in lymphoid progenitors whereas expression of ET2 in multipotent progenitors but not in committed myeloid progenitors blocks myeloid differentiation in culture (Cochrane et al., 2009; Dias, Mansson, Gurbuxani, Sigvardsson, & Kee, 2008). Interestingly, levels of Id1 expression is markedly elevated in granulocyte-macrophage progenitors, perhaps for the purpose of allowing these progenitors to proceed down the myeloid path (Cochrane et al., 2009). Indeed, overexpression of Id1 promotes the differentiation of multipotent progenitors into myeloid or dendritic cells (Cochrane et al., 2009). Overexpression of Id genes in human hematopoietic progenitors has also been shown to promote NK cell differentiation (Heemskerk et al., 1997).

In addition, Id2-deficient mice are devoid of NK cells, lymph nodes, Peyer's patch, and other secondary lymphoid tissues (Yokota et al., 1999). Recently, it has become recognized that these secondary lymphoid tissues are populated with innate lymphoid cells, which appear to be derived from related progenitors and are crucial for protecting the body from infection and for the formation of lymphoid tissues (Hwang & McKenzie, 2013; Spits & Di Santo, 2011). Id2 thus plays an important role in maintaining the capacity of multipotent progenitors to develop into NK cells, lymphoid tissue inducer cells, and innate lymphoid cells. Furthermore, Id2 is also highly expressed in dendritic cell progenitors, and ablation of the Id2 gene impairs the development of CD8-positive dendritic cells (Jackson et al., 2011; Kusunoki et al., 2003). Whether these diverse functions of Id2 are mediated through inhibition of E protein activity and prevention of the progenitors from adopting a lymphoid fate is an interesting question. Attempts have been made to address this question by crossing Id2-deficient mice with those lacking E2A (Boos, Yokota, Eberl, & Kee, 2007). Ablation of E2A rescued the defect of Id2 knockout mice with regard to bone marrow NK cell differentiation and secondary lymphoid tissue formation, suggesting that Id2-mediated inhibition of E2A is at least in part involved.

During T cell development in the thymus, committed T cell progenitors differentiate from the CD4 and CD8 double-negative stage to the double-positive (DP) stage, which depends on signaling from the pre-T cell receptor (pre-TCR). The Id3 gene is activated transiently and immediately after pre-TCR stimulation (Bain et al., 2001). This is thought to be mediated via the Ras/MAPK pathway which leads to activation of Egr transcription factors that are responsible for Id3 transcription. Downregulation of E protein activities by Id3 has been shown to be crucial for T cell progenitors to progress to the DP stage, because ablation of E2A or overexpression of Id1 enables the cells without pre-TCR to reach the DP stage (Engel & Murre, 2002; Kim et al., 2002). At the DP stage, developing T cells undergo positive selection, which allows T cells with appropriate affinities to self-antigens to progress to CD4 or CD8 single-positive stage. In this process, Id2 and Id3 have been shown to play important roles as deletion of both genes impairs positive selection and blocks the formation of CD4 or CD8 single-positive cells (Jones-Mason et al., 2012; Jones & Zhuang, 2007).

Moreover, a small fraction of developing T cells deviates from the pool of conventional T cells to become NKT cells which carry the invariant Vα14–Jα18 chain of TCRα paired with certain TCRβ chains and are selected by lipid antigens presented by CD1d (Bendelac, Savage, & Teyton, 2007). These cells exhibit the characteristics of innate immune cells and are capable of rapid activation upon pathogen recognition without prior antigen exposure. They have diverse functions in regulating immune responses by secreting various cytokines. NKT cell development is greatly influenced by the balance between E and Id proteins. Although ablation of the Id2 and Id3 genes prior to pre-TCR selection leads to increased differentiation towards the NKT lineage (Li, Wu, Jiang, & Zhuang, 2013), loss of these genes at the DP stage dramatically impairs the maturation of NKT cells in the thymus and reduces the output of NKT cells to the spleen and liver (Verykokakis et al., 2013). In contrast, increasing E protein activity by expressing ET2 at the DP stage not only blocks the maturation of NKT cells but also promotes the formation of alternative subsets of NKT cells known as NKT2 and NKT17 cells (Hu et al., 2013).

Collectively, the Id proteins appear to play crucial roles in steering various multipotent progenitors away from the conventional lymphoid lineages at different steps of hematopoiesis and lymphopoiesis, thus allowing them to differentiate into myeloid and dendritic cells, as well as a variety of innate lymphoid cells such as NK, NKT, and lymphoid tissue inducer cells. It will be interesting to determine if Id proteins fulfill their responsibilities by inhibiting lymphoid commitment directed by E proteins or by relieving the suppression of alternative cell fate mediated by E proteins. It also remains to be ruled out that Id proteins promote the differentiation of multipotent progenitors into nonlymphoid lineages by regulating gene expression independently of E proteins.

Despite the broad roles of Id proteins in diverse aspects of immune system development, the connection between Id proteins and immune disorders have been scarce. One exception is the Sjogren's syndrome-like phenotypes of Id3-deficient mice (Li, Dai, & Zhuang, 2004). These mice display typical symptoms such as dry eyes and mouth, lymphocyte infiltration in lachrymal and salivary glands, and development of anti-Ro and anti-La antibodies (Guo et al., 2011). Further investigation indicated that Id3 deficiency in T cells is responsible for the development of Sjogren's syndrome, because adoptive transfer of B cell-depleted Id3$^{-/-}$ bone marrow as well as T cell-specific ablation of Id3 triggered the disease (Li et al., 2004). Although the causes may be attributed to defective T cells, B cells play an important role in the manifestation of the disease as depletion of B cells with anti-CD20 alleviates the disease (Hayakawa, Tedder, & Zhuang, 2007). A T cell dependent over production of IL-13 in Id3$^{-/-}$ mice has also been linked to impaired salivary function or exocrinopathy (Mahlios & Zhuang, 2011). It is unclear which defect in T cell biology caused by Id3 deficiency leads to the development of Sjogren's syndrome. However, it is known that Id3-deficient mice exhibit impaired development of T regulatory cells and abnormal γδ T cell production (Maruyama et al., 2011; Ueda-Hayakawa, Mahlios, & Zhuang, 2009). Whether Id3 deficiency is associated with human primary Sjogren's syndrome has been investigated in a case-controlled study of 212 patients and 168 controls (Sellam et al., 2008). Id3 mRNA levels along with two SNPs in peripheral T cells, cultured salivary glandular epithelial cells, and in total minor salivary glands were measured. No significant differences were observed. However, the levels of Id3 in the cells tested might not be relevant to the cause of the disease.

Id2 has recently been recognized as an essential regulator for the formation of innate lymphoid cells which are distributed in secondary lymphoid organs throughout the body. These cells play protective roles in the innate immune responses to infectious microorganisms, in lymphoid tissue formation, in tissue remodeling after damage inflicted by injury or infection, and in the homeostasis of tissue stromal cells. As the investigation continues, it would not be surprising to find connections between Id2 and diseases such as asthma and inflammatory bowel diseases.

8. ID PROTEINS IN ADIPOGENESIS

Adipocyte differentiation has been shown to be controlled by a series of transcriptional events, and *in vitro* studies have identified important roles for PPARγ and C/EBPα in the development and maintenance of adipocytes (Barak et al., 1999; Rosen et al., 1999). Hormonal induction of adipogenesis *in vitro* leads to a transient rise in C/EBPδ and C/EBPβ levels which is followed by an increase in PPARγ expression (Wu, Bucher, & Farmer, 1996; Yeh, Cao, Classon, & McKnight, 1995). PPARγ can then go on to induce C/EBPα expression which in turn reinforces PPARγ expression leading to a positive feedback loop necessary for the maintenance of mature adipocytes (Wu et al., 1999).

Id proteins are widely known to play a role in regulating differentiation. Since the discovery of SREBP-1c, a bHLH-leucine-zipper protein important for lipogenesis, several studies have examined the role of Id proteins in adipocyte differentiation. Although Id proteins are typically thought to play an inhibitory role in differentiation, studies on Id2 and Id4 have interestingly suggested a role in promoting adipogenesis. Expression of Id2 and Id4 in preadipocytes is rapidly but transiently increased upon induction with the standard adipogenic cocktail of dexamethasone, 3-isobutyl-1-methylxanthine, and insulin. This increase is then closely followed by an upregulation of PPARγ expression suggesting that Id2 and Id4 may act upstream of PPARγ (Murad et al., 2010; Park et al., 2008). Knocking down Id2 or Id4 leads to a decrease in triglyceride accumulation as measured by Oil Red O staining as well as a reduction in PPARγ and aP2 expression, and $Id4^{-/-}$ mice show a significant reduction in fat mass compared to littermate controls (Bai et al., 2003; Park et al., 2008). Taken together, Id2 and Id4 appear to play a role in promoting adipocyte formation.

Id3, on the other hand, has been suggested to play an inhibitory role in adipocyte formation since preadipocytes expressing Id3 had reduced adipogenic potential (Moldes, Lasnier, Feve, Pairault, & Djian, 1997). It was initially thought that Id3 mediated its inhibitory effect by antagonizing SREBP-1c. However, these early *in vitro* studies relied heavily on the preadipocyte cell lines, 3T3-F442A and 3T3-L1, and it was later shown that SREBP-1c-deficient mice had no decrease in the amount of adipose tissue formed (Shimano et al., 1997). Moreover, it has also been suggested that Id3 does not bind to SREBP-1c directly but instead binds to E47 to interfere with the formation of an E47/SREBP-1c complex which can activate the adiponectin promoter (Doran et al., 2008). Adiponectin is an adipocyte specific hormone whose levels have been shown to be inversely related to obesity and has protective effects against metabolic disorders (Yamauchi & Kadowaki, 2013). Therefore, loss of Id3 might not interfere with adipocyte formation directly, but can instead lead to the upregulation of adiponectin which then improves the overall metabolic state.

More recently, it was shown that Id3 null MEFs and stromal vascular fraction taken from white adipose tissue had normal lipid accumulation as measured by Oil Red O staining when induced to differentiate *in vitro*, and the reduced fat mass found in Id3$^{-/-}$ mice placed on a high-fat diet was suggested to be due to an inhibition of visceral fat expansion (Cutchins et al., 2012). Adipose tissue growth is greatly dependent on the formation of new microvasculature as a means to deliver nutrients and growth factors to developing adipocytes, and Id3$^{-/-}$ mice were shown to have reduced VEGFA level in visceral fat compared to littermate controls (Cao, 2014; Cutchins et al., 2012). This suggests that Id3 does not function to regulate adipocyte differentiation directly, but instead promotes adipose development by controlling angiogenesis through VEGFA expression (Cutchins et al., 2012).

Id1 is also thought to not play a direct role in adipocyte differentiation, but might instead be important in maintaining energy homeostasis. Id1$^{-/-}$ mice was shown to have decreased body weight and fat mass, while lean body mass remained unchanged (Satyanarayana, Klarmann, Gavrilova, & Keller, 2012). However, *in vitro* analysis of Id1 null MEFs showed enhanced adipocyte differentiation (Satyanarayana et al., 2012). To account for the discrepancy, oxygen consumption and thermogenesis was measured. It was found that Id1$^{-/-}$ mice had increased oxygen consumption and an upregulation of thermogenic markers such as UCP1 and PGC1α. This led the authors to conclude that the decrease in fat mass was due to an increase in energy expenditure and thermogenesis and not due to deregulated adipogenesis in Id1$^{-/-}$ mice (Satyanarayana et al., 2012).

Obesity has been linked to an increased risk in diabetes, cardiovascular disease, and cancer, and given the current rise in obesity, research in metabolism is becoming increasingly important. Id proteins would make good targets for drug discovery especially since Id expression is usually low in adults under homeostatic conditions. However, expression of Id proteins in adipocytes seems to be associated with a lower metabolic rate as loss of Id1 can increase energy expenditure. Therefore, downregulating proteins such as Id1 could help increase an individual's overall metabolic state and could be a means in which to combat obesity with few unwanted side effects.

9. CONCLUDING REMARKS

Since the discovery of Id proteins over two decades ago, much has been learned about their biological functions. These are highlighted by the ability of Id proteins to delay the onset of cellular senescence, to stimulate cell cycle, and to promote vasculogenesis and angiogenesis, as well as to facilitate cell migration and invasion, all of which make Id proteins important player in tumorigenesis and metastasis. Consistently, increased expression of Id proteins has generally been associated with advanced stages of tumor progression and poor prognosis. In addition, the inhibitory effects of Id proteins on cell differentiation land them a critical role in stem cell maintenance. Id proteins also emerged as essential molecules for the formation innate immune cells, possibly due to their function in blocking lymphoid differentiation. In terms of the molecular mechanisms whereby Id proteins carry out these functions, it is easy to propose that they interfere with the transcriptional activity of bHLH E proteins and therefore, diminish the expression of E protein target genes. However, Id expression often stimulates the transcription of a number of genes, and whether these genes are controlled by E proteins has not been established. These caveats raise the question of whether Id proteins function independently of E proteins in some cellular settings, but the answer awaits more robust biochemical analyses.

REFERENCES

Alani, R. M., Young, A. Z., & Shifflett, C. B. (2001). Id1 regulation of cellular senescence through transcriptional repression of p16/Ink4a. *Proceedings of the National Academy of Sciences of United States of America*, *98*, 7812–7816.

Ao, J., Meng, J., Zhu, L., Nie, H., Yang, C., Li, J., et al. (2012). Activation of androgen receptor induces ID1 and promotes hepatocellular carcinoma cell migration and invasion. *Molecular Oncology*, *6*, 507–515.

Asirvatham, A. J., Carey, J. P., & Chaudhary, J. (2007). ID1-, ID2-, and ID3-regulated gene expression in E2A positive or negative prostate cancer cells. *Prostate*, *67*, 1411–1420.

Asirvatham, A. J., Schmidt, M. A., & Chaudhary, J. (2006). Non-redundant inhibitor of differentiation (Id) gene expression and function in human prostate epithelial cells. *Prostate*, *66*, 921–935.

Bai, X., Cerimele, F., Ushio-Fukai, M., Waqas, M., Campbell, P. M., Govindarajan, B., et al. (2003). Honokiol, a small molecular weight natural product, inhibits angiogenesis in vitro and tumor growth in vivo. *Journal of Biological Chemistry*, *278*, 35501–35507.

Bai, G., Sheng, N., Xie, Z., Bian, W., Yokota, Y., Benezra, R., et al. (2007). Id sustains Hes1 expression to inhibit precocious neurogenesis by releasing negative autoregulation of Hes1. *Developmental Cell, 13*, 283–297.

Bain, G., Cravatt, C. B., Loomans, C., Alberola-Ila, J., Hedrick, S. M., & Murre, C. (2001). Regulation of the helix-loop-helix proteins, E2A and Id3, by the Ras- ERK MAPK cascade. *Nature Immunology, 2*, 165–171.

Bain, G., Engel, I., Robanus Maandag, E. C., te Riele, H. P., Voland, J. R., Sharp, L. L., et al. (1997). E2A deficiency leads to abnormalities in alphabeta T-cell development and to rapid development of T-cell lymphomas. *Molecular and Cellular Biology, 17*, 4782–4791.

Ball, A. J., & Levine, F. (2005). Telomere-independent cellular senescence in human fetal cardiomyocytes. *Aging Cell, 4*, 21–30.

Barak, Y., Nelson, M. C., Ong, E. S., Jones, Y. Z., Ruiz-Lozano, P., Chien, K. R., et al. (1999). PPAR gamma is required for placental, cardiac, and adipose tissue development. *Molecular Cell, 4*, 585–595.

Barone, M. V., Pepperkok, R., Peverali, F. A., & Philipson, L. (1994). Id proteins control growth induction in mammalian cells. *Proceedings of the National Academy of Sciences of the United States of America, 91*, 4985–4988.

Bendelac, A., Savage, P. B., & Teyton, L. (2007). The biology of NKT cells. *Annual Review of Immunology, 25*, 297–336.

Benezra, R., Davis, R. L., Lockshon, D., Turner, D. L., & Weintraub, H. (1990). The protein Id: A negative regulator of helix-loop-helix DNA binding proteins. *Cell, 61*, 49–59.

Beyan, E., Beyan, C., & Turan, M. (2007). Hematological presentation in systemic lupus erythematosus and its relationship with disease activity. *Hematology, 12*, 257–261.

Boos, M. D., Yokota, Y., Eberl, G., & Kee, B. L. (2007). Mature natural killer cell and lymphoid tissue-inducing cell development requires Id2-mediated suppression of E protein activity. *Journal of Experimental Medicine, 204*, 1119–1130.

Cammenga, J., Mulloy, J. C., Berguido, F. J., MacGrogan, D., Viale, A., & Nimer, S. D. (2003). Induction of C/EBPalpha activity alters gene expression and differentiation of human CD34+ cells. *Blood, 101*, 2206–2214.

Cao, Y. (2014). Angiogenesis as a therapeutic target for obesity and metabolic diseases. *Chemical Immunology and Allergy, 99*, 170–179.

Carey, J. P., Knowell, A. E., Chinaranagari, S., & Chaudhary, J. (2013). Id4 promotes senescence and sensitivity to doxorubicin-induced apoptosis in DU145 prostate cancer cells. *Anticancer Research, 33*, 4271–4278.

Cheung, P. Y., Yip, Y. L., Tsao, S. W., Ching, Y. P., & Cheung, A. L. (2011). Id-1 induces cell invasiveness in immortalized epithelial cells by regulating cadherin switching and Rho GTPases. *Journal of Cellular Biochemistry, 112*, 157–168.

Christy, B. A., Sanders, L. K., Lau, L. F., Copeland, N. G., Jenkins, N. A., & Nathans, D. (1991). An Id-related helix-loop-helix protein encoded by a growth factor-inducible gene. *Proceedings of the National Academy of Sciences of the United States of America, 88*, 1815–1819.

Ciarrocchi, A., Piana, S., Valcavi, R., Gardini, G., & Casali, B. (2011). Inhibitor of DNA binding-1 induces mesenchymal features and promotes invasiveness in thyroid tumour cells. *European Journal of Cancer, 47*, 934–945.

Cochrane, S. W., Zhao, Y., Welner, R. S., & Sun, X. H. (2009). Balance between Id and E proteins regulates myeloid-versus-lymphoid lineage decisions. *Blood, 113*, 1016–1026.

Cutchins, A., Harmon, D. B., Kirby, J. L., Doran, A. C., Oldham, S. N., Skaflen, M., et al. (2012). Inhibitor of differentiation-3 mediates high fat diet-induced visceral fat expansion. *Arteriosclerosis, Thrombosis, and Vascular Biology, 32*, 317–324.

Deed, R. W., Hara, E., Atherton, G. T., Peters, G., & Norton, J. D. (1997). Regulation of Id3 cell cycle function by Cdk-2-dependent phosphorylation. *Molecular and Cellular Biology, 17*, 6815–6821.

Desprez, P. Y., Lin, C. Q., Thomasset, N., Sympson, C. J., Bissell, M. J., & Campisi, J. (1998). A novel pathway for mammary epithelial cell invasion induced by the helix-loop-helix protein Id-1. *Molecular and Cellular Biology, 18*, 4577–4588.

Dias, S., Mansson, R., Gurbuxani, S., Sigvardsson, M., & Kee, B. L. (2008). E2A proteins promote development of lymphoid-primed multipotent progenitors. *Immunity, 29*, 217–227.

Doran, A. C., Lehtinen, A. B., Meller, N., Lipinski, M. J., Slayton, R. P., Oldham, S. N., et al. (2010). Id3 is a novel atheroprotective factor containing a functionally significant single-nucleotide polymorphism associated with intima-media thickness in humans. *Circulation Research, 106*, 1303–1311.

Doran, A. C., Lipinski, M. J., Oldham, S. N., Garmey, J. C., Campbell, K. A., Skaflen, M. D., et al. (2012). B-cell aortic homing and atheroprotection depend on Id3. *Circulation Research, 110*, e1–e12.

Doran, A. C., Meller, N., Cutchins, A., Deliri, H., Slayton, R. P., Oldham, S. N., et al. (2008). The helix-loop-helix factors Id3 and E47 are novel regulators of adiponectin. *Circulation Research, 103*, 624–634.

Engel, I., & Murre, C. (2002). Disruption of pre-TCR expression accelerates lymphomagenesis in E2A- deficient mice. *Proceedings of the National Academy of the Sciences of the United States of America, 99*, 11322–11327.

Fong, S., Itahana, Y., Sumida, T., Singh, J., Coppe, J. P., Liu, Y., et al. (2003). Id-1 as a molecular target in therapy for breast cancer cell invasion and metastasis. *Proceedings of the National Academy of the Sciences of the United States of America, 100*, 13543–13548.

Guo, Z., Li, H., Han, M., Xu, T., Wu, X., & Zhuang, Y. (2011). Modeling Sjogren's syndrome with Id3 conditional knockout mice. *Immunology Letters, 135*, 34–42.

Gupta, G. P., Perk, J., Acharyya, S., De, C. P., Mittal, V., & Todorova-Manova, K. (2007). ID genes mediate tumor reinitiation during breast cancer lung metastasis. *Proceedings of the National Academy of the Sciences of the United States of America, 104*, 19506–19511.

Hara, E., Hall, M., & Peters, G. (1997). Cdk2-dependent phosphorylation of Id2 modulates activity of E2A-related transcription factors. *EMBO Journal, 16*, 332–342.

Hayakawa, I., Tedder, T. F., & Zhuang, Y. (2007). B-lymphocyte depletion ameliorates Sjogren's syndrome in Id3 knockout mice. *Immunology, 122*, 73–79.

Heemskerk, M. H., Blom, B., Oda, K., Stegmann, A. P., Bakker, A. Q., Weijer, K., et al. (1997). Inhibition of T cell and promotion of natural killer cell development by the dominant negative helix loop helix factor Id3. *Journal of Experimental Medicine, 186*, 1597–1602.

Hollnagel, A., Oehlmann, V., Heymer, J., Ruther, U., & Nordheim, A. (1999). Id genes are direct targets of bone morphogenetic protein induction in embryonic stem cells. *Journal of Biological Chemistry, 274*, 19838–19845.

Hu, T., Wang, H., Simmons, A., Bajana, S., Zhao, Y., Kovats, S., et al. (2013). Increased level of E protein activity during invariant NKT development promotes differentiation of invariant NKT2 and invariant NKT17 subsets. *Journal of Immunology, 191*, 5065–5073.

Hwang, Y. Y., & McKenzie, A. N. (2013). Innate lymphoid cells in immunity and disease. *Advances in Experimental Medicine & Biology, 785*, 9–26.

Ikuta, K., & Weissman, I. L. (1992). Evidence that hematopoietic stem cells express mouse c-kit but do not depend on steel factor for their generation. *Proceedings of the National Academy of the Sciences of the United States of America, 89*, 1502–1506.

Itahana, Y., Singh, J., Sumida, T., Coppe, J. P., Parrinello, S., Bennington, J. L., et al. (2003). Role of Id-2 in the maintenance of a differentiated and noninvasive phenotype in breast cancer cells. *Cancer Research, 63*, 7098–7105.

Jackson, J. T., Hu, Y., Liu, R., Masson, F., D'Amico, A., Carotta, S., et al. (2011). Id2 expression delineates differential checkpoints in the genetic program of CD8alpha+ and CD103+ dendritic cell lineages. *EMBO Journal, 30*, 2690–2704.

Jankovic, V., Ciarrocchi, A., Boccuni, P., Deblasio, T., Benezra, R., & Nimer, S. D. (2007). Id1 restrains myeloid commitment, maintaining the self-renewal capacity of hematopoietic stem cells. *Proceedings of the National Academy of the Sciences of the United States of America*, *104*, 1260–1265.

Ji, M., Li, H., Suh, H. C., Klarmann, K. D., Yokota, Y., & Keller, J. R. (2008). Id2 intrinsically regulates lymphoid and erythroid development via interaction with different target proteins. *Blood*, *112*, 1068–1077.

Jones, M. E., & Zhuang, Y. (2007). Acquisition of a functional T cell receptor during T lymphocyte development is enforced by HEB and E2A transcription factors. *Immunity*, *27*, 860–870.

Jones-Mason, M. E., Zhao, X., Kappes, D., Lasorella, A., Iavarone, A., & Zhuang, Y. (2012). E protein transcription factors are required for the development of CD4(+) lineage T cells. *Immunity*, *36*, 348–361.

Jung, S., Park, R. H., Kim, S., Jeon, Y. J., Ham, D. S., Jung, M. Y., et al. (2010). Id proteins facilitate self-renewal and proliferation of neural stem cells. *Stem Cells and Development*, *19*, 831–841.

Karaya, K., Mori, S., Kimoto, H., Shima, Y., Tsuji, Y., Kurooka, H., et al. (2005). Regulation of Id2 expression by CCAAT/enhancer binding protein beta. *Nucleic Acids Research*, *33*, 1924–1934.

Katagiri, T., Imada, M., Yanai, T., Suda, T., Takahashi, N., & Kamijo, R. (2002). Identification of a BMP-responsive element in Id1, the gene for inhibition of myogenesis. *Genes to Cells*, *7*, 949–960.

Kee, B. L. (2009). E and ID proteins branch out. *Nature Reviews Immunology*, *9*, 175–184.

Kim, D., Peng, X. C., & Sun, X. H. (1999). Massive apoptosis of thymocytes in T-cell-deficient Id1 transgenic mice. *Molecular and Cellular Biology*, *19*, 8240–8253.

Kim, D., Xu, M., Nie, L., Peng, X. C., Jimi, E., Voll, R. E., et al. (2002). Helix-loop-helix proteins regulate pre-TCR and TCR signaling through modulation of Rel/NF-kappaB activities. *Immunity*, *16*, 9–21.

Kusunoki, T., Sugai, M., Katakai, T., Omatsu, Y., Iyoda, T., Inaba, K., et al. (2003). TH2 dominance and defective development of a CD8+ dendritic cell subset in Id2-deficient mice. *Journal of Allergy and Clinical Immunology*, *111*, 136–142.

Lasorella, A., Boldrini, R., Dominici, C., Donfrancesco, A., Yokota, Y., Inserra, A., et al. (2002). Id2 is critical for cellular proliferation and is the oncogenic effector of N-myc in human neuroblastoma. *Cancer Research*, *62*, 301–306.

Lasorella, A., Iavarone, A., & Israel, M. A. (1996). Id2 specifically alters regulation of the cell cycle by tumor suppressor proteins. *Molecular and Cellular Biology*, *16*, 2570–2578.

Lasorella, A., Noseda, M., Beyna, M., Yokota, Y., & Iavarone, A. (2000). Id2 is a retinoblastoma protein target and mediates signalling by Myc oncoproteins. *Nature*, *407*, 592–598.

Li, H., Dai, M., & Zhuang, Y. (2004). A T cell intrinsic role of Id3 in a mouse model for primary Sjogren's syndrome. *Immunity*, *21*, 551–560.

Li, H., Ji, M., Klarmann, K. D., & Keller, J. R. (2010). Repression of Id2 expression by Gfi-1 is required for B-cell and myeloid development. *Blood*, *116*, 1060–1069.

Li, J., Wu, D., Jiang, N., & Zhuang, Y. (2013). Combined deletion of Id2 and Id3 genes reveals multiple roles for E proteins in invariant NKT cell development and expansion. *Journal of Immunology*, *191*, 5052–5064.

Lin, J., Guan, Z., Wang, C., Feng, L., Zheng, Y., Caicedo, E., et al. (2010). Inhibitor of differentiation 1 contributes to head and neck squamous cell carcinoma survival via the NF-kappaB/survivin and phosphoinositide 3-kinase/Akt signaling pathways. *Clinical Cancer Research*, *16*, 77–87.

Ling, M. T., Wang, X., Lee, D. T., Tam, P. C., Tsao, S. W., & Wong, Y. C. (2004). Id-1 expression induces androgen-independent prostate cancer cell growth through activation of epidermal growth factor receptor (EGF-R). *Carcinogenesis*, *25*, 517–525.

Ling, M. T., Wang, X., Ouyang, X. S., Xu, K., Tsao, S. W., & Wong, Y. C. (2003). Id-1 expression promotes cell survival through activation of NF-kappaB signalling pathway in prostate cancer cells. *Oncogene, 22*, 4498–4508.

Lipinski, M. J., Campbell, K. A., Duong, S. Q., Welch, T. J., Garmey, J. C., Doran, A. C., et al. (2012). Loss of Id3 increases VCAM-1 expression, macrophage accumulation, and atherogenesis in Ldlr−/− mice. *Arteriosclerosis, Thrombosis, and Vascular Biology, 32*, 2855–2861.

Liu, H., Jia, D., Li, A., Chau, J., He, D., Ruan, X., et al. (2013). p53 regulates neural stem cell proliferation and differentiation via BMP-Smad1 signaling and Id1. *Stem Cells and Development, 22*, 913–927.

Lowery, J. W., Frump, A. L., Anderson, L., DiCarlo, G. E., Jones, M. T., & de Caestecker, M. P. (2010). ID family protein expression and regulation in hypoxic pulmonary hypertension. *American Journal of Physiology. Regulatory, Integrative and Comparative Physiology, 299*, R1463–R1477.

Lyden, D., Hattori, K., Dias, S., Costa, C., Blaikie, P., Butros, L., et al. (2001). Impaired recruitment of bone-marrow-derived endothelial and hematopoietic precursor cells blocks tumor angiogenesis and growth. *Nature Medicine, 7*, 1194–1201.

Lyden, D., Young, A. Z., Zagzag, D., Yan, W., Gerald, W., O'Reilly, R., et al. (1999). Id1 and Id3 are required for neurogenesis, angiogenesis and vascularization of tumour xenografts. *Nature, 401*, 670–677.

Maeda, K., Malykhin, A., Teague-Weber, B. N., Sun, X. H., Farris, A. D., & Coggeshall, K. M. (2009). Interleukin-6 aborts lymphopoiesis and elevates production of myeloid cells in systemic lupus erythematosus-prone B6.Sle1.Yaa animals. *Blood, 113*, 4534–4540.

Mahlios, J., & Zhuang, Y. (2011). Contribution of IL-13 to early exocrinopathy in Id3−/− mice. *Molecular Immunology, 49*, 227–233.

Martini, M., Cenci, T., D'Alessandris, G. Q., Cesarini, V., Cocomazzi, A., Ricci-Vitiani, L., et al. (2013). Epigenetic silencing of Id4 identifies a glioblastoma subgroup with a better prognosis as a consequence of an inhibition of angiogenesis. *Cancer, 119*, 1004–1012.

Maruyama, T., Li, J., Vaque, J. P., Konkel, J. E., Wang, W., Zhang, B., et al. (2011). Control of the differentiation of regulatory T cells and T(H)17 cells by the DNA-binding inhibitor Id3. *Nature Immunology, 12*, 86–95.

Massari, M. E., & Murre, C. (2000). Helix-loop-helix proteins: Regulators of transcription in eucaryotic organisms. *Molecular and Cellular Biology, 20*, 429–440.

Mern, D. S., Hoppe-Seyler, K., Hoppe-Seyler, F., Hasskarl, J., & Burwinkel, B. (2010). Targeting Id1 and Id3 by a specific peptide aptamer induces E-box promoter activity, cell cycle arrest, and apoptosis in breast cancer cells. *Breast Cancer Research and Treatment, 124*, 623–633.

Miyazono, K., & Miyazawa, K. (2002). A target of BMP signaling. *Science's Signal Transduction Knowledge Environment, 2002*, E40.

Moldes, M., Lasnier, F., Feve, B., Pairault, J., & Djian, P. (1997). Id3 prevents differentiation of preadipose cells. *Molecular and Cellular Biology, 17*, 1796–1804.

Morrison, S. J., & Weissman, I. L. (1994). The long-term repopulating subset of hematopoietic stem cells is deterministic and isolatable by phenotype. *Immunity, 1*, 661–673.

Morrow, M. A., Mayer, E. W., Perez, C. A., Adlam, M., & Siu, G. (1999). Overexpression of the helix-loop-helix protein Id2 blocks T cell development at multiple stages. *Molecular Immunology, 36*, 491–503.

Mui, A. L., Wakao, H., Kinoshita, T., Kitamura, T., & Miyajima, A. (1996). Suppression of interleukin-3-induced gene expression by a C-terminal truncated Stat5: Role of Stat5 in proliferation. *EMBO Journal, 15*, 2425–2433.

Murad, J. M., Place, C. S., Ran, C., Hekmatyar, S. K., Watson, N. P., Kauppinen, R. A., et al. (2010). Inhibitor of DNA binding 4 (ID4) regulation of adipocyte differentiation and adipose tissue formation in mice. *Journal of Biological Chemistry*, *285*, 24164–24173.

Murre, C. (2005). Helix-loop-helix proteins and lymphocyte development. *Nature Immunology*, *6*, 1079–1086.

Nakahiro, T., Kurooka, H., Mori, K., Sano, K., & Yokota, Y. (2010). Identification of BMP-responsive elements in the mouse Id2 gene. *Biochemical and Biophysical Research Communications*, *399*, 416–421.

Nam, H. S., & Benezra, R. (2009). High levels of Id1 expression define B1 type adult neural stem cells. *Cell Stem Cell*, *5*, 515–526.

Nickoloff, B. J., Chaturvedi, V., Bacon, P., Qin, J. Z., Denning, M. F., & Diaz, M. O. (2000). Id-1 delays senescence but does not immortalize keratinocytes. *Journal of Biological Chemistry*, *275*, 27501–27504.

Nieborowska-Skorska, M., Hoser, G., Rink, L., Malecki, M., Kossev, P., Wasik, M. A., et al. (2006). Id1 transcription inhibitor-matrix metalloproteinase 9 axis enhances invasiveness of the breakpoint cluster region/abelson tyrosine kinase-transformed leukemia cells. *Cancer Research*, *66*, 4108–4116.

Nishiyama, K., Takaji, K., Kataoka, K., Kurihara, Y., Yoshimura, M., Kato, A., et al. (2005). Id1 gene transfer confers angiogenic property on fully differentiated endothelial cells and contributes to therapeutic angiogenesis. *Circulation*, *112*, 2840–2850.

O'Brien, C. A., Kreso, A., Ryan, P., Hermans, K. G., Gibson, L., Wang, Y., et al. (2012). ID1 and ID3 regulate the self-renewal capacity of human colon cancer-initiating cells through p21. *Cancer Cell*, *21*, 777–792.

Ohtani, N., Zebedee, Z., Huot, T. J., Stinson, J. A., Sugimoto, M., Ohashi, Y., et al. (2001). Opposing effects of Ets and Id proteins on p16INK4a expression during cellular senescence. *Nature*, *409*, 1067–1070.

Ouyang, X. S., Wang, X., Lee, D. T., Tsao, S. W., & Wong, Y. C. (2002). Over expression of ID-1 in prostate cancer. *Journal of Urology*, *167*, 2598–2602.

Owens, A. P., III, Subramanian, V., Moorleghen, J. J., Guo, Z., McNamara, C. A., Cassis, L. A., et al. (2010). Angiotensin II induces a region-specific hyperplasia of the ascending aorta through regulation of inhibitor of differentiation 3. *Circulation Research*, *106*, 611–619.

Paolella, B. R., Havrda, M. C., Mantani, A., Wray, C. M., Zhang, Z., & Israel, M. A. (2011). p53 directly represses Id2 to inhibit the proliferation of neural progenitor cells. *Stem Cells*, *29*, 1090–1101.

Park, S. T., Nolan, G. P., & Sun, X.-H. (1999). Growth inhibition and apoptosis due to restoration of E2A activity in T cell acute lymphoblastic leukemia cells. *Journal of Experimental Medicine*, *189*, 501–508.

Park, S. T., & Sun, X. H. (1998). The Tal1 oncoprotein inhibits E47-mediated transcription mechanism of inhibition. *Journal of Bioliological Chemistry*, *273*, 7030–7037.

Park, K. W., Waki, H., Villanueva, C. J., Monticelli, L. A., Hong, C., Kang, S., et al. (2008). Inhibitor of DNA binding 2 is a small molecule-inducible modulator of peroxisome proliferator-activated receptor-gamma expression and adipocyte differentiation. *Molecular Endocrinology*, *22*, 2038–2048.

Peng, X., Wang, Y., Kolli, S., Deng, J., Li, L., Wang, Z., et al. (2012). Physical and functional interaction between the ID1 and p65 for activation of NF-kappaB. *American Journal of Physiology—Cell Physiology*, *303*, C267–C277.

Perry, S. S., Zhao, Y., Nie, L., Cochrane, S. W., Huang, Z., & Sun, X. H. (2007). Id1, but not Id3, directs long-term repopulating hematopoietic stem-cell maintenance. *Blood*, *110*, 2351–2360.

Peverali, F. A., Ramqvist, T., Saffrich, R., Pepperkik, R., Barone, M. V., & Philipson, L. (1994). Regulation of G1 progression by E2A and Id helix-loop-helix proteins. *EMBO Journal, 13*, 4291–4301.

Pillai, S., Rizwani, W., Li, X., Rawal, B., Nair, S., Schell, M. J., et al. (2011). ID1 facilitates the growth and metastasis of non-small cell lung cancer in response to nicotinic acetylcholine receptor and epidermal growth factor receptor signaling. *Molecular and Cellular Biology, 31*, 3052–3067.

Prabhu, S., Ignatova, A., Park, S. T., & Sun, X.-H. (1997). Regulation of the expression of cyclin-dependent kinase inhibitor p21 by E2A and ID proteins. *Molecular and Cellular Biology, 17*, 5888–5896.

Riechmann, V., van Cruchten, I., & Sablitzky, F. (1994). The expression pattern of Id4, a novel dominant negative helix-loop-helix protein, is distinct from Id1, Id2 and Id3. *Nucleic Acids Research, 22*, 749–755.

Roberts, E. C., Deed, R. W., Inoue, T., Norton, J. D., & Sharrocks, A. D. (2001). Id helix-loop-helix proteins antagonize pax transcription factor activity by inhibiting DNA binding. *Molecular and Cellular Biology, 21*, 524–533.

Rosen, E. D., Sarraf, P., Troy, A. E., Bradwin, G., Moore, K., Milstone, D. S., et al. (1999). PPAR gamma is required for the differentiation of adipose tissue in vivo and in vitro. *Molecular Cell, 4*, 611–617.

Ruzinova, M. B., & Benezra, R. (2003). Id proteins in development, cell cycle and cancer. *Trends in Cell Biology, 13*, 410–418.

Ruzinova, M. B., Schoer, R. A., Gerald, W., Egan, J. E., Pandolfi, P. P., Rafii, S., et al. (2003). Effect of angiogenesis inhibition by Id loss and the contribution of bone-marrow-derived endothelial cells in spontaneous murine tumors. *Cancer Cell, 4*, 277–289.

Saisanit, S., & Sun, X. H. (1995). A novel enhancer, the pro-B enhancer, regulates Id1 gene expression in progenitor B cells. *Molecular and Cellular Biology, 15*, 1513–1521.

Saisanit, S., & Sun, X. H. (1997). Regulation of the pro-B-cell-specific enhancer of the Id1 gene involves the C/EBP family of proteins. *Molecular and Cellular Biology, 17*, 844–850.

Samanta, J., & Kessler, J. A. (2004). Interactions between ID and OLIG proteins mediate the inhibitory effects of BMP4 on oligodendroglial differentiation. *Development, 131*, 4131–4142.

Satyanarayana, A., Klarmann, K. D., Gavrilova, O., & Keller, J. R. (2012). Ablation of the transcriptional regulator Id1 enhances energy expenditure, increases insulin sensitivity, and protects against age and diet induced insulin resistance, and hepatosteatosis. *FASEB Journal, 26*, 309–323.

Sellam, J., Miceli-Richard, C., Gottenberg, J. E., Proust, A., Ittah, M., Lavie, F., et al. (2008). Is Inhibitor of differentiation 3 involved in human primary Sjogren's syndrome? *Rheumatology (Oxford), 47*, 437–441.

Sequerra, E. B., Costa, M. R., Menezes, J. R., & Hedin-Pereira, C. (2013). Adult neural stem cells: Plastic or restricted neuronal fates? *Development, 140*, 3303–3309.

Sharma, P., Patel, D., & Chaudhary, J. (2012). Id1 and Id3 expression is associated with increasing grade of prostate cancer: Id3 preferentially regulates CDKN1B. *Cancer Medicine, 1*, 187–197.

Shimano, H., Shimomura, I., Hammer, R. E., Herz, J., Goldstein, J. L., Brown, M. S., et al. (1997). Elevated levels of SREBP-2 and cholesterol synthesis in livers of mice homozygous for a targeted disruption of the SREBP-1 gene. *Journal of Clinical Investigation, 100*, 2115–2124.

Shuno, Y., Tsuno, N. H., Okaji, Y., Tsuchiya, T., Sakurai, D., Nishikawa, T., et al. (2010). Id1/Id3 knockdown inhibits metastatic potential of pancreatic cancer. *Journal of Surgical Research, 161*, 76–82.

Sikder, H., Huso, D. L., Zhang, H., Wang, B., Ryu, B., Hwang, S. T., et al. (2003). Disruption of Id1 reveals major differences in angiogenesis between transplanted and autochthonous tumors. *Cancer Cell, 4*, 291–299.

Spangrude, G. J., Heimfeld, S., & Weissman, I. L. (1988). Purification and characterization of mouse hematopoietic stem cells. *Science*, *241*, 58–62.

Spits, H., Couwenberg, F., Bakker, A. Q., Weijer, K., & Uittenbogaart, C. H. (2000). Id2 and Id3 inhibit development of CD34(+) stem cells into predendritic cell (pre-DC)2 but not into pre-DC1. Evidence for a lymphoid origin of pre-DC2. *Journal of Experimental Medicine*, *192*, 1775–1784.

Spits, H., & Di Santo, J. P. (2011). The expanding family of innate lymphoid cells: Regulators and effectors of immunity and tissue remodeling. *Nature Immunology*, *12*, 21–27.

Stankic, M., Pavlovic, S., Chin, Y., Brogi, E., Padua, D., Norton, L., et al. (2013). TGF-beta-Id1 signaling opposes Twist1 and promotes metastatic colonization via a mesenchymal-to-epithelial transition. *Cell Reports*, *5*, 1228–1242.

Stighall, M., Manetopoulos, C., Axelson, H., & Landberg, G. (2005). High ID2 protein expression correlates with a favourable prognosis in patients with primary breast cancer and reduces cellular invasiveness of breast cancer cells. *International Journal of Cancer*, *115*, 403–411.

Su, Y., Gao, L., Teng, L., Wang, Y., Cui, J., Peng, S., et al. (2013). Id1 enhances human ovarian cancer endothelial progenitor cell angiogenesis via PI3K/Akt and NF-kappaB/MMP-2 signaling pathways. *Journal of Translational Medicine*, *11*, 132.

Suh, H. C., Ji, M., Gooya, J., Lee, M., Klarmann, K. D., & Keller, J. R. (2009). Cell-nonautonomous function of Id1 in the hematopoietic progenitor cell niche. *Blood*, *114*, 1186–1195.

Sun, X.-H. (1994). Constitutive expression of the Id1 gene impairs mouse B cell development. *Cell*, *79*, 893–900.

Sun, X.-H., Copeland, N. G., Jenkins, N. A., & Baltimore, D. (1991). Id proteins, Id1 and Id2, selectively inhibit DNA binding by one class of helix-loop-helix proteins. *Molecular and Cellular Biology*, *11*, 5603–5611.

Swarbrick, A., Roy, E., Allen, T., & Bishop, J. M. (2008). Id1 cooperates with oncogenic Ras to induce metastatic mammary carcinoma by subversion of the cellular senescence response. *Proceedings of the National Academy of Sciences pf the United States of America*, *105*, 5402–5407.

Tang, J., Gordon, G. M., Nickoloff, B. J., & Foreman, K. E. (2002). The helix-loop-helix protein id-1 delays onset of replicative senescence in human endothelial cells. *Laboratory Investigation*, *82*, 1073–1079.

Taylor, A. M., Li, F., Thimmalapura, P., Gerrity, R. G., Sarembock, I. J., Forrest, S., et al. (2006). Hyperlipemia and oxidation of LDL induce vascular smooth muscle cell growth: An effect mediated by the HLH factor Id3. *Journal of Vascular Research*, *43*, 123–130.

Tobin, N. P., Sims, A. H., Lundgren, K. L., Lehn, S., & Landberg, G. (2011). Cyclin D1, Id1 and EMT in breast cancer. *BMC Cancer*, *11*, 417.

Ueda-Hayakawa, I., Mahlios, J., & Zhuang, Y. (2009). Id3 restricts the developmental potential of gamma delta lineage during thymopoiesis. *Journal of Immunology*, *182*, 5306–5316.

Valdimarsdottir, G., Goumans, M. J., Rosendahl, A., Brugman, M., Itoh, S., Lebrin, F., et al. (2002). Stimulation of Id1 expression by bone morphogenetic protein is sufficient and necessary for bone morphogenetic protein-induced activation of endothelial cells. *Circulation*, *106*, 2263–2270.

Verykokakis, M., Krishnamoorthy, V., Iavarone, A., Lasorella, A., Sigvardsson, M., & Kee, B. L. (2013). Essential functions for ID proteins at multiple checkpoints in invariant NKT cell development. *Journal of Immunology*, *191*, 5973–5983.

Volpert, O. V., Pili, R., Sikder, H. A., Nelius, T., Zaichuk, T., Morris, C., et al. (2002). Id1 regulates angiogenesis through transcriptional repression of thrombospondin-1. *Cancer Cell*, *2*, 473–483.

Vrijens, K., Lin, W., Cui, J., Farmer, D., Low, J., Pronier, E., et al. (2013). Identification of small molecule activators of BMP signaling. *PLoS One*, *8*, e59045.

Wagner, K., Zhang, P., Rosenbauer, F., Drescher, B., Kobayashi, S., Radomska, H. S., et al. (2006). Absence of the transcription factor CCAAT enhancer binding protein alpha results in loss of myeloid identity in bcr/abl-induced malignancy. *Proceedings of the National Academy of Sciences of the United States of America, 103*, 6338–6343.

Wazir, U., Jiang, W. G., Sharma, A. K., Newbold, R. F., & Mokbel, K. (2013). The mRNA expression of inhibitors of DNA binding-1 and −2 is associated with advanced tumour stage and adverse clinical outcome in human breast cancer. *Anticancer Research, 33*, 2179–2183.

Wu, Z., Bucher, N. L., & Farmer, S. R. (1996). Induction of peroxisome proliferator-activated receptor gamma during the conversion of 3T3 fibroblasts into adipocytes is mediated by C/EBPbeta, C/EBPdelta, and glucocorticoids. *Molecular and Cellular Biology, 16*, 4128–4136.

Wu, Z., Rosen, E. D., Brun, R., Hauser, S., Adelmant, G., Troy, A. E., et al. (1999). Cross-regulation of C/EBP alpha and PPAR gamma controls the transcriptional pathway of adipogenesis and insulin sensitivity. *Molecular Cell, 3*, 151–158.

Xu, M., Nie, L., Kim, S. H., & Sun, X. H. (2003). STAT5-induced Id-1 transcription involves recruitment of HDAC1 and deacetylation of C/EBPbeta. *EMBO Journal, 22*, 893–904.

Yamauchi, T., & Kadowaki, T. (2013). Adiponectin receptor as a key player in healthy longevity and obesity-related diseases. *Cell Metabolism, 17*, 185–196.

Yan, W., Young, A. Z., Soares, V. C., Kelley, R., Benezra, R., & Zhuang, Y. (1997). High incidence of T-cell tumors in E2A-null mice and E2A/Id1 double-knockout mice. *Molecular and Cellular Biology, 17*, 7317–7327.

Yang, J., Li, X., Li, Y., Southwood, M., Ye, L., Long, L., et al. (2013). Id proteins are critical downstream effectors of BMP signaling in human pulmonary arterial smooth muscle cells. *American Journal of Physiology. Lung Cellular and Molecular Physiology, 305*, L312–L321.

Yang, Y., Liou, H. C., & Sun, X. H. (2006). Id1 potentiates NF-kappaB activation upon T cell receptor signaling. *Journal of Biological Chemistry, 281*, 34989–34996.

Yeh, W. C., Cao, Z., Classon, M., & McKnight, S. L. (1995). Cascade regulation of terminal adipocyte differentiation by three members of the C/EBP family of leucine zipper proteins. *Genes and Development, 9*, 168–181.

Yokota, Y., Mansouri, A., Mori, S., Sugawara, S., Adachi, S., Nishikawa, S., et al. (1999). Development of peripheral lymphoid organs and natural killer cells depends on the helix-loop-helix inhibitor Id2. *Nature, 397*, 702–706.

Zeng, H., Yucel, R., Kosan, C., Klein-Hitpass, L., & Moroy, T. (2004). Transcription factor Gfi1 regulates self-renewal and engraftment of hematopoietic stem cells. *EMBO Journal, 23*, 4116–4125.

Zhang, X., Ling, M. T., Wang, Q., Lau, C. K., Leung, S. C., Lee, T. K., et al. (2007). Identification of a novel inhibitor of differentiation-1 (ID-1) binding partner, caveolin-1, and its role in epithelial-mesenchymal transition and resistance to apoptosis in prostate cancer cells. *Journal of Biological Chemistry, 282*, 33284–33294.

Zhang, P., Zhao, Y., & Sun, X. H. (2013). Notch-regulated periphery B cell differentiation involves suppression of E protein function. *Journal of Immunology, 191*, 726–736.

Zhao, Q., Beck, A. J., Vitale, J. M., Schneider, J. S., Gao, S., Chang, C., et al. (2011). Developmental ablation of Id1 and Id3 genes in the vasculature leads to postnatal cardiac phenotypes. *Developmental Biology, 349*, 53–64.

Zhao, Y., Ling, F., Wang, H. C., & Sun, X. H. (2013). Chronic TLR signaling impairs the long-term repopulating potential of hematopoietic stem cells of wild type but not Id1 deficient mice. *PLoS One, 8*, e55552.

Zheng, W., Wang, H., Xue, L., Zhang, Z., & Tong, T. (2004). Regulation of cellular senescence and p16(INK4a) expression by Id1 and E47 proteins in human diploid fibroblast. *Journal of Biological Chemistry, 279*, 31524–31532.

CHAPTER SIX

E(spl): Genetic, Developmental, and Evolutionary Aspects of a Group of Invertebrate Hes Proteins with Close Ties to Notch Signaling

Christos Delidakis*,†,1, Maria Monastirioti*, Srivathsa S. Magadi*,†

*Institute of Molecular Biology & Biotechnology, Foundation for Research & Technology Hellas, Heraklion, Greece
†Department of Biology, University of Crete, Heraklion, Crete, Greece
[1]Corresponding author: e-mail address: delidaki@imbb.forth.gr

Contents

Abstract

Enhancer-of-split (*E(spl)*) was genetically characterized in *Drosophila* as a dominant mutation that interacts with an allele of *Notch*, the receptor in a multipurpose signaling pathway throughout development. Although dominant mutations are often not informative

Current Topics in Developmental Biology, Volume 110
ISSN 0070-2153
http://dx.doi.org/10.1016/B978-0-12-405943-6.00006-3
© 2014 Elsevier Inc.
All rights reserved.

of the normal gene function, *E(spl)* turned out to encode a family of seven paralogous basic helix-loop-helix proteins of utmost importance in the implementation of the Notch signal in the receiving cell. They are transcriptionally induced by Notch in almost every instance where the signal is deployed, and they participate in numerous feedback circuits, where they interface with a panoply of additional more tissue-specific Notch targets to ensure the proper signaling outcome. Besides the bHLH domain, E(spl) contain a characteristic Orange domain and are classified in the Hes (hairy and enhancer-of-split) branch of the bHLH-Orange proteins. They act as DNA-binding repressors in close collaboration with the corepressor Groucho. In this review, we will focus on the regulation of E(spl) expression and on the function of E(spl) proteins. In the latter section, we will present some of the best-studied developmental events where E(spl) function has been analysed as well as the molecular mechanism of E(spl) activity that has transpired. Finally, we will review the evolution of this protein family, which, albeit of relatively recent origin, present only in insects and crustaceans, has undergone extensive diversification, including gene loss and duplication. Importantly, many of the characteristics of E(spl) proteins are more deeply rooted in the very ancient larger bHLH-O family, which seems to have forged a connection with the Notch pathway from the very beginning of multicellular animal life.

ABBREVIATIONS

AD activation domain
AMP adult midgut precursor
AS-C achaete–scute complex
BFM bearded family member
bHLH basic helix-loop-helix
bHLH-O basic helix-loop-helix-Orange
Brd-C Bearded complex
ChIP chromatin immunoprecipitation
CKII casein kinase II
CNS central nervous system
CSL CBF1, Su(H), Lag-1
Da daughterless
Dl Delta
Dpn deadpan
DV dorsoventral
E(spl) enhancer-of-split
E(spl)-C enhancer-of-split complex
EB enteroblast
EC enterocyte
ee enteroendocrine
Egfr epidermal growth factor receptor
GMC ganglion mother cell
gof gain-of-function
gro Groucho

Hes hairy and enhancer-of-split-like
Hey hairy and enhancer-of-split-like with a Y
ICP interstitial cell precursor
ISC intestinal stem cell
lof loss of function
MAPK mitogen-activated protein kinase
MeHg methylmercury
MF morphogenetic furrow
N Notch
NB neuroblast
OPC outer proliferation center
PMEC principal midgut epithelial cell
PNS peripheral nervous system
RTK receptor tyrosine kinase
SNS stomatogastric nervous system
SOP sensory organ precursor
SPS Su(H)-paired site
StUbl SUMO-targeted ubiquitin ligase
SUMO small ubiquitin-like modifier

1. E(SPL): FROM A SPONTANEOUS DOMINANT MUTATION TO A GROUP OF CORE DEVELOPMENTAL REGULATORS

Enhancer-of-split (*E(spl)*) was discovered by Mel Green (Welshons, 1956) as a spontaneous dominant mutation in *Drosophila* which enhanced the small eye phenotype caused by *split*, or N^{spl}, a recessive allele of *Notch* (Li, Lei, Irvine, & Baker, 2003), that was later shown to be a receptor involved in many developmental patterning events via contact-mediated cell communication. Dominant mutations often exert their effects by producing excessive or aberrant gene products (see Section 4.3 for the molecular analysis of $E(spl)^D$), and it is common practice among fly geneticists to induce revertants of a dominant mutation via deletion-promoting mutagens (e.g., X-rays) in order to generate gene knockouts and study the effects caused by the loss of function (lof) of the particular gene. When revertants of $E(spl)^D$ were isolated, it was realized that the lof phenotype for this locus was hyperplasia of the embryonic nervous system at the expense of larval epidermis, caused by the misspecification of most ectodermal cells as neural precursors, known as neuroblasts (NBs), early in embryogenesis (Knust, Bremer, et al., 1987; Lehmann, Jiménez, Dietrich, & Campos-Ortega, 1983).

This recessive phenotype, which was dubbed "neurogenic," was shared by lof alleles of a small number of genes, all of which were later found to participate in the Notch signaling pathway, Notch and its ligand Delta prominently featuring in this early neurogenic mutant collection.

Molecular cloning of the locus revealed not just a single gene but a gene cluster, the *Enhancer-of-split complex* (*E(spl)-C*), spanning some 60 kb and containing 13 genes (Delidakis & Artavanis-Tsakonas, 1992; Delidakis, Preiss, Hartley, & Artavanis-Tsakonas, 1991; Hartley, Preiss, & Artavanis-Tsakonas, 1988; Klämbt, Knust, Tietze, & Campos-Ortega, 1989; Knust, Schrons, Grawe, & Campos-Ortega, 1992; Knust, Tietze, & Campos-Ortega, 1987; Preiss, Hartley, & Artavanis-Tsakonas, 1988). These were originally named *E(spl)mδ, mγ, mβ, mα, m1, m2, ..., m8, m9/10* based on gene order along the chromosome. Seven of these, *E(spl)mδ, mγ, mβ, m3, m5, m7*, and *m8*, encode paralogous basic helix-loop-helix (bHLH) proteins; four, *BFM-mα, m6, m4*, and *m2*, encode paralogous proteins of the Bearded family (BFM: Bearded family member) (Lai, Bodner, & Posakony, 2000); one, *kazal-m1*, encodes a protease inhibitor (Wurmbach, Wech, & Preiss, 1999); and finally the distal-most gene of the complex, *m9/10*—later renamed *gro*—encodes the transcriptional corepressor Groucho. With such complexity, it comes as no surprise that pinpointing the molecular basis of the *E(spl)* lof and dominant phenotypes was not trivial. The dominant allele was eventually mapped to defects in the *E(spl)m8* gene, whereas most revertants were deletions that removed all classes of genes in the *E(spl)-C*. The embryonic neurogenic phenotype was attributed primarily to lof of the seven *bHLH* genes, although *gro* was also demonstrated to have a significant role. It was later shown that the Gro protein physically interacts with the E(spl) bHLH proteins and is needed for their function (Paroush et al., 1994). The genetic characterization of the *Bearded* family genes was even more challenging, as the genome contains an additional locus at a distant location, the bearded complex (*Brd-C*), which harbors another five *BFM* paralogs (Lai, Bodner, Kavaler, Freschi, & Posakony, 2000). Recently, these proteins were shown to be inhibitors of Neuralized, an E3 ubiquitin ligase which is needed to activate the Notch ligands, Delta, and Serrate, by ubiquitylation and subsequent endocytosis (Bardin & Schweisguth, 2006; De Renzis, Yu, Zinzen, & Wieschaus, 2006; Fontana & Posakony, 2009). Their complete lof, namely, their elimination from both *Brd-C* and *E(spl)-C* (while keeping the other classes of genes in the *E(spl)-C* intact), caused embryonic lethality, but this was not manifest as a neurogenic phenotype. Instead, it was due to increased Notch signaling and epithelial polarity disruption (Chanet & Schweisguth, 2012; Chanet, Vodovar, Mayau, & Schweisguth, 2009). Whereas loss of E(spl) bHLH function produces the neurogenic phenotype typical of *E(spl)-C* deficiencies, *gro* null alleles have more severe and pleiotropic defects, since this corepressor interacts with a multitude of repressors involved not only in neurogenesis but also in sex determination, embryonic pattern formation, and segmentation. This extreme phenotype was masked in the early genetic analysis, thanks to a strong maternal contribution in *gro* expression; it can only be revealed if *gro*$^{-/-}$ embryos are born from *gro*$^{-/-}$ mothers (actually −/− mosaic ovaries induced in −/+mothers, since *gro*$^{-/-}$ is lethal). The *kazal-m1* gene is still very little studied but appears to be nonessential for viability. The rest of this review will deal with the biology of the seven *E(spl) bHLH* genes and their diverse developmental functions. For simplicity, we will reserve the "E(spl)" term for the bHLH genes/proteins and refer to the Bearded family members as "BFM," although the official Flybase (http://flybase.org/) name for the *E(spl)-C*-encoded BFMs is E(spl)m4-BFM, E(spl)m6-BFM, and so on.

The seven E(spl) bHLH proteins belong to the "Orange" subfamily, as they contain an Orange domain just after their bHLH domain. Like the HLH domain, the Orange domain consists of two amphipathic alpha-helices and mediates protein–protein interactions, including dimerization of the basic helix-loop-helix-orange (bHLH-O) proteins, where it acts as an "extension" of the HLH domain (Dawson, Turner, Weintraub, & Parkhurst, 1995; Eastwood, Yin, Bandyopadhyay, & Bidwai, 2011; Knust et al., 1992; Taelman et al., 2004; Zarifi et al., 2012). bHLH-O proteins are subdivided into the Hes and Hey (hairy and enhancer-of-split-like with a Y) subgroups: E(spl) belong to the Hes subgroup, in fact Hes stands for "hairy and enhancer-of-split"-like. The *Drosophila* genome contains another four Hes genes (*hairy, deadpan* (*dpn*), *Similar-to-deadpan*, and *Her*) and two *Hey* genes (*Hey* and *clockwork-orange*); these genes have been implicated in a variety of functions, like body segmentation (*hairy*), asymmetric cell divisions (*Hey*), and circadian rhythm regulation (*cwo*), whereas others, like *Side* and *Her*, have hardly been studied at all (Duncan & Dearden, 2010; Moore, Barbel, Jan, & Jan, 2000). bHLH-O proteins show a high degree of similarity in their basic domains, and this correlates with their preference for B-type E-box-binding sites (CACGTG) as well as variants thereof, the C-boxes (CACGCG) and N-boxes (CACNAG) (Jennings, Tyler, & Bray, 1999;

Oellers, Dehio, & Knust, 1994; Ohsako, Hyer, Panganiban, Oliver, & Caudy, 1994; Tietze, Oellers, & Knust, 1992; Van Doren, Bailey, Esnayra, Ede, & Posakony, 1994). E(spl) proteins have been shown to bind all of these target sites with higher affinity for E_B- and C-boxes and lower for N-boxes; furthermore, the trinucleotides flanking these core hexanucleotides are also very important in determining binding affinity.

Hes (but not Hey) proteins, in *Drosophila* and other animal species, have two additional common characteristics besides the Orange domain and their overall sequence similarity within the bHLH domain. First, they all contain a KP dipeptide just before their basic domain; however, mutation of this proline to asparagine or threonine did not alter the functions of E(spl)m8 or Hairy, assayed either *in vitro* (DNA binding by electromobility shift assay) or *in vivo* (phenotypic effects in enhancement of the "split" eye phenotype or sex determination), so its significance remains elusive (Dawson et al., 1995; Oellers et al., 1994; Tietze et al., 1992). Their second common characteristic (with rare exceptions, like the *C. elegans* Hes protein Lin-22) is a WRPW tetrapeptide at, or very close to, the C-terminus, which binds very strongly to the WD40 repeats of the Groucho corepressor (and its vertebrate TLE homologs) (Jennings et al., 2006; Paroush et al., 1994). Consequently, Hes proteins act as repressors. E(spl) proteins need the Gro interaction in order to exert their activity, as WRPW mutants are completely inactive (Giagtzoglou, Alifragis, Koumbanakis, & Delidakis, 2003; Giagtzoglou, Koumbanakis, Fullard, Zarifi, & Delidakis, 2005; Giebel & Campos-Ortega, 1997; Kahali, Bose, Karandikar, Bishop, & Bidwai, 2009). In contrast, Hairy, another Hes protein, has additional motifs and can recruit three different cofactors, Gro, CtBP, and Sir2, in a target context-dependent manner (Bianchi-Frias et al., 2004; Poortinga, Watanabe, & Parkhurst, 1998; Rosenberg & Parkhurst, 2002). One of the E(spl) proteins, mδ, was also shown to interact with CtBP in biochemical assays (Poortinga et al., 1998), but the functional significance of this interaction remains to be demonstrated. Besides the common C-terminal motif, the broader C-terminal regions of Hes proteins do not show characteristic similarities. Many are enriched in proline, glutamate, serine, and threonine residues, which is known as the PEST motif and has been associated with intrinsically disordered regions displaying increased instability (Belizario, Alves, Garay-Malpartida, & Occhiucci, 2008).

The seven *E(spl) bHLH* genes and the four *BFM* genes in the *E(spl)-C* locus are all short intronless genes. This is a peculiarity of the *E(spl)* locus; all other bHLH-O genes in the fly contain introns and none of them are encountered in clusters with other *Hes* or *Hey* genes. A remarkable feature of *E(spl)* and *BFM* genes is that all of them are transcriptional targets of Notch signaling. This Notch responsiveness is consistent with the common lof phenotype (neurogenic defect) described for *Notch* and *E(spl)* and suggests that the E(spl) proteins mediate the effects of Notch signaling in the embryonic neuroectoderm. Besides the neuroectoderm, a great multitude of additional expression sites, both in the embryo and later stages, have been described, which will be detailed in Section 3. Unlike the dynamic Notch-responsive expression patterns of *E(spl)* and *BFM* genes, *gro* is ubiquitously expressed (Hartley et al., 1988; Knust, Tietze, et al., 1987). We shall next discuss in more detail how *E(spl)* genes are regulated, and how E(spl) proteins participate in various developmental processes; but first, we will take a look at the phylogenetic distribution of E(spl) proteins.

2. E(SPL) PROTEINS—WHERE DID THEY ALL COME FROM?

E(spl) are a particular type of bHLH-O proteins, specifically of the Hes subgroup. bHLH-O proteins are specific to animals. They are encountered in all animal genomes down to the basal Porifera and Cnidaria, but have never been found in plants or fungi, whose genomes contain numerous other classes of bHLH proteins (Ledent, Paquet, & Vervoort, 2002; Simionato et al., 2007). Phylogenetic analysis of bHLH-O protein sequences from animal species has been hampered by the rapid divergence of their sequence, especially outside the bHLH domain, yet with the increasing number of animal genomes sequenced, more robust evolutionary histories can be traced. It is now quite clear that Hes proteins from invertebrates cluster together in phylogenetic trees at the exclusion of those from vertebrates (Duncan & Dearden, 2010). This suggests that the duplications that generated Hes gene diversity happened independently after protostome–deuterostome divergence >600 Myr ago; in other words, the different clades of insect Hes proteins, such as Hairy, Dpn, or E(spl), do not have true orthologs in the vertebrates, hence the term Hes (Hairy *and* E(spl) homolog) describes this situation accurately.

True E(spl) orthologs may not even exist throughout the invertebrates, even if we limit our scope only to arthropods. Insects have three distinct Hes clades: Hairy/Dpn, Side, and E(spl), with clear orthologs identified in every queried genome (Duncan & Dearden, 2010). The same three Hes types have been found in crustaceans, a sister group of insects, whereas the situation in the other two large athropod groups, the chelicerates and myriapods, is less clear. Database searches have failed to uncover an E(spl) homolog from these groups. Instead, all bHLH-O proteins found are more similar to insect Dpn, Side, Hey, or Cwo. However, only limited genomic databases exist for chelicerates and myriapods. Until more genomic data have accrued, we can hypothesize that *E(spl)* genes arose in the common ancestor of insects and crustaceans and are absent from all other animal groups—they are, therefore, characteristic of the Pancrustacea group (crustaceans + insects) within the arthropods.

Is the *E(spl)* gene family as highly duplicated in other insects and crustaceans as it is in *Drosophila*? The answer is no. Already in three different species of mosquitoes, dipteran insects (like Drosophilids), that have diverged from flies by ~200 Myr, there is but a single *E(spl)* gene (Duncan & Dearden, 2010; Schlatter & Maier, 2005). However, in more distantly related insect species, like hymenopterans, lepidopterans, etc., clusters of two to three *E(spl)* genes have been found. A three-gene cluster was also found in *Daphnia pulex*, a crustacean. In all of the genomes tested, a single *BFM* gene was found in the proximity of the *E(spl)* clusters, even next to the single mosquito *E(spl)* gene, consistent with an ancient association of these two genes, perhaps reflecting their coregulation by Notch. Phylogenetic clustering analysis of the bHLH-O domains of E(spl) proteins from nine insects and one crustacean species grouped the E(spl) proteins into three clades, called E(spl)1, 2, and 3 (or Her) (Duncan & Dearden, 2010). It therefore appears that an ancestral *E(spl)* cluster of three *bHLH-O* genes and one *BFM* gene underwent multiple gene duplications and gene losses in the various descendant groups of the common ancestor of insects and crustaceans, which is estimated to have existed ~430 Myr ago. In dipterans, *E(spl)2* and *3* were lost, but in the Drosophilids, the remaining *E(spl)1* gene underwent a sevenfold duplication; this was accompanied by duplication of the *BFM* gene with concomitant migration of some of the duplicates to a distant chromosome locus, the *Brd-C*. Also in the *Drosophila* lineage, an eighth *E(spl)* bHLH copy arose and moved out of the *E(spl)-C* generating the *Her* (Hairy and E(spl)-related) gene, which must be a more recent event, as it is not even present in all Drosophilid species sequenced. As *Her* is ubiquitously expressed (Duncan & Dearden, 2010; Moore et al., 2000) and its function is unknown, we will not discuss it further in this review. In addition to genes moving out of the *E(spl)-C*, genes have also moved into the *E(spl)-C* over the course of evolution. For example, a tubulin tyrosine ligase gene is found in the *E(spl)-C* of the honeybee *Apis mellifera*, the Kazal protease inhibitor *m1* in flies and a number of novel genes in the beetle *Tribolium castaneum* (Kux, Kiparaki, & Delidakis, 2013).

At what point within the Dipteran lineage did the massive expansion of the *E(spl)-C* take place? Mosquitoes (1 *E(spl)* + 1 *BFM*) belong to a major Dipteran group, called Nematocera, whereas Drosophilid flies (7 *E(spl)* + 4 *BFM*) belong to the other major group, the Brachycera, which are estimated to have diverged by about 200 Myr. Drosophilids branched from the Brachycera lineage about 40–60 Myr ago. Whereas all *Drosophila* species sequenced show a perfect conservation in the composition and structure of the *E(spl)-C* (Duncan & Dearden, 2010; Maier, Marte, Schäfer, Yu, & Preiss, 1993; Schlatter & Maier, 2005), no information was available from non-Drosophilid Brachycera until recently. The Brachyceran family of Diopsidae is thought to have arisen no more than 76 Myr ago, and recently, the sequence of the *E(spl)-C* from the diopsid fly *Teleopsis dalmanni* was analyzed (Baker, Kuehl, & Wilkinson, 2011). The structure of the locus was remarkably conserved with that of the Drosophilids with the only difference that *E(spl)m5* had undergone a tandem inverse duplication, raising the number of *E(spl) bHLH* genes to eight. The remaining *bHLH*, *BFM*, and *kazal* genes were in the same relative location and the same orientation. So, the expansion of the *E(spl)* locus to a large gene cluster may be a characteristic of the Brachyceran lineage, something that was supported by limited expressed sequence tag sequences from other more distantly related Brachyceran flies. Interestingly, in the *T. dalmanni* cluster, the gene next to *E(spl)m8* is not *gro*, which was in fact not located at all within the sequenced ~150 kb contig. *gro* is also unlinked from the *E(spl)-C* in all non-Drosophilid-sequenced genomes. It therefore seems that the juxtaposition of *gro* with the *E(spl)-C* is a curious coincidence found only in Drosophilids, since it is neither conserved nor does it play a role in transcriptional regulation. In *Drosophila, gro* does not have a similar expression pattern with its neighboring *E(spl)* genes; in fact, there seems to be a chromatin boundary element separating *E(spl)m8* from *gro* (Schaaf, Misulovin, Gause, Koenig, & Dorsett, 2013).

3. REGULATION OF *E(SPL)* GENES

As *Drosophila E(spl)* genes duplicated relatively recently, not enough time may have elapsed to generate functional divergence. Analyzing their expression patterns should inform us to what extent they may show any specialization. Early embryonic expression provides no such evidence, as all seven *E(spl)* genes show robust expression in almost indistinguishable patterns (Knust et al., 1992; Wech, Bray, Delidakis, & Preiss, 1999). This contrasts with results in late embryos and postembryonic tissues, where different *E(spl)* genes have very distinct expression patterns (de Celis et al., 1996; Wech et al., 1999). Table 6.1 details the tissues where *E(spl)* genes are expressed at various stages of development, pointing out which subset of the *E(spl)* genes is detected in each case. It is a compilation of data from RNA *in situ* hybridization, antibody detection, and reporter genes and is by no means exhaustive.

Transcription of all *E(spl)* genes responds to Notch signaling mediated by the transcription factor Suppressor of Hairless [Su(H)], whose binding sites abound in the flanking regions of *E(spl)* genes. Su(H) belongs to the CSL proteins (CSL is an acronym for CBF1, Su(H), Lag-1, the mammalian, fly, and worm homolog, respectively), the DNA-binding factors that tether activated Notch (the cleaved intracellular domain of the Notch receptor, or Nicd) to its target enhancers—for recent reviews of Notch signaling, see Andersson, Sandberg, and Lendahl (2011), Hori, Sen, and Artavanis-Tsakonas (2013), Kopan and Ilagan (2009). *E(spl)* expression is lost if Notch signaling is compromised (de Celis et al., 1996; Jennings, de Celis, Delidakis, Preiss, & Bray, 1995; Jennings et al., 1994; Lecourtois & Schweisguth, 1995). Conversely, ectopic Notch activation induces ectopic expression of at least one *E(spl)* gene depending on tissue context. Interestingly, Notch responsiveness has also been documented for the *BFM* genes of the *E(spl)* locus (Bailey & Posakony, 1995; Bernard, Krejci, Housden, Adryan, & Bray, 2010; Lai, Bodner, & Posakony, 2000; Nellesen et al., 1999; Wurmbach et al., 1999) and may reflect an evolutionarily ancient connection between the *E(spl) Complex* and Notch signaling. Examples of Notch-independent *E(spl)* expression are rare. One case is the induction of *E(spl) mδ, mγ*, and *m7* in cultured cells by methylmercury (MeHg), a toxin which affects nervous system development. *E(spl)mδ, mγ, m5*, and *m8* were also induced in MeHg-exposed embryos but not in larvae or adults (Engel, Delwig, & Rand, 2012; Rand, Bland, & Bond, 2008). This toxin-induced *E(spl)* expression appears to occur by both a Notch-independent (immediate response) and a Notch-dependent (late response) mechanism (Bland & Rand, 2006). Another example is the expression of *E(spl)m5* in the dorsal cardiogenic mesoderm of the embryo, where it seems to be a target of Jak/STAT signaling (Johnson, Mokalled, Haden, & Olson, 2011). It is not, however, clear whether in this tissue there is cooperation between Jak/STAT and Notch to induce *E(spl)m5*.

Table 6.1 Summary of expression of *E(spl)* genes during *Drosophila* development

Developmental stage	Expression	References
Embryonic	Blastoderm/gastrulation: E(spl)m5, m7, m8 are expressed in mesectoderm, one cell stripe on either side of mesodermal anlage. Also in amnioserosa (dorsal midline). m3 is maternally expressed. Germband extension: At stage 8–11, all E(spl) genes are expressed similarly in clusters of cells in a grid-like pattern in the ventral ectoderm from which nonexpressing neuroblasts delaminate. From stage 10 onward, as NBs start proliferating, mγ is expressed in NBs. Mesoderm: At stage 10–12, E(spl)m3, m5, m7, and m8 are expressed throughout the mesoderm at uniformly low levels. Endoderm: At stages 9–11, all E(spl) genes except m5 are expressed in the entire anterior and posterior endoderm as ICPs and AMPs delaminate. SNS placodes: At stages 10–13, all E(spl) genes are expressed in the stomatogastric nervous system primordium as delaminating and invaginating (placode) SNS precursors segregate from the epithelium. Trachea: At stages 12–13, E(spl)mβ is expressed in all tracheal cells except fusion cells.	Hartenstein, Tepass, and Gruszynski-deFeo (1996), Jennings, Preiss, Delidakis, and Bray (1994), Knust et al. (1992), Knust, Tietze, et al. (1987) Kramatschek and Campos-Ortega (1994), Lecourtois and Schweisguth (1995), Llimargas (1999), Tepass and Hartenstein (1995), Wech et al. (1999), Zacharioudaki, Magadi, and Delidakis (2012)
Larval/pupal wing disk	E(spl) are expressed at high levels in the surrounding cells of SOPs: m7, m8 expressed in all proneural clusters, mγ and mδ are expressed in a subset. mβ is expressed more broadly in both the wing pouch and the notum regions of the disk with higher levels at the intervein regions. E(spl)m8, m7, mβ are expressed at the DV boundary (future wing margin). E(spl)mγ is expressed strongly in nonneuronal progeny of wing and notum SOPs.	Bailey and Posakony (1995), Cooper et al. (2000), de Celis et al. (1996), Nellesen, Lai, and Posakony (1999), Singson, Leviten, Bang, Hua, and Posakony (1994)

Continued

Table 6.1 Summary of expression of *E(spl)* genes during *Drosophila* development—cont'd

Developmental stage	Expression	References
Larval/pupal eye disk	E(spl)mδ, m7, m8, mβ, mγ are strongly expressed in the morphogenetic furrow. E(spl)mδ, mγ, mβ also accumulate posterior of the morphogenetic furrow in undifferentiated cells. E(spl)mδ is expressed in R4 and R7 photoreceptor cells of developing ommatidia. In the pupa, E(spl)mγ, mδ, and m8 are expressed in the progeny of olfactory sensilla.	Baker, Yu, and Han (1996), Cooper and Bray (1999, 2000), de Celis et al. (1996), Endo et al. (2012)
Larval trachea	E(spl)mβ is expressed in tracheal branch points.	Furriols and Bray (2001)
Larval/pupal leg disk	E(spl)mβ is expressed at the distal end of each leg segment and also in proneural clusters. mγ is expressed in femoral chordotonal organ proneural cluster.	de Celis, Tyler, de Celis, and Bray (1998), zur Lage and Jarman (1999)
Larval/pupal CNS	mγ, m8, and mβ are expressed in NBs of central brain, ventral nerve cord, and optic lobe m7 is expressed in OPC epithelium. mγ, m5, m7, and m8 are expressed in the medial transition zone of OPC of optic lobe. mβ expressed in subperineurial glia of OPC. mβ is expressed in dorsal cluster neurons in central brain.	Almeida and Bray (2005), Egger, Gold, and Brand (2010), Langen et al. (2013), Perez-Gomez, Slovakova, Rives-Quinto, Krejci, and Carmena (2013), Yasugi, Sugie, Umetsu, and Tabata (2010), Zacharioudaki et al. (2012)
Gut	E(spl)m8 is expressed in interstitial cells but not copper cells in the first larval instar. E(spl) mβ is expressed strongly in EB cells and weakly in ISCs in the adult midgut.	Tanaka et al. (2007), Bardin, Perdigoto, Southall, Brand, and Schweisguth (2010)
Ovary	E(spl)mβ is expressed in newly formed terminal filaments and cap cells during larval–pupal transition stage. E(spl)m7 and mβ are expressed in cap cells of adult ovary. At stages of oogenesis 1–6, E(spl)mβ and m7 are expressed in the polar follicle cells. At stages 7–10 of oogenesis, E(spl)mβ and m7 are expressed in endoreplicating follicle cells.	Assa-Kunik, Torres, Schejter, Johnston, and Shilo (2007), Grammont and Irvine (2001), Song, Call, Kirilly, and Xie (2007), Sun, Smith, Armento, and Deng (2008), Xie et al. (2008)

Other than these exceptional cases, *E(spl)* genes are not expressed in the absence of Notch signal. In fact, in the absence of Notch, CSL proteins can act as "default repressors" (Furriols & Bray, 2001). Su(H) represses its targets by binding to the adaptor protein Hairless (H), which recruits corepressors CtBP, Gro, Skip/Bx42, and perhaps others (Barolo, Stone, Bang, & Posakony, 2002; Morel et al., 2001; Nagel et al., 2005). Via interaction with Hairless and Skip, this repressive complex recruits histone chaperones Asf1 and Nap1, which in turn recruit complexes containing the H3K4 demethylase Lid and the histone deacetylase Rpd3 to establish a repressive chromatin state (Goodfellow et al., 2007; Moshkin et al., 2009). Activation is achieved by displacement of H/corepressors complexes by the intracellular domain of Notch (Nicd), which is released from the membrane by γ-secretase cleavage of the activated receptor. Recently, another Su(H) corepressor was discovered, the BEN domain protein Insensitive. The BEN domain of Insv interacts directly with Su(H) but at the same time is a DNA-binding domain (Dai et al., 2013; Duan et al., 2011). Insv repression of *E(spl)* genes does not require Hairless or Gro, suggesting that it acts via a parallel mechanism. However, unlike H, Insv is not essential for the repression of *E(spl)*, since $insv^{-/-}$ flies do not show significant derepression of *E(spl)*.

Upon Notch signaling, Nicd forms a complex with Su(H) and coactivator Mam (Nam, Sliz, Song, Aster, & Blacklow, 2006; Wilson & Kovall, 2006). Disruption of Su(H)-binding sites in enhancers isolated from the *E(spl)-C* abolished expression in most cell types, pointing to the importance of Su(H) binding in the induction of *E(spl)* genes by Notch (Bailey & Posakony, 1995; Castro, Barolo, Bailey, & Posakony, 2005; Lecourtois & Schweisguth, 1995; Nellesen et al., 1999). At the same time, it caused ectopic expression in cell types that do not receive Notch signaling, pointing to the importance of Su(H)-mediated default repression. The Su(H)/Nicd/Mam activator complex recruits HATs and other chromatin remodeling activities. Among the proteins known to associate with this activator complex on *E(spl)* regulatory regions are Putzig (Pzg) and Caf1-105 (Kugler & Nagel, 2010; Yu et al., 2013). Pzg seems to act as a component of the NURF nucleosome remodeling complex, while Caf1-105 is a subunit of a histone chaperone called chromatin assembly factor (CAF): both of these complexes are needed in the activation of *E(spl)mβ, m3*, and other Notch target genes in both cell lines and wing disks.

Many of the *E(spl)* enhancers display a particular configuration of two Su(H) sites in opposite orientation spaced by 17 bp, which has been called the Su(H)-paired site (SPS) (Bailey & Posakony, 1995; Nellesen et al., 1999). This motif enables cooperative binding of two Su(H)/Nicd/Mam complexes (Arnett et al., 2010) and is also encountered in some mammalian Notch targets, like *Hes1*. In most *E(spl)* enhancers, SPS or solo Su(H) sites are found in proximity to E_A-boxes, binding sites of proneural bHLH activators, like Da/Sc (see Section 4.2) (Cave, Loh, Surpris, Xia, & Caudy, 2005). These enhancers respond to dual input from Notch and the proneural activators, eliciting expression in proneural territories but only in those cells that also receive Notch signaling within these territories (Cave, Xia, & Caudy, 2009; Kramatschek & Campos-Ortega, 1994; Singson et al., 1994). Plenty of other TFs are predicted to bind *E(spl)-C cis* regulatory elements, and these sequences show conservation across many *Drosophila* species (Maeder, Polansky, Robson, & Eastman, 2007), attesting to the complex expression patterns of these genes.

Chromatin immunoprecipitation (ChIP) analysis of the binding of Su(H) to *E(spl)* genes has been performed in a number of cell lines and tissues. A major conclusion reached by these studies is that Su(H) is not always prebound onto its target enhancers before Notch signaling; even when it is prebound (and presumably acting as a repressor), its binding strength increases upon signaling (Krejci & Bray, 2007). As might be expected, not all *E(spl)* gene enhancers show the same Su(H) or PolII-binding profiles in all cell types. Consistently, in different cell types, a different subset of *E(spl)* genes is induced upon Notch signaling; the two more broadly induced genes are *E(spl)mβ* and *m3* (Krejci, Bernard, Housden, Collins, & Bray, 2009; Schaaf et al., 2013; Terriente-Felix et al., 2013). Cell line studies allow fine temporal resolution of *E(spl)* expression, as Notch can be acutely activated by a short calcium chelator treatment (Rand et al., 2000). *E(spl)* and *BFM* genes were the first to get activated after a 10 min pulse of Notch, peaking at 20–30 min and returning to basal at ~100 min after signaling termination (Housden et al., 2013). What is most remarkable is that *E(spl)* and *BFM* genes were the *only* genes that responded so rapidly to the signal, along with one more gene, encoding the apoptosis-promoting protein Hid. The remaining Notch targets, on the order of a few hundred up- or downregulated genes, peaked at later time intervals. This suggests that *E(spl)* and *BFM* genes play a central role in the Notch response, consistent with their broad deployment and justifying their use as markers of choice to reveal activity of the Notch pathway in *Drosophila*. It is not entirely clear mechanistically how E(spl) and (especially) BFM proteins ensure the proper unraveling of the Notch response. As far as the E(spl) bHLH-O proteins are concerned, there is evidence that they repress repressors of late-responding Notch target genes, which are more diverse and tissue-context dependent (Housden et al., 2013). They also autorepress and repress other early targets, in order to terminate the Notch signaling response. Autorepression coupled with a short protein half-life has been shown to be a prerequisite for oscillatory expression, which has been observed for certain vertebrate *Hes* genes (Kageyama, Niwa, Shimojo, Kobayashi, & Ohtsuka, 2010), although *E(spl)* mRNA oscillations have not yet been reported in *Drosophila* cells.

Additional ChIP analyses have identified a number of bound factors and chromatin marks on the *E(spl)-C*, like Polycomb complex subunits, Cohesin and Kollerin subunits, insulator-binding factors, H3K27Me3, H2A-Ub, etc. (Housden et al., 2013; Krejci & Bray, 2007; Schaaf et al., 2013, 2009), although the profile of these is not the same in all cell lines. What seems to be common is that the *E(spl)-C* constitutes a well-demarcated chromosomal domain whose chromatin state is different from that of adjacent domains. In the BG3 cell line, Cohesin and the PRC1/2 Polycomb complexes are needed to keep *E(spl)* gene expression low, even though there is a moderate level of Notch signaling going on, but whether such a mechanism operates in fly tissues is presently unclear (Schaaf et al., 2009, 2013).

In addition to their complex transcriptional regulation, *E(spl)* genes are heavily regulated posttranscriptionally by binding of microRNAs to their 3′-UTRs. All *E(spl)-C* genes, except *mβ*, contain three specific motifs for miRNA binding, the K-box (UGUGAU—recognized by miR2, miR6, miR11, and miR13), the GY-box (GUCUUCC, recognized by miR7), and the Brd-box (AGCUUUA, recognized by miR4 and miR79) (Bejarano et al., 2012; Lai, Tam, & Rubin, 2005). Overexpression of many of these microRNAs generates *Notch* lof phenotypes, whereas their down-regulation (using miR "sponges") can enhance Notch signaling, perhaps via modulating the levels of *E(spl)* gene expression. The importance of this regulation is mirrored in the conservation of these motifs in *E(spl)* homologs in distant insect species, such as the lepidopteran *Bombyx*, the aphid *Acyrthosiphon*, and even outside the insects, in the crustacean *Daphnia* (Duncan & Dearden, 2010; Lai et al., 2005).

4. FUNCTIONS OF E(SPL) PROTEINS IN *DROSOPHILA*

Genetic analysis of the *E(spl)* locus has shown that lof alleles of individual *E(spl)* genes are viable with no apparent phenotype (Nagel, Maier,

Krauss, Mezger, & Preiss, 2004; Nagel & Preiss, 1999). Deficiencies of increasing size cause increasingly more severe neurogenic phenotypes. Actually, small deficiencies removing as many as two *E(spl)* genes can be viable (Babaoglan, Housden, Furriols, & Bray, 2013; Delidakis & Artavanis-Tsakonas, 1992; Schrons, Knust, & Campos-Ortega, 1992). This shows that there is extensive functional redundancy among the seven *E(spl)* genes, not only in the embryo where all seven are coexpressed, but also at later stages, where expression patterns become more specialized. Therefore, in the discussion that follows, "*E(spl)* mutation," unless qualified, refers to deletion of the entire *E(spl)-C*. The examples we present will be limited to cases where genetic analysis has implicated *E(spl)* genes in a given process; many more processes are known to involve Notch signaling, but it has not been directly tested whether they also depend on E(spl) activity.

4.1. Lateral inhibition of neural precursors in the CNS

The earliest studied role of E(spl) proteins was that of restraining NB commitment in the embryonic neuroectoderm. NBs are stem cells that delaminate from the ventral trunk and the dorsal cephalic ectoderm and divide asymmetrically to generate all cells of the central nervous system (CNS). Elimination of the *E(spl)-C* produces embryonic lethality with a severe hyperplasia of the CNS, which stems from the misspecification of supernumerary ectodermal cells as NBs. This depletes the ectoderm of epithelial cells and no ventral or cephalic epidermis is formed. This misallocation of cell fates, producing nervous system hyperplasia with concomitant loss of ventral epidermis, is the classical "neurogenic" phenotype, which is shown by lof alleles of many of the Notch pathway genes (Knust, Bremer, et al., 1987; Lehmann et al., 1983). From early genetic and molecular analyses, it was concluded that nascent NBs send a Delta–Notch signal to laterally inhibit their ectodermal neighbors from adopting a NB fate, and this is implemented by E(spl) proteins. Indeed, all seven *E(spl)* genes are expressed strongly in the ventral and cephalic ectoderm at the time of NB delamination, where they are often seen to surround nascent NBs, which show low or no *E(spl)* expression (Jennings et al., 1994; Knust et al., 1992; Knust, Tietze, et al., 1987; Wech et al., 1999).

How do E(spl) proteins antagonize the differentiation of neuroectodermal cells to NBs? A first clue to the mechanism was provided by the demonstration that the *E(spl)* lof phenotype (the neurogenic phenotype) is the opposite of that of another class of genes, the so-called proneural genes in the *achaete–scute complex*, or *AS-C* (Garcia-Bellido & de Celis, 2009). Deletion of the *AS-C*, which contains three transcripts encoding paralogous bHLH activators, called Achaete (Ac), Scute (Sc), and Lethal of scute [L(1)sc], produces severe neural hypoplasia. This phenotypic contrast, supported with dosage-dependent genetic interactions between proneural and *E(spl)* genes, was instrumental for putting forward a hypothesis that proneural and E(spl) bHLH proteins act in a mutually antagonistic manner (Martin-Bermudo, Carmena, & Jiménez, 1995; Nakao & Campos-Ortega, 1996). However, this was more directly proven in the context of the peripheral nervous system (PNS), whose precursors (sensory organ precursors, SOPs) absolutely depend on proneural protein activity in order to acquire their fate (Culí & Modolell, 1998; Giagtzoglou et al., 2003; Giebel & Campos-Ortega, 1997). In the CNS NBs, on the contrary, the proneural genes play an important role, but additional factors seem to play a role, which makes genetic analysis less straightforward. Indeed, whereas null *E(spl)* alleles cause essentially all neuroectodermal cells to adopt a NB fate, null *AS-C* alleles do not eliminate the entirety of the NB complement. Instead, many NB persist and even divide to produce some differentiated neurons, although they often prematurely die by apoptosis (Jiménez & Campos-Ortega, 1990). Initially, it was hypothesized that other bHLH activators, perhaps homologues of the Neurogenin and Atonal families, could act redundantly with AS-C proteins in the initiation of CNS development. However, after characterization of the *Drosophila* genome, none of the other bHLH genes of this family was shown to be expressed in the embryonic neuroectoderm, making such a possibility unlikely. Instead, it appears that structurally distinct transcription factors, like the Snail family of zinc-finger transcription factors, may have a parallel role with the AS-C proteins in NB specification (Ip, Levine, & Bier, 1994). We will now turn to the description of cellular events in PNS generation, where the interplay between E(spl) and proneural proteins has been characterized in greater detail.

4.2. Lateral inhibition of neural precursors in the PNS—Inhibition of proneural proteins

Most peripheral sensory organs in *Drosophila* consist of a small number of cells, all of which are direct descendants of a single sensory organ precursor (SOP) cell per organ (Lai & Orgogozo, 2004). SOPs arise at different developmental times in various ectodermal epithelia, and their progeny neurons extend axons to the CNS ganglia. Unlike central NBs, which only partially depend on *AS-C* activity, peripheral SOPs are completely absent in lof conditions for *AS-C* (most mechano- and chemosensory organs of the adult fly) or other proneural genes, namely, atonal (*ato*) for chordotonal organs, *amos* for some of the antennal olfactory organs, and asense (*ase*) for some of the wing margin mechanosensory bristles (Campuzano & Modolell, 1992; Dominguez & Campuzano, 1993; Goulding, zur Lage, & Jarman, 2000; Jarman, Grau, Jan, & Jan, 1993; zur Lage, Prentice, Holohan, & Jarman, 2003). Like central NBs, SOPs laterally inhibit their neighbors via Notch signaling. Neural potential is initially bestowed not to the individual SOPs but to clusters of epithelial cells via the expression of proneural genes. Notch signaling takes place within these proneural clusters to single out the SOPs. In this process, the SOP remains devoid of E(spl) proteins, which are instead expressed in its neighbors, suggesting that the SOP somehow escapes from Notch signaling (Jennings et al., 1994). E(spl) expression in the cells surrounding the SOP is needed to prevent these cells from adopting a SOP fate; this is achieved by antagonizing the SOP-promoting activity of proneural proteins. This system explicitly illustrates the antagonism between AS-C proneural proteins and E(spl), for example, by their diametrically opposite lof and gain-of-function (gof) phenotypes. Whereas loss of *AS-C* results in clonal patches which lack all external sensory bristles, loss of *E(spl)* results in clonal patches with densely crowded supernumerary bristles (Heitzler, Bourouis, Ruel, Carteret, & Simpson, 1996). Reciprocally, overexpression of any of the three *AS-C* genes causes overproduction of bristles, whereas overexpression of any of the seven *E(spl)* genes causes reduction or complete elimination of bristles (Giebel & Campos-Ortega, 1997; Hinz, Giebel, & Campos-Ortega, 1994; Ligoxygakis, Bray, Apidianakis, & Delidakis, 1999; Tata & Hartley, 1995). These defects have been traced back to SOP formation, namely, supernumerary bristles arise from supernumerary SOPs, while loss of bristles is due to failure to generate SOPs. AS-C proteins act to promote SOP fate by turning on a number of genes, some of which are expressed throughout the proneural cluster, whereas others are expressed solely in SOPs (Reeves & Posakony, 2005). The simplest model, therefore, is that AS-C and E(spl) proteins target the same cohort of genes; AS-C-mediated activation of these target genes leads to the SOP fate, whereas repression of the same genes by E(spl) precludes the SOP fate.

Important insight into the mechanism of Ac/Sc-E(spl) antagonism was contributed by biochemical studies of the properties of these two groups of bHLH proteins. AS-C proteins act as transcriptional activators that bind a class A E-box, which contains a core of CAGSTG for Ac/Sc. They are poor binders by themselves and need to heterodimerize with the ubiquitous protein Da in order to bind their targets. Da is the only "E-protein" in *Drosophila*, homologous to the E2A, E2-2, and HEB proteins in mammals, and has a broad repertoire of tissue-specific bHLH interaction partners, each with its own preferred E_A-box-binding site. Unlike the proneural proteins, E(spl) proteins are repressors and prefer to bind E_B/C/N-boxes (see p. 219) as homodimers or heterodimers with other E(spl) proteins. Both E_A-and E_B/C/N-boxes are frequently encountered in enhancers that are activated by proneural genes (Culí & Modolell, 1998; Singson et al., 1994). This hints to the fact that genes activated by Da/AS-C heterodimers could be repressed by E(spl) dimers. However, this model could not be easily reconciled with an early observation that overexpression of basic domain mutants (incapable of DNA binding) of E(spl) proteins was still capable of repressing target genes and eliminating bristles (Giebel & Campos-Ortega, 1997; Nakao & Campos-Ortega, 1996). This was resolved by the demonstration that E(spl) proteins can interact with both members of proneural heterodimers, Da and Sc/Ac/Ato (Alifragis, Poortinga, Parkhurst, & Delidakis, 1997). Whereas the naturally basic-less HLH protein Extramacrochetae (Emc, homolog of vertebrate Id) interacts with Da and proneural proteins using the HLH helices and in doing so averts their DNA binding (Van Doren, Ellis, & Posakony, 1991), E(spl) proteins neither use their HLH helices to interact with proneurals nor do they abolish proneural DNA binding (Giagtzoglou et al., 2003, 2005; Zarifi et al., 2012). Instead, E(spl)–proneural interaction helps recruit E(spl) proteins to target enhancers. This was most convincingly shown by exchanging the WRPW motif of E(spl)m7 or mγ with the strong transactivation domain VP16. E(spl)-VP16 chimeric proteins were strong activators of target genes, suggesting that they act on, rather than off, DNA. In fact, E(spl)-VP16 fusions could activate transcription even if their basic domain had been disrupted. Interactions between proneural and E(spl) proteins have been mapped to the transactivation domains of the former and the N-terminal and Orange domains of the latter (Giagtzoglou et al., 2005; Zarifi et al., 2012). The ability of E(spl) proteins to be recruited on target enhancers both via DNA binding and via protein–protein interactions is summarized in Fig. 6.1.

Is the dispensability of the basic domain simply a consequence of E(spl) overexpression? Perhaps, this is so, since E(spl)-binding sites are widespread among true proneural target enhancers. Also minimal enhancers driven by tandem multimers of AS-C/Da sites (E_A-boxes) are not repressed by E(spl)

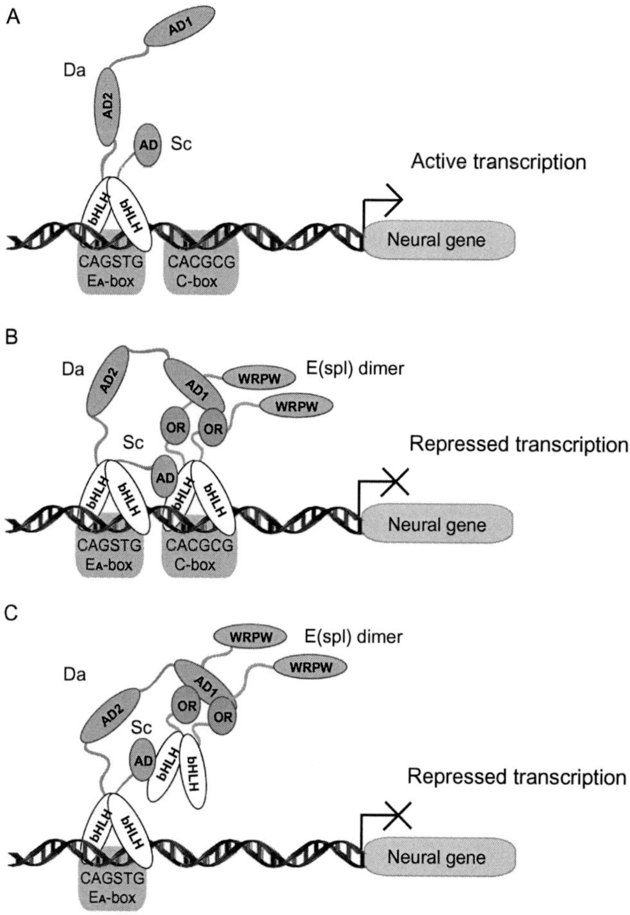

Figure 6.1 Graphic representation of the antagonism between proneural and E(spl) proteins. Proneural activators, such as the Da/Sc heterodimer depicted here, bind to E_A-boxes, whereas E(spl) repressors bind to E_B-, C-, and N-boxes. (A) In the absence of repressors, proneural proteins activate their target genes. Activation is mediated via two transactivation domains on Da, AD1 and AD2, as well as one transactivation domain on Sc. (B) In cells where E(spl) proteins are expressed, they are strongly bound by target enhancers thanks to DNA binding at the C-box as well as protein–protein interactions with two transactivation domains of the proneural heterodimer, the Da AD1 and the Sc AD. (C) If a cognate DNA target site is not present, E(spl) proteins can still be recruited onto proneural target enhancers via protein–protein interactions with Da and Sc. In this case, we hypothesize that higher levels of E(spl) will be needed to ensure recruitment and target gene repression, than in the case depicted in (B). *Adapted from a figure originally published in Zarifi et al. (2012), Copyright © American Society for Microbiology.* (See the color plate.)

proteins, unless the latter are overexpressed. However, artificial enhancers have been constructed that contain no E_B-, C-, or N-boxes yet seem to be efficiently repressed by E(spl) proteins in a wt background [no over-expression of E(spl)] (Culí & Modolell, 1998). This suggests that E(spl) proteins can be recruited to proneural proteins even when expressed at low endogenous levels, perhaps with the aid of additional tethering factors (still uncharacterized) that stabilize the proneural–E(spl) interactions even in the absence of a nearby DNA-E(spl)-binding site. Further characterization of the proneural and E(spl) interactomes as well as their enhancer targets will no doubt shed more light in the mechanism of their antagonistic interplay.

4.3. Lateral inhibition of neuronal precursors in the retina—Autoinhibition of E(spl)

In another neuroepithelium, the developing retina, E(spl) proteins play a similar role in inhibiting excessive neural differentiation. Unlike other sensory organs and CNS neurons, photoreceptor neurons arise directly from the neuroepithelium, without passing through an intermediate stage of a mitotically active NB or SOP, for a review of retinal morphogenesis, see Frankfort and Mardon (2002). The compound eye consists of identical ommatidia, each of which contains eight photoreceptor neurons (R1–8) plus accessory cells. In the undifferentiated neuroepithelium of the eye imaginal disk, the first cell to differentiate is R8, a process which takes place in the morphogenetic furrow (MF), an indentation that sweeps the retinal field from posterior to anterior, leaving differentiated cells behind it. Subsequently (behind the MF), R8 progressively induces surrounding undifferentiated cells to take on the R1–7 fates via epidermal growth factor receptor (Egfr) signaling. R8 cells arise from a large proneural field ahead of the MF, that expresses the proneural bHLH gene *atonal* (*ato*), another major difference from the CNS and bristles, which rely on the *AS-C*. Other photoreceptor cells, however, do not seem to directly require *ato* (or *AS-C*) activity, although *da* is autonomously needed in some (R2, 3, 4, and 5), but not others (R1, 6, and 7) (Brown et al., 1996). Lateral inhibition at the MF is mediated by Dl-N signaling and ensures the emergence of a well-spaced array of individual R8 photoreceptors. This is mediated via expression of *E(spl)* genes *m8, m7, mβ, mγ*, and *mδ* at the MF (de Celis et al., 1996); loss of *E(spl)* results in clusters of supernumerary R8 cells (Ligoxygakis, Yu, Delidakis, & Baker, 1998). As in SOP-specific enhancers, it is likely that R8-specific enhancers respond to activation by Ato/Da and repression by E(spl) in a mode similar to what we have depicted in Fig. 6.1. E(spl) recruitment probably occurs by synergistic DNA and protein-mediated binding via E(spl)-Da and E(spl)-Ato contacts.

The role of E(spl) proteins in R8 lateral inhibition is the basis of the original phenotype, which gave the name to these genes. *split* is a recessive allele of Notch (N^{spl}) that carries a point mutation that mildly increases Notch activity specifically in R8 cells (Li et al., 2003) and, as a result of aberrant R8 cell determination, produces small disorganized eyes. The first mutation isolated in the *E(spl)* locus, $E(spl)^D$, is a truncation of the 3′ part of the *E(spl) m8* gene, which removes its 3′-UTR and part of its C-terminal protein-coding sequence, leaving the bHLH-O domains intact. $E(spl)^D$ dominantly enhances N^{spl}: $N^{spl}/+$ (normally wildtype) now displays a split phenotype (becomes dominant), and N^{spl}/Y shows a more severe eye disorganization, almost complete elimination of the eye. As the *m8* 3′ UTR is targeted by microRNAs, its removal in the $E(spl)^D$ allele vastly increases m8 steady-state mRNA levels (Lai, Burks, & Posakony, 1998; Tietze et al., 1992), although the truncation of the protein also plays a role (Nagel, Yu, & Preiss, 1999). Interestingly, the overactive $E(spl)^D$ allele has no phenotype on its own, although it strongly suppresses R8 cell formation in the background of N^{spl}, perhaps pointing to the fact that R8 precursors are spared from Notch signaling and do not express *E(spl)* genes in the normal N^+ background. Indeed, forced overexpression of some *E(spl)* genes throughout the MF via the Gal4-UAS system can decrease the number of R8 cells and produce a smaller eye, although, strangely enough, *E(spl)mδ* but not *m8* is most active in this assay (Kahali et al., 2009; Ligoxygakis et al., 1998).

So, to a first approximation, $E(spl)^D$ represents a dominant gof (hypermorphic) allele of *E(spl)*. Does this fully account for the mutant phenotype, or has the protein gained some novel (neomorphic) features? There are actually two rather puzzling changes that have occured in the m8* protein (this is the notation that has been used to refer to the protein product of the $E(spl)^D$ allele). First of all, it has lost the C-terminal WRPW tetrapeptide and is incapable of interacting with Groucho (Nagel et al., 1999). This is the only instance where an E(spl) protein has been shown to act independently of Gro. Second, it has gained the ability to interact with Ato, Ac, and Sc, whereas wt m8 can only interact with Da but not with its proneural partners (Alifragis et al., 1997; Nagel & Preiss, 1999). This "unmasking" of a proneural interaction domain was eventually traced down to the presence of an autoinhibitory domain in the C-terminal third of E(spl)m8, which apparently interacts in cis with the N-terminal part (which includes the bHLH and Orange domains) to mask its ability to contact proneural proteins. This negative regulatory domain contains a casein kinase II (CKII) phosphorylation site and is inactivated upon phosphorylation (Kahali, Kim, Karandikar, Bishop, & Bidwai, 2010; Karandikar, Trott, Yin, Bishop, & Bidwai, 2004; Trott, Kalive, Paroush, & Bidwai, 2001). In this way, CKII activity enhances the repressive activity of E(spl)m8; this is also probably true for m5, m7, and mγ, which also contain similar CKII sites in their C-terminal PEST regions. Indeed a S>D phosphomimetic mutation of E(spl)m8 is more active than wt m8 in some overexpression assays, and RNAi for CKII can phenocopy weak E(spl) lof (Bose et al., 2006). However the increased affinity for Ato cannot in itself justify the increased activity of the m8* protein, as both E(spl)m8SD and E(spl)m8SDΔWRPW, m8 variants with a deactivated autoinhibitory domain (the latter also lacks the ability to interact with Gro, like m8*) have gained the ability to interact with Ato, but do not enhance the N^{spl} phenotype (Kahali et al., 2009). Sequencing of three revertants of the $E(spl)^D$ allele gave some insight to this puzzle. One of them has a small deletion overlapping the original deletion breakpoint, which causes a frameshift and only changes the nine aberrant aminoacids appended to the bHLH-O segment in m8* (into 14 new ones), suggesting that the novel nine amino acid tails significantly contributes to m8*'s aberrant activity (Nagel et al., 1999; Shepard, Broverman, & Muskavitch, 1989). In summary, analysis of the $E(spl)^D$ allele has revealed a complex "pathology," caused by both mRNA stabilization and changes in the protein structure of E(spl)m8 and helped reveal that E(spl)m8 and three more E(spl) proteins contain an autoinhibitory domain, which is conserved across Drosophilids.

The gof behavior of $E(spl)^D$ reflects the participation of *E(spl)* genes in a lateral inhibition process that takes place at the MF of the eye disk to generate regularly spaced R8 cells. Just before R8 singularization, Notch signaling is used to promote *ato* expression in the MF. This is achieved via repression of *hairy* and transcriptional activation of *da* and *ato* via a yet uncharacterized mechanism, which, apparently does not require *E(spl)* (Baonza & Freeman, 2001; Bhattacharya & Baker, 2009, 2012; Fu & Baker, 2003; Li & Baker, 2001; Ligoxygakis et al., 1998). Notch signaling is also reiteratively used in subsequent steps of retina differentiation, namely, the temporary inhibition of undifferentiated cells behind the MF to respond to Egfr differentiative signals (Overstreet, Fitch, & Fischer, 2004), the control of cell cycle behind the MF (Firth & Baker, 2005), the promotion of the R4 fate between the duet of simultaneously differentiating photoreceptors R3-R4 (Cooper & Bray, 1999; Fanto & Mlodzik, 1999), the promotion of R7 fate (Cooper & Bray, 2000), and the promotion of cone cell fate (Flores et al., 2000). In all these cases, E(spl) genes are variously expressed, but they seem not to play any role in the developmental outcome. It is possible, however, that their role has so far escaped classical genetic analysis because of functional overlap with other Hes, non-*E(spl)*, genes (see Section 4.7).

4.4. E(spl) in other instances of Notch-mediated lateral inhibition

Besides neural precursor specification in neuroepithelia, there are numerous other instances of lateral inhibition described during *Drosophila* development, and although Notch signaling has been implicated in all of these events, the role of E(spl) has been directly addressed only in some of these cell fate choices (Hartenstein, Rugendorff, Tepass, & Hartenstein, 1992). It should be noted that some cell types that display hyperplasia in Notch pathway mutant backgrounds may do so as a secondary effect, if they are induced by paracrine signals from another nearby cell type. One such example are the lipid-metabolizing cells of the larva called oenocytes. Although they become supernumerary in *Notch* and *E(spl)* lof mutations, this does not seem to result from a canonical lateral inhibition sorting out of oenocytes from equipotent progenitors. Instead, oenocytes are induced by receptor tyrosine kinase (RTK) signaling emanating from a SOP of the embryonic PNS, the dorsal-most SOP of the lateral chordotonal organ, also called SOP-1. SOP-1 is selected from a proneural cluster in a typical lateral inhibition mode and subsequently sends a signal mediated by Egfr RTK to neighboring cells to instruct them to adopt the oenocyte fate (Elstob, Brodu, & Gould, 2001; Rusten et al., 2001). Since failure of lateral inhibition in *E(spl)* mutations causes additional SOP-1s to form in the lateral chordotonal organ region, supernumerary surrounding cells are induced to the oenocyte fate.

One of the best characterized nonneural cases of lateral inhibition is the generation of muscle precursors in the embryo. *E(spl)* expression, in response to Notch signaling, is needed to restrict the number of Nautilus-positive muscle founder precursors (Corbin et al., 1991; Rusconi & Corbin, 1998). Nau is the only fly homologue of the vertebrate myogenic bHLH proteins and is expressed transiently in muscle founder precursor cells (Wei, Rong, & Paterson, 2007). These are selected from clusters of mesodermal cells that express the proneural gene L(1)sc via Notch-mediated lateral inhibition, a situation reminiscent of embryonic NB specification (Carmena, Bate, & Jiménez, 1995). Founder cell precursors then divide once to form the muscle founder cells, each of which is responsible for the generation of one syncytial embryonic muscle fiber, by initiating fusion with nearby mesodermal cells (reviewed in de Joussineau, Bataille, Jagla, & Jagla, 2012). In some of the L(1)sc-positive mesodermal clusters, Notch/E(spl)-mediated lateral inhibition is used to restrict the number not of somatic muscle precursor cells but of heart precursors, or cardioblasts (Grigorian, Mandal, Hakimi, Ortiz, & Hartenstein, 2011; Hartenstein et al., 1992).

Another nonneural case of E(spl)-mediated lateral inhibition takes place in the embryonic endoderm. At embryonic stages 9–11, the cells of the anterior and the posterior midgut primordia undergo a mesenchymal-to-epithelial transition and develop into either a principal cell type (PMECs, principal midgut epithelial cells), which give rise the future enterocytes (ECs), that form the absorptive epithelium of the midgut, or two other distinct cell types, the large basophilic interstitial cell precursors (ICPs) and the small adult midgut precursors (AMPs) (Tepass & Hartenstein, 1994). *E(spl)-C* genes are transiently expressed in the midgut primordia during these stages in response to Notch signaling, in order to suppress proneural genes that promote AMP/ICP fate thus allowing the development of the PMECs. *E(spl)-C* deletion mutants, similarly to mutations in other components of Notch pathway, result in increased AMP and ICP cell populations at the expense of the PMECs, thus a proper gut epithelium is not formed and midgut development is severely disturbed (Tepass & Hartenstein, 1995). As in other instances of lateral inhibition (within the neuroectodermal and mesodermal equivalence groups), E(spl) antagonizes the activity of proneural genes, which promote one fate (here AMP and ICP), and in doing so ensures the adoption of an alternative fate (here PMEC).

4.5. Involvement of E(spl) in other developmental processes—Interplay with the RTK/MAPK pathway

All processes we have described so far are examples where Notch signaling generates cell fate dichotomies in groups of equipotent cells; initial equipotence is the hallmark of lateral inhibition. However, Notch acts in numerous other instances where signal sending and signal receiving cells are already different from each other. In the majority of these instances, E(spl) proteins seem to not play a role, inasmuch as null *E(spl)* alleles do not perturb the signaling outcome. This is true even though at least one of the *E(spl)* genes has been reported to be expressed in most of these cases. Typical examples are mesectoderm specification (signaling between mesoderm and ectoderm) (Martin-Bermudo et al., 1995; Morel et al., 2001; Nagel, Wech, Schwinkendorf, & Preiss, 2007), wing dorsoventral (DV) boundary establishment (signaling between wing dorsal and ventral compartments) (de Celis et al., 1996), leg segment boundary establishment (signaling between consecutive leg segments) (de Celis et al., 1998), and ovarian follicle cell subtype specification (signaling between adjacent ovarian follicles) (Lopez-Schier & St Johnston, 2001). Cell fate adoption by progeny cells of SOP and NB lineages, which generally unfold via a series of asymmetric cell divisions causing the two cells to be different already at birth (Lai & Orgogozo, 2004; Skeath & Doe, 1998) are also instances where Notch signaling plays important roles without involving E(spl). There are, however, some notable exceptions, where E(spl) proteins mediate non lateral inhibition processes.

At the outer proliferation center (OPC) of the larval optic lobe, a spatiotemporally regulated epithelial-to-NB transition occurs gradually during larval life. The OPC is an epithelial structure on the larval brain that gives rise to NBs of the outer medulla, one of the visual processing centers in the adult brain. Similar to other neuroepithelia, certain OPC cells express the bHLH proneural gene *l(1)sc* before losing their epithelial characteristics and converting to asymmetrically dividing NBs. Different from other neuroepithelia, NB specification does not occur in a spaced pattern, instead *l(1)sc* expression initiates in a stripe spanning the medial edge of the OPC, where all cells switch to the NB fate en masse. E(spl) proteins (m8, m7, m5, and mγ) expressed in this medial OPC edge (also known as the transition zone) prevent premature NB specification (Egger et al., 2010; Perez-Gomez et al., 2013; Wang et al., 2011; Weng, Haenfler, & Lee, 2011), probably by antagonizing L(1)sc, as they do in the embryonic ventral neuroectoderm. E(spl) expression at the transition zone is mediated via a local peak in Notch signaling, which is downregulated after a while, concomitant with the massive switch to NB fate (Ngo et al., 2010; Orihara-Ono, Toriya, Nakao, & Okano, 2011). As in other cases, the mechanism whereby cells selectively shut down Notch signaling has not been adequately described, although in the OPC there is evidence for interplay with the Jak/Stat and Egfr pathways (Yasugi et al., 2010; Yasugi, Umetsu, Murakami, Sato, & Tabata, 2008). Egfr signaling promotes NB conversion by inducing the expression of *l(1)sc*. Egfr simultaneously induces Dl, which increases Notch signaling and E(spl) expression at the transition zone. Dl may also be involved in turning off E(spl) expression when it accumulates above a certain threshold, via a well-described *cis*-negative effect toward the Notch receptor (Jacobsen, Brennan, Arias, & Muskavitch, 1998; Weng et al., 2011). On the other hand, Jak/Stat signaling, which emanates from the opposite (lateral-most) edge of the OPC, inhibits NB conversion and maintains the OPC in the epithelial precursor state, probably by inhibiting Egfr signaling by a yet uncharacterized mechanism (Ngo et al., 2010; Yasugi et al., 2008). As a result, an asymmetric transition zone is established with L(1)sc-positive/E (spl)mγ-negative cells at its medial border (at the very edge of the OPC epithelium, where cells massively delaminate as NBs), doubly positive cells next to them in the lateral direction and L(1)sc-negative/E(spl)mγ-positive cells at the lateral border, next to the bulk of the OPC epithelium, which expresses lower levels of E(spl)mγ and m7. In summary, at the OPC transition zone, E(spl) regulate the timing of neuroepithelial cell differentiation in a field that has already been made asymmetric (consists of nonequivalent cells) through the action of Jak/Stat and Egfr signals. Besides the epithelial-to-NB transition, Notch signaling is also implicated in at least one other process in the larval optic lobe: it is required for OPC epithelial maintenance in cells away from the transition zone, i.e., before *l(1)sc* expression. Unlike other Notch pathway mutations, loss of *E(spl)* function does not result in loss of epithelial character at the more lateral OPC, suggesting that this activity of Notch does not require E(spl) proteins (Wang et al., 2011).

Another instance of interplay between Notch and Egfr signaling is that of wing veins. During late larval and early pupal wing development, Notch inhibits the outer cells of a wide provein primordium to adopt the vein fate and as a result ensures proper vein thickness (Biehs, Sturtevant, & Bier, 1998; de Celis, 2003; de Celis, Bray, & Garcia-Bellido, 1997; de Celis et al., 1996; Huppert, Jacobsen, & Muskavitch, 1997). In this context, *E(spl)mβ* is expressed in cells at the edges of the provein domains, whereas Egfr signaling is the highest in the middle vein cells, where *rhomboid*, an activator of Egfr ligands, is expressed. In *E(spl)* mutant clones, vein thickness is greatly increased. *E(spl)mβ* is not solely responsible for this process; additional *E(spl)* genes are also involved, illustrated by the fact that a small deletion removing *E(spl)mβ* and *mγ* does not have as severe a vein defect as that of a complete *E(spl)-C* deficiency (Babaoglan et al., 2013). In this context, it appears that E(spl) proteins repress expression of *rho* and in this way inhibit Egfr signaling, which promotes vein fate. It is not known whether E(spl) proteins interact with specific transcriptional activators of vein fate, similarly to their inhibition of proneural proteins during neural precursor formation.

Using this vein suppression activity of E(spl)mβ as an experimental paradigm, an interesting mechanism of E(spl) regulation by Egfr, and other RTKs, was revealed. It is known that E(spl) proteins require the Gro corepressor to exert their effects. Gro was found to be phosphorylated in two residues by MAP kinase, a key downstream kinase activated by the RTK/Ras pathway. A phosphomimetic form of Gro was found to relieve the vein loss phenotype caused by E(spl)mβ overexpression, whereas an unphosphorylatable form could by itself cause strong vein loss (Hasson et al., 2005). Thus, it is possible that one of the ways RTK signaling moderates the Notch response is by damping the activity of E(spl) effectors via phosphorylation of Gro, a mechanism which seems to be widespread (Helman et al., 2011) and could, for example, operate in the optic lobe OPC transition zone, in a manner analogous to the veins. Indeed, there are several instances where Notch and RTK signaling act antagonistically to each other, with Notch inhibiting progression to the next differentiation step and RTK promoting it. Besides wing veins and brain OPC, RTK involvement has been documented in the retina (recruitment of photoreceptors subsequent to R8), chordotonal organs of the PNS (recruitment of secondary SOPs by primary SOPs), stomatogastric nervous system (induction of secondary proneural epithelial folds by early delaminated neural precursors), and even in mechanosensory bristles (Culí, Martin-Blanco, & Modolell, 2001; Doroquez & Rebay, 2006; Gonzalez-Gaitan & Jackle, 2000; Hartenstein et al., 1996; zur Lage & Jarman, 1999; zur Lage, Powell, Prentice, McLaughlin, & Jarman, 2004). A common theme in all of these processes is that E(spl) proteins are expressed simultaneously with differentiation-promoting factors (e.g., L(1)sc in the OPC and Atonal in the chordotonal organs) without inhibiting the latter from exerting their prodifferentiation effects. Gro phosphorylation could mediate this reduction in E(spl) potency.

4.6. E(spl) in tissue maintenance and regeneration

Although Notch and E(spl) activity has been studied in many developmental contexts, their role in adult tissue homeostasis is just starting to be addressed. During adult life, the midgut epithelium, which consists of large polyploid absorptive EC and smaller enteroendocrine (ee) cells, suffers cell loss and is replenished by a small number of intestinal stem cells (ISCs), descendants of the embryonic AMPs (Mathur, Bost, Driver, & Ohlstein, 2010; Micchelli, Sudmeier, Perrimon, Tang, & Beehler-Evans, 2011). The ISCs divide asymmetrically to self-renew and to produce a postmitotic daughter cell, the enteroblast (EB), which subsequently differentiates either as an ee cell or as an EC. Both determination of the EB postmitotic state and its differentiation into an EC require Notch activity in apparently two consecutive events (Micchelli & Perrimon, 2006; Ohlstein & Spradling, 2006, 2007). Loss of *Notch* gives rise to stem cell "tumors," where all cells adopt the ISC fate and overproliferate. Loss of *E(spl)* has a somewhat different phenotype (Bardin et al., 2010): mutant clones contain an increased number of ISC-like cells (similarly to *Notch*), but they also contain a normal number of ECs and do not overproliferate. In the wildtype, expression of an *E(spl)mβ-lacZ* reporter is highly upregulated in EBs and is very low in ISCs or ECs; Repression of *E(spl)mβ* in the ISC is achieved via the Su(H)/H complex, as in other tissues. Bardin et al. (2010) concluded that E(spl) proteins mediate Notch signaling to promote EB fate and/or block ISC-like fate, thus ensuring the asymmetric outcome of the ISC mitosis. Interestingly, $da^{-/-}$ clones have the opposite phenotype of $E(spl)^{-/-}$, namely, loss of ISC by premature differentiation; clone consists of one or two ECs. Therefore, the asymmetric outcome of the ISC division seems to depend on the previously documented mutual antagonism between Da and E(spl) proteins. Since ECs persist in $E(spl)^{-/-}$ clones, the second Notch-mediated event that instructs the EB to differentiate to an EC is independent of E(spl) activity. So, in *E(spl)* lof clones, although the ISC division is defective (symmetric), the supernumerary ISCs continue to exchange Notch signals, and as a result, some of them differentiate into ECs; the Notch targets that mediate this latter event have not been characterized yet.

Midgut maintenance is a homeostatic mechanism which occurs even in the absence of pathological insults. Could E(spl) proteins also play a role in stress or pathological conditions? A recent report (Blanco et al., 2010) showed that many *E(spl)* genes (*m8, m7, m3, mβ, mδ*) are upregulated during regeneration of injured wing disks, which can happen after transplantation of the injured disk into adult hosts. More significantly, regenerating disks show reduced proliferation if the genetic dose of *Notch* or the *E(spl)-C* is halved, suggesting some involvement of these proteins in the response to injury. Another recent report showed that *E(spl)* expression is induced by exposure to toxic MeHg compounds, although whether *E(spl)* expression may have any role in resistance to such toxins is still not known (Engel et al., 2012; Rand et al., 2008). It would be interesting to determine whether *E(spl)* genes are expressed in other instances of response to injury, stress, or pathogens.

4.7. E(spl) can synergize with other bHLH-O proteins

Despite the ubiquitous response of *E(spl)-C* genes to Notch signaling, there is a large number of instances where the effect of Notch does not seem to be mediated exclusively, or at all, via E(spl), as the lof phenotype of *Notch* in these cases is much more severe than the lof phenotype of *E(spl)*. As mentioned before, this is more common in cases where Notch signaling takes place between nonequivalent cells. For example, in the signaling between dorsal and ventral populations of wing disk cells that establishes the DV wing boundary, four *E(spl)* genes are expressed, *m8, m7, mγ*, and *mβ*, as well as other targets of Notch, most importantly *wg* and *ct*. *Notch* null clones straddling the DV boundary cause deep wing scalloping, whereas *E(spl)* null clones in the same region do not affect wing integrity (de Celis et al., 1996). *Notch* clones also eliminate *wg* and *ct* from the DV boundary, whereas *E(spl)* clones only reduce *ct* expression (Diaz-Benjumea & Cohen, 1995; Ligoxygakis et al., 1999). This example illustrates the fact that the Notch pathway is not linear after Su(H) activation; rather a considerable number of transcriptional targets may synergize to effect the signaling outcome. The wing DV boundary analysis also raises a puzzling question of why *E(spl)* are expressed at all, since they appear to be dispensable. Could E(spl) proteins play a subtle role in this tissue, as suggested by the partial effect on *ct*? Alternatively, might *E(spl)* genes be rendered dispensable simply because another E(spl)-like molecule is expressed in response to Notch and acts redundantly to the four E(spl)?

Recent results (Babaoglan et al., 2013; Djiane et al., 2013; San Juan, Andrade-Zapata, & Baonza, 2012) showed that both of these hypotheses are true. These groups found that Dpn, another member of the Hes family of bHLH-O repressors is also regulated by Notch activity in the wing, where it is expressed at the DV boundary and in intervein territories, displaying a pattern similar to E(spl)mβ. Two separate Notch-responsive enhancers in the *dpn* gene were identified by ChIP-chip analysis for Su(H) binding. *dpn* mutant clones, much like *E(spl)* clones, did not reproduce Notch mutant phenotypes, such as elimination of *ct* and *wg* expression. However, *dpn E(spl)* double-mutant clones displayed significant (but still not complete) loss of *wg* and *ct*. It therefore seems that Dpn collaborates with E(spl) proteins to establish *ct* expression at the wing DV boundary, probably by repressing a repressor of *ct*. The realization that bHLH-O proteins do contribute to DV boundary establishment explained an older puzzling result (Ligoxygakis et al., 1999) where it had been shown that ectopic expression of *E(spl)mδ* or *mγ*, which were not supposed to mediate wing blade integrity, could revert the loss of Ct expression and the wing notching phenotypes caused by a temperature-sensitive allele of *Notch*. Besides the novel role of bHLH-O proteins in the wing DV boundary, Babaoglan et al. (2013)

detected two other examples of Dpn-E(spl) redundancy in wing vein and retina patterning.

Dpn and E(spl) seem to collaborate during CNS development as well. After their specification, NBs embark on a series of asymmetric cell divisions to self-renew and produce a transit progenitor, the GMC (ganglion mother cell), that divides once to produce two neurons (or glia) (Homem & Knoblich, 2012). *dpn* is expressed in all NBs since their birth and stays on until their terminal division or apoptosis (Bier, Vässin, Younger-Shepherd, Jan, & Jan, 1992). *E(spl)* expression, on the other hand, spares the NB during its birth, since the NB is the cell that escapes Notch-mediated lateral inhibition. However, *E(spl)mγ*, *m8*, and *mβ* begin to be expressed within the NB as soon as it commences cell divisions, as a result of a new round of Notch signaling, this time from GMCs to NBs (Almeida & Bray, 2005; Zacharioudaki et al., 2012). lof mutations of *E(spl)* or *dpn* (in mosaic clones) in larval NBs, show no, or at best subtle, defects. However, a dramatic phenotype was observed when double-mutant clones were induced (Zacharioudaki et al., 2012). In the majority of clones, the NB was absent, and only a few progeny neurons were present indicating that the NB had stopped dividing prematurely. These results indicate that Dpn and E(spl) act redundantly to ensure long-term maintenance of NBs during the larval proliferative phase.

It has been long appreciated that the seven *E(spl)* genes display a significant level of redundancy among themselves while having achieved a certain degree of specialization, e.g., in their expression patterns or in their ability to interact with proneural proteins (Alifragis et al., 1997; de Celis et al., 1996; Delidakis et al., 1991). It now appears that functional redundancy extends beyond the seven Hes paralogs of the *E(spl)* locus to encompass *dpn*, a more distantly related Hes gene (Babaoglan et al., 2013; San Juan et al., 2012; San Juan & Baonza, 2011; Zacharioudaki et al., 2012). There is no evidence to date for any other *Drosophila* Hes genes acting synergistically with *E(spl)*. In fact, *hairy* and *E(spl)* can act antagonistically. In DmD8-cultured cells as well as in larval wing disks, *hairy* and *E(spl)m3* or *mβ* repress each other. This is important in the regulation of the timing of response to Notch, since Hairy represses some of the later-responding Notch targets and early *E(spl)* expression is needed to repress *hairy* to allow late Notch targets to be expressed (Housden et al., 2013). The remaining fly Hes genes, *Side* and *Her*, have not been studied in depth, so the formal possibility exists that they may interface with *E(spl)* activity in some way. The more distant Hey proteins have also been shown to interact with Hes proteins in mammals (Iso et al., 2001).

This has not yet been addressed in *Drosophila*, where *Hey* is found to be expressed in a rather limited number of cell types (Monastirioti et al., 2010). One instance where Dpn may act in concert with Hey is in the repression of the *Sxl* gene, which is central in sex determination (Lu et al., 2008). It is possible that more Hes and Hey synergies exist that are yet to be discovered.

5. E(SPL) FUNCTION IN OTHER SPECIES

Although developmental mechanisms are being dissected with exquisite detail in the model systems of *Drosophila*, *C. elegans*, zebrafish, or mouse, we still know relatively little about the function of orthologs of important developmental regulators in nonmodel systems. These gaps in our knowledge are now beginning to be filled, thanks to the development of targeted knockdown technologies which can be applied to nonmodel systems, where traditional genetic tools are lacking.

E(spl) orthologs exist only in insects and crustaceans (see Section 2). Embryonic expression patterns of *E(spl)* orthologs have been described in the honeybee *A. mellifera* (three-gene *E(spl)* cluster), the beetle *T. castaneum* (two-gene cluster), and the water flea *Daphnia magna* (a crustacean; no genome available, but two *E(spl)* ortholog cDNAs cloned); one of the three orthologs of the pea aphid *Acyrthosiphon pisum* has also been examined. All these genes show complex expression patterns dominated by neuroectodermal expression, similar to *Drosophila* (Duncan & Dearden, 2010; Kux et al., 2013; Ungerer, Eriksson, & Stollewerk, 2012), with an additional prominent stripe pattern similar to segmentation genes in *Daphnia* (Eriksson, Ungerer, & Stollewerk, 2013). The *Apis* genes represent all three variants of the presumed ancestral *E(spl)* cluster, whereas the *Tribolium* and available *Daphnia* genes belong to the E(spl)1 and E(spl)3 families. These four *Tribolium* and *Daphnia* genes were found to be strongly responsive to Notch, as RNAi for Notch or pharmacological inhibition of the Notch pathway (via the γ-secretase inhibitor DAPT) strongly reduced their expression (Eriksson et al., 2013; Kux et al., 2013; Ungerer et al., 2012). In agreement with this result, a number of Su(H) consensus sites were identified in the vicinity of both *Tribolium* E(spl) genes and also in the three-gene *E(spl)* cluster of a different *Daphnia* species (*D. pulex*), for which genomic sequence is available. Furthermore, the 3′-UTRs of these genes were found to contain miRNA target sites of the K, GY, and Brd-box types, suggesting a similar mode of miRNA regulation as in *Drosophila*. Similar flanking domain architectures were detected in the *E(spl)* clusters of *Acyrthosiphon*, *Apis*, and a number of Diptera (including two non-Drosophilids, *T. dalmanni* and *G. morsitans*) (Baker et al., 2011; Duncan & Dearden, 2010), although experimental demonstration of Notch responsiveness is lacking for these species. The two *Tribolium* genes were also found to respond to activation by Ash, the proneural Ac/Sc homolog in the beetle, similar to the dual dependence of fly *E(spl)* neuroectodermal expression to both Notch signaling and proneural activators.

As arthropod and vertebrate *Hes* genes diverged independently, did their response to Notch exist in the ancestral *Hes* gene or was it independently acquired for some of the members (e.g., arthropod *E(spl)*, vertebrate *Hes1*) and not by others (e.g., *Drosophila hairy*)? Studies of basal metazoa suggest a very old connection between *Hes* genes and Notch. In two species of cnidaria, a sister group of bilateria, a Notch–*Hes* connection was shown by DAPT treatment (Marlow, Roettinger, Boekhout, & Martindale, 2012; Munder et al., 2013). In *Hydra*, the single *Hes* ortholog was shown to respond to Notch signaling and to be involved in tentacle patterning and head regeneration. In *Nematostella*, one *Hey* gene and at least three *Hes* genes were also shown to be downstream of Notch and to be involved in, among others, development of the simple neural cell network possessed by this animal. In fact, Notch inhibition resulted in increased neural cells and increased expression of *Nematostella* Ac/Sc orthologs, in addition to reducing *Hes* and *Hey* gene expression. In *C. elegans* (a nematode worm; belonging to the non-arthropod branch of the ecdysozoa), on the other hand, the sole *Hes* homolog, *lin-22*, does not seem to rely on Notch for its expression. Strangely enough, a family of six other pralogous bHLH proteins, the Ref-1 family, was shown to respond to Notch in early *C. elegans* embryogenesis (Neves & Priess, 2005). The Ref-1 proteins show a weak similarity to Hes proteins but contain no Orange domain and in its place have a second bHLH domain. As these genes are only found in nematodes, they have been termed "orphan" bHLH genes, but they may be highly divergent homologs of the *Hes* genes. The fact that *Drosophila dpn* was also recently implicated in several instances of Notch signaling, suggests that not only *E(spl)* genes, but other *Hes* genes can respond to Notch, and this was perhaps an ancestral connection that got lost in some cases, like fly *hairy*. Such an ancestral connection may have also existed for the *Hey* genes, since both *Drosophila* and mammalian *Hey* genes are targets of Notch signaling (Fischer & Gessler, 2007; Monastirioti et al., 2010); however, besides the recent *Nematostella* result, no studies exist for *Hey* orthologs in other metazoan branches.

Knockdown of *E(spl)* genes has only been reported in *Tribolium* (Bitra & Palli, 2010; Bitra, Tan, Dowling, & Palli, 2009; Kux et al., 2013) and led to a variety of defects. The most severe effect of pupal RNAi was female sterility or early arrest of embryonic development. Milder effects included a neurogenic phenotype of increased NBs in the progeny embryos, suggesting a conserved role in NB lateral inhibition. These defects were most robustly produced by dual *E(spl)1* + *E(spl)3* RNAi. Single *E(spl)1* knockdown gave no embryonic defects, whereas single *E(spl)3* RNAi showed milder sterility and neurogenic defects. *E(spl)1* RNAi at the larval stage, resulted in pupal lethality with wing and abdominal malformations (Bitra et al., 2009). *E(spl)3* RNAi at the same stage had no effect. The two *Tribolium E(spl)* orthologs were also tested by heterologous overexpression in transgenic flies. They were capable of eliminating bristles and wing veins, mimicking *Drosophila E(spl)* overexpression (Kux et al., 2013). Whereas both *Tribolium* E(spl) proteins could antagonize the bristle-promoting activity of overexpressed Sc, E(spl)1 was shown to interact with both Sc and Da but E(spl)3 could interact only with Sc. These functional studies showed that even though *Drosophila* contains no *E(spl)* genes of the *E(spl)3* class, a representative of the latter class behaves in a similar fashion to fly [*E(spl)1*-type] genes in many assays.

6. CLOSING REMARKS

In summary, E(spl) proteins act as chromatin-bound repressors in many cell types, where they regulate cell fate decisions in close connection with the Notch signaling pathway. Thanks to their duplication, *Drosophila* E(spl) can form many homo- and heterodimers with differential affinities for Da, Sc, Ato, and probably other proneural proteins; these preferences are partly mediated by their Orange domains. Whether subtle differences in affinities for DNA sites also exist among the various E(spl) dimers is an issue that has not yet been addressed. E(spl) proteins use their WRPW C-terminus to recruit the Gro corepressor in order to silence their target genes. Most certainly, the various protein–protein and protein–DNA interactions in which E(spl) proteins participate will be regulated by posttranslational modifications. CKII phosphorylation has been implicated in the relief of autoinhibition for a subset of the E(spl) proteins, whereas MAPK phosphorylation of their corepressor, Gro, can affect their transcriptional activity. Small ubiquitin-like modifier (SUMO) modifications could also affect E(spl) function. Although E(spl) proteins themselves are not known to be

SUMOylated, they interact with two SUMO-modified proteins, Gro and Sens. Most importantly, E(spl) proteins also interact with a StUbl (SUMO-targeted ubiquitin ligase), called Degringolade, and their ubiquitylation seems to decrease their affinity for Gro (Abed et al., 2011; Barry et al., 2011; Powell et al., 2012). Further modifications and protein interactors will most likely be discovered which may influence E(spl) activities in different tissue contexts. Since E(spl) seem to collaborate with at least one other Hes protein, Dpn, we can speculate that the molecular mechanisms that have been described for E(spl) are not a peculiarity of these highly duplicated and diversified fly proteins but may be of more general relevance. Already many of the mechanisms elucidated by *Drosophila* studies, such as Gro interactions, also hold true for vertebrate Hes proteins. It is possible that in the future, many more conserved features will be recognized within the family of Hes proteins, or even the broader bHLH-O family, which seem to have a prominent place in the evolution of multicellularity in the animal kingdom.

ACKNOWLEDGMENTS

The authors are grateful to Vasso Theodorou for drawing Fig. 6.1 and to Anette Preiss for comments on the chapter.

REFERENCES

Abed, M., Barry, K. C., Kenyagin, D., Koltun, B., Phippen, T. M., Delrow, J. J., et al. (2011). Degringolade, a SUMO-targeted ubiquitin ligase, inhibits Hairy/Groucho-mediated repression. *The EMBO Journal*, *30*, 1289–1301.

Alifragis, P., Poortinga, G., Parkhurst, S. M., & Delidakis, C. (1997). A network of interacting transcriptional regulators involved in *Drosophila* neural fate specification revealed by the yeast two-hybrid system. *Proceedings of the National Academy of Sciences of the United States of America*, *94*, 13099–13104.

Almeida, M. S., & Bray, S. J. (2005). Regulation of post-embryonic neuroblasts by Drosophila Grainyhead. *Mechanisms of Development*, *122*, 1282–1293.

Andersson, E. R., Sandberg, R., & Lendahl, U. (2011). Notch signaling: Simplicity in design, versatility in function. *Development*, *138*, 3593–3612.

Arnett, K. L., Hass, M., McArthur, D. G., Ilagan, M. X., Aster, J. C., Kopan, R., et al. (2010). Structural and mechanistic insights into cooperative assembly of dimeric Notch transcription complexes. *Nature Structural and Molecular Biology*, *17*, 1312–1317.

Assa-Kunik, E., Torres, I. L., Schejter, E. D., Johnston, D. S., & Shilo, B. Z. (2007). Drosophila follicle cells are patterned by multiple levels of Notch signaling and antagonism between the Notch and JAK/STAT pathways. *Development*, *134*, 1161–1169.

Babaoglan, A. B., Housden, B. E., Furriols, M., & Bray, S. J. (2013). Deadpan contributes to the robustness of the notch response. *PloS One*, *8*, e75632.

Bailey, A. M., & Posakony, J. W. (1995). Suppressor of Hairless directly activates transcription of *Enhancer of split* Complex genes in response to Notch receptor activity. *Genes and Development*, *9*, 2609–2622.

Baker, R. H., Kuehl, J. V., & Wilkinson, G. S. (2011). The Enhancer of split complex arose prior to the diversification of schizophoran flies and is strongly conserved between Drosophila and stalk-eyed flies (Diopsidae). *BMC Evolutionary Biology*, *11*, 354.

Baker, N. E., Yu, S., & Han, D. (1996). Evolution of proneural atonal expression during distinct regulatory phases in the developing Drosophila eye. *Current Biology*, *6*, 1290–1301.

Baonza, A., & Freeman, M. (2001). Notch signalling and the initiation of neural development in the Drosophila eye. *Development*, *128*, 3889–3898.

Bardin, A. J., Perdigoto, C. N., Southall, T. D., Brand, A. H., & Schweisguth, F. (2010). Transcriptional control of stem cell maintenance in the Drosophila intestine. *Development*, *137*, 705–714.

Bardin, A. J., & Schweisguth, F. (2006). Bearded family members inhibit Neuralized-mediated endocytosis and signaling activity of Delta in Drosophila. *Developmental Cell*, *10*, 245–255.

Barolo, S., Stone, T., Bang, A. G., & Posakony, J. W. (2002). Default repression and Notch signaling: Hairless acts as an adaptor to recruit the corepressors Groucho and dCtBP to Suppressor of Hairless. *Genes and Development*, *16*, 1964–1976.

Barry, K. C., Abed, M., Kenyagin, D., Werwie, T. R., Boico, O., Orian, A., et al. (2011). The Drosophila STUbL protein Degringolade limits HES functions during embryogenesis. *Development*, *138*, 1759–1769.

Bejarano, F., Bortolamiol-Becet, D., Dai, Q., Sun, K., Saj, A., Chou, Y. T., et al. (2012). A genome-wide transgenic resource for conditional expression of Drosophila microRNAs. *Development*, *139*, 2821–2831.

Belizario, J. E., Alves, J., Garay-Malpartida, M., & Occhiucci, J. M. (2008). Coupling caspase cleavage and proteasomal degradation of proteins carrying PEST motif. *Current Protein and Peptide Science*, *9*, 210–220.

Bernard, F., Krejci, A., Housden, B., Adryan, B., & Bray, S. J. (2010). Specificity of Notch pathway activation: Twist controls the transcriptional output in adult muscle progenitors. *Development*, *137*, 2633–2642.

Bhattacharya, A., & Baker, N. E. (2009). The HLH protein Extramacrochaetae is required for R7 cell and cone cell fates in the Drosophila eye. *Developmental Biology*, *327*, 288–300.

Bhattacharya, A., & Baker, N. E. (2012). The role of the bHLH protein hairy in morphogenetic furrow progression in the developing Drosophila eye. *PLoS One*, 7, e47503.

Bianchi-Frias, D., Orian, A., Delrow, J. J., Vazquez, J., Rosales-Nieves, A. E., & Parkhurst, S. M. (2004). Hairy transcriptional repression targets and cofactor recruitment in Drosophila. *PLoS Biology*, *2*, E178.

Biehs, B., Sturtevant, M. A., & Bier, E. (1998). Boundaries in the Drosophila wing imaginal disc organize vein-specific genetic programs. *Development*, *125*, 4245–4257.

Bier, E., Vässin, H., Younger-Shepherd, S., Jan, L. Y., & Jan, Y. N. (1992). *deadpan*, an essential pan-neural gene in *Drosophila*, encodes a helix-loop-helix protein similar to the *hairy* gene product. *Genes and Development*, *6*, 2137–2151.

Bitra, K., & Palli, S. R. (2010). The members of bHLH transcription factor superfamily are required for female reproduction in the red flour beetle, Tribolium castaneum. *Journal of Insect Physiology*, *56*, 1481–1489.

Bitra, K., Tan, A., Dowling, A., & Palli, S. R. (2009). Functional characterization of PAS and HES family bHLH transcription factors during the metamorphosis of the red flour beetle, Tribolium castaneum. *Gene*, *448*, 74–87.

Blanco, E., Ruiz-Romero, M., Beltran, S., Bosch, M., Punset, A., Serras, F., et al. (2010). Gene expression following induction of regeneration in Drosophila wing imaginal discs. Expression profile of regenerating wing discs. *BMC Developmental Biology*, *10*, 94.

Bland, C., & Rand, M. D. (2006). Methylmercury induces activation of Notch signaling. *Neurotoxicology*, *27*, 982–991.

Bose, A., Kahali, B., Zhang, S., Lin, J. M., Allada, R., Karandikar, U., et al. (2006). Drosophila CK2 regulates lateral-inhibition during eye and bristle development. *Mechanisms of Development*, *123*, 649–664.

Brown, N. L., Paddock, S. W., Sattler, C. A., Cronmiller, C., Thomas, B. J., & Carroll, S. B. (1996). daughterless is required for Drosophila photoreceptor cell determination, eye morphogenesis, and cell cycle progression. *Developmental Biology*, *179*, 65–78.

Campuzano, S., & Modolell, J. (1992). Patterning of the *Drosophila* nervous system: The *achaete-scute* gene complex. *Trends in Genetics*, *8*, 202–208.

Carmena, A., Bate, M., & Jiménez, F. (1995). Lethal of scute, a proneural gene, participates in the specification of muscle progenitors during Drosophila embryogenesis. *Genes and Development*, *9*, 2373–2383.

Castro, B., Barolo, S., Bailey, A. M., & Posakony, J. W. (2005). Lateral inhibition in proneural clusters: Cis-regulatory logic and default repression by Suppressor of Hairless. *Development*, *132*, 3333–3344.

Cave, J. W., Loh, F., Surpris, J. W., Xia, L., & Caudy, M. A. (2005). A DNA transcription code for cell-specific gene activation by notch signaling. *Current Biology*, *15*, 94–104.

Cave, J. W., Xia, L., & Caudy, M. A. (2009). The Daughterless N-terminus directly mediates synergistic interactions with Notch transcription complexes via the SPS+A DNA transcription code. *BMC Research Notes*, *2*, 65.

Chanet, S., & Schweisguth, F. (2012). Regulation of epithelial polarity by the E3 ubiquitin ligase Neuralized and the Bearded inhibitors in Drosophila. *Nature Cell Biology*, *14*, 467–476.

Chanet, S., Vodovar, N., Mayau, V., & Schweisguth, F. (2009). Genome engineering-based analysis of Bearded family genes reveals both functional redundancy and a nonessential function in lateral inhibition in Drosophila. *Genetics*, *182*, 1101–1108.

Cooper, M. T., & Bray, S. J. (1999). Frizzled regulation of Notch signalling polarizes cell fate in the Drosophila eye. *Nature*, *397*, 526–530.

Cooper, M. T., & Bray, S. J. (2000). R7 photoreceptor specification requires Notch activity. *Current Biology*, *10*, 1507–1510.

Cooper, M. T., Tyler, D. M., Furriols, M., Chalkiadaki, A., Delidakis, C., & Bray, S. (2000). Spatially restricted factors cooperate with Notch in the regulation of enhancer of split genes. *Developmental Biology*, *221*, 390–403.

Corbin, V., Michelson, A. M., Abmayr, S. M., Neel, V., Alcamo, E., Maniatis, T., et al. (1991). A role for the Drosophila neurogenic genes in mesoderm differentiation. *Cell*, *67*, 311–323.

Culí, J., Martin-Blanco, E., & Modolell, J. (2001). The EGF receptor and N signalling pathways act antagonistically in Drosophila mesothorax bristle patterning. *Development*, *128*, 299–308.

Culí, J., & Modolell, J. (1998). Proneural gene self-stimulation in neural precursors: An essential mechanism for sense organ development that is regulated by Notch signaling. *Genes and Development*, *12*, 2036–2047.

Dai, Q., Ren, A., Westholm, J. O., Serganov, A. A., Patel, D. J., & Lai, E. C. (2013). The BEN domain is a novel sequence-specific DNA-binding domain conserved in neural transcriptional repressors. *Genes and Development*, *27*, 602–614.

Dawson, S. R., Turner, D. L., Weintraub, H., & Parkhurst, S. M. (1995). Specificity for the Hairy/Enhancer of split basic helix-loop-helix (bHLH) proteins maps outside the bHLH domain and suggests two separable modes of transcriptional repression. *Molecular and Cellular Biology*, *15*, 6923–6931.

de Celis, J. F. (2003). Pattern formation in the Drosophila wing: The development of the veins. *Bioessays*, *25*, 443–451.

de Celis, J. F., Bray, S., & Garcia-Bellido, A. (1997). Notch signalling regulates veinlet expression and establishes boundaries between veins and interveins in the Drosophila wing. *Development, 124*, 1919–1928.

de Celis, J., de Celis, J., Ligoxygakis, P., Preiss, A., Delidakis, C., & Bray, S. (1996). Functional relationships between *Notch, Su(H)* and the bHLH genes of the *E(spl) complex*: The *E(spl)* genes mediate only a subset of *Notch* activity during imaginal development. *Development, 122*, 2719–2728.

de Celis, J. F., Tyler, D. M., de Celis, J., & Bray, S. J. (1998). Notch signalling mediates segmentation of the Drosophila leg. *Development, 125*, 4617–4626.

de Joussineau, C., Bataille, L., Jagla, T., & Jagla, K. (2012). Diversification of muscle types in Drosophila: Upstream and downstream of identity genes. *Current Topics in Developmental Biology, 98*, 277–301.

De Renzis, S., Yu, J., Zinzen, R., & Wieschaus, E. (2006). Dorsal-ventral pattern of Delta trafficking is established by a Snail-Tom-Neuralized pathway. *Developmental Cell, 10*, 257–264.

Delidakis, C., & Artavanis-Tsakonas, S. (1992). The Enhancer of split [E(spl)] locus of Drosophila encodes seven independent helix-loop-helix proteins. *Proceedings of the National Academy of Sciences of the United States of America, 89*, 8731–8735.

Delidakis, C., Preiss, A., Hartley, D. A., & Artavanis-Tsakonas, S. (1991). Two genetically and molecularly distinct functions involved in early neurogenesis reside within the Enhancer of split locus of Drosophila melanogaster. *Genetics, 129*, 803–823.

Diaz-Benjumea, F. J., & Cohen, S. M. (1995). Serrate signals through Notch to establish a Wingless-dependent organizer at the dorsal/ventral compartment boundary of the *Drosophila* wing. *Development, 121*, 4215–4225.

Djiane, A., Krejci, A., Bernard, F., Fexova, S., Millen, K., & Bray, S. J. (2013). Dissecting the mechanisms of Notch induced hyperplasia. *The EMBO Journal, 32*, 60–71.

Dominguez, M., & Campuzano, S. (1993). *asense*, a member of the *Drosophila achaete-scute* complex, is a proneural and neural differentiation gene. *The EMBO Journal, 12*, 2049–2060.

Doroquez, D. B., & Rebay, I. (2006). Signal integration during development: Mechanisms of EGFR and Notch pathway function and cross-talk. *Critical Reviews in Biochemistry and Molecular Biology, 41*, 339–385.

Duan, H., Dai, Q., Kavaler, J., Bejarano, F., Medranda, G., Negre, N., et al. (2011). Insensitive is a corepressor for Suppressor of Hairless and regulates Notch signalling during neural development. *The EMBO Journal, 30*, 3120–3133.

Duncan, E. J., & Dearden, P. K. (2010). Evolution of a genomic regulatory domain: The role of gene co-option and gene duplication in the Enhancer of split complex. *Genome Research, 20*, 917–928.

Eastwood, K., Yin, C., Bandyopadhyay, M., & Bidwai, A. (2011). New insights into the Orange domain of E(spl)-M8, and the roles of the C-terminal domain in autoinhibition and Groucho recruitment. *Molecular and Cellular Biochemistry, 356*, 217–225.

Egger, B., Gold, K. S., & Brand, A. H. (2010). Notch regulates the switch from symmetric to asymmetric neural stem cell division in the Drosophila optic lobe. *Development, 137*, 2981–2987.

Elstob, P. R., Brodu, V., & Gould, A. P. (2001). spalt-dependent switching between two cell fates that are induced by the Drosophila EGF receptor. *Development, 128*, 723–732.

Endo, K., Karim, M. R., Taniguchi, H., Krejci, A., Kinameri, E., Siebert, M., et al. (2012). Chromatin modification of Notch targets in olfactory receptor neuron diversification. *Nature Neuroscience, 15*, 224–233.

Engel, G. L., Delwig, A., & Rand, M. D. (2012). The effects of methylmercury on Notch signaling during embryonic neural development in Drosophila melanogaster. *Toxicology In Vitro, 26*, 485–492.

Eriksson, B. J., Ungerer, P., & Stollewerk, A. (2013). The function of Notch signalling in segment formation in the crustacean Daphnia magna (Branchiopoda). *Developmental Biology, 383*, 321–330.
Fanto, M., & Mlodzik, M. (1999). Asymmetric Notch activation specifies photoreceptors R3 and R4 and planar polarity in the Drosophila eye. *Nature, 397*, 523–526.
Firth, L. C., & Baker, N. E. (2005). Extracellular signals responsible for spatially regulated proliferation in the differentiating Drosophila eye. *Developmental Cell, 8*, 541–551.
Fischer, A., & Gessler, M. (2007). Delta-Notch—and then? Protein interactions and proposed modes of repression by Hes and Hey bHLH factors. *Nucleic Acids Research, 35*, 4583–4596.
Flores, G. V., Duan, H., Yan, H., Nagaraj, R., Fu, W., Zou, Y., et al. (2000). Combinatorial signaling in the specification of unique cell fates. *Cell, 103*, 75–85.
Fontana, J. R., & Posakony, J. W. (2009). Both inhibition and activation of Notch signaling rely on a conserved Neuralized-binding motif in Bearded proteins and the Notch ligand Delta. *Developmental Biology, 333*, 373–385.
Frankfort, B. J., & Mardon, G. (2002). R8 development in the Drosophila eye: A paradigm for neural selection and differentiation. *Development, 129*, 1295–1306.
Fu, W., & Baker, N. E. (2003). Deciphering synergistic and redundant roles of Hedgehog, Decapentaplegic and Delta that drive the wave of differentiation in Drosophila eye development. *Development, 130*, 5229–5239.
Furriols, M., & Bray, S. (2001). A model Notch response element detects Suppressor of Hairless-dependent molecular switch. *Current Biology, 11*, 60–64.
Garcia-Bellido, A., & de Celis, J. F. (2009). The complex tale of the achaete-scute complex: A paradigmatic case in the analysis of gene organization and function during development. *Genetics, 182*, 631–639.
Giagtzoglou, N., Alifragis, P., Koumbanakis, K. A., & Delidakis, C. (2003). Two modes of recruitment of E(spl) repressors onto target genes. *Development, 130*, 259–270.
Giagtzoglou, N., Koumbanakis, K. A., Fullard, J., Zarifi, I., & Delidakis, C. (2005). Role of the Sc C terminus in transcriptional activation and E(spl) repressor recruitment. *The Journal of Biological Chemistry, 280*, 1299–1305.
Giebel, B., & Campos-Ortega, J. A. (1997). Functional dissection of the *Drosophila* Enhancer of split protein, a suppressor of neurogenesis. *Proceedings of the National Academy of Sciences of the United States of America, 94*, 6250–6254.
Gonzalez-Gaitan, M., & Jackle, H. (2000). Tip cell-derived RTK signaling initiates cell movements in the Drosophila stomatogastric nervous system anlage. *EMBO Reports, 1*, 366–371.
Goodfellow, H., Krejci, A., Moshkin, Y., Verrijzer, C. P., Karch, F., & Bray, S. J. (2007). Gene-specific targeting of the histone chaperone asf1 to mediate silencing. *Developmental Cell, 13*, 593–600.
Goulding, S. E., zur Lage, P., & Jarman, A. P. (2000). amos, a proneural gene for Drosophila olfactory sense organs that is regulated by lozenge. *Neuron, 25*, 69–78.
Grammont, M., & Irvine, K. D. (2001). Fringe and Notch specify polar cell fate during Drosophila oogenesis. *Development, 128*, 2243–2253.
Grigorian, M., Mandal, L., Hakimi, M., Ortiz, I., & Hartenstein, V. (2011). The convergence of Notch and MAPK signaling specifies the blood progenitor fate in the Drosophila mesoderm. *Developmental Biology, 353*, 105–118.
Hartenstein, A. Y., Rugendorff, A., Tepass, U., & Hartenstein, V. (1992). The function of the neurogenic genes during epithelial development in the Drosophila embryo. *Development, 116*, 1203–1220.
Hartenstein, V., Tepass, U., & Gruszynski-deFeo, E. (1996). Proneural and neurogenic genes control specification and Morphogenesis of stomatogastric nerve cell precursors in Drosophila. *Developmental Biology, 173*, 213–227.

Hartley, D. A., Preiss, A., & Artavanis-Tsakonas, S. (1988). A deduced gene product from the Drosophila neurogenic locus, enhancer of split, shows homology to mammalian G-protein beta subunit. *Cell, 55*, 785–795.

Hasson, P., Egoz, N., Winkler, C., Volohonsky, G., Jia, S. T., Dinur, T., et al. (2005). EGFR signaling attenuates Groucho-dependent repression to antagonize Notch transcriptional output. *Nature Genetics, 37*, 101–105.

Heitzler, P., Bourouis, M., Ruel, L., Carteret, C., & Simpson, P. (1996). Genes of the *Enhancer of split* and *achaete-scute* complexes are required for a regulatory loop between *Notch* and *Delta* during lateral signalling in *Drosophila*. *Development, 122*, 161–171.

Helman, A., Cinnamon, E., Mezuman, S., Hayouka, Z., Von Ohlen, T., Orian, A., et al. (2011). Phosphorylation of Groucho mediates RTK feedback inhibition and prolonged pathway target gene expression. *Current Biology, 21*, 1102–1110.

Hinz, U., Giebel, B., & Campos-Ortega, J. A. (1994). The basic-helix-loop-helix domain of *Drosophila* Lethal of Scute protein is sufficient for proneural function and activates neurogenic genes. *Cell, 76*, 77–87.

Homem, C. C., & Knoblich, J. A. (2012). Drosophila neuroblasts: A model for stem cell biology. *Development, 139*, 4297–4310.

Hori, K., Sen, A., & Artavanis-Tsakonas, S. (2013). Notch signaling at a glance. *Journal of Cell Science, 126*, 2135–2140.

Housden, B. E., Fu, A. Q., Krejci, A., Bernard, F., Fischer, B., Tavare, S., et al. (2013). Transcriptional dynamics elicited by a short pulse of notch activation involves feed-forward regulation by E(spl)/Hes genes. *PLoS Genetics, 9*, e1003162.

Huppert, S. S., Jacobsen, T. L., & Muskavitch, M. A. (1997). Feedback regulation is central to Delta-Notch signalling required for *Drosophila* wing vein morphogenesis. *Development, 124*, 3283–3291.

Ip, Y. T., Levine, M., & Bier, E. (1994). Neurogenic expression of snail is controlled by separable CNS and PNS promoter elements. *Development, 120*, 199–207.

Iso, T., Sartorelli, V., Poizat, C., Iezzi, S., Wu, H. Y., Chung, G., et al. (2001). HERP, a novel heterodimer partner of HES/E(spl) in Notch signaling. *Molecular and Cellular Biology, 21*, 6080–6089.

Jacobsen, T. L., Brennan, K., Arias, A. M., & Muskavitch, M. A. (1998). Cis-interactions between Delta and Notch modulate neurogenic signalling in Drosophila. *Development, 125*, 4531–4540.

Jarman, A. P., Grau, Y., Jan, L. Y., & Jan, Y. N. (1993). *atonal* is a proneural gene that directs chordotonal organ formation in the Drosophila peripheral nervous system. *Cell, 73*, 1307–1321.

Jennings, B., de Celis, J., Delidakis, C., Preiss, A., & Bray, S. (1995). Role of *Notch* and *achaete-scute* complex in the expression of *Enhancer of split* bHLH proteins. *Development, 121*, 3745–3752.

Jennings, B. H., Pickles, L. M., Wainwright, S. M., Roe, S. M., Pearl, L. H., & Ish-Horowicz, D. (2006). Molecular recognition of transcriptional repressor motifs by the WD domain of the Groucho/TLE corepressor. *Molecular Cell, 22*, 645–655.

Jennings, B., Preiss, A., Delidakis, C., & Bray, S. (1994). The Notch signalling pathway is required for *Enhancer of split* bHLH protein expression during neurogenesis in the *Drosophila* embryo. *Development, 120*, 3537–3548.

Jennings, B. H., Tyler, D. M., & Bray, S. J. (1999). Target specificities of Drosophila enhancer of split basic helix-loop-helix proteins. *Molecular and Cellular Biology, 19*, 4600–4610.

Jiménez, F., & Campos-Ortega, J. A. (1990). Defective neuroblast commitment in mutants of the *achete*-scute complex and adjacent genes of *D. melanogaster*. *Neuron, 5*, 81–89.

Johnson, A. N., Mokalled, M. H., Haden, T. N., & Olson, E. N. (2011). JAK/Stat signaling regulates heart precursor diversification in Drosophila. *Development, 138*, 4627–4638.

Kageyama, R., Niwa, Y., Shimojo, H., Kobayashi, T., & Ohtsuka, T. (2010). Ultradian oscillations in Notch signaling regulate dynamic biological events. *Current Topics in Developmental Biology*, *92*, 311–331.

Kahali, B., Bose, A., Karandikar, U., Bishop, C. P., & Bidwai, A. P. (2009). On the mechanism underlying the divergent retinal and bristle defects of M8* (E(spl)D) in Drosophila. *Genesis*, *47*, 456–468.

Kahali, B., Kim, J. E., Karandikar, U., Bishop, C. P., & Bidwai, A. P. (2010). Evidence that the C-terminal domain (CtD) autoinhibits neural repression by Drosophila E(spl)M8. *Genesis*, *48*, 44–55.

Karandikar, U. C., Trott, R. L., Yin, J., Bishop, C. P., & Bidwai, A. P. (2004). Drosophila CK2 regulates eye morphogenesis via phosphorylation of E(spl)M8. *Mechanisms of Development*, *121*, 273–286.

Klämbt, C., Knust, E., Tietze, K., & Campos-Ortega, J. A. (1989). Closely related transcripts encoded by the neurogenic gene complex *Enhancer* of *split* of *Drosophila melanogaster*. *The EMBO Journal*, *8*, 203–210.

Knust, E., Bremer, K. A., Vässin, H., Ziemer, A., Tepass, U., & Campos-Ortega, J. A. (1987). The enhancer of split locus and neurogenesis in Drosophila melanogaster. *Developmental Biology*, *122*, 262–273.

Knust, E., Schrons, H., Grawe, F., & Campos-Ortega, J. A. (1992). Seven genes of the *Enhancer of split* Complex of *Drosophila melanogaster* encode helix-loop-helix proteins. *Genetics*, *132*, 505–518.

Knust, E., Tietze, K., & Campos-Ortega, J. A. (1987). Molecular analysis of the neurogenic locus *Enhancer* of *split* of *Drosophila melanogaster*. *The EMBO Journal*, *6*, 4113–4123.

Kopan, R., & Ilagan, M. X. (2009). The canonical Notch signaling pathway: Unfolding the activation mechanism. *Cell*, *137*, 216–233.

Kramatschek, B., & Campos-Ortega, J. A. (1994). Neuroectodermal transcription of the *Drosophila* neurogenic genes *E(spl)* and *HLH-m5* is regulated by proneural genes. *Development*, *120*, 815–826.

Krejci, A., Bernard, F., Housden, B. E., Collins, S., & Bray, S. J. (2009). Direct response to Notch activation: Signaling crosstalk and incoherent logic. *Science Signaling*, *2*, 1.

Krejci, A., & Bray, S. (2007). Notch activation stimulates transient and selective binding of Su(H)/CSL to target enhancers. *Genes and Development*, *21*, 1322–1327.

Kugler, S. J., & Nagel, A. C. (2010). A novel Pzg-NURF complex regulates Notch target gene activity. *Molecular Biology of the Cell*, *21*, 3443–3448.

Kux, K., Kiparaki, M., & Delidakis, C. (2013). The two Tribolium E(spl) genes show evolutionarily conserved expression and function during embryonic neurogenesis. *Mechanisms of Development*, *130*, 207–225.

Lai, E. C., Bodner, R., Kavaler, J., Freschi, G., & Posakony, J. W. (2000). Antagonism of notch signaling activity by members of a novel protein family encoded by the bearded and enhancer of split gene complexes. *Development*, *127*, 291–306.

Lai, E. C., Bodner, R., & Posakony, J. W. (2000). The enhancer of split complex of Drosophila includes four Notch-regulated members of the bearded gene family. *Development*, *127*, 3441–3455.

Lai, E. C., Burks, C., & Posakony, J. W. (1998). The K box, a conserved 3' UTR sequence motif, negatively regulates accumulation of Enhancer of split Complex transcripts. *Development*, *125*, 4077–4088.

Lai, E. C., & Orgogozo, V. (2004). A hidden program in Drosophila peripheral neurogenesis revealed: Fundamental principles underlying sensory organ diversity. *Developmental Biology*, *269*, 1–17.

Lai, E. C., Tam, B., & Rubin, G. M. (2005). Pervasive regulation of Drosophila Notch target genes by GY-box-, Brd-box-, and K-box-class microRNAs. *Genes and Development*, *19*, 1067–1080.

Langen, M., Koch, M., Yan, J., De Geest, N., Erfurth, M. L., Pfeiffer, B. D., et al. (2013). Mutual inhibition among postmitotic neurons regulates robustness of brain wiring in Drosophila. *eLife*, *2*, e00337.

Lecourtois, M., & Schweisguth, F. (1995). The neurogenic Suppressor of Hairless DNA-binding protein mediates the transcriptional activation of the *Enhancer* of split Complex genes triggered by Notch signaling. *Genes and Development*, *9*, 2598–2608.

Ledent, V., Paquet, O., & Vervoort, M. (2002). Phylogenetic analysis of the human basic helix-loop-helix proteins. *Genome Biology*, *3* Research0030.1–research0030.18.

Lehmann, R., Jiménez, F., Dietrich, U., & Campos-Ortega, J. A. (1983). On the phenotype and development of mutants of early neurogenesis in *Drosophila melanogaster. Roux's Archives of Developmental Biology*, *192*, 62–74.

Li, Y., & Baker, N. E. (2001). Proneural enhancement by Notch overcomes Suppressor-of-Hairless repressor function in the developing Drosophila eye. *Current Biology*, *11*, 330–338.

Li, Y., Lei, L., Irvine, K. D., & Baker, N. E. (2003). Notch activity in neural cells triggered by a mutant allele with altered glycosylation. *Development*, *130*, 2829–2840.

Ligoxygakis, P., Bray, S. J., Apidianakis, Y., & Delidakis, C. (1999). Ectopic expression of individual *E(spl)* genes has differential effects on different cell fate decisions and underscores the biphasic requirement for notch activity in wing margin establishment in *Drosophila. Development*, *126*, 2205–2214.

Ligoxygakis, P., Yu, S. Y., Delidakis, C., & Baker, N. E. (1998). A subset of Notch functions during *Drosophila* eye development require *Su(H)* and the *E(spl)* gene complex. *Development*, *125*, 2893–2900.

Llimargas, M. (1999). The Notch pathway helps to pattern the tips of the Drosophila tracheal branches by selecting cell fates. *Development*, *126*, 2355–2364.

Lopez-Schier, H., & St Johnston, D. (2001). Delta signaling from the germ line controls the proliferation and differentiation of the somatic follicle cells during Drosophila oogenesis. *Genes and Development*, *15*, 1393–1405.

Lu, H., Kozhina, E., Mahadevaraju, S., Yang, D., Avila, F. W., & Erickson, J. W. (2008). Maternal Groucho and bHLH repressors amplify the dose-sensitive X chromosome signal in Drosophila sex determination. *Developmental Biology*, *323*, 248–260.

Maeder, M. L., Polansky, B. J., Robson, B. E., & Eastman, D. A. (2007). Phylogenetic footprinting analysis in the upstream regulatory regions of the Drosophila enhancer of split genes. *Genetics*, *177*, 1377–1394.

Maier, D., Marte, B. M., Schäfer, W., Yu, Y., & Preiss, A. (1993). *Drosophila* evolution challenges postulated redundancy in the *E(spl)* gene complex. *Proceedings of the National Academy of Sciences of the United States of America*, *90*, 5464–5468.

Marlow, H., Roettinger, E., Boekhout, M., & Martindale, M. Q. (2012). Functional roles of Notch signaling in the cnidarian Nematostella vectensis. *Developmental Biology*, *362*, 295–308.

Martin-Bermudo, M. D., Carmena, A., & Jiménez, F. (1995). Neurogenic genes control gene expression at the transcriptional level in early neurogenesis and in mesectoderm specification. *Development*, *121*, 219–224.

Mathur, D., Bost, A., Driver, I., & Ohlstein, B. (2010). A transient niche regulates the specification of Drosophila intestinal stem cells. *Science*, *327*, 210–213.

Micchelli, C. A., & Perrimon, N. (2006). Evidence that stem cells reside in the adult Drosophila midgut epithelium. *Nature*, *439*, 475–479.

Micchelli, C. A., Sudmeier, L., Perrimon, N., Tang, S., & Beehler-Evans, R. (2011). Identification of adult midgut precursors in Drosophila. *Gene Expression Patterns*, *11*, 12–21.

Monastirioti, M., Giagtzoglou, N., Koumbanakis, K. A., Zacharioudaki, E., Deligiannaki, M., Wech, I., et al. (2010). Drosophila Hey is a target of Notch in asymmetric divisions during embryonic and larval neurogenesis. *Development*, *137*, 191–201.

Moore, A. W., Barbel, S., Jan, L. Y., & Jan, Y. N. (2000). A genomewide survey of basic helix-loop-helix factors in Drosophila. *Proceedings of the National Academy of Sciences of the United States of America*, *97*, 10436–10441.

Morel, V., Lecourtois, M., Massiani, O., Maier, D., Preiss, A., & Schweisguth, F. (2001). Transcriptional repression by suppressor of hairless involves the binding of a hairless-dCtBP complex in Drosophila. *Current Biology*, *11*, 789–792.

Moshkin, Y. M., Kan, T. W., Goodfellow, H., Bezstarosti, K., Maeda, R. K., Pilyugin, M., et al. (2009). Histone chaperones ASF1 and NAP1 differentially modulate removal of active histone marks by LID-RPD3 complexes during NOTCH silencing. *Molecular Cell*, *35*, 782–793.

Munder, S., Tischer, S., Grundhuber, M., Buchels, N., Bruckmeier, N., Eckert, S., et al. (2013). Notch-signalling is required for head regeneration and tentacle patterning in Hydra. *Developmental Biology*, *383*, 146–157.

Nagel, A. C., Krejci, A., Tenin, G., Bravo-Patino, A., Bray, S., Maier, D., et al. (2005). Hairless-mediated repression of notch target genes requires the combined activity of Groucho and CtBP corepressors. *Molecular and Cellular Biology*, *25*, 10433–10441.

Nagel, A. C., Maier, D., Krauss, S., Mezger, M., & Preiss, A. (2004). Neurogenic phenotypes induced by RNA interference with bHLH genes of the Enhancer of split complex of Drosophila melanogaster. *Genesis*, *39*, 105–114.

Nagel, A. C., & Preiss, A. (1999). $Notch^{spl}$ is deficient for inductive processes in the eye, and $E(spl)^{D}$ enhances it by interfering with proneural activity. *Developmental Biology*, *208*, 406–415.

Nagel, A. C., Wech, I., Schwinkendorf, D., & Preiss, A. (2007). Involvement of co-repressors Groucho and CtBP in the regulation of single-minded in Drosophila. *Hereditas*, *144*, 195–205.

Nagel, A. C., Yu, Y., & Preiss, A. (1999). Enhancer of split [E(spl)(D)] is a gro-independent, hypermorphic mutation in Drosophila. *Developmental Genetics*, *25*, 168–179.

Nakao, K., & Campos-Ortega, J. A. (1996). Persistent expression of genes of the *Enhancer of split* Complex suppresses neural development in *Drosophila*. *Neuron*, *16*, 275–286.

Nam, Y., Sliz, P., Song, L., Aster, J. C., & Blacklow, S. C. (2006). Structural basis for cooperativity in recruitment of MAML coactivators to Notch transcription complexes. *Cell*, *124*, 973–983.

Nellesen, D. T., Lai, E. C., & Posakony, J. W. (1999). Discrete enhancer elements mediate selective responsiveness of *Enhancer of split Complex* genes to common transcriptional activators. *Developmental Biology*, *213*, 33–53.

Neves, A., & Priess, J. R. (2005). The REF-1 family of bHLH transcription factors pattern C. elegans embryos through Notch-dependent and Notch-independent pathways. *Developmental Cell*, *8*, 867–879.

Ngo, K. T., Wang, J., Junker, M., Kriz, S., Vo, G., Asem, B., et al. (2010). Concomitant requirement for Notch and Jak/Stat signaling during neuro-epithelial differentiation in the Drosophila optic lobe. *Developmental Biology*, *346*, 284–295.

Oellers, N., Dehio, M., & Knust, E. (1994). bHLH proteins encoded by the *Enhancer of split* complex of *Drosophila* negatively interfere with transcriptional activation mediated by proneural genes. *Molecular and General Genetics*, *244*, 465–473.

Ohlstein, B., & Spradling, A. (2006). The adult Drosophila posterior midgut is maintained by pluripotent stem cells. *Nature*, *439*, 470–474.

Ohlstein, B., & Spradling, A. (2007). Multipotent Drosophila intestinal stem cells specify daughter cell fates by differential notch signaling. *Science*, *315*, 988–992.

Ohsako, S., Hyer, J., Panganiban, G., Oliver, I., & Caudy, M. (1994). hairy function as a DNA binding helix-loop-helix repressor of *Drosophila* sensory organ formation. *Genes and Development*, *8*, 2743–2755.

Orihara-Ono, M., Toriya, M., Nakao, K., & Okano, H. (2011). Downregulation of Notch mediates the seamless transition of individual Drosophila neuroepithelial progenitors into optic medullar neuroblasts during prolonged G1. *Developmental Biology, 351*, 163–175.

Overstreet, E., Fitch, E., & Fischer, J. A. (2004). Fat facets and Liquid facets promote Delta endocytosis and Delta signaling in the signaling cells. *Development, 131*, 5355–5366.

Paroush, Z., Finley, R. L., Jr., Kidd, T., Wainwright, S. M., Ingham, P. W., Brent, R., et al. (1994). Groucho is required for Drosophila neurogenesis, segmentation, and sex determination and interacts directly with hairy-related bHLH proteins. *Cell, 79*, 805–815.

Perez-Gomez, R., Slovakova, J., Rives-Quinto, N., Krejci, A., & Carmena, A. (2013). A Serrate-Notch-Canoe complex mediates essential interactions between glia and neuroepithelial cells during Drosophila optic lobe development. *Journal of Cell Science, 126*, 4873–4884.

Poortinga, G., Watanabe, M., & Parkhurst, S. M. (1998). Drosophila CtBP: A Hairy-interacting protein required for embryonic segmentation and hairy-mediated transcriptional repression. *The EMBO Journal, 17*, 2067–2078.

Powell, L. M., Chen, A., Huang, Y. C., Wang, P. Y., Kemp, S. E., & Jarman, A. P. (2012). The SUMO pathway promotes basic helix-loop-helix proneural factor activity via a direct effect on the Zn finger protein senseless. *Molecular and Cellular Biology, 32*, 2849–2860.

Preiss, A., Hartley, D. A., & Artavanis-Tsakonas, S. (1988). The molecular genetics of Enhancer of split, a gene required for embryonic neural development in Drosophila. *The EMBO Journal, 7*, 3917–3927.

Rand, M. D., Bland, C. E., & Bond, J. (2008). Methylmercury activates enhancer-of-split and bearded complex genes independent of the notch receptor. *Toxicological Sciences, 104*, 163–176.

Rand, M. D., Grimm, L. M., Artavanis-Tsakonas, S., Patriub, V., Blacklow, S. C., Sklar, J., et al. (2000). Calcium depletion dissociates and activates heterodimeric notch receptors. *Molecular and Cellular Biology, 20*, 1825–1835.

Reeves, N., & Posakony, J. W. (2005). Genetic programs activated by proneural proteins in the developing Drosophila PNS. *Developmental Cell, 8*, 413–425.

Rosenberg, M. I., & Parkhurst, S. M. (2002). Drosophila Sir2 is required for heterochromatic silencing and by euchromatic Hairy/E(Spl) bHLH repressors in segmentation and sex determination. *Cell, 109*, 447–458.

Rusconi, J. C., & Corbin, V. (1998). Evidence for a novel Notch pathway required for muscle precursor selection in Drosophila. *Mechanisms of Development, 79*, 39–50.

Rusten, T. E., Cantera, R., Urban, J., Technau, G., Kafatos, F. C., & Barrio, R. (2001). Spalt modifies EGFR-mediated induction of chordotonal precursors in the embryonic PNS of Drosophila promoting the development of oenocytes. *Development, 128*, 711–722.

San Juan, B. P., Andrade-Zapata, I., & Baonza, A. (2012). The bHLH factors Dpn and members of the E(spl) complex mediate the function of Notch signalling regulating cell proliferation during wing disc development. *Biology Open, 1*, 667–676.

San Juan, B. P., & Baonza, A. (2011). The bHLH factor deadpan is a direct target of Notch signaling and regulates neuroblast self-renewal in Drosophila. *Developmental Biology, 352*, 70–82.

Schaaf, C. A., Misulovin, Z., Gause, M., Koenig, A., & Dorsett, D. (2013). The Drosophila enhancer of split gene complex: Architecture and coordinate regulation by notch, cohesin, and polycomb group proteins. *G3 (Bethesda), 3*, 1785–1794.

Schaaf, C. A., Misulovin, Z., Sahota, G., Siddiqui, A. M., Schwartz, Y. B., Kahn, T. G., et al. (2009). Regulation of the Drosophila Enhancer of split and invected-engrailed gene complexes by sister chromatid cohesion proteins. *PloS One, 4*, e6202.

Schlatter, R., & Maier, D. (2005). The Enhancer of split and Achaete-Scute complexes of Drosophilids derived from simple ur-complexes preserved in mosquito and honeybee. *BMC Evolutionary Biology*, *5*, 67.

Schrons, H., Knust, E., & Campos-Ortega, J. A. (1992). The *Enhancer* of *split* Complex and adjacent genes in the 96F region of *Drosophila melanogaster* are required for segregation of neural and epidermal progenitor cells. *Genetics*, *132*, 481–503.

Shepard, S. B., Broverman, S. A., & Muskavitch, M. A. T. (1989). A tripartite interaction among alleles of *Notch*, *Delta*, and *Enhancer* of *split* during imaginal development of *Drosophila melanogaster*. *Genetics*, *122*, 429–438.

Simionato, E., Ledent, V., Richards, G., Thomas-Chollier, M., Kerner, P., Coornaert, D., et al. (2007). Origin and diversification of the basic helix-loop-helix gene family in metazoans: Insights from comparative genomics. *BMC Evolutionary Biology*, *7*, 33.

Singson, A., Leviten, M. W., Bang, A. G., Hua, X. H., & Posakony, J. W. (1994). Direct downstream targets of proneural activators in the imaginal disc include genes involved in lateral inhibitory signaling. *Genes and Development*, *8*, 2058–2071.

Skeath, J. B., & Doe, C. Q. (1998). Sanpodo and Notch act in opposition to Numb to distinguish sibling neuron fates in the Drosophila CNS. *Development*, *125*, 1857–1865.

Song, X., Call, G. B., Kirilly, D., & Xie, T. (2007). Notch signaling controls germline stem cell niche formation in the Drosophila ovary. *Development*, *134*, 1071–1080.

Sun, J., Smith, L., Armento, A., & Deng, W. M. (2008). Regulation of the endocycle/gene amplification switch by Notch and ecdysone signaling. *Journal of Cell Biology*, *182*, 885–896.

Taelman, V., Van Wayenbergh, R., Solter, M., Pichon, B., Pieler, T., Christophe, D., et al. (2004). Sequences downstream of the bHLH domain of the Xenopus hairy-related transcription factor-1 act as an extended dimerization domain that contributes to the selection of the partners. *Developmental Biology*, *276*, 47–63.

Tanaka, R., Takase, Y., Kanachi, M., Enomoto-Katayama, R., Shirai, T., & Nakagoshi, H. (2007). Notch-, Wingless-, and Dpp-mediated signaling pathways are required for functional specification of Drosophila midgut cells. *Developmental Biology*, *304*, 53–61.

Tata, F., & Hartley, D. A. (1995). Inhibition of cell fate in *Drosophila* by *Enhancer* of *split* genes. *Mechanisms of Development*, *51*, 305–315.

Tepass, U., & Hartenstein, V. (1994). Epithelium formation in the Drosophila midgut depends on the interaction of endoderm and mesoderm. *Development*, *120*, 579–590.

Tepass, U., & Hartenstein, V. (1995). Neurogenic and proneural genes control cell fate specification in the Drosophila endoderm. *Development*, *121*, 393–405.

Terriente-Felix, A., Li, J., Collins, S., Mulligan, A., Reekie, I., Bernard, F., et al. (2013). Notch cooperates with Lozenge/Runx to lock haemocytes into a differentiation programme. *Development*, *140*, 926–937.

Tietze, K., Oellers, N., & Knust, E. (1992). *Enhancer of split*^{D}, a dominant mutation of *Drosophila*, and its use in the study of functional domains of a helix-loop-helix protein. *Proceedings of the National Academy of Sciences of the United States of America*, *89*, 6152–6156.

Trott, R. L., Kalive, M., Paroush, Z., & Bidwai, A. P. (2001). Drosophila melanogaster casein kinase II interacts with and phosphorylates the basic helix-loop-helix proteins m5, m7, and m8 derived from the Enhancer of split complex. *The Journal of Biological Chemistry*, *276*, 2159–2167.

Ungerer, P., Eriksson, B. J., & Stollewerk, A. (2012). Unravelling the evolution of neural stem cells in arthropods: Notch signalling in neural stem cell development in the crustacean Daphnia magna. *Developmental Biology*, *371*, 302–311.

Van Doren, M., Bailey, A. M., Esnayra, J., Ede, K., & Posakony, J. W. (1994). Negative regulation of proneural gene activity: Hairy is a direct transcriptional repressor of *achaete*. *Genes and Development*, *8*, 2729–2742.

Van Doren, M., Ellis, H. M., & Posakony, J. W. (1991). The Drosophila extramacrochaetae protein antagonizes sequence-specific DNA binding by daughterless/achaete-scute protein complexes. *Development, 113*, 245–255.

Wang, W., Liu, W., Wang, Y., Zhou, L., Tang, X., & Luo, H. (2011). Notch signaling regulates neuroepithelial stem cell maintenance and neuroblast formation in Drosophila optic lobe development. *Developmental Biology, 350*, 414–428.

Wech, I., Bray, S., Delidakis, C., & Preiss, A. (1999). Distinct expression patterns of different enhancer of split bHLH genes during embryogenesis of Drosophila melanogaster. *Development Genes and Evolution, 209*, 370–375.

Wei, Q., Rong, Y., & Paterson, B. M. (2007). Stereotypic founder cell patterning and embryonic muscle formation in Drosophila require nautilus (MyoD) gene function. *Proceedings of the National Academy of Sciences of the United States of America, 104*, 5461–5466.

Welshons, W. J. (1956). Dosage experiments with split mutants in the presence of an enhancer of split. *Drosophila Information Service, 30*, 157–158.

Weng, M., Haenfler, J. M., & Lee, C. Y. (2011). Changes in Notch signaling coordinates maintenance and differentiation of the Drosophila larval optic lobe neuroepithelia. *Developmental Neurobiology, 72*, 1376–1390.

Wilson, J. J., & Kovall, R. A. (2006). Crystal structure of the CSL-Notch-Mastermind ternary complex bound to DNA. *Cell, 124*, 985–996.

Wurmbach, E., Wech, I., & Preiss, A. (1999). The *Enhancer of split* complex of *Drosophila melanogaster* harbours three classes of Notch responsive genes. *Mechanisms of Development, 80*, 171–180.

Xie, T., Song, X., Jin, Z., Pan, L., Weng, C., Chen, S., et al. (2008). Interactions between stem cells and their niche in the Drosophila ovary. *Cold Spring Harbor Symposia on Quantitative Biology, 73*, 39–47.

Yasugi, T., Sugie, A., Umetsu, D., & Tabata, T. (2010). Coordinated sequential action of EGFR and Notch signaling pathways regulates proneural wave progression in the Drosophila optic lobe. *Development, 137*, 3193–3203.

Yasugi, T., Umetsu, D., Murakami, S., Sato, M., & Tabata, T. (2008). Drosophila optic lobe neuroblasts triggered by a wave of proneural gene expression that is negatively regulated by JAK/STAT. *Development, 135*, 1471–1480.

Yu, Z., Wu, H., Chen, H., Wang, R., Liang, X., Liu, J., et al. (2013). CAF-1 promotes Notch signaling through epigenetic control of target gene expression during Drosophila development. *Development, 140*, 3635–3644.

Zacharioudaki, E., Magadi, S. S., & Delidakis, C. (2012). bHLH-O proteins are crucial for Drosophila neuroblast self-renewal and mediate Notch-induced overproliferation. *Development, 139*, 1258–1269.

Zarifi, I., Kiparaki, M., Koumbanakis, K. A., Giagtzoglou, N., Zacharioudaki, E., Alexiadis, A., et al. (2012). Essential roles of Da transactivation domains in neurogenesis and in E(spl)-mediated repression. *Molecular and Cellular Biology, 32*, 4534–4548. http://dx.doi.org/10.1128/MCB.00827-12.

zur Lage, P., & Jarman, A. P. (1999). Antagonism of EGFR and notch signalling in the reiterative recruitment of Drosophila adult chordotonal sense organ precursors. *Development, 126*, 3149–3157.

zur Lage, P. I., Powell, L. M., Prentice, D. R. A., McLaughlin, P., & Jarman, A. P. (2004). EGF receptor signaling triggers recruitment of Drosophila sense organ precursors by stimulating proneural gene autoregulation. *Developmental Cell, 7*, 687–696.

zur Lage, P. I., Prentice, D. R., Holohan, E. E., & Jarman, A. P. (2003). The Drosophila proneural gene amos promotes olfactory sensillum formation and suppresses bristle formation. *Development, 130*, 4683–4693.

CHAPTER SEVEN

Expression Dynamics and Functions of Hes Factors in Development and Diseases

Taeko Kobayashi[*,†,1], **Ryoichiro Kageyama**[*,†,‡,1]

[*]Institute for Virus Research, Kyoto University, Kyoto, Japan
[†]Japan Science and Technology Agency, Core Research for Evolutional Science and Technology (CREST), Saitama, Japan
[‡]World Premier International Research Initiative–Institute for Integrated Cell-Material Sciences (WPI-iCeMS), Kyoto University, Kyoto, Japan
[1]Corresponding authors: e-mail address: tkobayas@virus.kyoto-u.ac.jp; rkageyam@virus.kyoto-u.ac.jp

Contents

Abstract

Hes genes, encoding basic helix–loop–helix (HLH) transcriptional repressors, are mammalian homologues of Drosophila hairy and Enhancer of split genes, both of which are required for normal neurogenesis in Drosophila. There are seven members in the human Hes family, Hes1–7, which are expressed in many tissues and play various roles mainly in development. All Hes proteins have three conserved domains: basic HLH (bHLH), Orange, and WRPW domains. The basic region binds to target DNA sequences, while the HLH region forms homo- and heterodimers with other bHLH proteins, the Orange domain is responsible for the selection of partners during heterodimer formation, and the WRPW domain recruits corepressors. Hes1, Hes5, and Hes7 are known as downstream effectors of canonical Notch signaling, which regulates cell differentiation

Current Topics in Developmental Biology, Volume 110
ISSN 0070-2153
http://dx.doi.org/10.1016/B978-0-12-405943-6.00007-5

© 2014 Elsevier Inc.
All rights reserved.

via cell–cell interaction. Hes factors regulate many events in development by repressing the expression of target genes, many of which encode transcriptional activators that promote cell differentiation. For example, Hes1, Hes3, and Hes5 are highly expressed by neural stem cells, and inactivation of these genes results in insufficient maintenance of stem cell proliferation and prematurely promotes neuronal differentiation. Recently, it was shown that the expression dynamics of Hes1 plays crucial roles in proper developmental timings and fate-determination steps of embryonic stem cells and neural progenitor cells. Here, we discuss some key features of Hes factors in development and diseases.

1. INTRODUCTION

Development is a highly organized process that depends on the timely proliferation of stem cells and their differentiation into multiple cell types. The balance between cell proliferation and cell fate determination is critical to forming organs with the right shape and size and the right cell composition. Hes factors (hairy and Enhancer of split (E(spl)) homologs) play important roles in the development of many organs to expand progenitor cell pools and regulate cell fate decisions. Some Hes factors, Hes1, Hes5, and Hes7, are dynamically expressed by auto-negative feedback loops and also function as effectors of the Notch signaling pathway, which coordinates cell proliferation and differentiation via cell–cell interaction. Recent studies have shown that the expression dynamics of Hes1 contribute to its multiple functions. In this review, we describe the features of Hes factors, their functions in development and diseases, and the significance of Hes1 oscillation, including its expression dynamics in embryonic stem (ES) cells and neural progenitor/stem (NS) cells.

2. HES FAMILY MEMBERS AND PROTEIN STRUCTURES

The protein products of the Hes gene family comprise seven members in human, Hes1–7. Hes1, Hes2, and Hes4 are more similar to Drosophila hairy in their amino acid sequences than are the other members (Fig. 7.1A). Hes4 is the most similar to Hes1 and is not present in mouse or rat. *Hes3* encodes two isoforms due to alternative splicing: Hes3 variant 1 lacks the N-terminal half of the basic region for DNA binding, while Hes3 variant 2 has a complete basic region (Hirata, Ohtsuka, Bessho, & Kageyama, 2000; Katoh & Katoh, 2007). Hes factors have three conserved

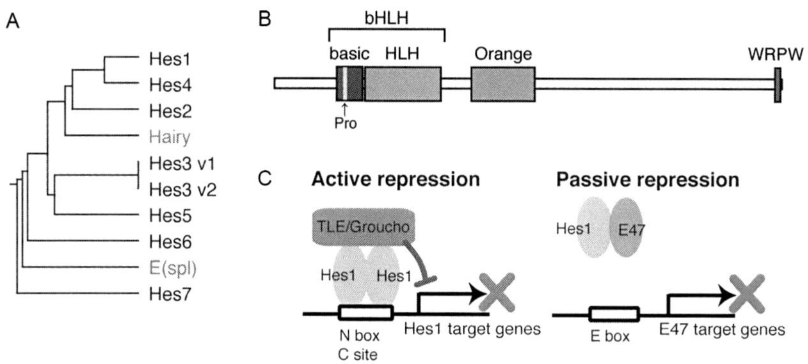

Figure 7.1 Features of Hes factors and their functions. (A) Comparison of protein homology in human Hes factors (Hes1–7, black) and their Drosophila homologues (hairy and E(spl), gray). Phylogenetic tree was generated by ClustalW software from the amino acid sequence alignment. (B) Structure of Hes1 protein. (C) Dual repressor functions of Hes factors: active repression and passive repression. During active repression, Hes factors bind to the N-box or the class C site in the promoter regions of target genes and repress their transcription. During passive repression, Hes factors form heterodimers with bHLH activators and inhibit their binding to the E box in the promoter regions of target genes. (See the color plate.)

domains: the basic helix–loop–helix (bHLH) domain, the Orange domain, and the WRPW domain, harboring a WRPW motif (Trp-Arg-Pro-Trp) (Fig. 7.1B; Kageyama, Ohtsuka, & Kobayashi, 2007). There are also other Hes-related bHLH factors, such as Hey1, Hey2, HeyL, and HELT (Hes/Hey-like), that do not have the WRPW motif.

The bHLH domain is essential for DNA binding and homo- and heterodimer formation. The N-terminal basic region plays an important role in the specific binding to target DNA sequences, that is, the N-box (CACNAG) and the class C site (CACG(C/A)G) (Sasai, Kageyama, Tagawa, Shigemoto, & Nakanishi, 1992). It has been suggested that a proline residue in the middle of the basic region (Fig. 7.1B), which is conserved in all Hes factors as well as Drosophila hairy and E(spl) proteins, may be involved in the target sequence specificity, providing a higher affinity for the N-box and the class C site than for the E box (CANNTG), a consensus binding sequence for activator-type bHLH factors. The HLH domain is characterized by two alpha-helix motifs linked by a flexible loop structure and interacts with other HLH motifs of binding partner proteins to form homo- or heterodimers. Hes factors can form homo- and heterodimers with Hes-related bHLH repressors, such as Hey factors, as well as with other bHLH factors, such as Mash1, E47, and Id. The Orange domain has two amphipathic helices that are involved in the selection of partners to form heterodimers. The WRPW domain, located at the C-terminal end, is essential for the repressor activity via recruitment of the corepressor Transducin-like E(spl) (TLE1–4)/Groucho-related gene (Grg), a homologue of Drosophila Groucho. The Hey family factors contain a YRPW (Tyr-Arg-Pro-Trp) motif instead of the WRPW motif, which contributes to the weaker binding of Hey factors to TLE than the Hes factors (Fischer & Gessler, 2007; Iso et al., 2001).

3. HES GENERAL FUNCTIONS

Hes factors repress transcription by two different mechanisms, active repression and passive repression (Fig. 7.1C; Kageyama et al., 2007). During active repression, Hes factors form homo- or heterodimers with other Hes/Hey factors and bind to the N-box or the class C site sequences in the promoter region of target genes. In the case of the heterodimers between Hes1 and Hey1 or Hey2, binding to the class C site exhibits a higher affinity than Hes1 or Hey1 homodimers, and such heterodimers repress transcription more efficiently than the homodimers. The C-terminal WRPW domain in Hes factors interacts with TLE/Grg corepressors (Buscarlet, Perin, Laing, Brickman, & Stifani, 2008), which recruit histone deacetylase and repress the target gene expression by inducing changes in the chromatin structures. Hes1 can also repress target genes in a TLE-independent manner, by recruiting Sirt1, a homologue of silent information regulator 1 and a class III histone deacetylase, via its bHLH domain, and thereby repressing target genes (Takata & Ishikawa, 2003). Although Hes family genes encode transcriptional repressors, there is one exceptional gene, *Hes6*. The Hes6 protein has no DNA-binding ability and inhibits Hes1 function by forming heterodimers with it.

During passive repression, Hes factors form heterodimers with bHLH-type activators, such as Mash1 and E47, and inhibit their binding to the E box in the promoter region of target genes (Akazawa, Sasai, Nakanishi, & Kageyama, 1992). Thus, Hes factors function as dominant negative regulators to these bHLH activators. Id proteins, which lack the basic DNA-binding domain, also negatively regulate bHLH activators by inhibiting their DNA binding through heterodimerization. Hes factors can tightly repress the expression of target genes via these two types of repression. For example, Hes1 represses *Mash1*, a mouse homologue of Drosophila achate-scute genes that encode transcriptional activators and promote neural differentiation, both via active repression by directly binding to the class C site on the *Mash1* promoter and via passive repression by capturing E-protein activators such as E47 (Fig. 7.1C).

4. ROLES OF HES FACTORS IN VARIOUS TISSUES

Hes gene expression is regulated by the Notch-dependent signaling pathway (Fig. 7.2). Notch signaling is activated when ligands such as Delta or Jagged expressed on the surface of neighboring cells bind to Notch receptors. Upon activation of Notch signaling, the Notch intracellular domain (ICD) is released from the cell surface and is translocated into the nucleus, where it forms a complex with the DNA-binding protein RBPj/CSL. The RBPj–ICD complex recruits additional transcriptional coactivators and induces the expression of downstream genes, including *Hes* and *Hey*. This pathway is often called the canonical pathway of Notch signaling. Mouse *Hes1*, *Hes5*, and *Hes7*, which are all activated by Notch signaling, have RBPj-binding sites on their promoters (Bessho, Miyoshi, Sakata, & Kageyama, 2001; Katoh & Katoh, 2007; Ohtsuka et al., 1999). On the other

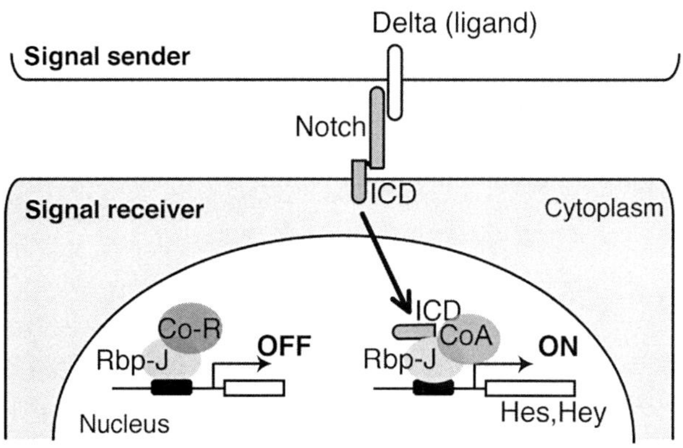

Figure 7.2 Notch signaling pathway. The Notch receptor is a transmembrane protein activated by the binding of Notch ligands expressed on the surface of neighboring cells (signal sender). Upon ligand binding, the Notch receptor on the signal receiver cell is sequentially cleaved by the ADAM protease and the gamma-secretase, releasing the intracellular domain (ICD) of Notch from the plasma membrane. ICD translocates into the nucleus and forms a complex with the DNA-binding transcription factor RBP-J. Without the ICD, RBP-J represses target gene expression, but is converted to a transcriptional activator when complexed with the ICD and the coactivators (CoA). Finally, Notch signaling activation induces the expression of target genes such as *Hes* and *Hey*. (See the color plate.)

hand, other signaling pathways are also reported to induce Hes expression. For example, Hes1 expression is induced by many signaling pathways, including bone morphogenetic protein (BMP), fibroblast growth factor (FGF), leukemia inhibitory factor (LIF), NF-κB, Sonic hedgehog (Shh), and Wnt signaling pathways (Kageyama, Hatakeyama, & Ohtsuka, 2006; Nakayama, Satoh, Igari, Kageyama, & Nishida, 2008). Hes1 regulates the proliferation of NS cells in the central (see Sections 5.2 and 7.3) and peripheral nervous systems under the control of these pathways. *Hes7*, a segmentation clock gene involved in somitogenesis, is regulated by the Notch and FGF pathways during somite segmentation (Niwa et al., 2007). Thus, many pathways regulate Hes expression in a tissue-specific manner.

Hes genes are involved in regulation of many different cell types, including hematopoietic stem cells (HSCs) in embryos (Kageyama et al., 2007), digestive organs, lung, cochlea, skeletal tissues, testis, and vascular endothelial cells, as discussed below. In digestive organs, Hes1 and Hes5 are expressed by proliferating progenitor/stem cells and play an important role in the maintenance of these cells by preventing their differentiation. Hes1 is detected in the developing pancreas, where it represses the expression of the bHLH-type activators Neurogenin3 (Ngn3) and pancreas transcription factor 1 (Ptf1a), which enhance the differentiation of endocrine and exocrine cells, respectively. Hes1 maintains progenitors by inhibiting the expression of these bHLH activators. Genetic inactivation of *Hes1* upregulates the expression of these bHLH activators, resulting in, for example, severe pancreatic hypoplasia and the formation of an ectopic pancreas (Fukuda et al., 2006; Jensen et al., 2000; Sumazaki et al., 2004). Hes1 is also expressed in the intestinal crypt, where stem cells and proliferating progenitors are located, and represses the expression of bHLH-type activator genes, *Math1* and *Ngn3*, whose translational products promote the differentiation of secretory cells. Math1 regulates the development of all types of secretory cells, while Ngn3 regulates enteroendocrine cell development. *Hes1* knockout (KO) mice showed an overproduction of secretory cells (Jensen et al., 2000). Importantly, *Hes1* inactivation markedly reduces tumor cell proliferation and induces their differentiation into intestinal epithelial cells without affecting the homeostasis of normal intestinal crypts in adenomatous polyposis coli mutation-induced intestinal tumors, suggesting that Hes1 is a promising target for intestinal cancer therapy (Ueo et al., 2012).

In sensory organs, it has been reported that Hes and Hes-related genes are expressed in the developing inner ear and the developing retina. Hes1, Hes5, and Hey1 are expressed in the developing cochlea of the inner ear, and they cooperatively inhibit the formation of hair cells, which are sensory neurons. Inactivation of these genes leads to disorganized cell alignment and polarity and to hearing loss due to hair cell overproduction (Tateya, Imayoshi, Tateya, Ito, & Kageyama, 2011). In the developing retina, Hes1 and Hes5 function downstream of both Shh and Notch signaling, inhibit the activity of proneural bHLH genes such as *Mash1* and *Math3*, and maintain progenitor cell populations (Ohtsuka, Sakamoto, Guillemot, & Kageyama, 2001; Wall et al., 2009). Hes6 is exceptionally expressed in the differentiated neural retina and functions as an inhibitor of Hes1, thereby promoting neuronal differentiation (Bae, Bessho, Hojo, & Kageyama, 2000).

In HSCs, both Hes1 and Hes5 are induced by Notch signaling and regulate several blood lineages, including T cells, B cells, and stem cells. Forced expression of Hes1 results in increased numbers of quiescent HSCs, and *Hes1*-deficient HSCs are unable to differentiate into T cells, suggesting that Hes1 is an important regulator for HSC functions (Tomita et al., 1999). It has been shown that, although *Hes1*-deficient mice do not show clear abnormalities in their hematopoietic system, the depletion of both *Hes1* and *Hes5* induces overproduction of nonfunctional HSCs, in which Hes target genes such as *Gata2* are abnormally upregulated (Guiu et al., 2013). Hes1 is also reported to be an important regulator of the development of T cell acute lymphoblastic leukemia (Espinosa et al., 2010; Wendorff et al., 2010).

Hes7 is a segmentation clock gene involved in somitogenesis of embryos (Bessho, Miyoshi, et al., 2001; Bessho, Sakata, et al., 2001). Somites are metameric structures periodically segmented from the anterior parts of the presomitic mesoderm. These segmented, paired blocks of mesoderm form vertebrae, ribs, skeletal muscles, and subcutaneous tissues. During somitogenesis, the oscillatory expression of Hes7 regulates the periodic segmentation events of the presomitic mesoderm. *Hes7* is a causative gene in the inheritable diseases, spondylocostal dysostosis type 4 and congenital scoliosis, which exhibit developmental bone and vertebral defects (Sparrow, GuillŽn-Navarro, Fatkin, & Dunwoodie, 2008; Sparrow et al., 2012).

5. HES FUNCTIONS IN THE CENTRAL NERVOUS SYSTEM

5.1. Hes genes in NS cells

Hes genes play crucial roles in inhibiting neuronal differentiation and promoting the proliferation of NS cells in the developing central nervous system (Ohtsuka et al., 2001). During neural development, neuroepithelial cells appear initially and undergo extensive self-renewal. As development proceeds, neuroepithelial cells are gradually elongated and become radial glial cells, which have a cell body in the inner region, called the ventricular zone, and a radial fiber extending from the internal surface to the pial (outer) surface of the brain. Both neuroepithelial cells and radial glial cells are considered to be NS cells. NS cells are known to change their competency over time to generate distinct types of neurons. Radial glial cells divide asymmetrically, forming one radial glial cell and one neuron or basal progenitor from each cell division. Basal progenitors proliferate and produce more neurons in the subventricular zone. After producing neurons, NS cells give rise to oligodendrocytes and astrocytes (Fig. 7.3A). Hes1 and Hes3 are widely expressed by neuroepithelial cells, whereas Hes1 and Hes5 are mainly expressed by radial glial cells (Hatakeyama et al., 2004).

Hes1, Hes3, and Hes5 are important for maintaining NS cells. Inactivation of these Hes genes leads to failure of maintenance of the undifferentiated state, accelerating neuronal differentiation. As a result, virtually all NS cells become neurons and are therefore depleted prematurely (Hatakeyama et al., 2004). In the absence of these Hes genes, proneural bHLH genes, such as *Mash1*, *Neurogenin2* (*Ngn2*), and others, are highly upregulated, leading to premature neuronal differentiation. On the other hand, miss-expression of Hes1 or Hes5 inhibits neuronal differentiation and maintains NS cells (Ohtsuka et al., 2001). Based on their expression patterns, Hes1 and Hes3 are important for the maintenance of neuroepithelial cells, while Hes1 and Hes5 are required for most radial glial cells (Hatakeyama et al., 2004), although these Hes factors are mostly redundant and compensate each other.

5.2. Hes genes in brain morphogenesis

Hes genes play important roles in brain morphogenesis (Kageyama et al., 2007). The developing nervous system is partitioned into many compartments by boundary structures, such as the zona limitans intrathalamica, the isthmus, the floor plate, and the roof plate (Fig. 7.3B). The zona limitans intrathalamica is the boundary between the thalamus and the prethalamus; the isthmus is the boundary between the midbrain and the hindbrain; and the floor and roof plates are the boundaries between the right and the left parts of the neural tube. Cells in these boundary structures proliferate poorly and do not give rise to neurons, suggesting that the boundary cells are mostly quiescent, whereas cells in compartments between boundaries proliferate extensively, maintain the NS state, and differentiate into neurons. Several

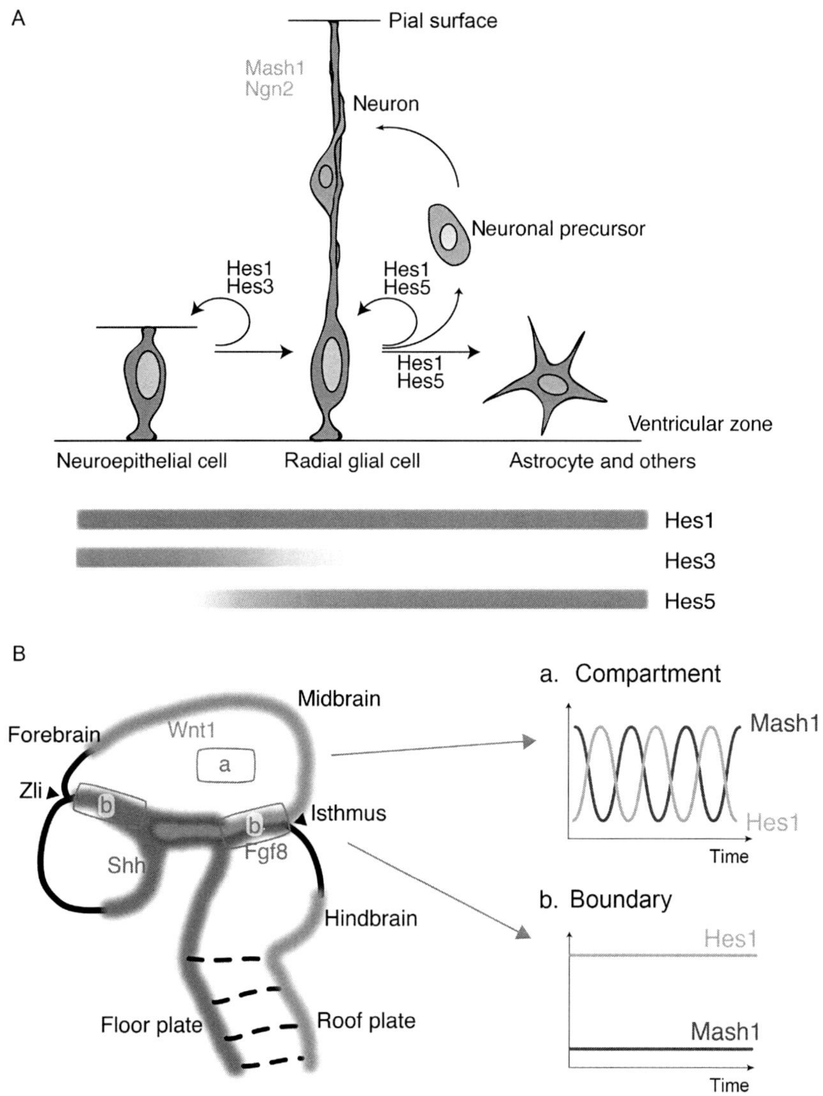

Figure 7.3 Hes functions in the central nervous system. (A) Neural stem cells and their differentiation during neural development (upper panel). Neuroepithelial cells initially proliferate by symmetric cell division and then become radial glial cells, which sequentially generate neurons by asymmetric cell division. Both neuroepithelial cells and radial glial cells are considered to be neural stem (NS) cells. Radial glial cells eventually differentiate into astrocytes, oligodendrocytes, and ependymal cells. Hes factors, expressed in these NS cells, maintain these NS cells and then promote astrocyte differentiation. Proneural bHLH genes, such as *Mash1* and *Ngn2*, are activated in differentiating neurons, leading to the promotion of neurogenesis. Expression patterns of Hes genes in the ventricular zone are indicated (lower panel). (B) The structure of the developing

(Continued)

knockout mouse studies showed that the zona limitans intrathalamica is lost in the absence of *Hes1* and *Hes5*, and that the isthmus and the floor and roof plates are not formed in *Hes1/3/5* triple knockout mice. In the absence of Hes genes, ectopic neuronal differentiation occurs in these defective boundary regions, collapsing the boundary structures (Hirata, Tomita, Bessho, & Kageyama, 2001). Forced expression of Hes1 in compartment cells inhibits both cell proliferation and neuronal differentiation by repressing cell cycle regulator genes and proneural genes (Baek, Hatakeyama, Sakamoto, Ohtsuka, & Kageyama, 2006; Shimojo, Ohtsuka, & Kageyama, 2008). These results demonstrate that Hes genes are crucial to maintaining these boundary structures in the developing brain by inhibiting differentiation and proliferation of their cells.

Hes1 regulates the maintenance of both quiescent boundary cells and proliferating compartment NS cells, but how does it control two opposite cell cycle states (quiescent vs. proliferating)? It has been shown that the expression of Hes1 is different in these two regions (Baek et al., 2006). Hes1 protein levels are persistent and high in boundary cells, whereas they are variable in individual compartment cells (Fig. 7.3B). It was recently found that this variability in Hes1 levels in compartments is due to oscillatory expression (see below) (Shimojo et al., 2008). These different expression patterns of Hes1 may be responsible for its different regulatory influences in compartment and boundary cells: repressed proliferation and differentiation in boundary cells, and intense proliferation and neuronal differentiation in compartment cells.

6. HES1 OSCILLATION AND CELL PROLIFERATION IN CULTURED CELLS

Hes1 expression oscillates in many types of cultured cells, including fibroblasts, myoblasts, and neuroblasts, and the oscillatory expression can be induced following serum stimulation or Notch activation (Hirata et al., 2002; Masamizu et al., 2006). The period of Hes1 oscillation is about

Figure 7.3—Cont'd central nervous system (CNS) and differential expression of Hes1 gene. The CNS in the mouse at E10.5 (left) is partitioned into two regions, compartments (a) and boundaries (b). Boundary cells function as organizing centers by expressing several signaling ligands, including Shh, Wnt1, or Fgf8. The temporal expression patterns of Hes1 and Mash1 (right panels) are different in compartment and boundary cells. Hes1 expression is variable and oscillating in NS cells in compartments, but is sustained in boundary cells. *Adapted from Kageyama, Ohtsuka, & Kobayashi, 2007.* (See the color plate.)

2 h in many mouse cell lines. Hes7, which also oscillates with a 2-h periodicity in the presomitic mesoderm of mouse embryos, acts as a molecular clock to regulate the periodic somite segmentation (Pourquié, 2011). This oscillatory expression is regulated by a negative feedback loop (Fig. 7.4A). After the activation of *Hes1* transcription by Notch signaling or others, translated Hes1 protein represses its own expression by directly binding to N-box sequences in its promoter. Hes1 gene products, both mRNA and protein, are very unstable, with a very short half-life of about 20 min and therefore disappear rapidly, indicating that this autorepression is short

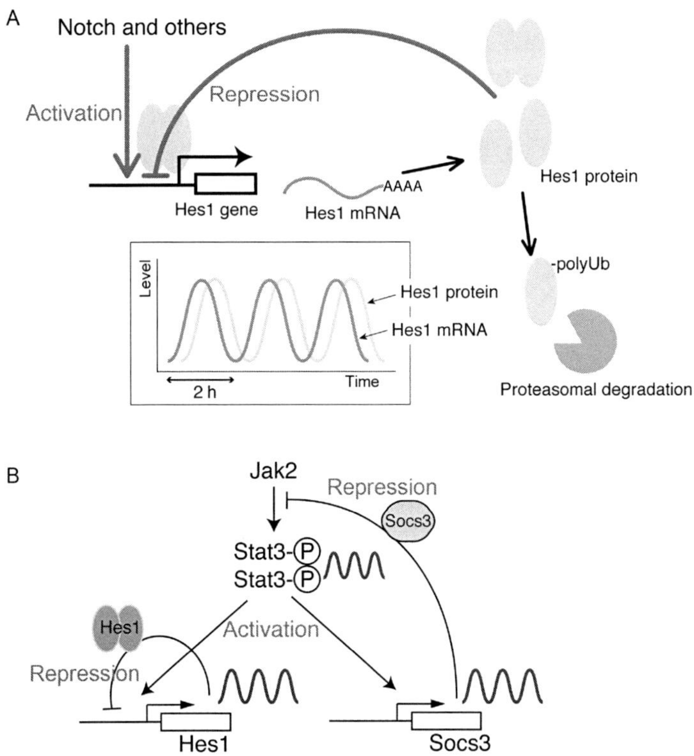

Figure 7.4 Oscillatory expression of Hes1. (A) Hes1 oscillation is regulated by negative feedback: Hes1 protein represses its own expression by directly binding to N-box sequences in its promoter. Both Hes1 mRNA and protein are unstable and disappear quickly when the expression is repressed. By this mechanism, Hes1 autonomously initiates oscillatory expression with a periodicity of about 2 h. (B) Jak–Stat signaling regulates Hes1 oscillation. Jak2 activity is regulated via negative feedback by *Socs3*, a downstream gene of Jak–Stat signaling. The levels of phosphorylated Stat3 (Stat3-P) and Socs3 are oscillating, which regulate Hes1 oscillation and cell proliferation. (See the color plate.)

and transient. Disappearance of Hes1 protein allows the next round of expression. In this way, Hes1 autonomously initiates oscillatory expression with a short periodicity. Real-time imaging analyses of Hes1 expression in single cells revealed that the oscillatory expression of Hes1 is intrinsic and surely occurs in normal culture conditions even in the absence of special stimulations such as serum shock, which synchronizes Hes1 induction in cell populations (Masamizu et al., 2006).

It has been reported that the oscillatory expression of Hes1 is regulated by the Jak–Stat signaling pathway in mouse fibroblast cells (Fig. 7.4B; Yoshiura et al., 2007). Jak2 kinase activates Stat3 by phosphorylation, and activated Stat3 induces the expression of downstream genes, including *Socs3*, an inhibitor of Jak2 kinase. Stat3 phosphorylation oscillates in an approximately 2-h cycle, depending on the negative feedback regulation of Jak2 through Socs3 expression. The blockade of Jak2–Stat3–Socs3 oscillation inhibits Hes1 oscillation and stabilizes Hes1 protein levels by a still unknown mechanism.

Results of several studies suggest that Hes1 oscillation and cell proliferation are correlated with each other. Sustained Hes1 expression leads to G1 phase retardation of the cell cycle (Yoshiura et al., 2007). Hes1 can directly or indirectly repress the transcription of cell cycle-related genes, including *E2f1*, *p21Cip1*, *p27Kip1*, and *p57Kip2* (Castella, Sawai, Nakao, Wagner, & Caudy, 2000; Georgia, Soliz, Li, Zhang, & Bhushana, 2006; Hartman et al., 2004; Murata et al., 2005). Hes1 oscillation may be important for efficient cell proliferation. Hes1 is identified as one of the factors that protects the quiescent state in fibroblast cells by preventing these cells from undergoing differentiation or senescence (Sang, Coller, & Roberts, 2008). High and sustained expression of Hes1 protects fibroblasts against senescence or differentiation, but the expression of dominant negative-type Hes1 mutants allows quiescent fibroblasts to undergo differentiation. Increased expression of Hes1 has been reported in malignant tumors, including ovarian carcinomas and breast cancers (Sang, Roberts, & Coller, 2010; Wang et al., 2010), but the patterns of its expression in these cells remain unknown. It is important to understand the more detailed relationships between Hes1 expression patterns or levels and cell proliferation or quiescent state.

The molecular mechanism by which cells exhibit different (oscillatory vs. sustained) expression modes of Hes1 is still unknown. One candidate is the regulation of microRNA-9 (miR-9) (Bonev, Stanley, & Papalopulu, 2012; Tan, Ohtsuka, González, & Kageyama, 2012). miR-9 negatively regulates Hes1 mRNA stability by interacting with the 3′-UTR of Hes1 mRNA, enabling oscillatory expression of Hes1. miR-9 expression is absent in the roof plate and floor plate where Hes1 expression is high and sustained. Overexpression of miR-9 induces cell cycle exit and neuronal differentiation, while miR-9 knockdown inhibits neuronal differentiation of NS cells and dampens Hes1 oscillation.

7. HES1 OSCILLATION IN ES AND NS CELLS

7.1. Hes1 oscillation in ES cells

ES cells are pluripotent stem cells derived from the inner cell mass of blastocyst-stage embryos. Due to their pluripotency, these cells have the ability to differentiate into various cell types belonging to all three germ layers: ectoderm, mesoderm, and endoderm (Smith, 2001). ES cells are known to exhibit heterogeneous gene inductions in response to the same differentiation cues, but the basic mechanism governing such heterogeneity in ES cells is not fully understood. Hes1 oscillation is one of the candidates for regulating heterogeneous differentiation of ES cells (Kobayashi & Kageyama, 2010a; Kobayashi et al., 2009). Hes1 is highly expressed in mouse ES cells under the control of LIF and BMP, two essential factors for mouse ES cell culture, and the expression oscillates in individual mouse ES cells with a 3–5-h periodicity, a longer period than that observed in fibroblast cells. This Hes1 oscillation contributes to the heterogeneous properties of ES cell differentiation: Hes1 protein-low cells are prone to differentiate into neuroectodermal cells, while Hes1 protein-high cells differentiate into early mesodermal cells (Fig. 7.5). Although Hes1 oscillation is dispensable for ES cell pluripotency (*Hes1*-null ES cells maintain their pluripotent states in cultured ES cells), Hes1 oscillation regulates the fate-determination step

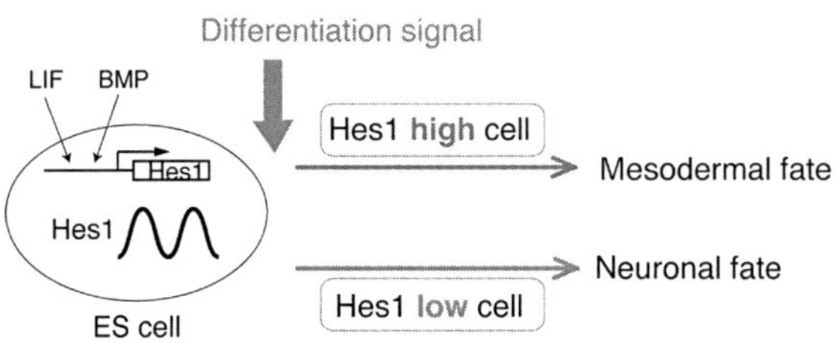

Figure 7.5 Hes1 oscillation in ES cells. Hes1 expression oscillates in embryonic stem (ES) cells via LIF and BMP stimulation. When differentiation signals activate ES cell differentiation, cells low in Hes1 protein levels differentiate into neuroectodermal cells (neuronal fate), and those high in Hes1 protein levels tend to differentiate into early mesodermal cells. (See the color plate.)

during ES cell differentiation. In ES cells, Hes1 regulates Notch signaling molecules and cell cycle-related molecules, both of which are expressed in an oscillating manner in ES cells, suggesting that Hes1 oscillation sets the heterogeneous activation of Notch signaling and the variable timing of cell cycle exit in genetically identical ES cells (Kobayashi & Kageyama, 2010b). Other fluctuating molecules, including Nanog, Rex1, and Stella, have been identified in ES cells, so they may also contribute more diverse responses in cells to produce various kinds of cells at once even in the same environment (Chambers et al., 2007; Hayashi, Lopes, Tang, & Surani, 2008; Kobayashi & Kageyama, 2011; Toyooka, Shimosato, Murakami, Takahashi, & Niwa, 2008). One attractive hypothesis is that such fluctuating and oscillatory expression of genes is a common feature of stem cells to endow them with their specific properties (Chang, Hemberg, Barahona, Ingber, & Huang, 2008; Furusawa & Kaneko, 2001, 2012).

7.2. Hes1 oscillation in developing mouse brain

Hes1 expression oscillates dynamically in NS cells both in mouse embryonic brain and in *in vitro* cultured mouse NS cells with a periodicity of about 2–3 h (Shimojo et al., 2008). The period and amplitude are variable from cell to cell and from cycle to cycle, suggesting that this oscillation does not serve as an accurate molecular clock but rather enables various (not synchronizing) responses to be individually timed. In NS cells, Hes1 represses the expression of proneural genes, Notch signaling molecules, and cell cycle regulators. Interestingly, time-lapse imaging analyses show that both the proneural gene *Ngn2* and the gene for the Notch ligand *Delta-like 1* (*Dll1*) are expressed in an oscillatory manner in NS cells (Fig. 7.6A). These oscillations are specific to undifferentiated NS cells and are not observed in differentiating neurons. It was previously thought that *Ngn2* and *Dll1* expression did not occur in NS cells, but were specific to differentiating neurons. Ngn2 upregulates the expression of Dll1, which activates Hes1 via Notch signaling in neighboring cells to repress Ngn2 and Dll1 expression and maintain these cells in an undifferentiated state. Thus, Notch signaling is unidirectional, from neurons to NS cells, in this context. However, *Ngn2* and *Dll1* are actually expressed by NS cells in an oscillatory manner, which is probably dependent on periodic repression by Hes1 oscillation. This oscillatory expression provides another interesting strategy to maintain NS cells in their undifferentiated state. Oscillatory expression of Ngn2 activates only early-response genes such as *Dll1* without activation of the neuronal differentiation program. The oscillatory expressions of *Dll1* and *Ngn2* maintain

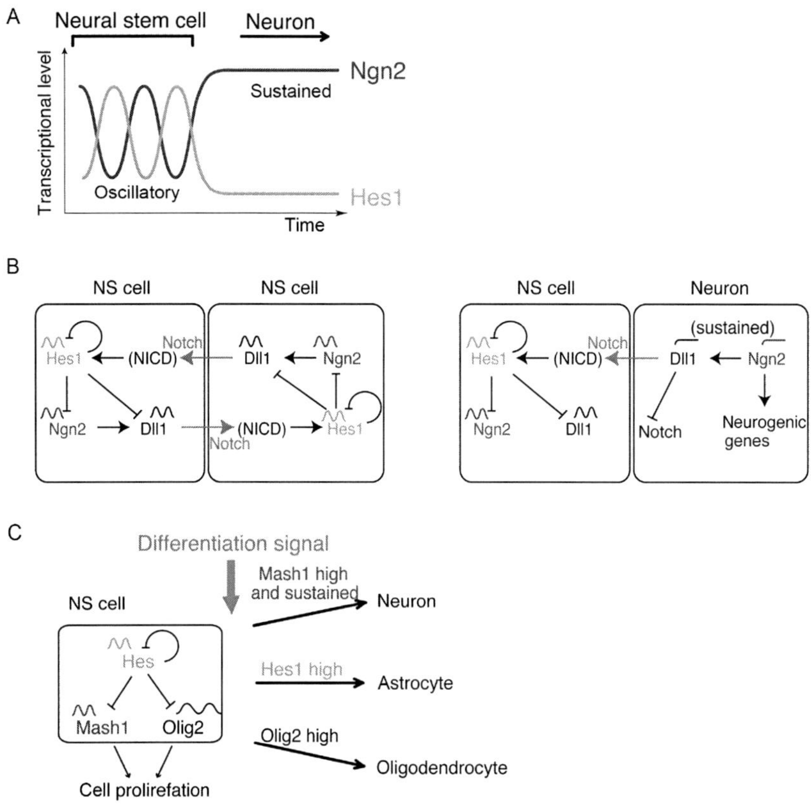

Figure 7.6 Hes1 oscillation in NS cells. (A) The levels of both Hes1 and *Ngn2*, a downstream gene of Hes1, oscillate in neural stem (NS) cells. When Hes1 expression is repressed in differentiating neurons, *Ngn2* expression becomes sustained. (B) Hes1 oscillation in NS cells induces the oscillatory expression of *Ngn2* and *Dll1* by periodic repression. *Dll1* oscillation mutually activates Notch signaling between neighboring NS cells, which maintain their proliferation without any help from neurons (left). When Hes1 is repressed, *Ngn2* and *Dll1* expression becomes sustained in differentiating neurons (right). Notch signaling is unidirectional from differentiating neurons to NS cells. Sustained Dll1 expression might prevent neurons from receiving Notch signaling by the *cis*-inhibitory function. (C) In NS cells, oscillatory expression of Hes1, Mash1, and Olig2 is important for cell proliferation and stem cell maintenance: oscillatory expression of several fate-determination factors contributes to the multipotent state of NS cells. When differentiation signals arrive, the expression of a single factor becomes sustained or high and determines the cell fate. (See the color plate.)

proliferating NS cell populations by mutual activation of Notch signaling (Fig. 7.6B, left). The expression of *Ngn2* becomes sustained when Hes1 expression is repressed, thereby promoting neuronal differentiation (Fig. 7.6B, right).

7.3. Hes1 oscillation in NS cells and the optogenetic control

NS cells are multipotent, generating neurons, oligodendrocytes, and astrocytes. NS cells, derived from the ventral telencephalon of mouse embryos, express Hes1, Mash1, and Olig2. Mash1 promotes cell cycle exit and neuronal fate determination; Olig2 regulates oligodendrocyte specification, and Hes1 regulates the maintenance and proliferation of NS cells. Importantly, these three factors are also reported to function in other events: Mash1 upregulates the gene expression for cell cycle progression (Castro et al., 2011), Olig2 functions in NS cell proliferation (Ligon et al., 2007), and Hes1 promotes astrocyte differentiation (Ohtsuka et al., 2001). Thus, each of these factors has two contradictory functions: maintenance and proliferation of NS cells and determination of a specific cell fate. Time-lapse imaging of these proteins in single NS cells revealed that their gene expression oscillates: the levels of both Hes1 and Mash1 proteins oscillate with a periodicity of 2–3 h, and that of Olig2 oscillates with a longer periodicity (5–8 h) (Imayoshi et al., 2013). Moreover, these gene expressions change dynamically when NS cells are differentiating. During neuronal differentiation, Hes1 expression is repressed before asymmetric cell division, when one NS cell produces one neuron and one NS cell. Mash1 protein gradually accumulates and expression becomes sustained in differentiating neurons, whereas Hes1 and Mash1 oscillations resume in daughter NS cells. During astrocyte differentiation, Hes1 protein continues to oscillate but the expression level becomes higher than in NS cells, and then Mash1 and Olig2 expressions become undetectable. During oligodendrocyte differentiation, Olig2 protein expression oscillates at high levels, and Mash1 and Hes1 are downregulated. These observations reveal that one of the bHLH factors, which are fluctuating in NS cells, becomes dominant and sustained during differentiation (Fig. 7.6C).

A light-switchable gene expression system is able to generate the oscillatory gene expression with short periodicity. The system makes use of the *Neurospora crassa* photoreceptor Vivid that is fused with the Gal4 DNA-binding domain and the p65 activation domain (GAVPO) (Wang, Chen, & Yang, 2012). Blue light illumination induces the formation of GAVPO homodimers, which bind to UAS sequences and activate transcription of target genes under the control of UAS sequences. A Mash1-inducible system with both Mash1 coding region and its 3′-UTR under the control of UAS sequences in *Mash1*-null NS cells reveals the importance of gene expression modes (Imayoshi et al., 2013). Repeated exposure to blue light at 3-h intervals successfully generates oscillatory expression of Mash1 with a 3-h period, which enhances the cell cycle progression of NS cells. On the other hand, repeated exposure at 30-min intervals, which generates sustained expression of Mash1, increases neuronal differentiation. These results clearly demonstrate that the differences in expression modes of these factors are important for a cell's switch between proliferation and differentiation (Imayoshi et al., 2013). The detailed molecular mechanism by which the oscillatory expression upregulates the cell cycle progression genes remains unclear.

8. CONCLUSIONS AND PERSPECTIVES

Hes factors regulate the maintenance of stem cells and the cell fate decision into multiple cell types in various tissues under multiple signaling pathways, including Notch signaling as described here. These functions are sometimes contradictory to one another. Different expression dynamics is clearly one of the biological strategies to resolve such complex regulations. We mainly focused on Hes1 oscillation in cultured cells, ES cells, and NS cells. It will be interesting to study the role of Hes factor oscillations in other (periodic) developmental processes using imaging techniques. It is also important to address the detailed molecular mechanisms maintaining Hes1 oscillation with biochemical studies and those that switch the expression modes between oscillatory and sustained expressions.

Oscillations were also reported in some signaling pathways regulating Hes factors, such as Notch, Jak–Stat, and Fgf signalings (Kageyama, Niwa, Shimojo, Kobayashi, & Ohtsuka, 2010; Niwa et al., 2007; Yoshiura et al., 2007). For example, Notch and Fgf signaling cooperatively regulate Hes7 oscillation and achieve a spatiotemporal regulation during somite segmentation events (Niwa et al., 2011). Oscillatory Hes7 expression is essential for maintaining periodic somite segmentation. Fgf signaling is important for initiating periodic Hes7 expression in the posterior end of the presomitic mesoderm, and Notch signaling synchronizes and amplifies Hes7 expression from the posterior to the anterior end, thereby inducing periodic somite segmentation at the anterior end. At the same time, Hes7 connects two signaling pathways by regulating the expression of signaling factors involved in both pathways, leading to coupled oscillation of the two pathways. Such oscillatory coupling between multiple signals via Hes factors may be important for other kinds of tissue development as well.

In mouse embryos, Hes7 expression oscillates with a 2-h cycle, and somite segmentation events also occur with a 2-h cycle. It is known that the period of somite segmentation is different among different species: 30 min in zebrafish, 90 min in chick, and 6–8 h in human. In the case of Hes1 oscillation in cultured cells, the period is variable between cell types derived from the same species. It is 2–3 h in NS cells and 3–5 h in ES cells. It is not known how different cells or cell types set the lengths of these cycles and what the importance of the differential cycle lengths may be. Reconstitution and regulation of the oscillatory expressions using light-inducible systems such as the GAVPO system is an attractive tool not only to reveal the function of oscillation itself but also to understand the significance of the oscillation period.

REFERENCES

Akazawa, C., Sasai, Y., Nakanishi, S., & Kageyama, R. (1992). Molecular characterization of a rat negative regulator with a basic helix-loop-helix structure predominantly expressed in the developing nervous system. *The Journal of Biological Chemistry, 267*, 21879–21885.

Bae, S.-K., Bessho, Y., Hojo, M., & Kageyama, R. (2000). The bHLH gene Hes6, an inhibitor of Hes1, promotes neuronal differentiation. *Development, 127*, 2933–2943.

Baek, J. H., Hatakeyama, J., Sakamoto, S., Ohtsuka, T., & Kageyama, R. (2006). Persistent and high levels of Hes1 expression regulate boundary formation in the developing central nervous system. *Development, 133*, 2467–2476.

Bessho, Y., Miyoshi, G., Sakata, R., & Kageyama, R. (2001). Hes7: A bHLH-type repressor gene regulated by Notch and expressed in the presomitic mesoderm. *Genes to Cells, 6*, 175–185.

Bessho, Y., Sakata, R., Komatsu, S., Shiota, K., Yamada, S., & Kageyama, R. (2001). Dynamic expression and essential functions of Hes7 in somite segmentation. *Genes & Development, 15*, 2642–2647.

Bonev, B., Stanley, P., & Papalopulu, N. (2012). MicroRNA-9 modulates Hes1 ultradian oscillations by forming a double-negative feedback loop. *Cell Reports, 2*, 10–18.

Buscarlet, M., Perin, A., Laing, A., Brickman, J. M., & Stifani, S. (2008). Inhibition of cortical neuron differentiation by Groucho/TLE1 requires interaction with WRPW, but not Eh1, repressor peptides. *The Journal of Biological Chemistry, 283*, 24881–24888.

Castella, P., Sawai, S., Nakao, K., Wagner, J. A., & Caudy, M. (2000). HES-1 repression of differentiation and proliferation in PC12 cells: Role for the helix 3-helix 4 domain in transcription repression. *Molecular and Cellular Biology, 20*, 6170–6183.

Castro, D. S., Martynoga, B., Parras, C., Ramesh, V., Pacary, E., Johnston, C., et al. (2011). A novel function of the proneural factor Ascl1 in progenitor proliferation identified by genome-wide characterization of its targets. *Genes & Development, 25*, 930–945.

Chambers, I., Silva, J., Colby, D., Nichols, J., Nijmeijer, B., Robertson, M., et al. (2007). Nanog safeguards pluripotency and mediates germline development. *Nature, 450*, 1230–1234.

Chang, H. H., Hemberg, M., Barahona, M., Ingber, D. E., & Huang, S. (2008). Transcriptome-wide noise controls lineage choice in mammalian progenitor cells. *Nature, 453*, 544–547.

Espinosa, L., Cathelin, S., D'Altri, T., Trimarchi, T., Statnikov, A., & Guiu, J. (2010). The Notch/Hes1 pathway sustains NF-kB activation through CYLD repression in T cell leukemia. *Cancer Cell, 18*, 268–281.

Fischer, A., & Gessler, M. (2007). Delta-Notch—And then? Protein interactions and proposed modes of repression by Hes and Hey bHLH factors. *Nucleic Acids Research*, *35*, 4583–4596.

Fukuda, A., Kawaguchi, Y., Furuyama, K., Kodama, S., Horiguchi, M., Kuhara, T., et al. (2006). Ectopic pancreas formation in Hes1-knockout mice reveals plasticity of endodermal progenitors of the gut, bile duct, and pancreas. *The Journal of Clinical Investigation*, *116*, 1484–1493.

Furusawa, C., & Kaneko, K. (2001). Theory of robustness of irreversible differentiation in a stem cell system: Chaos hypothesis. *Journal of Theoretical Biology*, *209*, 395–416.

Furusawa, C., & Kaneko, K. (2012). A dynamical-systems view of stem cell biology. *Science*, *338*, 215–217.

Georgia, S., Soliz, R., Li, M., Zhang, P., & Bhushana, A. (2006). p57 and Hes1 coordinate cell cycle exit with self-renewal of pancreatic progenitors. *Developmental Biology*, *298*, 22–31.

Guiu, J., Shimizu, R., D'Altri, T., Fraser, S. T., Hatakeyama, J., Bresnick, E. H., et al. (2013). Hes repressors are essential regulators of hematopoietic stem cell development downstream of Notch signaling. *The Journal of Experimental Medicine*, *210*, 71–84.

Hartman, J., Müller, P., Foster, J. S., Wimalasena, J., Gustafsson, J.-A., & Ström, A. (2004). HES-1 inhibits 17beta-estradiol and heregulin-beta1-mediated upregulation of E2F-1. *Oncogene*, *23*, 8826–8833.

Hatakeyama, J., Bessho, Y., Katoh, K., Ookawara, S., Fujioka, M., Guillemot, F., et al. (2004). Hes genes regulate size, shape and histogenesis of the nervous system by control of the timing of neural stem cell differentiation. *Development*, *131*, 5539–5550.

Hayashi, K., Lopes, S. M. C. d. S., Tang, F., & Surani, M. A. (2008). Dynamic equilibrium and heterogeneity of mouse pluripotent stem cells with distinct functional and epigenetic states. *Cell Stem Cell*, *3*, 391–401.

Hirata, H., Ohtsuka, T., Bessho, Y., & Kageyama, R. (2000). Generation of structurally and functionally distinct factors from the basic helix-loop-helix gene Hes3 by alternative first exons. *The Journal of Biological Chemistry*, *275*, 19083–19089.

Hirata, H., Tomita, K., Bessho, Y., & Kageyama, R. (2001). Hes1 and Hes3 regulate maintenance of the isthmic organizer and development of the mid/hindbrain. *The EMBO Journal*, *20*, 4454–4466.

Hirata, H., Yoshiura, S., Ohtsuka, T., Bessho, Y., Harada, T., Yoshikawa, K., et al. (2002). Oscillatory expression of the bHLH factor Hes1 regulated by a negative feedback loop. *Science*, *298*, 840–843.

Imayoshi, I., Isomura, A., Harima, Y., Kawaguchi, K., Kori, H., Miyachi, H., et al. (2013). Oscillatory control of factors determining multipotency and fate in mouse neural progenitors. *Science*, *342*, 1203–1208.

Iso, T., Sartorelli, V., Chung, G., Shichinohe, T., Kedes, L., & Hamamori, Y. (2001). HERP, a new primary target of Notch regulated by ligand binding. *Molecular and Cellular Biology*, *21*, 6071–6079.

Jensen, J., Pedersen, E. E., Galante, P., Hald, J., Heller, R. S., Ishibashi, M., et al. (2000). Control of endodermal endocrine development by Hes-1. *Nature Genetics*, *24*, 36–44.

Kageyama, R., Hatakeyama, J., & Ohtsuka, T. (2006). Roles of Hes bHLH factors in neural development. In G. Thiel (Ed.), *Transcription factors in the nervous system* (pp. 3–22). Weinheim: Wiley-VDH.

Kageyama, R., Niwa, Y., Shimojo, H., Kobayashi, T., & Ohtsuka, T. (2010). Ultradian oscillations in Notch signaling regulate dynamic biological events. *Current Topics in Developmental Biology*, *92*, 311–331.

Kageyama, R., Ohtsuka, T., & Kobayashi, T. (2007). The Hes gene family: Repressors and oscillators that orchestrate embryogenesis. *Development*, *134*, 1243–1251.

Katoh, M., & Katoh, M. (2007). Integrative genomic analyses on HES/HEY family: Notch-independent HES1, HES3 transcription in undifferentiated ES cells, and

Notch-dependent HES1, HES5, HEY1, HEY2, HEYL transcription in fetal tissues, adult tissues, or cancer. *International Journal of Oncology, 31*, 461–466.

Kobayashi, T., & Kageyama, R. (2010a). Hes1 oscillation: Making variable choices for stem cell differentiation. *Cell Cycle, 9*, 207–208.

Kobayashi, T., & Kageyama, R. (2010b). Hes1 regulates embryonic stem cell differentiation by suppressing Notch signaling. *Genes to Cells, 15*, 689–698.

Kobayashi, T., & Kageyama, R. (2011). Hes1 oscillations contribute to heterogeneous differentiation responses in embryonic stem cells. *Genes, 2*, 219–228.

Kobayashi, T., Mizuno, H., Imayoshi, I., Furusawa, C., Shirahige, K., & Kageyama, R. (2009). The cyclic gene Hes1 contributes to diverse differentiation responses of embryonic stem cells. *Genes & Development, 23*, 1870–1875.

Ligon, K. L., Huillard, E., Mehta, S., Kesari, S., Liu, H., Alberta, J. A., et al. (2007). Olig2-regulated lineage-restricted pathway controls replication competence in neural stem cells and malignant glioma. *Neuron, 53*, 503–517.

Masamizu, Y., Ohtsuka, T., Takashima, Y., Nagahara, H., Takenaka, Y., Yoshikawa, K., et al. (2006). Real-time imaging of the somite segmentation clock: Revelation of unstable oscillators in the individual presomitic mesoderm cells. *Proceedings of the National Academy of Sciences of the United States of America, 103*, 1313–1318.

Murata, K., Hattori, M., Hirai, N., Shinozuka, Y., Hirata, H., Kageyama, R., et al. (2005). Hes1 directly controls cell proliferation through the transcriptional repression of p27Kip1. *Molecular and Cellular Biology, 25*, 4262–4271.

Nakayama, K., Satoh, T., Igari, A., Kageyama, R., & Nishida, E. (2008). FGF induces oscillations of Hes1 expression and Ras/ERK activation. *Current Biology, 18*, R332–R334.

Niwa, Y., Masamizu, Y., Liu, T., Nakayama, R., Deng, C.-X., & Kageyama, R. (2007). The Initiation and propagation of Hes7 oscillation are cooperatively regulated by Fgf and Notch signaling in the somite segmentation clock. *Developmental Cell, 13*, 298–304.

Niwa, Y., Shimojo, H., Isomura, A., González, A., Miyachi, H., & Kageyama, R. (2011). Different types of oscillations in Notch and Fgf signaling regulate the spatiotemporal periodicity of somitogenesis. *Genes & Development, 25*, 1115–1120.

Ohtsuka, T., Ishibashi, M., Gradwohl, G., Nakanishi, S., Guillemot, F., & Kageyama, R. (1999). Hes1 and Hes5 as Notch effectors in mammalian neuronal differentiation. *The EMBO Journal, 18*, 2196–2207.

Ohtsuka, T., Sakamoto, M., Guillemot, F., & Kageyama, R. (2001). Roles of the basic helix-loop-helix genes Hes1 and Hes5 in expansion of neural stem cells of the developing brain. *The Journal of Biological Chemistry, 276*, 30467–30474.

Pourquié, O. (2011). Vertebrate segmentation: From cyclic gene networks to scoliosis. *Cell, 145*, 650–663.

Sang, L., Coller, H. A., & Roberts, J. M. (2008). Control of the reversibility of cellular quiescence by the transcriptional repressor Hes1. *Science, 321*, 1095–1100.

Sang, L., Roberts, J. M., & Coller, H. A. (2010). Hijacking HES1: How tumors co-opt the anti-differentiation strategies of quiescent cells. *Trends in Molecular Medicine, 16*, 17–26.

Sasai, Y., Kageyama, R., Tagawa, Y., Shigemoto, R., & Nakanishi, S. (1992). Two mammalian helix-loop-helix factors structurally related to Drosophila hairy and Enhancer of split. *Genes & Development, 6*, 2620–2634.

Shimojo, H., Ohtsuka, T., & Kageyama, R. (2008). Oscillations in Notch signaling regulate maintenance of neural progenitors. *Neuron, 58*, 52–64.

Smith, A. G. (2001). Embryo-derived stem cells: Of mice and men. *Annual Review of Cell and Developmental Biology, 17*, 435–462.

Sparrow, Duncan B., Chapman, G., Smith, A. J., Mattar, M. Z., Major, J. A., & O'Reilly, V. C. (2012). A mechanism for gene-environment interaction in the etiology of congenital scoliosis. *Cell, 149*, 295–306.

Sparrow, D. B., GuillŽn-Navarro, E., Fatkin, D., & Dunwoodie, S. L. (2008). Mutation of HAIRY-AND-ENHANCER-OF-SPLIT-7 in humans causes spondylocostal dysostosis. *Human Molecular Genetics*, *17*, 3761–3766.
Sumazaki, R., Shiojiri, N., Isoyama, S., Masu, M., Keino-Masu, K., Osawa, M., et al. (2004). Conversion of biliary system to pancreatic tissue in Hes1-deficient mice. *Nature Genetics*, *36*, 83–87.
Takata, T., & Ishikawa, F. (2003). Human Sir2-related protein SIRT1 associates with the bHLH repressors HES1 and HEY2 and is involved in HES1- and HEY2-mediated transcriptional repression. *Biochemical and Biophysical Research Communications*, *301*, 250–257.
Tan, S.-L., Ohtsuka, T., González, A., & Kageyama, R. (2012). MicroRNA9 regulates neural stem cell differentiation by controlling Hes1 expression dynamics in the developing brain. *Genes to Cells*, *17*, 952–961.
Tateya, T., Imayoshi, I., Tateya, I., Ito, J., & Kageyama, R. (2011). Cooperative functions of Hes/Hey genes in auditory hair cell and supporting cell development. *Developmental Biology*, *352*, 329–340.
Tomita, K., Hattori, M., Nakamura, E., Nakanishi, S., Minato, N., & Kageyama, R. (1999). The bHLH gene Hes1 is essential for expansion of early T cell precursors. *Genes & Development*, *13*, 1203–1210.
Toyooka, Y., Shimosato, D., Murakami, K., Takahashi, K., & Niwa, H. (2008). Identification and characterization of subpopulations in undifferentiated ES cell culture. *Development*, *135*, 909–918.
Ueo, T., Imayoshi, I., Kobayashi, T., Ohtsuka, T., Seno, H., Nakase, H., et al. (2012). The role of Hes genes in intestinal development, homeostasis and tumor formation. *Development*, *139*, 1071–1082.
Wall, D. S., Mears, A. J., McNeill, B., Mazerolle, C., Thurig, S., Wang, Y., et al. (2009). Progenitor cell proliferation in the retina is dependent on Notch-independent Sonic hedgehog/Hes1 activity. *The Journal of Cell Biology*, *184*, 101–112.
Wang, X., Chen, X., & Yang, Y. (2012). Spatiotemporal control of gene expression by a light-switchable transgene system. *Nature Methods*, *9*, 266–269.
Wang, X., Fu, Y., Chen, X., Ye, J., Lü, B., Lü, F. Y. W., et al. (2010). The expressions of bHLH gene HES1 and HES5 in advanced ovarian serous adenocarcinomas and their prognostic significance: A retrospective clinical study. *Journal of Cancer Research and Clinical Oncology*, *136*, 989–996.
Wendorff, A. A., Koch, U., Wunderlich, F. T., Wirth, S., Dubey, C., & Brüning, J. C. (2010). Hes1 is a critical but context-dependent mediator of canonical Notch signaling in lymphocyte development and transformation. *Immunity*, *33*, 671–684.
Yoshiura, S., Ohtsuka, T., Takenaka, Y., Nagahara, H., Yoshikawa, K., & Kageyama, R. (2007). Ultradian oscillations of Stat, Smad, and Hes1 expression in response to serum. *Proceedings of the National Academy of Sciences of the United States of America*, *104*, 11292–11297.

CHAPTER EIGHT

Hey bHLH Transcription Factors

David Weber*, Cornelia Wiese*, Manfred Gessler*,†,1

*Developmental Biochemistry, Theodor-Boveri-Institute/Biocenter, Wuerzburg University, Wuerzburg, Germany

†Comprehensive Cancer Center Mainfranken, Wuerzburg University, Wuerzburg, Germany

1Corresponding author: e-mail address: gessler@biozentrum.uni-wuerzburg.de

Contents

Abstract

Hey bHLH transcription factors are direct targets of canonical Notch signaling. The three mammalian Hey proteins are closely related to Hes proteins and they primarily repress target genes by either directly binding to core promoters or by inhibiting other transcriptional activators. Individual candidate gene approaches and systematic screens identified a number of Hey target genes, which often encode other transcription factors involved in various developmental processes. Here, we review data on interaction partners and target genes and conclude with a model for Hey target gene regulation. Furthermore, we discuss how expression of Hey proteins affects processes like cell fate

Current Topics in Developmental Biology, Volume 110
ISSN 0070-2153
http://dx.doi.org/10.1016/B978-0-12-405943-6.00008-7

© 2014 Elsevier Inc.
All rights reserved.

decisions and differentiation, e.g., in cardiovascular, skeletal, and neural development or oncogenesis and how this relates to the observed developmental defects and phenotypes observed in various knockout mice.

1. INTRODUCTION

The vertebrate Hey proteins belong to the class E basic helix-loop-helix (bHLH) transcription factors and are closely related to the Hes family of proteins and the *Drosophila hairy* and *Enhancer of split* genes. They are integral members of the Notch signaling pathway, a central signaling system in embryonic development, differentiation processes, and later tissue homeostasis (Fig. 8.1A).

Three mammalian Hey proteins Hey1, Hey2, and HeyL (also known as Hrt1/2/3, Hesr1/2/3, Herp2/1/3, or Chf2/1/3) have been identified (Chin et al., 2000; Iso, Sartorelli, Chung, et al., 2001; Kokubo, Lun, &

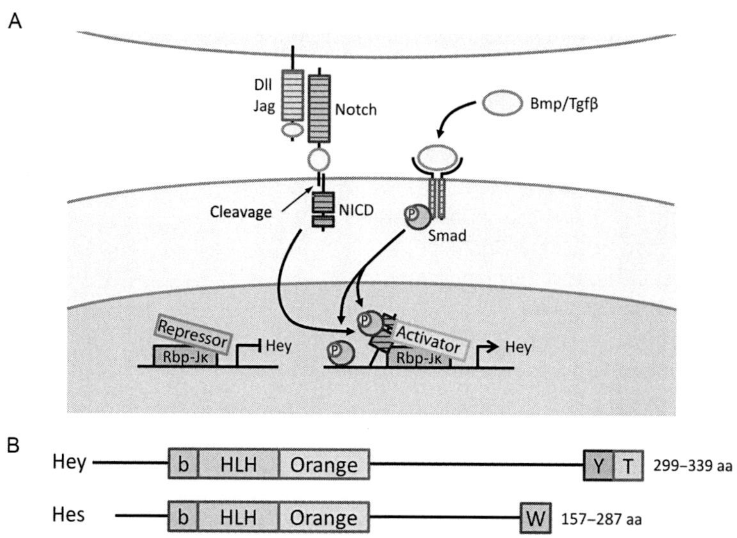

Figure 8.1 Hey genes are targets of canonical Notch signaling. (A) Ligands of the Delta (Dll) or Jagged (Jag) family induce intramembrane cleavage of the Notch receptor. The Notch intracellular domain (NICD) replaces transcriptional corepressors with activators in the Rbp-Jκ complex enabling transcription of Hey genes. Hey transcription can also be induced by Bmp/Tgfβ signaling: Bmp or Tgfβ ligands bind to their receptors leading to a phosphorylation of Smad proteins, which either directly bind to the Hey promoter or interact with the NICD. (B) Domain organization of Hey and Hes proteins. Amino acid (aa) numbers indicate the length of the proteins. (See the color plate.)

Johnson, 1999; Leimeister, Externbrink, Klamt, & Gessler, 1999; Nakagawa et al., 2000; Sartorelli et al., 1999), which are directly induced by Notch signaling and function as transcriptional repressors (Iso, Sartorelli, Chung, et al., 2001; Leimeister, Schumacher, Steidl, & Gessler, 2000; Maier & Gessler, 2000; Nakagawa et al., 2000). Hey proteins are conserved throughout vertebrate evolution (Winkler, Elmasri, Klamt, Volff, & Gessler, 2003) and there is also a single Hey *Drosophila* homolog (Kokubo et al., 1999; Leimeister et al., 1999; Monastirioti et al., 2010). Common features are a basic, a HLH, and an Orange domain that are similar to those of Hes proteins and two conserved C-terminal motifs (Fig. 8.1B). The basic domain directly contacts DNA and preferentially binds E-box sequences (CACGTG, CACGCG) (Heisig et al., 2012; Iso, Sartorelli, Poizat, et al., 2001; van den Berg et al., 2003). The same sequence is bound by Hes proteins, despite the exchange of a conserved and functionally relevant proline residue within the basic domain of Hes proteins to glycine in Hey proteins, which might affect DNA binding (Iso, Sartorelli, Poizat, et al., 2001; Popovic et al., 2014). The HLH domain serves as a platform for homo- and heterodimerization with Hey and Hes proteins as well as for other protein interactions (Iso, Chung, Hamamori, & Kedes, 2002; Leimeister, Dale, et al., 2000). Two further α-helices form the Orange domain, which functions as an additional interface for protein interactions. These are followed by conserved YRPW (YXXW for HeyL) and GTE (I/V) GAF peptides at the C-terminus. Contrary to the C-terminal WRPW peptide of Hes proteins, YRPW cannot bind TLE corepressors (Monastirioti et al., 2010; van den Berg et al., 2003). The function of both C-terminal Hey motifs is currently unknown.

In the following, we describe the pathways that are involved in the regulation of Hey expression and summarize currently known Hey interaction partners, crosstalk with other signaling pathways, and downstream targets in different cell types. We then discuss the role of Hey proteins in different developmental processes, such as cardiac development, angiogenesis, bone formation, and neural development as well as in tumor formation.

2. NOTCH SIGNALING AND CROSSTALK WITH OTHER PATHWAYS

Hey and *Hes* genes are the most prominent targets of canonical Notch signaling in vertebrates and they often convey a significant part of the Notch signal. They are activated upon juxtacrine binding of Jagged or Delta-like ligands (Jag1, Jag2, Dll1, Dll3, and Dll4) to the extracellular domain of Notch receptors (Notch1–4) (Fig. 8.1A). This interaction leads to two consecutive proteolytic cleavages, releasing the Notch intracellular domain (NICD), which then translocates into the nucleus and interacts with the DNA-binding protein Rbp-Jκ. It displaces corepressors (e.g., NcoR, Sharp, CtBP) and allows recruitment of coactivators (e.g., Mastermind/MAML, p300/CBP) converting the Rbp-Jκ complex from a repressor to an activator.

For *Hey* genes, a differential expression was seen in several Notch pathway mutants : While *Notch1* deletion leads to a loss of *HeyL* expression in somites or the endocardium, overexpression of constitutively active Notch1 (N1ICD) in the mesodermal lineage leads to induction of *Hey1* (Firulli, Hadzic, McDaid, & Firulli, 2000; Leimeister, Schumacher, et al., 2000; Watanabe et al., 2006). However, similar deletion and overexpression experiments with Notch2 (N2ICD) in the heart or Notch3 in vascular smooth muscle cells did not identify obvious alterations in *Hey* gene expression, suggesting a Notch2/3 independent expression in other instances (King et al., 2006; Kokubo, Tomita-Miyagawa, Hamada, & Saga, 2007). Reporter gene assays and mutation analysis of *Hey* promoters clearly supported the direct mode of activation via Rbp-Jκ (Iso, Sartorelli, Chung, et al., 2001; Iso et al., 2002; Maier & Gessler, 2000). In these experiments, there is little evidence for selectivity of certain Notch receptors toward an individual *Hey* gene leaving the question of specificity unanswered.

In addition to activation by Notch, there is crosstalk with other signaling pathways to enhance *Hey* transcription: Jak2–Stat3 and Jak2–Erk1/2 signaling has been implicated in enhancing Notch activity in neuroepithelia cells (Kamakura et al., 2004). Foxc1 and Foxc2 directly induce *Hey2* in endothelial cells via Foxc-binding elements within the *Hey2* promoter and via interaction with NICD (Hayashi & Kume, 2008). In zebrafish, the *Hey2* ortholog *gridlock* is induced by Sox7 and Sox18 during arteriovenous differentiation (Pendeville et al., 2008). However, most data are available on Tgfβ/Bmp signaling and regulation by the *Hey* antagonist CoupTF-II.

2.1. Tgfβ/Bmp signaling induces *Hey* genes

Hey induction by Tgfβ/Bmp seems to occur through two different mechanisms (Fig. 8.1A): On one hand, activation by Bmp is driven by direct binding of Smad proteins to *Hey* promoters and was still observed when activation of Notch receptors was abrogated by chemical inhibition of γ-secretase activity (Sharff et al., 2009; Zavadil, Cermak, Soto-Nieves, & Bottinger, 2004). On the other hand, activation can be enhanced or even depend on interaction of NICD with Smad1/5 (Dahlqvist et al., 2003; Itoh et al., 2004; Zavadil et al., 2004). Notch-dependent activation was also observed by Tgfβ-induced Smad3 (Blokzijl et al., 2003).

Tgfβ/Bmp and Notch signaling tightly interact to control several cellular differentiation processes: during osteoblast differentiation, Hey proteins are activated by Bmp2 (de Jong, Steegenga, et al., 2004; Liu et al., 2007), Bmp7 (Lavery et al., 2009), and Bmp9 (Sharff et al., 2009). Tgfβ counteracted activation by Bmp2 in this case (de Jong, Steegenga, et al., 2004). Bmp signaling is conveyed by Smad1/5/8, which together with Smad4 directly bind to the *Hey1* promoter, e.g., upon Bmp9 induction (Sharff et al., 2009). In myogenic differentiation, *Hey1* as well as *Hes1* are induced by Bmp4 via Smad1 activation and binding to the NICD (Dahlqvist et al., 2003).

Hey proteins were shown to mediate Tgfβ-induced epithelial–mesenchymal transition (EMT) in keratinocytes. In this case, initial activation of *Hey1* was Notch-independent driven by Smad3/4 binding to the *Hey1* promoter, while a prolonged induction required a direct interaction of NICD with Smad3 (Zavadil et al., 2004). In endothelia cells, *Hey1* is similarly induced in a Notch-dependent manner by Bmp9 and Bmp10 through Smad1/5 binding to the NICD to control cell migration and sprouting angiogenesis (Itoh et al., 2004; Larrivee et al., 2012; Moya et al., 2012; Ricard et al., 2012).

2.2. CoupTF-II and Hif1 control cardiovascular *Hey* expression

CoupTF-II (Nr2f2) is another key regulator of *Hey* expression. It represses *Hey2* in atrial cardiomyocytes (Wu et al., 2013). In venous endothelial cells, CoupTF-II homodimers directly repress *Hey1* and *Hey2* contributing to the dichotomy of arterial versus venous fate decisions (Aranguren, Beerens, et al., 2013; Diez et al., 2007; Korten et al., 2013; Popovic et al., 2014). In lymphatic endothelia cells, however, Hey1/2 and CoupTF-II are coexpressed. This appears to be facilitated by Prox1 forming heterodimers with CoupTF-II, which lack repressive capacity toward *Hey1* and *Hey2* (Korten et al., 2013; Yoo et al., 2012).

The strong expression of Hey proteins in sprouting angiogenesis suggests that they are part of the hypoxia response. Indeed, hypoxia-sensing Hif1 has been shown to induce *Hey* expression directly and via induction of *Vegf*,

Dll4, and *Jag2* (Diez et al., 2007; Patel et al., 2005; Pietras, von Stedingk, Lindgren, Pahlman, & Axelson, 2011; Williams, Li, Murga, Harris, & Tosato, 2006). These data clearly show that besides conveying Notch signals *Hey* genes are under the control of multiple other signaling pathways and seem to integrate these into a unified cellular response.

3. INTERACTION PARTNERS OF HEY PROTEINS

Hey proteins form homo- and heterodimers via the HLH domain with Hey and Hes proteins as well as several other HLH factors. They can also interact with additional transcription factors and they bind transcriptional corepressors like histone deacetylases, which may mediate the repressive capacity of Hey proteins. Table 8.1 gives an overview of known Hey-interacting proteins.

3.1. Interaction with cofactors

Hey1 has been shown to directly interact with NcoR and Sin3a *in vitro* via the bHLH domain. Both are part of the Sin3 corepressor complex, which recruits histone deacetylase-1 (Hdac1) (Gould, Harrison, Hewitt, & Whitehouse, 2009; Iso, Sartorelli, Poizat, et al., 2001). For the Hey2 bHLH domain, a direct interaction has been described with Sirt1, a member of another histone deacetylase family (Takata & Ishikawa, 2003). Trichostatin A treatment that inhibits Hdacs but not Sirt1 partially blocks Hey-mediated repression (Nakagawa et al., 2000; Takata & Ishikawa, 2003). These results suggest that Hey proteins repress their target promoters via recruitment of different histone deacetylases.

3.2. Interaction with other transcription factors

There are various examples of Hey-interacting transcription factors where Hey proteins tend to abolish the activating capabilities of these factors (Table 8.1). All three Hey proteins were shown to bind the androgen receptor and its coactivator Src1 and to repress transcription from androgen-dependent promoters, possibly via recruitment of trichostatin-insensitive deacetylases such as Sirt1 (Belandia et al., 2005; Lavery et al., 2011).

Hey proteins also interact with Gata factors to inhibit transcriptional activation of, e.g., the *Itga2b* promoter by Gata1/2 (Elagib et al., 2004), the *Anf* promoter by Gata4/6 (Fischer et al., 2005; Kathiriya et al., 2004), and the *Myh11* promoter by Gata6 (Shirvani, Xiang, Koibuchi, & Chin, 2006).

Table 8.1 Hey interaction partners

Interaction partner	Hey protein	Comment	References
Cofactors			
Sin3a, NcoR	Hey1,2	Augments Hey repression, Hdac1 recruitment	Iso, Sartorelli, Poizat, et al. (2001)
Sirt1	Hey2	Augments Hey repression	Takata and Ishikawa (2003)
Other HLH factors			
Arnt	Hey1/2	Repression of Arnt/Epas induction of *Vegf*	Chin et al. (2000)
Hand1/2	Hey1/2/L		Firulli et al. (2000)
Id1	Hey1	Reduced half-life of Id1	Itoh et al. (2004)
Id4	Hey2	Releases Hes1 from Hey2–Hes1 complex	Tokuzawa et al. (2010)
E2A/E2-2	Hey1/2		Leimeister, Dale, et al. (2000)
MyoD	Hey1/2	Repression of transcriptional activity	Sun et al. (2001)
Ptf1-p48	Hey1/2	Repression of transcriptional activity	Ghosh and Leach (2006)
Other TFs			
Ar/Src1	Hey1	Repression of Ar/Src1-induced gene expression	Belandia et al. (2005)
Crebzf, p53	Hey1	Activation of p53 activity	Lopez-Mateo, Villaronga, Llanos, and Belandia (2012)
Gata1,2	Hey1,2	Repression of transcriptional activity	Elagib et al. (2004)

Continued

Table 8.1 Hey interaction partners—cont'd

Interaction partner	Hey protein	Comment	References
Gata4,6	Hey1,2,L	Repression of transcriptional activity	Fischer et al. (2005), Kathiriya et al. (2004), Shirvani et al. (2006), Xiang et al. (2006)
Rbp-Jκ	Hey1,2,L	Repression of transcriptional activity	King et al. (2006)
Runx2	Hey2	Repression of transcriptional activity	Garg et al. (2005), Zamurovic et al. (2004)
Srf	Hey2	Prevents Srf interaction with CArG box	Doi et al. (2005), Proweller, Pear, and Parmacek (2005)
Stat3	Hey1,2	Enhances transcriptional activity	Kamakura et al. (2004)
Others			
Boip	Hey1	No interaction with Hey2	Van Wayenbergh et al. (2003)
KSHV Rta	Hey1	Hey ubiquitination and proteasomal degradation	Gould et al. (2009)

Hey1 also interacts with MyoD preventing MyoD target gene activation to inhibit normal muscle cell differentiation (Sun et al., 2001). In a similar way, Hey1 binding of Srf interferes with Srf-CArG box binding, preventing Srf-dependent target gene activation (Doi et al., 2005).

These interactions indicate that Hey proteins can function as corepressors, although the precise mechanism has not always been elucidated. Most of these factors require acetylation to be fully functional: MyoD is only active in its acetylated form and similarly, acetylation of Gata proteins and the androgen receptor were shown to be important for DNA binding and target gene regulation (Fu et al., 2003; Hayakawa et al., 2004; Mal, Sturniolo, Schiltz, Ghosh, & Harter, 2001; Ozawa et al., 2001). As histone deacetylases not only act on histones, but also other proteins including MyoD, Gata factors, and the androgen receptor, it is conceivable that

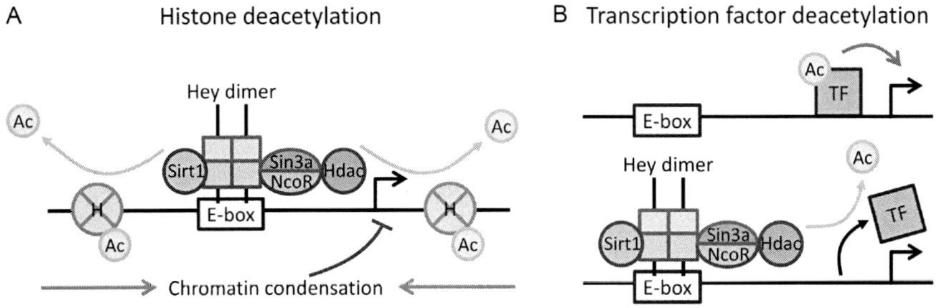

Figure 8.2 Model for Hey target gene regulation: (A) Hey dimers bound to target sequences (E-box) interact with histone deacetylases either directly (Sirt1) or indirectly (Hdac) through the Sin3a/NcoR complex. This leads to deacetylation of histones (H) inducing chromatin condensation and a reduction of transcription. (B) Hey proteins with attached Sirt1/Hdac may deacetylate other transcriptional activators either bound in proximity to Hey or via direct protein–protein interaction and abolish their activating capabilities, thereby reducing transcription. Alternatively, inactive heteromeric complexes may be formed. (See the color plate.)

Hey proteins may interfere with the activating capabilities of these factors via recruitment of histone deacetylases.

Based on the protein interactions described earlier, a dual mode of repression of target genes can be envisaged: (A) directly by E-box-mediated DNA binding and histone deacetylation or (B) indirectly by interfering with activating transcription factors, either by facilitating their deacetylation or through formation of inactive heteromeric complexes (Fig. 8.2).

4. DOWNSTREAM TARGETS OF HEY PROTEINS

Most data for Hey-regulated genes comes from individual target gene approaches. In the cardiovascular system, Hey proteins were shown to directly repress, e.g., *Gata4/6* and *Tbx2* (Fischer et al., 2005; Kokubo et al., 2007). Several papers describe Hey repression of *Nfatc1* and *Runx2*, thereby inhibiting osteoblast differentiation (Jing et al., 2010; Salie et al., 2010; Tu et al., 2012). For a more comprehensive overview of such examples, see Fischer and Gessler (2007).

There have been few genome and transcriptome wide screens to identify Hey target genes. A first screen was performed in human umbilical vein endothelial cells (HUVECs) overexpressing Hey2, where Hey2 was found to direct an artery-specific gene expression profile (Chi et al., 2003). More recently, however, it was shown that Hey2 overexpression alone is not sufficient to induce an arterial program, but needs additional cooperating factors (Aranguren, Agirre, et al., 2013).

Hey target genes were also sought in smooth muscle cells, where smooth muscle α-actin was found to be repressed by Hey1 and Hey2 (Tang, Urs, & Liaw, 2008). Time-course analysis of platelet-derived growth factor (PDGF)-stimulated *Hey2*$^{-/-}$ and wild-type mouse aortic smooth muscle cells identified a surprisingly large number of 9827 transcripts with altered expression, which makes it difficult to identify meaningful targets (Shirvani, Mookanamparambil, Ramoni, & Chin, 2007). Another screen used primary neonatal cardiomyocytes from wild-type and transgenic mice overexpressing Hey2, the latter being protected from phenylephrine-induced hypertrophy. Here, genes involved in phosphatase and adenylate cyclase activity, fluid transport, and autophagy were found to be repressed by Hey2 (Yu et al., 2010).

Finally, a screen with inducible overexpression of Hey1 or Hey2 in HEK293 cells identified a total of 507 Hey1- and 2313 Hey2-regulated genes (Heisig et al., 2012). Hey proteins were found to modulate gene expression levels to a rather limited extent, but with strong overlap between Hey factors. Target genes were mainly other transcription factors involved in developmental processes including skeletal, neural, and vascular development (Fig. 8.3). Chromatin immunoprecipitation revealed a much greater number of largely overlapping Hey1- and Hey2-binding sites (Heisig et al., 2012). These cluster around transcription start sites, especially of downregulated genes, supporting the notion that Hey proteins primarily act as repressors. Canonical E-box motifs were identified as preferred, but not essential binding sites. These data establish Hey proteins as highly redundant transcriptional repressors that act directly on target promoters, although additional functions through other transcription factors certainly exist.

5. HEY PROTEINS IN HEART DEVELOPMENT AND DISEASE

The broad and dynamic expression of *Hey* genes, controlled by Notch signaling in crosstalk with other signaling pathways, and the broad spectrum of downstream targets result in multiple functions of Hey proteins in development and tissues homeostasis. For a more concise overview of *Hey* knockout mice, see also Wiese, Heisig, and Gessler (2010).

In the developing heart, all three members of the Hey family are expressed in site- and cell type-specific patterns (Fig. 8.4). In the

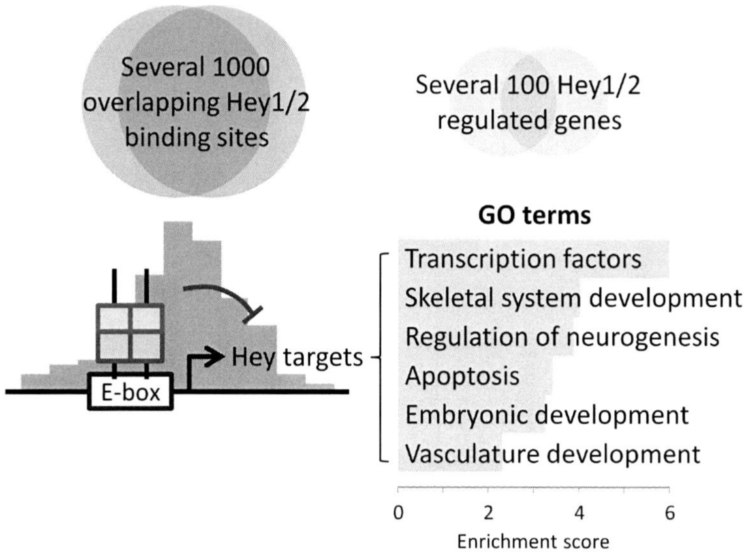

Figure 8.3 Gene regulation by Hey proteins. Hey proteins preferentially bind to E-box motifs in close proximity to transcription start sites of several 1000 promoters. In most screens, the number of regulated genes was limited to several hundred. These are mainly classified as other transcription factors involved in various developmental processes. *Taken from Heisig et al. (2012).* (See the color plate.)

myocardium, *Hey1* and *Hey2* are expressed complementary: *Hey1* is found in atrial progenitor cells, the developing inflow and outflow tract, and atrial myocardium, whereas *Hey2* expression is restricted to ventricular progenitor cells, ventricular myocardium, outflow tract, and the cushion mesenchyme of the outflow tract (Donovan, Kordylewska, Jan, & Utset, 2002; Leimeister et al., 1999; Nakagawa, Nakagawa, Richardson, Olson, & Srivastava, 1999). In contrast, *HeyL* is only expressed in endocardial cells of the atrioventricular canal, together with *Hey1* and *Hey2* (Fischer et al., 2007).

The myocardial expression of *Hey1* and *Hey2* plays a role in establishing the atrioventricular canal boundary by regulating *Tbx2* and *Bmp2* expression. This was demonstrated first in mouse embryos with forced expression of *Hey1* and *Hey2* abrogating *Tbx2* and *Bmp2* expression (Kokubo et al., 2007) and confirmed by N1ICD-driven ectopic expression of *Hey1* (Luna-Zurita et al., 2010).

Ventricular myocardial *Hey2* expression has further been shown to be important in suppressing atrial gene expression in the left ventricular myocardium, thereby allowing proper maturation (Koibuchi & Chin, 2007; Xin et al., 2007). This repressor function seems to be evolutionary conserved

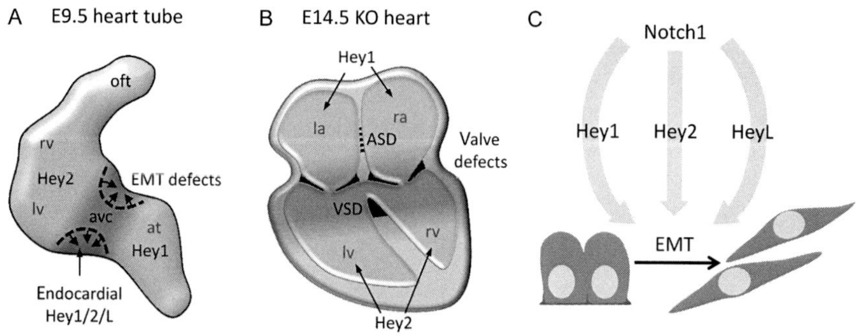

Figure 8.4 Hey expression in the developing heart and defects in knockout mice: (A) At embryonic day E9.5, Hey1 is expressed in cardiomyocytes of the developing atrium (at), while Hey2 is expressed in cardiomyocytes of the future left and right ventricle (rv, lv). All three Hey proteins are expressed in the endocardium. Hey2 and Hey1/L knockout mice exhibit epithelia–mesenchymal transition (EMT) defects in the endocardium of the atrioventricular canal (avc). (B) At embryonic day E14.5, Hey1 and Hey2 are expressed in atrial and ventricular cardiomyocytes, respectively. *Hey2* knockout mice have ventricular septum defects (VSDs), valve defects, and atrial septum defects (ASDs). These are the result of the earlier EMT defect, as these structures develop from mesenchymal cells, which have undergone EMT. (C) *Notch1* as well as *Hey2* and *Hey1/L* knockout mice exhibit similar cardiac EMT defects suggesting a cascade with Notch1 activating *Hey* genes in a combinatorial manner with partial redundancy to facilitate EMT. (See the color plate.)

since the *Hey2* ortholog *gridlock* in zebrafish also acts as a negative regulator in this case of *Gata5*-mediated transcription (Jia et al., 2007).

In the atria, *Hey2* expression is suppressed by the transcription factor CoupTF-II (Wu et al., 2013). Consistently, cardiomyocyte-specific *CoupTF-II* knockout mice developed ventricularized atria as demonstrated by altered cardiomyocyte morphology, electrical activities, and expression of ventricular genes including *Hey2*.

Mouse knockout studies revealed no apparent developmental defect in *Hey1*$^{-/-}$ and *HeyL*$^{-/-}$ mice (Fischer, Schumacher, Maier, Sendtner, & Gessler, 2004; Fischer et al., 2007), in contrast to a *Hey2* knockout that is lethal within the first 10 days after birth due to ventricular septum defect (VSD) and cardiac valve defects (Donovan et al., 2002; Gessler et al., 2002; Kokubo et al., 2004; Sakata et al., 2002). Combined loss of *Hey1* and *HeyL* leads to similar deficiencies as seen in *Hey2*$^{-/-}$ mice with membranous VSD and dysplastic atrioventricular and pulmonary valves (Fischer et al., 2007). Mice lacking *Hey1* and *Hey2* have even more severe cardiac defects (Kokubo, Miyagawa-Tomita, Nakazawa, Saga, & Johnson, 2005).

The observed cardiac phenotype can be explained by the combined expression of all three *Hey* genes in endocardial cells of the atrioventricular canal to mediate Notch signaling (Fig. 8.4C). These cells are involved in septation and cardiac valve formation after EMT. In this process, myocardial factors like Bmp2 induce delamination of epithelial cells into the atrioventricular cushion matrix and phenotypic conversion. These mesenchymal cells subsequently proliferate and differentiate to develop valvular leaflets and the membranous septum. Atrioventricular explants of *Hey1/HeyL* and *Hey2* knockout mice indeed exhibited strongly reduced EMT (Fischer et al., 2007). In line with this, a Notch1-induced myocardial expression of *Hey1* in the atrioventricular canal leads to *Bmp2* repression, which subsequently impairs endocardial EMT (Luna-Zurita et al., 2010).

Hey2$^{-/-}$ mice exhibit an increase in heart size with enlarged ventricles and an abnormal pear-like shape (Donovan et al., 2002; Gessler et al., 2002; Kokubo et al., 2004; Sakata et al., 2002). Morphologically, they develop a membranous VSD with overriding aorta and in few cases, defects of the muscular septum. Also atresia or abnormalities of the tricuspid valve and atrial septal defects (Donovan et al., 2002), in particular, secundum atrial septal defects (Kokubo et al., 2004) or persistent foramen ovale were observed (Fischer, Klamt, et al., 2004). Myocardial *Hey2* overexpression in *Hey2*$^{-/-}$ mice could rescue septation and myocardial defects, but not the aortic vascular, and to a lesser extent valve defects (Sakata et al., 2006), suggesting a combinatorial effect of Hey2 in different cell types.

In surviving adult *Hey2*$^{-/-}$ mice, dysfunctions indicative of aortic valve anomalies, stenosis, and regurgitation as well as fibrosis in the ventricle have been detected, suggesting an additional role of *Hey2* in maintaining the homeostasis of the aortic valve and ventricle (Kokubo et al., 2013).

Furthermore, the severity and extent of the observed malformations in *Hey2*$^{-/-}$ mice are clearly dependent on the mouse genetic background. In CD1 outbred mice, e.g., fewer VSDs and a reduced mortality were found compared to C57BL/6 mice, pointing to genetic modifiers (Fischer, Klamt, et al., 2004). *Hey2* itself was found to be a modifier gene for congenital heart diseases in a trisomy 21 mouse model (Li et al., 2012).

Hey2 deficiency has been reported to also impair coronary vessel development with reduced recruitment of vascular smooth muscle cells into coronary arteries and abnormally remodeled vein networks (Watanabe, Koibuchi, & Chin, 2010). Since *Hey2* is not expressed in the epicardium, which contributes to coronary vessel formation, the authors speculate on a myocardial influence, perhaps via altered PDGF and angiopoietin signals.

5.1. *Hey* genes in adult cardiac disease

Besides the developmental septum, valve, and coronary vessel defects, *Hey2* deficiency causes cardiomyopathy in mice that reach adulthood in a mixed genetic background (Sakata et al., 2002) with cardiac hypertrophy being a possible reason for heart failure. Interestingly, overexpression of *Hey2* in adult myocardium was shown to be protective against phenylephrine-induced hypertrophy *in vivo* and *in vitro*. This antihypertrophic function is possibly mediated via suppression of hypertrophic genes such as *Anf*, *Bnp*, and *Gata4* (Xiang et al., 2006). Hey2 also impacts on the adenylate cyclase pathway, cellular phosphatases, and genes involved in autophagy, some of which may impact on hypertrophy (Yu et al., 2010). In a mouse model for pressure overload hypertrophy, *Hey2* overexpression promoted physiological over pathological hypertrophy. Here, the authors also implicated suppression of apoptosis and regulation of multiple transcriptional pathways (Yu et al., 2009). Heterozygous deletion of *Hey2* in turn enhanced cardiac hypertrophy and progression to heart failure with increased apoptosis and elevated Gata4 levels (Liu, Yu, Wu, & Chin, 2010). Further analysis demonstrated abnormalities in calcium cycling mediated by the ryanodine receptor-binding protein Fkbp12.6, whose expression was increased in myocardial specific *Hey2*$^{-/-}$ mice after aortic banding (Liu et al., 2012).

Recently, a genome-wide association study described genetic polymorphisms in the *HEY2* gene as a regulator of cardiac electrical function involved in Brugada syndrome (Bezzina et al., 2013). In *Hey2*$^{-/-}$ mice, the expression of the Brugada syndrome-associated cardiac sodium channel *Nav1.5* in the compact ventricular layer was higher compared to wild-type hearts and conduction velocity in the right ventricular outflow tract was increased. This suggests an additional regulatory role for Hey2 in repolarizing currents (Bezzina et al., 2013).

Taken together, it has become clear that Hey genes control heart development and subsequent function through their expression in different combinations in several cell types and at multiple time points.

6. HEY PROTEINS IN VASCULAR DEVELOPMENT

Arterial, venous, and lymphatic vessels have a distinct molecular signature that makes them morphologically and functionally different. Hey factors are specifically expressed in arteries, but absent from veins (Leimeister, Schumacher, et al., 2000; Nakagawa et al., 1999). They are associated with processes like embryonic vasculogenesis, angiogenesis, endothelial cell fate decisions, maintenance of vessel homeostasis, and recovery from vascular injury.

The first involvement of Hey proteins in vascular biology was found in zebrafish *gridlock* mutants (Zhong, Rosenberg, Mohideen, Weinstein, & Fishman, 2000). During early vasculogenesis in zebrafish embryos, *gridlock*, the *Hey2* ortholog, is necessary for proliferation of vascular progenitor cells (Chun et al., 2011). Thereafter, diminished *gridlock* expression leads to arterial reduction and expansion of the vein region (Zhong, Childs, Leu, & Fishman, 2001). Despite aortic occlusion, blood flow can be restored during development by recruitment of collateral vessels (arteriogenesis) independent of hypoxia or ischemia suggesting that gridlock has no fundamental function in arteriogenesis during development (Gray et al., 2007).

In contrast, neither *Hey1* nor *Hey2* are essential for embryonic vessel development in the mouse on their own. Loss of both factors, however, is lethal after embryonic day 9.5 due to a global lack of vascular remodeling and hemorrhage, whereas initial vasculogenesis appeared unaffected (Fischer, Schumacher, et al., 2004; Kokubo et al., 2005). This affects all vascular beds, i.e., embryonic, yolk sac, and placental blood vessels. In addition, expression of arterial endothelial markers like ephrin B2 and markers for smooth muscle cells like smooth muscle actin and Sm22 is reduced indicating that Hey1 and Hey2 activity is required for arterial specification.

Human arterial endothelial cells soon lose their specific molecular signature during *in vitro* cultivation due to gradual loss of canonical Notch activity and *Hey2* expression (Aranguren, Agirre, et al., 2013). However, *Hey2* overexpression only partially restored an arterial gene expression program, and only overexpression of eight different transcription factors including Hey2 in HUVECs robustly converted these into arterial endothelial cells. This is in line with prior data showing that Hey1 and Hey2 can suppress expression of the venous endothelial marker EphB4 in HUVECs, but are unable to induce arterial ephrin B2, which suggests a major role for Hey1/2 in repressing venous fate rather than to induce an arterial phenotype (Iso et al., 2006).

CoupTF-II is a key antagonist of Notch–Hey signaling in endothelial cells and determines vein identity (Fig. 8.5). Hypoxia-induced *Hey1* and *Hey2* expression in embryonic endothelial cells were shown to repress *CoupTF-II* and to induce arterial cell fate (Diez et al., 2007). On the other hand, downregulation or overexpression of *Hey2* in human arterial or venous endothelial cells, respectively, did not influence *CoupTF-II*

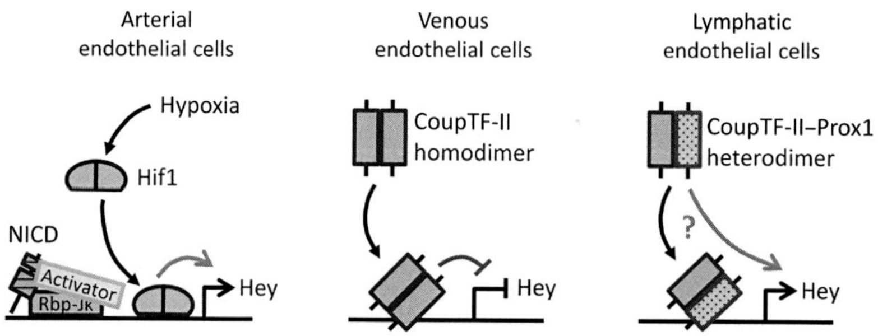

Figure 8.5 Regulation of Hey expression in endothelial cells: In arterial endothelial cells, Hey is directly induced by hypoxia-sensing Hif1 and the Delta–Notch pathway. In venous endothelial cells, transcription of Hey is repressed by CoupTF-II homodimers that directly bind to the Hey promoter. In lymphatic endothelial cells, repression by CoupTF-II is abolished by Prox1, which forms heterodimers with CoupTF-II and blocks its repressive capacity toward Hey genes. (See the color plate.)

expression (Korten et al., 2013). These discrepancies may be due to methodical differences or cellular context-dependent function of Hey factors in regulating *CoupTF-II*, but warrant further study. Vice versa, downregulation of CoupTF-II in human venous endothelial cells was found to elevate *Hey2* (Chen, Qin, Cheng, Tsai, & Tsai, 2012; Korten et al., 2013). CoupTF-II can directly bind to the *Hey1* and *Hey2* promoter and overexpression in human arterial endothelial cells led to downregulation of *Hey2* and other arterial markers (Aranguren, Beerens, et al., 2013; Korten et al., 2013). Therefore, it appears that CoupTF-II can directly repress *Hey* in endothelial cells, but the reverse may only function indirectly or cell type-dependent.

In lymphatic endothelial cells, in contrast, both Hey1 and Hey2 repress *Prox1* and *CoupTF-II* by regulating their promoter activity (Kang et al., 2010). Interestingly, CoupTF-II/Prox1 heterodimers lack the inhibitory function on *Hey* expression (Fig. 8.5). Expression of *Hey* and some lymphatic endothelial cell-specific genes seems to be dependent on this interaction (Aranguren, Beerens, et al., 2013). Hence, this suggests a regulatory feedback loop of Hey, CoupTF-II and Prox1 in lymphatic endothelial cells.

The lateral inhibition paradigm of Notch signaling seems to apply in angiogenesis, where the formation of tip cells that direct sprouting and guide adjacent stalk cells is highly regulated. Notch signaling in stalk cells, which is induced by Dll4 present on tip cells, can balance tip versus stalk cell formation. In this scenario, activation of *Hey1* and *Hey2* in stalk cells is regulated by Notch signaling but also via Bmp9–Alk1 (Larrivee et al., 2012). Hey2 inhibits tip cell formation and endothelial sprouting and represses Vegf signaling.

In concordance with that, there is a reduction of *Hey1* in migrating and proliferating human capillary endothelial cells, but an induction in capillary-like networks and mature quiescent vessels (Henderson, Wang, Taylor, Aitkenhead, & Hughes, 2001). Overexpression of *Hey1* in human retinal capillary endothelial cells repressed Vegfr2, inhibited migration and proliferation, and reduced generation of new tube-like structures, pointing to a stabilizing homeostatic function (Li et al., 2006).

6.1. Role of *Hey2* in smooth muscle cells during vascular injury

Hey2 has been further described as an important regulator of vascular smooth muscle cell accumulation during vascular remodeling (Li et al., 2009; Sakata et al., 2004). In an arterial wire injury model, *Hey2*$^{-/-}$ mice revealed decreased neointima formation, most probably due to reduced proliferation (Sakata et al., 2004). Furthermore, *Hey2*-deficient vascular smooth muscle cells proliferated and migrated slowly *in vitro* (Li et al., 2009; Sakata et al., 2004). Consistently, *Hey2* overexpression in a balloon injury carotid rat artery model could partially rescue smooth muscle cell proliferation (Caolo et al., 2011).

Besides its function in smooth muscle proliferation after injury, Hey2 can also inhibit myocardin-dependent smooth muscle cell differentiation by inhibiting the function of myocardin (Doi et al., 2005). This was also demonstrated after wire-mediated vascular injury, where induction of *Hey2* in the adventitia contributed to the inhibition of differentiation of bone marrow-derived cells into smooth muscle cells (Doi et al., 2009).

7. CONTROL OF MYOGENESIS AND MUSCLE STEM CELLS

One of the first functional tests for *Hey* genes already uncovered an antimyogenic function in C2C12 and 10T1/2 cells (Sun et al., 2001). Hey1 blocked MyoD-dependent activation of the *myogenin* promoter in 10T1/2 cells, presumably by interfering with the activity of MyoD/E47 heterodimers, and this repressive and antimyogenic activity was dependent on the C-terminus of Hey1. In C2C12 cells as well as in primary muscle satellite cells, *Hey1* and *Hes1* apparently mediate the effect of Bmp4 to inhibit muscle differentiation (Dahlqvist et al., 2003). In this case, a synergistic action of Notch and Bmp4/Smad1 signaling could be shown. The inhibitory effect of Hey1 on myogenesis was subsequently ascribed to its binding and inhibition of core myogenic promoters such as *myogenin* and *Mef2c* (Buas, Kabak, & Kadesch, 2010).

The most clear-cut data were provided by the elegant analysis of *Hey1/HeyL* knockout mice (Fukada et al., 2011). Both genes are expressed in satellite cells, which are skeletal muscle stem cells that contribute to the regenerative potential in later life. *Hey1/HeyL* knockout myoblasts generate fewer quiescent satellite cells during development and will then lead to gradual decrease in number with age lead to severe impairment of adult muscle homeostasis with a loss of regenerative capacity.

A related mechanism may be at work in embryonal rhabdomyosarcoma, which is characterized by strong Notch pathway activity, where blockade of *Notch1* or *Hey1* allowed partial myogenic differentiation of tumor cells at the expense of proliferation (Belyea, Naini, Bentley, & Linardic, 2011). Taken together, the main function of *Hey* genes in myogenic cells seems to be in the blockade of further differentiation to keep cells in an immature state—either as satellite (stem) cells or still proliferative in the context of rhabdomyosarcoma (Fig. 8.6).

8. BONE DEVELOPMENT AND HOMEOSTASIS

In osteogenesis, all three *Hey* genes appear to have related roles and are likely able to partially compensate for each other. Initial experiments showed that *Hey1* is a downstream target of Bmp2 in osteoblast differentiation of C2C12 and MC3T3 cells (de Jong, Vaes, et al., 2004; Zamurovic, Cappellen, Rohner, & Susa, 2004). Similarly, the Bmp9-induced osteogenic differentiation of mesenchymal stem cells is partly dependent on *Hey1*, which appears to convey some of the osteogenic effects (Sharff et al., 2009). Very similar data were obtained for Bmp7 that can independently promote osteogenesis, again mediated through *Hey1* (Lavery et al., 2009). These observations have been extended by analysis of *Hey* mutant mice: transgenic *Hey1* overexpression leads to progressive osteopenia in adults (Salie et al., 2010). On the contrary, mice deficient in *Hey1* and *HeyL* exhibited high bone mass due to higher numbers of osteoblasts. These findings agree with the effects ascribed to *Notch* in bone biology (Tu et al., 2012): In this context, it appears that Notch signaling functions in part through a Hey1-mediated inhibition of *Nfatc1* to suppress osteoblastogenesis (Fig. 8.6). Consistent with this notion, *Rbp-Jκ* deletion increased *Nfatc1* expression in bone and pharmacological inhibition of Nfat alleviated the high-bone-mass phenotype caused by *Rbp-Jκ* deletion (Tu et al., 2012). In similar

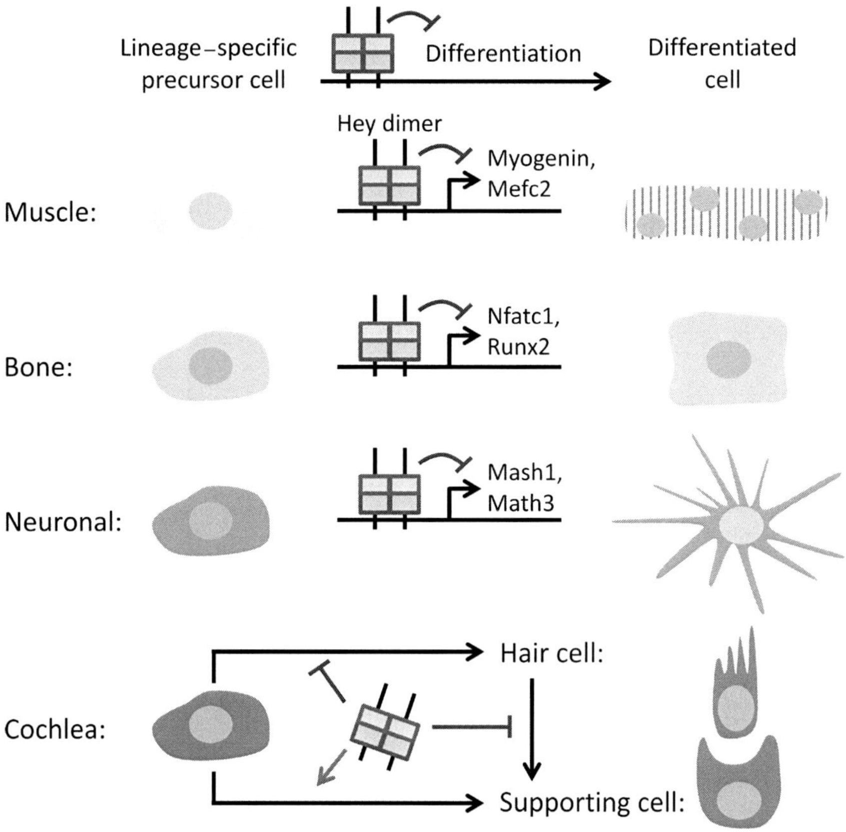

Figure 8.6 Hey proteins play a critical role in differentiation processes. They repress target genes, which drive differentiation within certain lineage and thereby affect myogenic, osteogenic, and neuronal differentiation. (See the color plate.)

experiments, Zanotti and Canalis (2013) could show that *Hey2* deletion in the osteoblast lineage also leads to a high-bone-mass phenotype, while *Hey2* overexpression in osteoblasts decreases skeletal mass and regulates bone remodeling, especially in male mice. Even in cardiac valves, related mechanisms seem to be at play since osteogenic genes, such as *osteopontin* and *sclerostin*, were upregulated in dysfunctional and sclerosing valves of *Hey2*$^{-/-}$ mice (Kokubo et al., 2013).

The antiosteogenic action of Hey proteins seems to be part of a balanced system with Hes1 and Id4. In mesenchymal stem cells, *Id4* knockdown suppresses osteoblast differentiation, whereas forced expression is promoting (Tokuzawa et al., 2010). This is recapitulated *in vivo* since *Id4*$^{-/-}$ mice suffer from reduced osteoblast differentiation, which was suggested to be due to unrestrained activity of Hey2/Hes1 heterodimers. Notably, Id4 interacted only with Hey2, but not Hes1, leading to the speculation that Id4 expression may liberate Hes1 from heterodimeric complexes to act on its own and to stabilize the osteogenic factor Runx2.

The main Hey/Hes function seems to be in limiting overshooting osteoblast differentiation, which is consistent with the role of Notch signaling in numerous other cell types and organs. There are additional effects on the counteracting osteoclast differentiation, but these seem to be on a smaller scale. Important mediators of Hey/Hes effects are *Nfatc* and *Runx2* genes, while Id4 may have a role as a partial antagonist targeting Hey2.

9. HEY PROTEINS CONTROL NEURAL DEVELOPMENT

A multitude of effects has been documented for *Hey* genes in the nervous system. *Hey1* and *Hey2* are expressed in the ventricular zone of the developing neural tube. Misexpression in the mouse brain at embryonic day 13.5 leads to a preferential maintenance of neural precursor cells with an increase in late-born neurons populating outer layers of the brain. In contrast, misexpression at later stages inhibits neurogenesis and promotes generation of astroglia (Sakamoto, Hirata, Ohtsuka, Bessho, & Kageyama, 2003). The authors concluded that *Hey1* and *Hey2* negatively regulate neuronal bHLH genes like *Mash1* and *Math3* and promote maintenance of neural precursor cells (Fig. 8.6). The developing retina also adds to the classical Notch paradigm of binary cell fate decision by *Hey* genes: overexpression of *Hey2*, but not *Hey1* or *HeyL* promotes gliogenesis while inhibiting sensory rod formation (Satow et al., 2001).

Hey2 and *HeyL* are expressed in neural crest cells and later in boundary caps of the dorsal root ganglia, giving rise to specialized Schwann cells, which would provide a scenario for comparable cell fate decisions (Coulpier et al., 2009). Furthermore, deletion of *HeyL* leads to a reduction in TrkC-positive neurons in dorsal root ganglia, while their number is increased in *Hey1*$^{-/-}$ mice, suggestive of an antagonistic function (Mukhopadhyay, Jarrett, Chlon, & Kessler, 2009). More surprisingly, a differentiation-promoting effect was reported for *HeyL* in neural progenitor cells, antagonistic to *Hey1/2* and *Hes1* (Jalali et al., 2011). This is currently still difficult to reconcile with the known similarities and interchangeable functions of the three Hey proteins and will await for further study.

Hey2 has even been implicated in the maintenance of adult neural stem cells. Loss of *TAp73*, a member of the *p53* family of genes, leads to a reduction of stem cells in the dentate gyrus of the hippocampus and this effect appears to be mediated by *Hey2*, promoting a neural precursor maintenance program and preventing premature differentiation (Fujitani et al., 2010).

Beyond these effects on cell fate decisions and proliferation control, *Hey* genes may also affect neural functions: Fuke et al. (2006) described a regulation of the dopamine transporter gene *Dat1* by Hey1 and Hey2 via a 3′-noncoding VNTR region and concomitantly altered behavioral traits in knockout mice.

10. *HEY* FUNCTIONS IN EAR DEVELOPMENT

Hey and *Hes* genes are expressed in different combinations in the various types of supporting cells of the organ of Corti. Specialized pillar cells are characterized by *Hey2* expression, which is activated by Notch and FGF signals and blocks hair cell differentiation, while mutation of *Hey2* results in differentiation into hair cells (Doetzlhofer et al., 2009). This concept has been extended since various combinations of *Hey* and *Hes* genes could restrain the number of hair cells in the developing cochlea, while reduced dosage led to an increase of hair cells with concomitantly increased supporting cells. Only upon their complete loss, the balance tips toward excessive numbers of hair cells in disorganized and nonfunctional arrays (Scott et al., 2008; Tateya, Imayoshi, Tateya, Ito, & Kageyama, 2011). Even postnatally, Hes1 and Hey proteins control transdifferentiation of supporting cells to sensory hair cells after cochlear damage (Korrapati, Roux, Glowatzki, & Doetzlhofer, 2013).

11. IMMUNE FUNCTIONS OF HEY PROTEINS

Hey genes are dynamically expressed in the developing thymus and different subsets of immune cells (Leimeister, Schumacher, et al., 2000, and our own unpublished data). In bone marrow-derived macrophages, *Hey1* and *Hes1* limit the expression of IL-6- and IL-12-type cytokines and modulate Toll-like receptor responses and macrophage effector functions in this way (Hu et al., 2008). These data indicate that *Hey* genes participate in immune cell functions, but they again seem to be partly redundant with each other or additional *Hes* genes.

12. *HEY* GENES IN CANCER

12.1. An oncogenic HEY1–NCOA2 fusion protein

There is a single tumor entity, where *HEY1* appears to have a direct involvement in oncogenesis: a large fraction of mesenchymal chondrosarcomas harbor a recurrent *HEY1–NCOA2* fusion that links *HEY1* exon 4 in frame with *NCOA2* exon13 (Wang et al., 2012). Both genes map within less than 10 Mb on the long arm of chromosome 8 (8q21, 8q13) and the fusion event is likely due to a deletion of intervening DNA sequences. The resultant fusion protein is comprised of the HEY1 bHLH domain fused to the C-terminal activation domains of NCOA2. Interestingly, very similar *NCOA2* breakpoints are present in other oncogenic fusions such as *MYST3–NCOA2* and *ETV6–NCOA2* fusion in leukemia (van den Berg et al., 2003), the *PAX3–NCOA2* fusions in alveolar rhabdomyosarcoma (Kokubo et al., 1999), and the *AHRR–NCOA2* fusions in soft tissue angiosarcoma (Jin et al., 2012). The HEY1–NCOA2 fusion protein is expected not only to bind typical HEY1 target sites but also to act as a transcriptional activator instead that seems to keep immature chondroblasts in a proliferative state, which is the hallmark of this rare, fast-growing bone tumor of children and young adults. It remains to be seen if typical HEY target genes involved in osteoblast differentiation are the relevant targets for oncogenesis in this tumor type.

12.2. Deregulated *Hey* genes in diverse malignancies

There is no further evidence for a direct *HEY* gene involvement in additional cancer types due to mutational activation or inactivation, but there are multiple reports on conspicuous expression patterns. Such aberrant expression of *HEY* genes in tumors may indicate the activity of upstream DLL/JAG ligands or the four NOTCH receptors and their glycosylation partners. In addition, elevated *HEY* gene expression may reflect the presence of activated endothelia and ongoing angiogenesis that involves NOTCH signaling though *HEY* genes (Parker et al., 2004).

Tumor types of special interest are colorectal, pancreatic, and prostate cancers as well as glioblastoma and rhabdomyosarcoma. In colorectal cancer, *HEY1* overexpression correlates with poor outcome and expression of *HEY1*, as well as *HES1* and *SOX9*, and is predictive of survival with

5-fluorouracil chemotherapy (Candy et al., 2013). In pancreatic cancer, nuclear staining of HEY1 (and NOTCH3) is significantly associated with reduced disease-free and overall survival, where the authors proposed a future potential for HEY1 as a biomarker for diagnosis, prognosis, and treatment efficacy (Mann et al., 2012). For glioblastoma, a strong correlation of *HEY1* expression with tumor-grade and survival has been reported (Hulleman et al., 2009). As already mentioned earlier, embryonal rhabdomyosarcoma is also characterized by high HEY1 levels *in vivo* and a reduced proliferation *in vitro* upon HEY1 knockdown by shRNA (Belyea et al., 2011).

In normal prostate cells, HEY1 interacts with the androgen receptor and functions as a corepressor, but in prostate cancer, this interaction is no longer detectable and HEY1 is excluded from the nucleus (Belandia et al., 2005). Similarly, HEYL was found to repress androgen receptor-induced genes and it exhibits the same pattern of nuclear exclusion in prostate cancer cells (Lavery et al., 2011). Furthermore, HEY2 expression was identified as a statistically significant and independent predictor for recurrence and outcome in prostate carcinoma (Tradonsky et al., 2012). Although the mechanistic details may well be quite different in the tumor entities described, HEY expression status may become relevant as a biomarker and this may hint to functionally relevant signaling pathways.

13. CONCLUSION

The scope of physiological and pathological processes that involve *Hey* genes has been expanded dramatically in recent years. While the strongest evidence may still point to a role in developmental processes affecting multiple cell types and organs, it has become clear that tissue and organ homeostasis and especially the maintenance of a local supply of reserve cells as a tissue-specific regeneration resource are key functions of Hey proteins. As known from Notch signaling, this may involve different mechanisms from binary cell fate decisions to inductive processes and control of proliferation. More sophisticated cell type-specific deletions and appropriate phenotyping may reveal additional functions of Hey proteins in future. Furthermore, the still limited and fragmented knowledge on Hey interaction partners and effectors holds promise that novel and relevant insights are still to be gained from this type of studies.

REFERENCES

Aranguren, X. L., Agirre, X., Beerens, M., Coppiello, G., Uriz, M., Vandersmissen, I., et al. (2013). Unraveling a novel transcription factor code determining the human arterial-specific endothelial cell signature. *Blood, 122*, 3982–3992.

Aranguren, X. L., Beerens, M., Coppiello, G., Wiese, C., Vandersmissen, I., Lo Nigro, A., et al. (2013). COUP-TFII orchestrates venous and lymphatic endothelial identity by homo- or hetero-dimerisation with PROX1. *Journal of Cell Science, 126*, 1164–1175.

Belandia, B., Powell, S. M., Garcia-Pedrero, J. M., Walker, M. M., Bevan, C. L., & Parker, M. G. (2005). Hey1, a mediator of notch signaling, is an androgen receptor corepressor. *Molecular and Cellular Biology, 25*, 1425–1436.

Belyea, B. C., Naini, S., Bentley, R. C., & Linardic, C. M. (2011). Inhibition of the Notch-Hey1 axis blocks embryonal rhabdomyosarcoma tumorigenesis. *Clinical Cancer Research, 17*, 7324–7336.

Bezzina, C. R., Barc, J., Mizusawa, Y., Remme, C. A., Gourraud, J. B., Simonet, F., et al. (2013). Common variants at SCN5A-SCN10A and HEY2 are associated with Brugada syndrome, a rare disease with high risk of sudden cardiac death. *Nature Genetics, 45*, 1044–1049.

Blokzijl, A., Dahlqvist, C., Reissmann, E., Falk, A., Moliner, A., Lendahl, U., et al. (2003). Cross-talk between the Notch and TGF-beta signaling pathways mediated by interaction of the Notch intracellular domain with Smad3. *Journal of Cell Biology, 163*, 723–728.

Buas, M. F., Kabak, S., & Kadesch, T. (2010). The Notch effector Hey1 associates with myogenic target genes to repress myogenesis. *Journal of Biological Chemistry, 285*, 1249–1258.

Candy, P. A., Phillips, M. R., Redfern, A. D., Colley, S. M., Davidson, J. A., Stuart, L. M., et al. (2013). Notch-induced transcription factors are predictive of survival and 5-fluorouracil response in colorectal cancer patients. *British Journal of Cancer, 109*, 1023–1030.

Caolo, V., Schulten, H. M., Zhuang, Z. W., Murakami, M., Wagenaar, A., Verbruggen, S., et al. (2011). Soluble Jagged-1 inhibits neointima formation by attenuating Notch-Herp2 signaling. *Arteriosclerosis, Thrombosis, and Vascular Biology, 31*, 1059–1065.

Chen, X., Qin, J., Cheng, C. M., Tsai, M. J., & Tsai, S. Y. (2012). COUP-TFII is a major regulator of cell cycle and Notch signaling pathways. *Molecular Endocrinology, 26*, 1268–1277.

Chi, J. T., Chang, H. Y., Haraldsen, G., Jahnsen, F. L., Troyanskaya, O. G., Chang, D. S., et al. (2003). Endothelial cell diversity revealed by global expression profiling. *Proceedings of the National Academy of Sciences of the United States of America, 100*, 10623–10628.

Chin, M. T., Maemura, K., Fukumoto, S., Jain, M. K., Layne, M. D., Watanabe, M., et al. (2000). Cardiovascular basic helix loop helix factor 1, a novel transcriptional repressor expressed preferentially in the developing and adult cardiovascular system. *Journal of Biological Chemistry, 275*, 6381–6387.

Chun, C. Z., Remadevi, I., Schupp, M. O., Samant, G. V., Pramanik, K., Wilkinson, G. A., et al. (2011). Fli + etsrp + hemato-vascular progenitor cells proliferate at the lateral plate mesoderm during vasculogenesis in zebrafish. *PloS One, 6*, e14732.

Coulpier, F., Le Crom, S., Maro, G. S., Manent, J., Giovannini, M., Maciorowski, Z., et al. (2009). Novel features of boundary cap cells revealed by the analysis of newly identified molecular markers. *Glia, 57*, 1450–1457.

Dahlqvist, C., Blokzijl, A., Chapman, G., Falk, A., Dannaeus, K., Ibanez, C. F., et al. (2003). Functional Notch signaling is required for BMP4-induced inhibition of myogenic differentiation. *Development (Cambridge, England), 130*, 6089–6099.

de Jong, D. S., Steegenga, W. T., Hendriks, J. M., van Zoelen, E. J., Olijve, W., & Dechering, K. J. (2004). Regulation of Notch signaling genes during BMP2-induced differentiation of osteoblast precursor cells. *Biochemical and Biophysical Research Communications, 320*, 100–107.

de Jong, D. S., Vaes, B. L., Dechering, K. J., Feijen, A., Hendriks, J. M., Wehrens, R., et al. (2004). Identification of novel regulators associated with early-phase osteoblast differentiation. *Journal of Bone and Mineral Research*, *19*, 947–958.

Diez, H., Fischer, A., Winkler, A., Hu, C. J., Hatzopoulos, A. K., Breier, G., et al. (2007). Hypoxia-mediated activation of Dll4-Notch-Hey2 signaling in endothelial progenitor cells and adoption of arterial cell fate. *Experimental Cell Research*, *313*, 1–9.

Doetzlhofer, A., Basch, M. L., Ohyama, T., Gessler, M., Groves, A. K., & Segil, N. (2009). Hey2 regulation by FGF provides a Notch-independent mechanism for maintaining pillar cell fate in the organ of Corti. *Developmental Cell*, *16*, 58–69.

Doi, H., Iso, T., Shiba, Y., Sato, H., Yamazaki, M., Oyama, Y., et al. (2009). Notch signaling regulates the differentiation of bone marrow-derived cells into smooth muscle-like cells during arterial lesion formation. *Biochemical and Biophysical Research Communications*, *381*, 654–659.

Doi, H., Iso, T., Yamazaki, M., Akiyama, H., Kanai, H., Sato, H., et al. (2005). HERP1 inhibits myocardin-induced vascular smooth muscle cell differentiation by interfering with SRF binding to CArG box. *Arteriosclerosis, Thrombosis, and Vascular Biology*, *25*, 2328–2334.

Donovan, J., Kordylewska, A., Jan, Y. N., & Utset, M. F. (2002). Tetralogy of fallot and other congenital heart defects in Hey2 mutant mice. *Current Biology*, *12*, 1605–1610.

Elagib, K. E., Xiao, M., Hussaini, I. M., Delehanty, L. L., Palmer, L. A., Racke, F. K., et al. (2004). Jun blockade of erythropoiesis: Role for repression of GATA-1 by HERP2. *Molecular and Cellular Biology*, *24*, 7779–7794.

Firulli, B. A., Hadzic, D. B., McDaid, J. R., & Firulli, A. B. (2000). The basic helix-loop-helix transcription factors dHAND and eHAND exhibit dimerization characteristics that suggest complex regulation of function. *Journal of Biological Chemistry*, *275*, 33567–33573.

Fischer, A., & Gessler, M. (2007). Delta-Notch—And then? Protein interactions and proposed modes of repression by Hes and Hey bHLH factors. *Nucleic Acids Research*, *35*, 4583–4596.

Fischer, A., Klamt, B., Schumacher, N., Glaeser, C., Hansmann, I., Fenge, H., et al. (2004). Phenotypic variability in Hey2 −/−mice and absence of HEY2 mutations in patients with congenital heart defects or Alagille syndrome. *Mammalian Genome*, *15*, 711–716.

Fischer, A., Klattig, J., Kneitz, B., Diez, H., Maier, M., Holtmann, B., et al. (2005). Hey basic helix-loop-helix transcription factors are repressors of GATA4 and GATA6 and restrict expression of the GATA target gene ANF in fetal hearts. *Molecular and Cellular Biology*, *25*, 8960–8970.

Fischer, A., Schumacher, N., Maier, M., Sendtner, M., & Gessler, M. (2004). The Notch target genes Hey1 and Hey2 are required for embryonic vascular development. *Genes & Development*, *18*, 901–911.

Fischer, A., Steidl, C., Wagner, T. U., Lang, E., Jakob, P. M., Friedl, P., et al. (2007). Combined loss of Hey1 and HeyL causes congenital heart defects because of impaired epithelial to mesenchymal transition. *Circulation Research*, *100*, 856–863.

Fu, M., Rao, M., Wang, C., Sakamaki, T., Wang, J., Di Vizio, D., et al. (2003). Acetylation of androgen receptor enhances coactivator binding and promotes prostate cancer cell growth. *Molecular and Cellular Biology*, *23*, 8563–8575.

Fujitani, M., Cancino, G. I., Dugani, C. B., Weaver, I. C., Gauthier-Fisher, A., Paquin, A., et al. (2010). TAp73 acts via the bHLH Hey2 to promote long-term maintenance of neural precursors. *Current Biology*, *20*, 2058–2065.

Fukada, S., Yamaguchi, M., Kokubo, H., Ogawa, R., Uezumi, A., Yoneda, T., et al. (2011). Hesr1 and Hesr3 are essential to generate undifferentiated quiescent satellite cells and to maintain satellite cell numbers. *Development*, *138*, 4609–4619.

Fuke, S., Minami, N., Kokubo, H., Yoshikawa, A., Yasumatsu, H., Sasagawa, N., et al. (2006). Hesr1 knockout mice exhibit behavioral alterations through the dopaminergic nervous system. *Journal of Neuroscience Research*, *84*, 1555–1563.

Garg, V., Muth, A. N., Ransom, J. F., Schluterman, M. K., Barnes, R., King, I. N., et al. (2005). Mutations in NOTCH1 cause aortic valve disease. *Nature*, *437*, 270–274.

Gessler, M., Knobeloch, K. P., Helisch, A., Amann, K., Schumacher, N., Rohde, E., et al. (2002). Mouse gridlock: No aortic coarctation or deficiency, but fatal cardiac defects in Hey2$^{-/-}$ mice. *Current Biology*, *12*, 1601–1604.

Ghosh, B., & Leach, S. D. (2006). Interactions between hairy/enhancer of split-related proteins and the pancreatic transcription factor Ptf1-p48 modulate function of the PTF1 transcriptional complex. *Biochemical Journal*, *393*, 679–685.

Gould, F., Harrison, S. M., Hewitt, E. W., & Whitehouse, A. (2009). Kaposi's sarcoma-associated herpesvirus RTA promotes degradation of the Hey1 repressor protein through the ubiquitin proteasome pathway. *Journal of Virology*, *83*, 6727–6738.

Gray, C., Packham, I. M., Wurmser, F., Eastley, N. C., Hellewell, P. G., Ingham, P. W., et al. (2007). Ischemia is not required for arteriogenesis in zebrafish embryos. *Arteriosclerosis, Thrombosis, and Vascular Biology*, *27*, 2135–2141.

Hayakawa, F., Towatari, M., Ozawa, Y., Tomita, A., Privalsky, M. L., & Saito, H. (2004). Functional regulation of GATA-2 by acetylation. *Journal of Leukocyte Biology*, *75*, 529–540.

Hayashi, H., & Kume, T. (2008). Foxc transcription factors directly regulate Dll4 and Hey2 expression by interacting with the VEGF-Notch signaling pathways in endothelial cells. *PloS One*, *3*, e2401.

Heisig, J., Weber, D., Englberger, E., Winkler, A., Kneitz, S., Sung, W. K., et al. (2012). Target gene analysis by microarrays and chromatin immunoprecipitation identifies HEY proteins as highly redundant bHLH repressors. *PLoS Genetics*, *8*, e1002728.

Henderson, A. M., Wang, S. J., Taylor, A. C., Aitkenhead, M., & Hughes, C. C. (2001). The basic helix-loop-helix transcription factor HESR1 regulates endothelial cell tube formation. *Journal of Biological Chemistry*, *276*, 6169–6176.

Hu, X., Chung, A. Y., Wu, I., Foldi, J., Chen, J., Ji, J. D., et al. (2008). Integrated regulation of Toll-like receptor responses by Notch and interferon-gamma pathways. *Immunity*, *29*, 691–703.

Hulleman, E., Quarto, M., Vernell, R., Masserdotti, G., Colli, E., Kros, J. M., et al. (2009). A role for the transcription factor HEY1 in glioblastoma. *Journal of Cellular and Molecular Medicine*, *13*, 136–146.

Iso, T., Chung, G., Hamamori, Y., & Kedes, L. (2002). HERP1 is a cell type-specific primary target of Notch. *Journal of Biological Chemistry*, *277*, 6598–6607.

Iso, T., Maeno, T., Oike, Y., Yamazaki, M., Doi, H., Arai, M., et al. (2006). Dll4-selective Notch signaling induces ephrinB2 gene expression in endothelial cells. *Biochemical and Biophysical Research Communications*, *341*, 708–714.

Iso, T., Sartorelli, V., Chung, G., Shichinohe, T., Kedes, L., & Hamamori, Y. (2001). HERP, a new primary target of Notch regulated by ligand binding. *Molecular and Cellular Biology*, *21*, 6071–6079.

Iso, T., Sartorelli, V., Poizat, C., Iezzi, S., Wu, H. Y., Chung, G., et al. (2001). HERP, a novel heterodimer partner of HES/E(spl) in Notch signaling. *Molecular and Cellular Biology*, *21*, 6080–6089.

Itoh, F., Itoh, S., Goumans, M. J., Valdimarsdottir, G., Iso, T., Dotto, G. P., et al. (2004). Synergy and antagonism between Notch and BMP receptor signaling pathways in endothelial cells. *EMBO Journal*, *23*, 541–551.

Jalali, A., Bassuk, A. G., Kan, L., Israsena, N., Mukhopadhyay, A., McGuire, T., et al. (2011). HeyL promotes neuronal differentiation of neural progenitor cells. *Journal of Neuroscience Research*, *89*, 299–309.

Jia, H., King, I. N., Chopra, S. S., Wan, H., Ni, T. T., Jiang, C., et al. (2007). Vertebrate heart growth is regulated by functional antagonism between Gridlock and Gata5. *Proceedings of the National Academy of Sciences of the United States of America*, *104*, 14008–14013.

Jin, Y., Moller, E., Nord, K. H., Mandahl, N., Von Steyern, F. V., Domanski, H. A., et al. (2012). Fusion of the AHRR and NCOA2 genes through a recurrent translocation t(5;8) (p15;q13) in soft tissue angiofibroma results in upregulation of aryl hydrocarbon receptor target genes. *Genes, Chromosomes & Cancer*, *51*, 510–520.

Jing, W., Xiong, Z., Cai, X., Huang, Y., Li, X., Yang, X., et al. (2010). Effects of gamma-secretase inhibition on the proliferation and vitamin D(3) induced osteogenesis in adipose derived stem cells. *Biochemical and Biophysical Research Communications*, *392*, 442–447.

Kamakura, S., Oishi, K., Yoshimatsu, T., Nakafuku, M., Masuyama, N., & Gotoh, Y. (2004). Hes binding to STAT3 mediates crosstalk between Notch and JAK-STAT signalling. *Nature Cell Biology*, *6*, 547–554.

Kang, J., Yoo, J., Lee, S., Tang, W., Aguilar, B., Ramu, S., et al. (2010). An exquisite cross-control mechanism among endothelial cell fate regulators directs the plasticity and heterogeneity of lymphatic endothelial cells. *Blood*, *116*, 140–150.

Kathiriya, I. S., King, I. N., Murakami, M., Nakagawa, M., Astle, J. M., Gardner, K. A., et al. (2004). Hairy-related transcription factors inhibit GATA-dependent cardiac gene expression through a signal-responsive mechanism. *Journal of Biological Chemistry*, *279*, 54937–54943.

King, I. N., Kathiriya, I. S., Murakami, M., Nakagawa, M., Gardner, K. A., Srivastava, D., et al. (2006). Hrt and Hes negatively regulate Notch signaling through interactions with RBP-Jkappa. *Biochemical and Biophysical Research Communications*, *345*, 446–452.

Koibuchi, N., & Chin, M. T. (2007). CHF1/Hey2 plays a pivotal role in left ventricular maturation through suppression of ectopic atrial gene expression. *Circulation Research*, *100*, 850–855.

Kokubo, H., Lun, Y., & Johnson, R. L. (1999). Identification and expression of a novel family of bHLH cDNAs related to Drosophila hairy and enhancer of split. *Biochemical and Biophysical Research Communications*, *260*, 459–465.

Kokubo, H., Miyagawa-Tomita, S., Nakashima, Y., Kume, T., Yoshizumi, M., Nakanishi, T., et al. (2013). Hesr2 knockout mice develop aortic valve disease with advancing age. *Arteriosclerosis, Thrombosis, and Vascular Biology*, *33*, e84–e92.

Kokubo, H., Miyagawa-Tomita, S., Nakazawa, M., Saga, Y., & Johnson, R. L. (2005). Mouse hesr1 and hesr2 genes are redundantly required to mediate Notch signaling in the developing cardiovascular system. *Developmental Biology*, *278*, 301–309.

Kokubo, H., Miyagawa-Tomita, S., Tomimatsu, H., Nakashima, Y., Nakazawa, M., Saga, Y., et al. (2004). Targeted disruption of hesr2 results in atrioventricular valve anomalies that lead to heart dysfunction. *Circulation Research*, *95*, 540–547.

Kokubo, H., Tomita-Miyagawa, S., Hamada, Y., & Saga, Y. (2007). Hesr1 and Hesr2 regulate atrioventricular boundary formation in the developing heart through the repression of Tbx2. *Development*, *134*, 747–755.

Korrapati, S., Roux, I., Glowatzki, E., & Doetzlhofer, A. (2013). Notch signaling limits supporting cell plasticity in the hair cell-damaged early postnatal murine cochlea. *PloS One*, *8*, e73276.

Korten, S., Brunssen, C., Poitz, D. M., Grossklaus, S., Brux, M., Schnittler, H. J., et al. (2013). Impact of Hey2 and COUP-TFII on genes involved in arteriovenous differentiation in primary human arterial and venous endothelial cells. *Basic Research in Cardiology*, *108*, 362.

Larrivee, B., Prahst, C., Gordon, E., del Toro, R., Mathivet, T., Duarte, A., et al. (2012). ALK1 signaling inhibits angiogenesis by cooperating with the Notch pathway. *Developmental Cell*, *22*, 489–500.

Lavery, K., Hawley, S., Swain, P., Rooney, R., Falb, D., & Alaoui-Ismaili, M. H. (2009). New insights into BMP-7 mediated osteoblastic differentiation of primary human mesenchymal stem cells. *Bone*, *45*, 27–41.

Lavery, D. N., Villaronga, M. A., Walker, M. M., Patel, A., Belandia, B., & Bevan, C. L. (2011). Repression of androgen receptor activity by HEYL, a third member of the Hairy/Enhancer-of-split-related family of Notch effectors. *Journal of Biological Chemistry, 286*, 17796–17808.

Leimeister, C., Dale, K., Fischer, A., Klamt, B., Hrabe de Angelis, M., Radtke, F., et al. (2000). Oscillating expression of c-Hey2 in the presomitic mesoderm suggests that the segmentation clock may use combinatorial signaling through multiple interacting bHLH factors. *Developmental Biology, 227*, 91–103.

Leimeister, C., Externbrink, A., Klamt, B., & Gessler, M. (1999). Hey genes: A novel subfamily of hairy- and Enhancer of split related genes specifically expressed during mouse embryogenesis. *Mechanisms of Development, 85*, 173–177.

Leimeister, C., Schumacher, N., Steidl, C., & Gessler, M. (2000). Analysis of HeyL expression in wild-type and Notch pathway mutant mouse embryos. *Mechanisms of Development, 98*, 175–178.

Li, H., Cherry, S., Klinedinst, D., DeLeon, V., Redig, J., Reshey, B., et al. (2012). Genetic modifiers predisposing to congenital heart disease in the sensitized Down syndrome population. *Circulation. Cardiovascular Genetics, 5*, 301–308.

Li, Y., Takeshita, K., Liu, P. Y., Satoh, M., Oyama, N., Mukai, Y., et al. (2009). Smooth muscle Notch1 mediates neointimal formation after vascular injury. *Circulation, 119*, 2686–2692.

Li, B., Tang, S. B., Hu, J., Gao, Y., Zhang, G., Lin, S. F., et al. (2006). Protective effects of transcription factor HESR1 on retinal vasculature. *Microvascular Research, 72*, 146–152.

Liu, T., Gao, Y., Sakamoto, K., Minamizato, T., Furukawa, K., Tsukazaki, T., et al. (2007). BMP-2 promotes differentiation of osteoblasts and chondroblasts in Runx2-deficient cell lines. *Journal of Cellular Physiology, 211*, 728–735.

Liu, Y., Korte, F. S., Moussavi-Harami, F., Yu, M., Razumova, M., Regnier, M., et al. (2012). Transcription factor CHF1/Hey2 regulates EC coupling and heart failure in mice through regulation of FKBP12.6. *American Journal of Physiology. Heart and Circulatory Physiology, 302*, H1860–H1870.

Liu, Y., Yu, M., Wu, L., & Chin, M. T. (2010). The bHLH transcription factor CHF1/Hey2 regulates susceptibility to apoptosis and heart failure after pressure overload. *American Journal of Physiology. Heart and Circulatory Physiology, 298*, H2082–H2092.

Lopez-Mateo, I., Villaronga, M. A., Llanos, S., & Belandia, B. (2012). The transcription factor CREBZF is a novel positive regulator of p53. *Cell Cycle, 11*, 3887–3895.

Luna-Zurita, L., Prados, B., Grego-Bessa, J., Luxan, G., del Monte, G., Benguria, A., et al. (2010). Integration of a Notch-dependent mesenchymal gene program and Bmp2-driven cell invasiveness regulates murine cardiac valve formation. *Journal of Clinical Investigation, 120*, 3493–3507.

Maier, M. M., & Gessler, M. (2000). Comparative analysis of the human and mouse Hey1 promoter: Hey genes are new Notch target genes. *Biochemical and Biophysical Research Communications, 275*, 652–660.

Mal, A., Sturniolo, M., Schiltz, R. L., Ghosh, M. K., & Harter, M. L. (2001). A role for histone deacetylase HDAC1 in modulating the transcriptional activity of MyoD: Inhibition of the myogenic program. *EMBO Journal, 20*, 1739–1753.

Mann, C. D., Bastianpillai, C., Neal, C. P., Masood, M. M., Jones, D. J., Teichert, F., et al. (2012). Notch3 and HEY-1 as prognostic biomarkers in pancreatic adenocarcinoma. *PloS One*, 7, e51119.

Monastirioti, M., Giagtzoglou, N., Koumbanakis, K. A., Zacharioudaki, E., Deligiannaki, M., Wech, I., et al. (2010). Drosophila Hey is a target of Notch in asymmetric divisions during embryonic and larval neurogenesis. *Development (Cambridge, England), 137*, 191–201.

Moya, I. M., Umans, L., Maas, E., Pereira, P. N., Beets, K., Francis, A., et al. (2012). Stalk cell phenotype depends on integration of Notch and Smad1/5 signaling cascades. *Developmental Cell*, *22*, 501–514.

Mukhopadhyay, A., Jarrett, J., Chlon, T., & Kessler, J. A. (2009). HeyL regulates the number of TrkC neurons in dorsal root ganglia. *Developmental Biology*, *334*, 142–151.

Nakagawa, O., McFadden, D. G., Nakagawa, M., Yanagisawa, H., Hu, T., Srivastava, D., et al. (2000). Members of the HRT family of basic helix-loop-helix proteins act as transcriptional repressors downstream of Notch signaling. *Proceedings of the National Academy of Sciences of the United States of America*, *97*, 13655–13660.

Nakagawa, O., Nakagawa, M., Richardson, J. A., Olson, E. N., & Srivastava, D. (1999). HRT1, HRT2, and HRT3: A new subclass of bHLH transcription factors marking specific cardiac, somitic, and pharyngeal arch segments. *Developmental Biology*, *216*, 72–84.

Ozawa, Y., Towatari, M., Tsuzuki, S., Hayakawa, F., Maeda, T., Miyata, Y., et al. (2001). Histone deacetylase 3 associates with and represses the transcription factor GATA-2. *Blood*, *98*, 2116–2123.

Parker, B. S., Argani, P., Cook, B. P., Liangfeng, H., Chartrand, S. D., Zhang, M., et al. (2004). Alterations in vascular gene expression in invasive breast carcinoma. *Cancer Research*, *64*, 7857–7866.

Patel, N. S., Li, J. L., Generali, D., Poulsom, R., Cranston, D. W., & Harris, A. L. (2005). Up-regulation of delta-like 4 ligand in human tumor vasculature and the role of basal expression in endothelial cell function. *Cancer Research*, *65*, 8690–8697.

Pendeville, H., Winandy, M., Manfroid, I., Nivelles, O., Motte, P., Pasque, V., et al. (2008). Zebrafish Sox7 and Sox18 function together to control arterial-venous identity. *Developmental Biology*, *317*, 405–416.

Pietras, A., von Stedingk, K., Lindgren, D., Pahlman, S., & Axelson, H. (2011). JAG2 induction in hypoxic tumor cells alters Notch signaling and enhances endothelial cell tube formation. *Molecular Cancer Research*, *9*, 626–636.

Popovic, M., Wienk, H., Coglievina, M., Boelens, R., Pongor, S., & Pintar, A. (2014). The basic-helix-loop-helix region of the transcriptional repressor HES-1 is preorganized to bind DNA. *Proteins*, *82*(4), 537–545.

Proweller, A., Pear, W. S., & Parmacek, M. S. (2005). Notch signaling represses myocardin-induced smooth muscle cell differentiation. *Journal of Biological Chemistry*, *280*, 8994–9004.

Ricard, N., Ciais, D., Levet, S., Subileau, M., Mallet, C., Zimmers, T. A., et al. (2012). BMP9 and BMP10 are critical for postnatal retinal vascular remodeling. *Blood*, *119*, 6162–6171.

Sakamoto, M., Hirata, H., Ohtsuka, T., Bessho, Y., & Kageyama, R. (2003). The basic helix-loop-helix genes Hesr1/Hey1 and Hesr2/Hey2 regulate maintenance of neural precursor cells in the brain. *Journal of Biological Chemistry*, *278*, 44808–44815.

Sakata, Y., Kamei, C. N., Nakagami, H., Bronson, R., Liao, J. K., & Chin, M. T. (2002). Ventricular septal defect and cardiomyopathy in mice lacking the transcription factor CHF1/Hey2. *Proceedings of the National Academy of Sciences of the United States of America*, *99*, 16197–16202.

Sakata, Y., Koibuchi, N., Xiang, F., Youngblood, J. M., Kamei, C. N., & Chin, M. T. (2006). The spectrum of cardiovascular anomalies in CHF1/Hey2 deficient mice reveals roles in endocardial cushion, myocardial and vascular maturation. *Journal of Molecular and Cellular Cardiology*, *40*, 267–273.

Sakata, Y., Xiang, F., Chen, Z., Kiriyama, Y., Kamei, C. N., Simon, D. I., et al. (2004). Transcription factor CHF1/Hey2 regulates neointimal formation in vivo and vascular smooth muscle proliferation and migration in vitro. *Arteriosclerosis, Thrombosis, and Vascular Biology*, *24*, 2069–2074.

Salie, R., Kneissel, M., Vukevic, M., Zamurovic, N., Kramer, I., Evans, G., et al. (2010). Ubiquitous overexpression of Hey1 transcription factor leads to osteopenia and chondrocyte hypertrophy in bone. *Bone, 46*, 680–694.

Sartorelli, V., Puri, P. L., Hamamori, Y., Ogryzko, V., Chung, G., Nakatani, Y., et al. (1999). Acetylation of MyoD directed by PCAF is necessary for the execution of the muscle program. *Molecular Cell, 4*, 725–734.

Satow, T., Bae, S. K., Inoue, T., Inoue, C., Miyoshi, G., Tomita, K., et al. (2001). The basic helix-loop-helix gene hesr2 promotes gliogenesis in mouse retina. *Journal of Neuroimmune Pharmacology, 21*, 1265–1273.

Scott, R. H., Douglas, J., Baskcomb, L., Huxter, N., Barker, K., Hanks, S., et al. (2008). Constitutional 11p15 abnormalities, including heritable imprinting center mutations, cause nonsyndromic Wilms tumor. *Nature Genetics, 40*, 1329–1334.

Sharff, K. A., Song, W. X., Luo, X., Tang, N., Luo, J., Chen, J., et al. (2009). Hey1 basic helix-loop-helix protein plays an important role in mediating BMP9-induced osteogenic differentiation of mesenchymal progenitor cells. *Journal of Biological Chemistry, 284*, 649–659.

Shirvani, S. M., Mookanamparambil, L., Ramoni, M. F., & Chin, M. T. (2007). Transcription factor CHF1/Hey2 regulates the global transcriptional response to platelet-derived growth factor in vascular smooth muscle cells. *Physiological Genomics, 30*, 61–68.

Shirvani, S., Xiang, F., Koibuchi, N., & Chin, M. T. (2006). CHF1/Hey2 suppresses SM-MHC promoter activity through an interaction with GATA-6. *Biochemical and Biophysical Research Communications, 339*, 151–156.

Sun, J., Kamei, C. N., Layne, M. D., Jain, M. K., Liao, J. K., Lee, M. E., et al. (2001). Regulation of myogenic terminal differentiation by the hairy-related transcription factor CHF2. *Journal of Biological Chemistry, 276*, 18591–18596.

Takata, T., & Ishikawa, F. (2003). Human Sir2-related protein SIRT1 associates with the bHLH repressors HES1 and HEY2 and is involved in HES1- and HEY2-mediated transcriptional repression. *Biochemical and Biophysical Research Communications, 301*, 250–257.

Tang, Y., Urs, S., & Liaw, L. (2008). Hairy-related transcription factors inhibit Notch-induced smooth muscle alpha-actin expression by interfering with Notch intracellular domain/CBF-1 complex interaction with the CBF-1-binding site. *Circulation Research, 102*, 661–668.

Tateya, T., Imayoshi, I., Tateya, I., Ito, J., & Kageyama, R. (2011). Cooperative functions of Hes/Hey genes in auditory hair cell and supporting cell development. *Developmental Biology, 352*, 329–340.

Tokuzawa, Y., Yagi, K., Yamashita, Y., Nakachi, Y., Nikaido, I., Bono, H., et al. (2010). Id4, a new candidate gene for senile osteoporosis, acts as a molecular switch promoting osteoblast differentiation. *PLoS Genetics, 6*, e1001019.

Tradonsky, A., Rubin, T., Beck, R., Ring, B., Seitz, R., & Mair, S. (2012). A search for reliable molecular markers of prognosis in prostate cancer: A study of 240 cases. *American Journal of Clinical Pathology, 137*, 918–930.

Tu, X., Chen, J., Lim, J., Karner, C. M., Lee, S. Y., Heisig, J., et al. (2012). Physiological notch signaling maintains bone homeostasis via RBPjk and Hey upstream of NFATc1. *PLoS Genetics, 8*, e1002577.

van den Berg, A., Kroesen, B. J., Kooistra, K., de Jong, D., Briggs, J., Blokzijl, T., et al. (2003). High expression of B-cell receptor inducible gene BIC in all subtypes of Hodgkin lymphoma. *Genes, Chromosomes & Cancer, 37*, 20–28.

Van Wayenbergh, R., Taelman, V., Pichon, B., Fischer, A., Kricha, S., Gessler, M., et al. (2003). Identification of BOIP, a novel cDNA highly expressed during spermatogenesis that encodes a protein interacting with the orange domain of the hairy-related transcription factor HRT1/Hey1 in Xenopus and mouse. *Developmental Dynamics, 228*, 716–725.

Wang, L., Motoi, T., Khanin, R., Olshen, A., Mertens, F., Bridge, J., et al. (2012). Identification of a novel, recurrent HEY1-NCOA2 fusion in mesenchymal chondrosarcoma based on a genome-wide screen of exon-level expression data. *Genes, Chromosomes & Cancer, 51*, 127–139.

Watanabe, T., Koibuchi, N., & Chin, M. T. (2010). Transcription factor CHF1/Hey2 regulates coronary vascular maturation. *Mechanisms of Development, 127*, 418–427.

Watanabe, Y., Kokubo, H., Miyagawa-Tomita, S., Endo, M., Igarashi, K., Aisaki, K., et al. (2006). Activation of Notch1 signaling in cardiogenic mesoderm induces abnormal heart morphogenesis in mouse. *Development (Cambridge, England), 133*, 1625–1634.

Wiese, C., Heisig, J., & Gessler, M. (2010). Hey bHLH factors in cardiovascular development. *Pediatric Cardiology, 31*, 363–370.

Williams, C. K., Li, J. L., Murga, M., Harris, A. L., & Tosato, G. (2006). Up-regulation of the Notch ligand delta-like 4 inhibits VEGF-induced endothelial cell function. *Blood, 107*, 931–939.

Winkler, C., Elmasri, H., Klamt, B., Volff, J. N., & Gessler, M. (2003). Characterization of hey bHLH genes in teleost fish. *Development Genes and Evolution, 213*, 541–553.

Wu, S. P., Cheng, C. M., Lanz, R. B., Wang, T., Respress, J. L., Ather, S., et al. (2013). Atrial identity is determined by a COUP-TFII regulatory network. *Developmental Cell, 25*, 417–426.

Xiang, F., Sakata, Y., Cui, L., Youngblood, J. M., Nakagami, H., Liao, J. K., et al. (2006). Transcription factor CHF1/Hey2 suppresses cardiac hypertrophy through an inhibitory interaction with GATA4. *American Journal of Physiology Heart and Circulatory Physiology, 290*, H1997–H2006.

Xin, M., Small, E. M., van Rooij, E., Qi, X., Richardson, J. A., Srivastava, D., et al. (2007). Essential roles of the bHLH transcription factor Hrt2 in repression of atrial gene expression and maintenance of postnatal cardiac function. *Proceedings of the National Academy of Sciences of the United States of America, 104*, 7975–7980.

Yoo, J., Lee, H. N., Choi, I., Choi, D., Chung, H. K., Kim, K. E., et al. (2012). Opposing regulation of PROX1 by interleukin-3 receptor and NOTCH directs differential host cell fate reprogramming by Kaposi sarcoma herpes virus. *PLoS Pathogens, 8*, e1002770.

Yu, M., Liu, Y., Xiang, F., Li, Y., Cullen, D., Liao, R., et al. (2009). CHF1/Hey2 promotes physiological hypertrophy in response to pressure overload through selective repression and activation of specific transcriptional pathways. *OMICS, 13*, 501–511.

Yu, M., Xiang, F., Beyer, R. P., Farin, F. M., Bammler, T. K., & Chin, M. T. (2010). Transcription factor CHF1/Hey2 regulates specific pathways in serum stimulated primary cardiac myocytes: Implications for cardiac hypertrophy. *Current Genomics, 11*, 287–296.

Zamurovic, N., Cappellen, D., Rohner, D., & Susa, M. (2004). Coordinated activation of notch, Wnt, and transforming growth factor-beta signaling pathways in bone morphogenic protein 2-induced osteogenesis. Notch target gene Hey1 inhibits mineralization and Runx2 transcriptional activity. *Journal of Biological Chemistry, 279*, 37704–37715.

Zanotti, S., & Canalis, E. (2013). Hairy and Enhancer of Split-related with YRPW motif (HEY)2 regulates bone remodeling in mice. *Journal of Biological Chemistry, 288*, 21547–21557.

Zavadil, J., Cermak, L., Soto-Nieves, N., & Bottinger, E. P. (2004). Integration of TGF-beta/Smad and Jagged1/Notch signalling in epithelial-to-mesenchymal transition. *EMBO Journal, 23*, 1155–1165.

Zhong, T. P., Childs, S., Leu, J. P., & Fishman, M. C. (2001). Gridlock signalling pathway fashions the first embryonic artery. *Nature, 414*, 216–220.

Zhong, T. P., Rosenberg, M., Mohideen, M. A., Weinstein, B., & Fishman, M. C. (2000). Gridlock, an HLH gene required for assembly of the aorta in zebrafish. *Science (New York), 287*, 1820–1824.

CHAPTER NINE

Stra13 and Sharp-1, the Non-Grouchy Regulators of Development and Disease

Jin Rong Ow[*,†], **Yong Hua Tan**[*], **Yu Jin**[*], **Avinash G. Bahirvani**[*], **Reshma Taneja**[*,†,1]

[*]Department of Physiology, Yong Loo Lin School of Medicine, National University of Singapore, Singapore, Singapore

[†]National University of Singapore Graduate School for Integrative Sciences and Engineering, National University of Singapore, Singapore, Singapore

[1]Corresponding author: e-mail address: phsrt@nus.edu.sg

Contents

Abstract

Stra13 and Sharp-1 are transcriptional repressors that share domain structure similarity with members of the basic helix-loop-helix-Orange subfamily. In contrast to other members that include Hes and Hey proteins, transcriptional repression mediated by Stra13 and Sharp-1 does not involve recruitment of the corepressor Groucho. Both proteins undergo sumoylation at evolutionarily conserved sites, and this posttranslational modification serves as a platform for association with chromatin-modifying enzymes including histone deacetylases and histone methyltransferases. In addition to being widely expressed during embryonic development and in adult tissues, the expression of both genes is induced by a number of stimuli. Loss-of-function and gain-of-function studies have demonstrated their function in cellular differentiation and regeneration, in regulation of circadian rhythms, immune homeostasis, and metabolism. Given their diverse

Current Topics in Developmental Biology, Volume 110
ISSN 0070-2153
http://dx.doi.org/10.1016/B978-0-12-405943-6.00009-9

© 2014 Elsevier Inc.
All rights reserved.

physiological functions in several tissues, it is not surprising that deregulated expression of Stra13 and Sharp-1 is apparent in human pathologies. Here, we review our current understanding of their cellular functions that suggest a requirement in maintenance of tissue homeostasis.

1. INTRODUCTION

A subtractive hybridization screen aimed at identifying retinoic acid-responsive genes in P19 embryonal carcinoma cells led to the identification of a novel gene *Stra13* (stimulated with retinoic acid 13) (Boudjelal et al., 1997). Around the same time, orthologues were discovered from a human chondrocyte cell line cultured in presence of dibutyryl cyclic AMP, and from rat brain, that were named *Dec1* and *Sharp-2*, respectively (Rossner, Dörr, Gass, Schwab, & Nave, 1997; Shen et al., 1997). A highly related gene *Sharp-1* was also isolated from rat brain (Rossner et al., 1997). Subsequent studies led to the cloning of its human and mouse counterparts named *Dec2* and *Sharp-1* (Azmi & Taneja, 2002; Fujimoto et al., 2001). Orthologues for *Stra13/Dec1/Sharp-2* and *Sharp-1/Dec2* have also been identified in zebrafish (Chen, Zhou, Xu, Xu, & Xue, 2010; Yao et al., 2006) as well as in *Drosophila melanogaster*, which contains a single gene named *clockwork orange (cwo)* (Kadener, Stoleru, McDonald, Nawathean, & Rosbash, 2007). To avoid confusion, in this review, we shall refer to Stra13/Sharp-2/Dec1 as Stra13 and Sharp-1/Dec2 as Sharp-1 regardless of species.

2. MECHANISMS OF TRANSCRIPTIONAL REPRESSION

Based on their domain structure and transcriptional properties, Stra13 and Sharp-1 have been classified as transcriptional repressors (Davis & Turner, 2001). Both proteins share the highest sequence homology with each other relative to other members of the basic helix-loop-helix (bHLH) transcription repressor subfamilies that include the Hes and Hey proteins in vertebrates (Fig. 9.1).

The two proteins share 98% identity in the conserved bHLH domain and bind to class B E-box sites [CACGTG] with high affinity as homodimers (Azmi, Sun, Ozog, & Taneja, 2003; St-Pierre, Flock, Zacksenhaus, & Egan, 2002). Transcriptional repression by these factors is multifaceted and involves several mechanisms. Repression has been shown to be dependent on DNA binding at target gene promoters (Li et al., 2003; Rossner

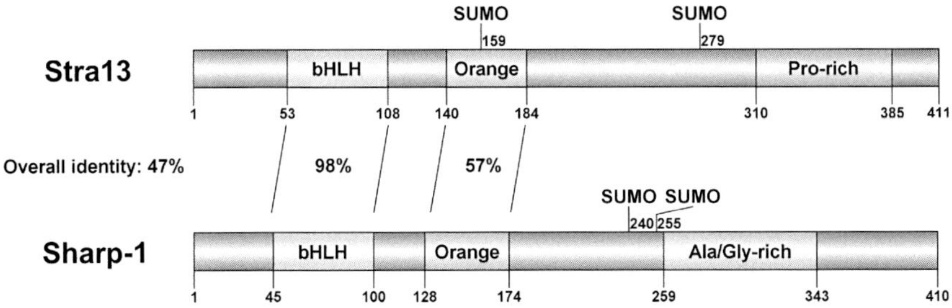

Figure 9.1 Domain structure and posttranslational modifications of mouse Stra13 and Sharp-1. The conserved bHLH and Orange domains as well as sumoylation sites (SUMO) in both proteins are shown. Numbers reflect amino acid residues. The image was made using the domain visualization software D.O.G. 2.0 (Ren et al., 2009). (See the color plate.)

et al., 2008). In addition, various cofactors have been shown to interact with Stra13 and Sharp-1. In contrast to other transcriptional repressor subfamilies, Stra13 and Sharp-1 lack a WRPW/YRPW motif in the C-terminus that is required to recruit the corepressor Groucho (Fisher, Ohsako, & Caudy, 1996). Early studies showed that transcriptional repression by Stra13 is mediated by recruitment of histone deacetylase 1 (HDAC1) that plays a role in its own transcription through an autoregulatory loop (Sun & Taneja, 2000). Subsequent studies have shown that HDAC1 recruitment is also important for repression of a number of target genes such as STAT1 (Ivanov, Salnikow, Ivanova, Bai, & Lerman, 2007), RXRα (Cho et al., 2009), SREBP1c (Cho et al., 2009), and PPARγ2 (Park & Park, 2012). Similarly, Sharp-1 was also found to interact with HDAC1 (Fujimoto et al., 2007; Garriga-Canut, Roopra, & Buckley, 2001; Li et al., 2011) and more recently with the histone methyltransferase G9a (Ling, Gopinadhan, et al., 2012). Interestingly, recruitment of HDAC1 by Stra13, and G9a by Sharp-1 is dependent on sumoylation. Stra13 has canonical sumoylation sites at lysine (K) 159 and K279 (Hong et al., 2011). Mutation of these sites abrogates recruitment of HDAC1. Similarly, Sharp-1 is sumoylated at K240 and K255, which regulates association with G9a (Wang, Shankar, Kher, Ling, & Taneja, 2013). The recruitment of these cofactors is important for Stra13 and Sharp-1-dependent functions. For instance, sumoylation of Stra13 is essential for mediating cellular growth arrest, whereas Sharp-1 sumoylation regulates its ability to inhibit myogenic differentiation (Wang et al., 2012, 2013).

In addition to recruitment of corepressors, Stra13 mediates repression through interaction with components of the basal transcription machinery (Sun & Taneja, 2000). While both proteins bind DNA as homodimers, they readily dimerize with several other bHLH proteins including MyoD,

E12/E47, USF, as well as with each other. In addition, both proteins interact with other transcription factors such as C/EBP and MyoD to repress their transcriptional activity and function (Azmi, Ozog, & Taneja, 2004; Dhar & Taneja, 2001; Fujimoto et al., 2007; Gulbagci et al., 2009; Ling, Bharathy, et al., 2012). Thus, transcriptional repression is mediated by both active and passive mechanisms. This includes (1) direct DNA binding to E-box sites of target promoters; (2) recruitment of co-repressors that mediate repressive chromatin marks or methylate non-histone proteins; (3) dimerization with bHLH factors to form nonfunctional complexes; (4) competition for binding to E-box sites; (5) interaction with non-bHLH transcription factors to inhibit their transcriptional activity and; (6) promotion of proteasome-mediated degradation (Fig. 9.2).

3. EMBRYONIC AND ADULT TISSUE EXPRESSION

In developing mouse embryos, *Stra13* is first seen in the extra-embryonic trophectoderm at embryonic day (E) 8.5, and from E9.5 dpc, it is expressed in the developing neural system, heart, skeletal muscles, thymus, kidney, and the gastrointestinal tract (Boudjelal et al., 1997). Similarly, in zebrafish, *Stra13* is expressed in the blastula and gastrula and subsequently restricted to various parts of the nervous system, heart, skeletal muscle, and kidneys as organogenesis proceeds (Yao et al., 2006), highlighting a conservation of the developmental expression of Stra13 in these tissues.

Sharp-1 expression at E8.5 is apparent in the developing forebrain and heart and later becomes prominent in the central nervous system and in sensory organs. From E10.5, *Sharp-1* is expressed in the myotome as well as in the developing limbs and expands to liver and prevertebrae from E12.5 (Azmi & Taneja, 2002). This expression profile is recapitulated during zebrafish development, with expression in the nervous system, sensory organs, somites, and heart (Chen et al., 2010). A comparison of *Stra13* and *Sharp-1* spatiotemporal expression in zebrafish embryogenesis shows that while both genes share certain spatial expression, *Sharp-1* eventually becomes more tissue specific while *Stra13* maintains ubiquitous expression. Temporally, however, these two proteins are expressed at different developmental stages. Thus, despite similarities in their overall structure, DNA-binding specificities and, thereby, target genes, their expression may be distinct in some cell types.

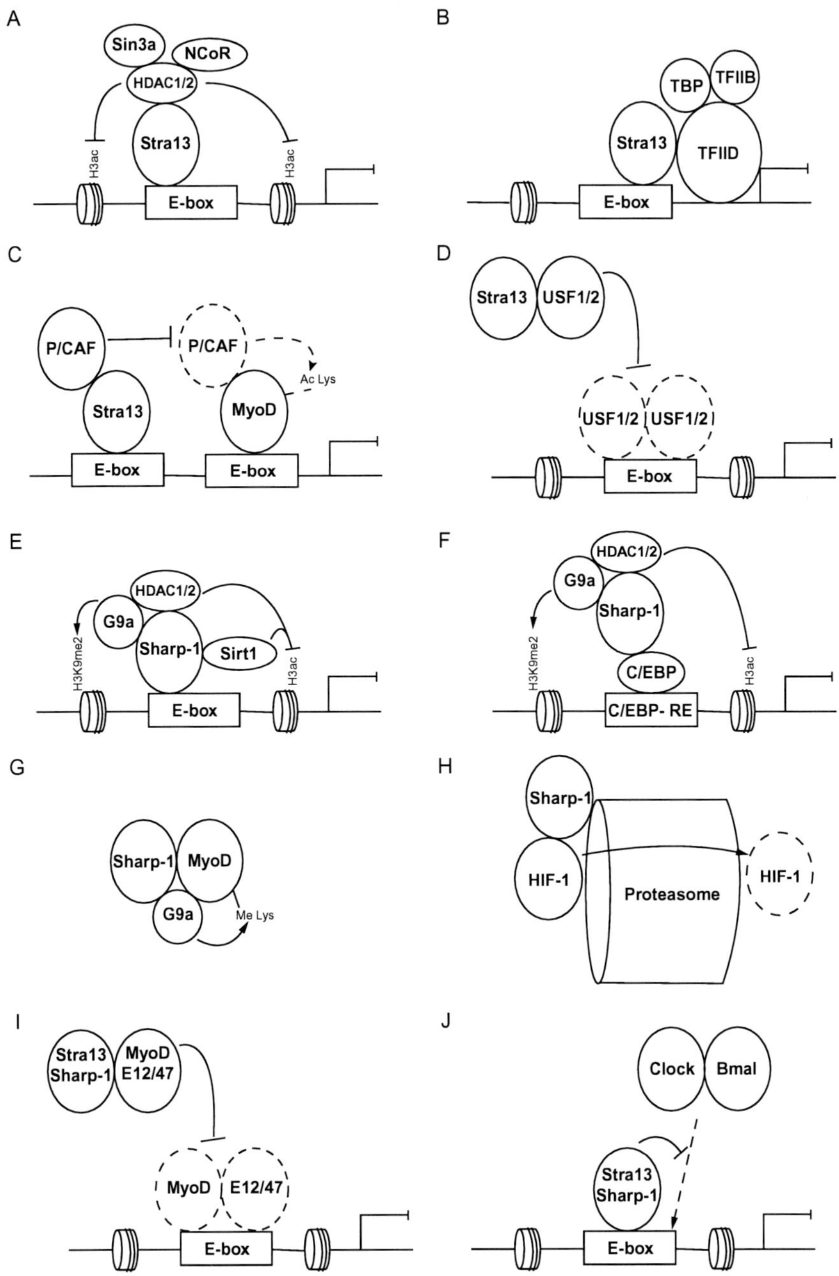

Figure 9.2 Stra13 and Sharp-1 mediate transcriptional repression by multiple mechanisms. Stra13 can repress target genes by (A) recruiting the HDAC1/Sin3A/NCoR corepressor complex or (B) interacting with components of the TFIID complex. (C) In skeletal

(*Continued*)

In adult tissues, *Sharp-1* is expressed in skeletal muscle and brain, and at lower levels in the heart, pancreas, lung, placenta, liver, and kidney (Azmi et al., 2003; Fujimoto et al., 2001; Rossner et al., 1997). On the other hand, *Stra13* is expressed more ubiquitously (Boudjelal et al., 1997; Rossner et al., 1997), suggesting a possible divergence in the role of these proteins in tissue function and maintenance. Moreover, the expression of both genes is upregulated by numerous stimuli, including retinoic acid, serum deprivation, TGF-β, cyclic AMP, light, cytokines, insulin, and hypoxia (Boudjelal et al., 1997; Chen, Kang, & Massagué, 2001; Honma et al., 2002; Miyazaki et al., 2002; Seimiya et al., 2002; Shen et al., 2001; Sun & Taneja, 2000; Yamada et al., 2003; Zawel et al., 2002).

4. ROLES IN DIFFERENTIATION AND DEVELOPMENT

4.1. Musculoskeletal system

The first direct implication that Stra13 regulates differentiation of several cell types came from gain-of-function studies in P19 cells that resulted in repression of mesodermal and endodermal markers. These studies suggested that Stra13 may function as a transcriptional repressor of mesodermal differentiation pathway (Boudjelal et al., 1997).

Indeed, *in vitro*, both Stra13 and Sharp-1 inhibit differentiation of skeletal muscle precursor cells. Stra13 represses MyoD activity, resulting in reduced expression of myogenic genes including *PGC-1α*, *MCK*, and *Myog*. This inhibition is relieved by PCAF (Hsiao, Huang, Chang, & Chen, 2009). Since *PGC-1α*, which is preferentially expressed in slow-twitch muscles (Hsiao et al., 2009), is regulated by Stra13, it is possible that Stra13 may control fiber-type switching, although such phenotype has not been

Figure 9.2—Cont'd muscles cells, Stra13 sequesters P/CAF, thereby prevents MyoD activity. (D) Stra13 interacts with USF and prevents its ability to bind DNA. (E) In a manner similar to Stra13, Sharp-1 can recruit the histone deacetylases HDAC1 and Sirt1, as well as the histone methyltransferase G9a and mediate repressive chromatin modifications that repress target gene expression. (F) Sharp-1 interacts with transcription factors such as C/EBP to inhibit their transcriptional activity likely by recruitment of HDAC1 and G9a. (G) Sharp-1 represses MyoD activity by forming a complex with G9a and MyoD to bring about lysine methylation of MyoD. (H) Sharp-1 is induced by hypoxia but can block hypoxic responses by presenting HIF-1α to the proteasome complex to induce its degradation. (I, J) Mechanisms utilized by both Stra13 and Sharp-1 involve sequestering MyoD or E12/47 to prevent formation of functional MyoD-E47 DNA-binding heterodimers, and regulation of the molecular clock by repressing Clock/Bmal-induced activation of the Per1 promoter through competition for binding to E-box sites.

reported in *Stra13*$^{-/-}$ mutants. Similar to Stra13, Sharp-1 also represses myogenic differentiation via inhibition of MyoD activity. This occurs via multiple mechanisms. Sharp-1 efficiently dimerizes with MyoD and E47, thus preventing formation of functional MyoD–E47 heterodimers. Sharp-1 also recruits G9a, which catalyzes repressive histone 3 lysine 9 dimethylation (H3K9me2) at promoters of MyoD-target genes (Ling, Gopinadhan, et al., 2012) and also methylates MyoD to inhibit its transcriptional activity (Ling, Bharathy, et al., 2012).

Despite their expression in embryonic muscle, neither *Stra13*$^{-/-}$ nor *Sharp-1*$^{-/-}$ mice exhibit overt developmental defects in skeletal muscle. Neither the size of muscle fibers nor is the number of Pax7^{+} satellite cells altered. However, both null mutants show a remarkably similar phenotype with an impaired regenerative response to injury (Acharjee et al., 2014; Sun et al., 2007). Interestingly, consistent with their ability to inhibit myogenic differentiation, the early stages of muscle regeneration are accelerated, likely due to a relief of MyoD repression in both null mutant mice. At late stages, however, sustained Notch and TGFβ signaling result in muscle necrosis and degeneration, resulting in a dystrophy-like phenotype.

The role of Stra13 in muscular dystrophy has been further evaluated using a mouse model for Duchenne muscular dystrophy (mdx). Double mutants lacking both Stra13 and dystrophin show enhanced dystrophic pathology and degeneration of muscle fibers, which is ameliorated with the antioxidant *N*-acetylcysteine. These data establish a link between Stra13 and oxidative stress in skeletal muscle (Vercherat et al., 2009). Consistent with its potent ability to inhibit myogenesis, Sharp-1 is overexpressed in inclusion body myositis mesangioblasts that do not differentiate into skeletal muscle (Morosetti et al., 2006). It is, however, unknown whether Sharp-1 plays a direct pathological role in this context.

Besides skeletal muscle, Stra13 is important in other tissues of the musculoskeletal system. *Stra13* was cloned as a gene induced with dibutyryl cAMP during chondrocyte differentiation (Shen et al., 1997). When ectopically expressed, Stra13 accelerates chondrogenic differentiation in mesenchymal stem cells (MSCs) (Shen et al., 2001, 2002). Nevertheless, the fact that overexpression of Stra13 in MSCs also leads to osteogenesis with increased expression of osteopontin and alkaline phosphatase (Iwata et al., 2006) questions the lineage-determining function of Stra13. Rather, Stra13 may act in cooperation with other lineage-specific transcription factors that direct chondrogenesis versus osteogenesis while repressing the adipogenic lineage in MSCs.

4.2. Metabolism

Stra13 was first shown to regulate adipogenesis as an effector of hypoxia-inducible factor (Yun, Maecker, Johnson, & Giaccia, 2002), and subsequent studies showed that it recruits HDAC1 to repress C/EBPβ transcriptional activity (Park & Park, 2012). In a similar mechanism of action, Sharp-1 also interacts with C/EBPβ to form a repressive complex containing HDAC1 and G9a at the promoters of *C/EBPα* and *PPARγ* (Gulbagci et al., 2009). These studies were the first to suggest that Sharp-1 may repress adipogenic differentiation via epigenetic mechanisms that have been subsequently validated in skeletal myogenesis (Ling, Bharathy, et al., 2012; Ling, Gopinadhan, et al., 2012). Stra13 and Sharp-1 are also regulated by RORα and mediate RORα-dependent repression of adipogenesis (Ozaki et al., 2012).

Under hypoxic conditions, Stra13 and Sharp-1 repress Sterol regulatory element-binding protein 1c (SREBP1c), involved in inducing several enzymes of the lipogenesis pathway, including fatty acid synthase. Stra13 inhibits SREBP1c DNA binding which may occur by competition for binding to the E-box site in the *SREBP1c* promoter or by direct interaction that reduces the affinity of SREBP1c for DNA (Choi et al., 2008).

Besides regulating adipogenesis and lipogenesis in adipocytes, Stra13 and Sharp-1 are also involved in regulation of lipogenesis in the liver. *Stra13* is an insulin-responsive gene that is upregulated through the PI3K pathway (Yamada et al., 2003) and regulates other insulin-responsive genes such as phosphoenolpyruvate carboxykinase (Yamada, Ogata-Kawata, Matsuura, & Miyamoto, 2005). Independently, glucose was also discovered to activate the expression of Stra13 through carbohydrate response element-binding protein (ChREBP) binding at the carbohydrate response element (ChoRE) in the *Stra13* promoter. Interestingly, in a negative feedback loop, Stra13 represses ChREBP activity on the ChoRE of lipogenic genes, likely to prevent excessive lipogenesis in the presence of high glucose (Iizuka & Horikawa, 2008). Along with Stra13, Sharp-1 also acts as repressor of the retinoic acid receptor (RXR) and retinoic acid receptor–liver X receptor (RXR–LXR) by interacting with RXR and repressing target genes mediated by the RXR–LXR heterodimer (Cho et al., 2009; Noshiro et al., 2009). The repression of RXRα transactivation by Stra13 is HDAC dependent, but HDAC independent for Sharp-1, and plays an important role in regulation of SREBP1c expression in liver (Cho et al., 2009). Together, these studies highlight the multiple pathways that Stra13 and Sharp-1 regulate metabolism by modulating adipogenesis and lipogenesis in adipocytes and hepatocytes. In view of this, dysregulation of these proteins may lead to or contribute to metabolic syndromes resulting from diabetes or obesity.

4.3. Immune system

A key role for Stra13 and Sharp-1 is in regulation of the immune system. Initial studies demonstrated that Stra13 is upregulated upon T-cell activation, and *Stra13*$^{-/-}$ mice exhibit lymphoid organ hyperplasia upon aging in a mixed genetic background (Sun, Lu, Li, Flavell, & Taneja, 2001). Independently, Stra13 was found to play a role in homeostasis of regulatory T (T_{reg}) cells as a marked reduction of these cells was observed in aged *Stra13*$^{-/-}$ mice (Miyazaki et al., 2010). Stra13 was found to bind to the CD25 gene [the alpha chain for IL-2 receptor (IL-2R)] together with Runx-1, which is required for long-term maintenance of T_{reg} cells in the periphery to prevent effector T cell-mediated autoimmunity. Recently, using a mouse model of autoimmune encephalomyelitis (EAE), Stra13 was found to be required for expression of proinflammatory cytokines such as IL-2, which mediate survival and proliferation of activated T cells, and thus is necessary for efficient T cell priming during the development of EAE (Lin et al., 2014; Martínez-Llordella et al., 2013).

In a gain-of-function approach, Stra13 transgenic mice (*Stra13-Tg*) that ectopically express Stra13 in B and T cells were shown to have impaired bone marrow and thymus development, with reduction in pro-B cell and pro-T cell populations (Seimiya et al., 2004). Mechanistically, it has been proposed that Stra13 may act by delaying B cell entry into S phase of the cell cycle (Seimiya et al., 2002), thus explaining the lower numbers of immune cell progenitors in the developing immune organs of *Stra13-Tg* mice.

Like Stra13, Sharp-1 is also a regulator of T-cell activation. Sharp-1 selectively promotes Th2 differentiation, as opposed to Th1 and Th17, by controlling the expression of IL-4 indirectly through Gata3 and Junb. In a reciprocal manner, Gata3 further regulates Sharp-1 expression in an autoregulatory loop that enhances Th2 differentiation signals. Besides involvement in IL-4 signaling pathway, Sharp-1 also influences IL-2 signal transduction. Sharp-1 upregulates the expression of CD25 in a Stat6-dependent manner, which promotes Th2 differentiation through enhancement of the IL-2R-mediated signaling pathway. Despite the above-mentioned roles of Sharp-1 in immune cells, Sharp-1 transgenic and knockout mice do not appear to exhibit any overt phenotype. However, when challenged with chicken ovalbumin or

Schistosoma mansoni eggs, there is increased Th2 response in the Sharp-1 transgenic mice and a corresponding defective Th2 response in Sharp-1 knockout mice (Liu et al., 2009; Yang et al., 2009). Together, these data indicate that Sharp-1 is required for Th2 responses *in vivo*.

Other than T cells, Sharp-1 is expressed in cells of the innate immune system (Yang et al., 2009). Mesangial cells, which express pattern recognition receptors, are innate immune cells that mediate immune surveillance in the kidneys. Upon challenge with polyinosinic–polycytidylic acid (poly-IC) to simulate viral infection, Sharp-1 is upregulated downstream of Toll-like receptor 3. While Sharp-1 does not appear to regulate the acute immune response, knockdown of Sharp-1 resulted in a retarded decline of IFN-β expression. As such, Sharp-1 may function in a feedback loop that prevents sustained antiviral immune responses that may result in inflammatory renal disease such as lupus nephritis (Imaizumi et al., 2011).

5. DEREGULATION OF STRA13 AND SHARP-1 IN CANCER

5.1. Roles in cell cycle arrest senescence

Stra13 and Sharp-1 expression is deregulated in a wide spectrum of cancers. This association may be consequent to their functions in cellular proliferation, senescence, apoptosis, DNA repair, and hypoxia responses (Table 9.1).

Many studies have shown that Stra13 mediates cell cycle arrest and promotes cellular senescence. Several mechanisms and pathways may account for growth suppression induced by Stra13 which include downregulation of c-Myc expression (Sun & Taneja, 2000), as regulator of cyclin D1 expression and as a p53 target gene (Qian, Zhang, Yan, & Chen, 2008). Consistent with this, Stra13 knockdown in lung cancer cell lines increased growth rates accompanied by an increase in cyclin D1 expression (Liu, Wang, et al., 2013). This inverse correlation of Stra13 and cyclin D1 was also seen in fibroblast cells (Wang et al., 2012) and NSCLC samples (Liu, Wang, et al., 2013), while gastric cancers revealed the same trend between Stra13 and Ki67 (Jia et al., 2013). Additionally, irradiation of human oral squamous cell carcinoma cell line HSC-2 upregulated Stra13 expression, with *in vivo* studies on irradiated HSC-2 xenografts displaying high levels of Stra13 and low cyclin D1 in tumors compared to controls (Bhawal et al., 2011). In addition to its ability to cause G_1 cell cycle arrest, Stra13 is upregulated during premature senescence induced by K-rasV12 (Collado et al., 2005). Overexpression of Stra13 is sufficient to mediate senescence that is partially p53 dependent in MCF7 cells as p53 knockdown reduces, but does not

Table 9.1 Types of cancers Stra13 and Sharp-1 are associated with

Type of cancer	References
STRA13	
Breast	Chakrabarti et al. (2004), Turley et al. (2004), Ehata et al. (2007), Giusiano et al. (2010, 2011), Wu et al. (2011), Liu, Miao, et al. (2013)
Lung	Giatromanolaki et al. (2003), Zhang and Li (2007), Phadke, Krynetskaia, Mishra, and Krynetskiy (2011), Liu, Wang, et al. (2013)
Liver	Shi et al. (2011), Peng et al. (2012), Ma, Shi, Lu, Wu, and Wang (2013), Yi et al. (2013)
Pancreatic	Yoon et al. (2001), Wu, Sato, Yamada, et al. (2012)
Stomach	Zheng et al. (2009), Zheng et al. (2012), Jia et al. (2013)
Colorectal	Li et al. (2002), Koukourakis et al. (2006)
Oropharyngeal	Bhawal et al. (2011), Xu et al. (2012)
Melanoma	Shen et al. (2001), Olbryt et al. (2006), Feige et al. (2011)
Kidney	Zhu et al. (2013)
Brain	Preusser et al. (2005)
SHARP-1	
Breast	Wu et al. (2002), Liu, Chung, Li, and Taneja (2010), Liu, Sato, et al. (2010), Wu et al. (2011), Montagner et al. (2012)
Lung	Falvella et al. (2008)
Pancreatic	Sato et al. (2012), Inaguma et al. (2013)
Uterus	Yunokawa et al. (2007)
Oropharyngeal	Wu, Sato, Bhawal, et al. (2012)

completely abrogate Stra13-dependent senescence (Qian et al., 2008). On the other hand, p21 is dispensable for Stra13-induced senescence. Stra13 expression is positively correlated to senescence in tumor samples (Xu et al., 2012) and has been used as a senescence marker for hepatoma cell lines (Yi et al., 2013). In addition, glyceraldehyde 3-phosphate dehydrogenase knockdown, which led to defective glycolysis that resulted in energy crisis, drives A549 cells into senescence by upregulating endogenous Stra13 (Phadke et al., 2011).

Like Stra13, Sharp-1 exhibits antiproliferative effects and causes growth arrest (Liu, Chung, et al., 2010; Liu, Sato, et al., 2010). Treatment of rexinoid bexarotene, a potential chemopreventive agent, significantly suppressed hyperplasia in mammary gland from MMTV-erbB2 transgenic mice (Wu et al., 2002). Upon bexarotene treatment, Sharp-1 expression was induced both *in vitro* and *in vivo*, causing the downregulation of cyclin D1, which may underlie its ability to inhibit proliferation of mammary epithelial cells (Li et al., 2008). In agreement, Sharp-1 overexpression downregulates cyclin D1 expression in lung cancer, suggesting Sharp-1 works as a potential tumor suppressor (Falvella et al., 2008).

Intriguingly, despite its antiproliferative and senescence-inducing abilities that are hallmarks of tumor-suppressive mechanisms, Stra13 expression is elevated in several cancers. For instance, Stra13 expression is high in carcinomas compared to normal breast tissue (Chakrabarti et al., 2004; Turley et al., 2004) and is increased with the grade of breast carcinoma (Chakrabarti et al., 2004). Moreover, Stra13 expression positively correlated with casein kinase II (CK-2), a potential protumorigenic protein in breast cancer (Giusiano et al., 2010, 2011; Liu, Miao, et al., 2013). Similar observations were made for gastric tumors where Stra13 expression was elevated in gastric cancer samples as compared to adjacent normal tissues (Jia et al., 2013; Zheng et al., 2009). In these instances, Stra13 may act as a repressor of differentiation not only in development but in pathology as well, where it may serve to maintain a more proliferative stem cell-like phenotype in cancerous cells. In all, these data indicate that *Stra13* may act as an oncogene in particular types of cancer.

While there is less known on the association of Sharp-1 in cancer, recent studies suggest a tumor-suppressive role. Sharp-1 expression is low in lung cancer cell lines and its overexpression inhibits colony formation (Falvella et al., 2008). Similarly, Sharp-1 expression is low in pancreatic cancer tissues but high in adjacent noncancerous tissues, suggesting that Sharp-1 may have an inhibitory effect on the tumor progression of pancreatic cell lines (Sato et al., 2012). More recent studies have shown that *Sharp-1* is a target of p63, a central regulator of metastasis and tumor invasiveness (Adorno et al., 2009). Interestingly, although *Sharp-1* is a hypoxia-inducible gene (see below), in triple negative breast cancers it counteracts the effect of hypoxia and HIF-mediated metastasis by targeting HIF1 to the proteasome (Montagner et al., 2012) and mediates the antimetastatic function of p63.

5.2. Apoptosis

Stra13 has been shown to both induce and block apoptosis in different cell types. For instance, in response to ionizing radiation, *Stra13*$^{-/-}$ thymocytes exhibit reduced apoptosis and p53 levels (Thin, Li, Chung, Sun, & Taneja, 2007) and Stra13 has been shown to interact with p53 (Qian, Jung, & Chen 2012; Thin et al., 2007). Consistently, Stra13 has a proapoptotic effect in paclitaxel-treated MCF7 cells (Wu et al., 2011), and sunitinib treatment of human renal cell carcinoma lines causes p53 activation, with an increase in Stra13 expression (Zhu et al., 2013). In contrast, in some cells, Stra13 appears to repress apoptosis and promote survival (Ehata et al., 2007 Li, Bi, Yan, & Wan, 2006; Li et al., 2002; Peng et al., 2012). Caspases 3, 7, and 9 are downregulated upon Stra13 overexpression in serum deprived cells, and survivin, an antiapoptotic factor belonging to the IAP family, is regulated by it. *Stra13* is a p53 target gene and its overexpression results in growth arrest and senescence similar to p53. However, it inhibits apoptosis by association with p53 and preventing upregulation of a p53 target gene *MIC-1* (Qian et al., 2012).

Like Stra13, Sharp-1 plays an antiapoptotic function in tumor cells. Knockdown of Sharp-1 by siRNA in MCF-7 human breast cancer cells in the presence or the absence of paclitaxel, an antitumor drug, led to enhancement of apoptosis by upregulating the expressions of apoptosis-related genes such as Fas, c-Myc, caspases-8, poly (ADP-ribose) polymerase (PARP), and Bax (Liu, Chung, et al., 2010; Liu, Sato, et al., 2010; Wu et al., 2011). Sharp-1 expression is upregulated in human oral cancer HSC-3 cell line upon cisplatin treatment. In addition, overexpression of Sharp-1 in HSC-3 cells inhibits proapoptotic factor Bim and cisplatin-induced apoptosis (Wu, Sato, Bhawal, et al., 2012).

5.3. Hypoxia signaling and DNA repair

A hypoxic microenvironment is characteristic of many solid tumors and triggers the induction of genes involved in angiogenesis, tumor growth, and metastasis. A primary mediator of hypoxic responses is the transcription factor hypoxia-inducible factor-1 (HIF-1), which is a heterodimer of HIF-1α and HIF-1β (Adams et al., 2009). Target genes of HIF-1 can be categorized into angiogenesis-related factors, glucose transporters and glycolytic enzymes, tumor invasion and metastasis-related factors, and cell proliferation and apoptosis-related factors (Semenza, 2009; Yang, Sun, Wang, & Jiao, 2013). *Stra13* and

Sharp-1 are target genes of HIF-1 and are induced by hypoxia (Feige et al., 2011; Koukourakis et al., 2006; Ma et al., 2013; Miyazaki et al., 2002; Olbryt et al., 2006; Yoon et al., 2001; Zheng et al., 2012). Under normoxic conditions, Stra13 expression is downregulated by von Hippel–Lindau (VHL), a component of VHL/E3 ligase complex which ubiquitinates HIF-1 and targets it for degradation (Ivanova, Ivanov, Danilkovitch-Miagkova, & Lerman, 2001; Wykoff, Pugh, Maxwell, Harris, & Ratcliffe, 2000). Stra13 expression is correlated with HIF-1 levels in various tumors (Chakrabarti et al., 2004; Giatromanolaki et al., 2003; Jia et al., 2013; Turley et al., 2004). However, recent studies also indicate Stra13 as being a regulator of HIF-1 and hypoxia responses (Li et al., 2006; Zhang & Li, 2007). Stra13 overexpression was shown to inhibit the expression of microphthalmia-associated transcription factor (MITF), a lineage-selective oncogene implicated in melanoma and clear-cell sarcoma. Stra13 knockdown was able to prevent HIF-1α-mediated inhibition of MITF expression in hypoxia-treated human melanoma UACC62 cells. Chromatin immunoprecipitation and promoter reporter analysis also further revealed that Stra13 is a transcriptional repressor of MITF, establishing the hypoxia-HIF-1α-Stra13 pathway of MITF regulation (Feige et al., 2011).

Similarly, Sharp-1 negatively regulates vascular endothelial growth factor (VEGF) expression by physically interacting with HIF-1α and disrupting the binding between HIF-1α and the HRE at the *VEGF* promoter under hypoxic conditions. Separately, overexpression of Sharp-1 was found to significantly reduce endogenous HIF-1α protein level in both normoxic and hypoxic conditions through proteasome-mediated degradation (Montagner et al., 2012). Thus, while both Stra13 and Sharp-1 are induced by hypoxia and are direct targets of HIF, they appear to negatively regulate hypoxic signaling through distinct mechanisms.

Stra13 and Sharp-1 impair DNA mismatch repair by binding to the E-box-like motif at *MLH-1* promoter region and resulting in transcription repression of this key human *mismatch repair (MMR)* gene (Nakamura et al., 2008). Downregulation of MLH-1 hinders MMR function, which subsequently causes genomic instability in cancer cells (Yuan, Narayanan, Rockwell, & Glazer, 2000), proposing a potential mechanism for oncogenic roles for these bHLH proteins.

6. CONCLUSION

The past decade has shed light on various aspects of Stra13 and Sharp-1 biology. Insights have been gained into the mechanisms by which these factors function to coordinate transcriptional responses through a multitude of molecular mechanisms. Many gain- and loss-of-function studies have identified their roles in cellular differentiation of various cell types, including skeletal muscle, fat, bone, and cartilage. Both genes are also important in the regeneration of skeletal muscle and homeostasis in the immune system.

Unlike their functions in these physiological processes, less is presently understood on the mechanisms by which their deregulated expression contributes to disease. As discussed above, a growing number of reports have shown deregulated Stra13 and Sharp-1 expression in a wide spectrum of tumors. The role of Stra13 and Sharp-1 in tumorigenesis is complex. Correlative studies of Stra13 and Sharp-1 with tumor markers, differentiation status, and metastatic potential have pointed both proteins to possess tumor-promoting and -suppressive characteristics. Whether the changes in Stra13 and Sharp-1 expression in most of these cancers are causal or consequent to the tumor onset and the tumor's hypoxic environment is yet to be elucidated. Neither Stra13 nor Sharp-1 null mutants have been reported to develop spontaneous tumors, and it would be important to test whether the mutants exhibit differences in tumor initiation, progression and metastasis in cancers of various origins. It is also unclear whether the two proteins are mutated in different types of cancers, exhibit altered subcellular localization, or are differentially modified by posttranslational modifications allowing them to associate with different proteins in different cell types or cell contexts that may account for the seemingly discrepant observations in tumorigenesis. Despite the limited number of mechanistic studies, the functions of Stra13 and Sharp-1 in regulation of the cell cycle, DNA repair, and apoptosis, as well as the intersection with various signaling pathways suggest a primary role in tumorigenesis. In many instances, Stra13 and Sharp-1 participate in negative feedback loops such as autoregulation of their own expression, as well as inhibition of signaling pathways that stimulate their expression indicating a central role in maintenance of tissue homeostasis. The multitude of roles Stra13 and Sharp-1 play in development, differentiation, and pathology of a wide array of tissue types strongly reinforces the value of future investigations in understanding the role of these critically important proteins in human biology and medicine.

REFERENCES

Acharjee, S., Chung, T.-K., Gopinadhan, S., Shankar, S. R., Wang, Y., Li, L., et al. (2014). Sharp-1 regulates TGF-β signaling and skeletal muscle regeneration. *Journal of Cell Science*, *127*, 599–608.

Adams, J. M., Difazio, L. T., Rolandelli, R. H., Luján, J. J., Haskó, G., Csóka, B., et al. (2009). HIF-1: A key mediator in hypoxia. *Acta Physiologica Hungarica*, *96*, 19–28.

Adorno, M., Cordenonsi, M., Montagner, M., Dupont, S., Wong, C., Hann, B., et al. (2009). A mutant-p53/Smad complex opposes p63 to empower TGFbeta-induced metastasis. *Cell, 137*, 87–98.

Azmi, S., Ozog, A., & Taneja, R. (2004). Sharp-1/DEC2 inhibits skeletal muscle differentiation through repression of myogenic transcription factors. *Journal of Biological Chemistry, 279*, 52643–52652.

Azmi, S., Sun, H., Ozog, A., & Taneja, R. (2003). mSharp-1/DEC2, a basic helix-loop-helix protein functions as a transcriptional repressor of E box activity and Stra13 expression. *Journal of Biological Chemistry, 278*, 20098–20109.

Azmi, S., & Taneja, R. (2002). Embryonic expression of mSharp-1/mDEC2, which encodes a basic helix-loop-helix transcription factor. *Mechanisms of Development, 114*, 181–185.

Bhawal, U. K., Sato, F., Arakawa, Y., Fujimoto, K., Kawamoto, T., Tanimoto, K., et al. (2011). Basic helix-loop-helix transcription factor DEC1 negatively regulates cyclin D1. *Journal of Pathology, 224*, 420–429.

Boudjelal, M., Taneja, R., Matsubara, S., Bouillet, P., Dolle, P., & Chambon, P. (1997). Overexpression of Stra13, a novel retinoic acid-inducible gene of the basic helix-loop-helix family, inhibits mesodermal and promotes neuronal differentiation of P19 cells. *Genes & Development, 11*, 2052–2065.

Chakrabarti, J., Turley, H., Campo, L., Han, C., Harris, A. L., Gatter, K. C., et al. (2004). The transcription factor DEC1 (stra13, SHARP2) is associated with the hypoxic response and high tumour grade in human breast cancers. *British Journal of Cancer, 91*, 954–958.

Chen, C. R., Kang, Y., & Massagué, J. (2001). Defective repression of c-myc in breast cancer cells: A loss at the core of the transforming growth factor beta growth arrest program. *Proceedings of the National Academy of Sciences of the United States of America, 98*, 992–995.

Chen, L., Zhou, J., Xu, H., Xu, G., & Xue, J. (2010). Identification and developmental expression of Dec2 in zebrafish. *Fish Physiology and Biochemistry, 36*, 667–675.

Cho, Y., Noshiro, M., Choi, M., Morita, K., Kawamoto, T., Fujimoto, K., et al. (2009). The basic helix-loop-helix proteins differentiated embryo chondrocyte (DEC) 1 and DEC2 function as corepressors of retinoid X receptors. *Molecular Pharmacology, 76*, 1360–1369.

Choi, S. M., Cho, H.-J., Cho, H., Kim, K. H., Kim, J. B., & Park, H. (2008). Stra13/DEC1 and DEC2 inhibit sterol regulatory element binding protein-1c in a hypoxia-inducible factor-dependent mechanism. *Nucleic Acids Research, 36*, 6372–6385.

Collado, M., Gil, J., Efeyan, A., Guerra, C., Schuhmacher, A. J., Barradas, M., et al. (2005). Tumour biology: senescence in premalignant tumours. *Nature, 436*, 624.

Davis, R. L., & Turner, D. L. (2001). Vertebrate hairy and enhancer of split related proteins: Transcriptional repressors regulating cellular differentiation and embryonic patterning. *Oncogene, 20*, 8342–8357.

Dhar, M., & Taneja, R. (2001). Cross-regulatory interaction between Stra13 and USF results in functional antagonism. *Oncogene, 20*, 4750–4756.

Ehata, S., Hanyu, A., Hayashi, M., Aburatani, H., Kato, Y., Fujime, M., et al. (2007). Transforming growth factor-beta promotes survival of mammary carcinoma cells through induction of antiapoptotic transcription factor DEC1. *Cancer Research, 67*, 9694–9703.

Falvella, F. S., Colombo, F., Spinola, M., Campiglio, M., Pastorino, U., & Dragani, T. A. (2008). BHLHB3: A candidate tumor suppressor in lung cancer. *Oncogene, 27*, 3761–3764.

Feige, E., Yokoyama, S., Levy, C., Khaled, M., Igras, V., Lin, R. J., et al. (2011). Hypoxia-induced transcriptional repression of the melanoma-associated oncogene MITF. *Proceedings of the National Academy of Sciences of the United States of America, 108*, E924–E933.

Fisher, A. L., Ohsako, S., & Caudy, M. (1996). The WRPW motif of the hairy-related basic helix-loop-helix repressor proteins acts as a 4-amino-acid transcription repression and protein-protein interaction domain. *Molecular and Cellular Biology*, *16*, 2670–2677.

Fujimoto, K., Hamaguchi, H., Hashiba, T., Nakamura, T., Kawamoto, T., Sato, F., et al. (2007). Transcriptional repression by the basic helix-loop-helix protein Dec2: Multiple mechanisms through E-box elements. *International Journal of Molecular Medicine*, *19*, 925–932.

Fujimoto, K., Shen, M., Noshiro, M., Matsubara, K., Shingu, S., Honda, K., et al. (2001). Molecular cloning and characterization of DEC2, a new member of basic helix-loop-helix proteins. *Biochemical and Biophysical Research Communications*, *280*, 164–171.

Garriga-Canut, M., Roopra, A., & Buckley, N. J. (2001). The basic helix-loop-helix protein, sharp-1, represses transcription by a histone deacetylase-dependent and histone deacetylase-independent mechanism. *Journal of Biological Chemistry*, *276*, 14821–14828.

Giatromanolaki, A., Koukourakis, M. I., Sivridis, E., Turley, H., Wykoff, C. C., Gatter, K. C., et al. (2003). DEC1 (STRA13) protein expression relates to hypoxia-inducible factor 1-alpha and carbonic anhydrase-9 overexpression in non-small cell lung cancer. *Journal of Pathology*, *200*, 222–228.

Giusiano, S., Cochet, C., Filhol, O., Duchemin-Pelletier, E., Secq, V., Bonnier, P., et al. (2011). Protein kinase CK2α subunit over-expression correlates with metastatic risk in breast carcinomas: Quantitative immunohistochemistry in tissue microarrays. *European Journal of Cancer*, *47*, 792–801.

Giusiano, S., Secq, V., Carcopino, X., Carpentier, S., Andrac, L., Lavaut, M.-N., et al. (2010). Immunohistochemical profiling of node negative breast carcinomas allows prediction of metastatic risk. *International Journal of Oncology*, *36*, 889–898.

Gulbagci, N. T., Li, L., Ling, B., Gopinadhan, S., Walsh, M., Rossner, M., et al. (2009). SHARP1/DEC2 inhibits adipogenic differentiation by regulating the activity of C/EBP. *EMBO Reports*, *10*, 79–86.

Hong, Y., Xing, X., Li, S., Bi, H., Yang, C., Zhao, F., et al. (2011). SUMOylation of DEC1 protein regulates its transcriptional activity and enhances its stability. *PLoS One*, *6*, e23046.

Honma, S., Kawamoto, T., Takagi, Y., Fujimoto, K., Sato, F., Noshiro, M., et al. (2002). Dec1 and Dec2 are regulators of the mammalian molecular clock. *Nature*, *419*, 841–844.

Hsiao, S. P., Huang, K. M., Chang, H. Y., & Chen, S. L. (2009). P/CAF rescues the Bhlhe40-mediated repression of MyoD transactivation. *Biochemical Journal*, *422*, 343–352.

Iizuka, K., & Horikawa, Y. (2008). Regulation of lipogenesis via BHLHB2/DEC1 and ChREBP feedback looping. *Biochemical and Biophysical Research Communications*, *374*, 95–100.

Imaizumi, T., Sato, F., Tanaka, H., Matsumiya, T., Yoshida, H., Yashiro-Aizawa, T., et al. (2011). Basic-helix-loop-helix transcription factor DEC2 constitutes negative feedback loop in IFN-β-mediated inflammatory responses in human mesangial cells. *Immunology Letters*, *136*, 37–43.

Inaguma, S., Riku, M., Hashimoto, M., Murakami, H., Saga, S., Ikeda, H., et al. (2013). GLI1 interferes with the DNA mismatch repair system in pancreatic cancer through BHLHE41-mediated suppression of MLH1. *Cancer Research*, *73*, 7313–7323.

Ivanov, S. V., Salnikow, K., Ivanova, A. V., Bai, L., & Lerman, M. I. (2007). Hypoxic repression of STAT1 and its downstream genes by a pVHL/HIF-1 target DEC1/STRA13. *Oncogene*, *26*, 802–812.

Ivanova, A. V., Ivanov, S. V., Danilkovitch-Miagkova, A., & Lerman, M. I. (2001). Regulation of STRA13 by the von Hippel-Lindau tumor suppressor protein, hypoxia, and the UBC9/ubiquitin proteasome degradation pathway. *Journal of Biological Chemistry*, *276*, 15306–15315.

Iwata, T., Kawamoto, T., Sasabe, E., Miyazaki, K., Fujimoto, K., Noshiro, M., et al. (2006). Effects of overexpression of basic helix-loop-helix transcription factor Dec1 on osteogenic and adipogenic differentiation of mesenchymal stem cells. *European Journal of Cell Biology, 85*, 423–431.

Jia, Y.-F., Xiao, D.-J., Ma, X.-L., Song, Y.-Y., Hu, R., Kong, Y., et al. (2013). Differentiated embryonic chondrocyte-expressed gene 1 is associated with hypoxia-inducible factor 1α and Ki67 in human gastric cancer. *Diagnostic Pathology, 8*, 37.

Kadener, S., Stoleru, D., McDonald, M., Nawathean, P., & Rosbash, M. (2007). Clockwork Orange is a transcriptional repressor and a new Drosophila circadian pacemaker component. *Genes & Development, 21*, 1675–1686.

Koukourakis, M. I., Giatromanolaki, A., Polychronidis, A., Simopoulos, C., Gatter, K. C., Harris, A. L., et al. (2006). Endogenous markers of hypoxia/anaerobic metabolism and anemia in primary colorectal cancer. *Cancer Science, 97*, 582–588.

Li, Y., Bi, Z., Yan, B., & Wan, Y. (2006). UVB radiation induces expression of HIF-1alpha and VEGF through the EGFR/PI3K/DEC1 pathway. *International Journal of Molecular Medicine, 18*, 713–719.

Li, Y., Shen, Q., Kim, H.-T., Bissonnette, R. P., Lamph, W. W., Yan, B., et al. (2011). The rexinoid bexarotene represses cyclin D1 transcription by inducing the DEC2 transcriptional repressor. *Breast Cancer Research and Treatment, 128*, 667–677.

Li, Y., Xie, M., Song, X., Gragen, S., Sachdeva, K., Wan, Y., et al. (2003). DEC1 negatively regulates the expression of DEC2 through binding to the E-box in the proximal promoter. *Journal of Biological Chemistry, 278*, 16899–16907.

Li, Y., Zhang, Y., Hill, J., Kim, H.-T., Shen, Q., Bissonnette, R. P., et al. (2008). The rexinoid, bexarotene, prevents the development of premalignant lesions in MMTV-erbB2 mice. *British Journal of Cancer, 98*, 1380–1388.

Li, Y., Zhang, H., Xie, M., Hu, M., Ge, S., Yang, D., et al. (2002). Abundant expression of Dec1/stra13/sharp2 in colon carcinoma: Its antagonizing role in serum deprivation-induced apoptosis and selective inhibition of procaspase activation. *Biochemical Journal, 367*, 413–422.

Lin, C.-C., Bradstreet, T. R., Schwarzkopf, E. A., Sim, J., Carrero, J. A., Chou, C., et al. (2014). Bhlhe40 controls cytokine production by T cells and is essential for pathogenicity in autoimmune neuroinflammation. *Nature Communications, 5*, 3551.

Ling, B. M. T., Bharathy, N., Chung, T.-K., Kok, W. K., Li, S., Tan, Y. H., et al. (2012). Lysine methyltransferase G9a methylates the transcription factor MyoD and regulates skeletal muscle differentiation. *Proceedings of the National Academy of Sciences of the United States of America, 109*, 841–846.

Ling, B. M. T., Gopinadhan, S., Kok, W. K., Shankar, S. R., Gopal, P., Bharathy, N., et al. (2012). G9a mediates Sharp-1-dependent inhibition of skeletal muscle differentiation. *Molecular Biology of the Cell, 23*, 4778–4785.

Liu, J. J., Chung, T. K., Li, J., & Taneja, R. (2010). Sharp-1 modulates the cellular response to DNA damage. *FEBS Letters, 584*, 619–624.

Liu, Z., Li, Z., Mao, K., Zou, J., Wang, Y., Tao, Z., et al. (2009). Dec2 promotes Th2 cell differentiation by enhancing IL-2R signaling. *Journal of Immunology, 183*, 6320–6329.

Liu, Y., Miao, Y., Wang, J., Lin, X., Wang, L., Xu, H.-T., et al. (2013). DEC1 is positively associated with the malignant phenotype of invasive breast cancers and negatively correlated with the expression of claudin-1. *International Journal of Molecular Medicine, 31*, 855–860.

Liu, Y., Sato, F., Kawamoto, T., Fujimoto, K., Morohashi, S., Akasaka, H., et al. (2010). Anti-apoptotic effect of the basic helix-loop-helix (bHLH) transcription factor DEC2 in human breast cancer cells. *Genes to Cells, 15*, 315–325.

Liu, Y., Wang, L., Lin, X.-Y., Wang, J., Yu, J.-H., Miao, Y., et al. (2013). The transcription factor DEC1 (BHLHE40/STRA13/SHARP-2) is negatively associated with TNM stage in non-small-cell lung cancer and inhibits the proliferation through cyclin D1 in A549 and BE1 cells. *Tumour Biology*, *34*, 1641–1650.

Ma, W., Shi, X., Lu, S., Wu, L., & Wang, Y. (2013). Hypoxia-induced overexpression of DEC1 is regulated by HIF-1α in hepatocellular carcinoma. *Oncology Reports*, *30*, 2957–2962.

Martínez-Llordella, M., Esensten, J. H., Bailey-Bucktrout, S. L., Lipsky, R. H., Marini, A., Chen, J., et al. (2013). CD28-inducible transcription factor DEC1 is required for efficient autoreactive CD4+ T cell response. *Journal of Experimental Medicine*, *210*, 1603–1619.

Miyazaki, K., Kawamoto, T., Tanimoto, K., Nishiyama, M., Honda, H., & Kato, Y. (2002). Identification of functional hypoxia response elements in the promoter region of the DEC1 and DEC2 genes. *Journal of Biological Chemistry*, *277*, 47014–47021.

Miyazaki, K., Miyazaki, M., Guo, Y., Yamasaki, N., Kanno, M., Honda, Z., et al. (2010). The role of the basic helix-loop-helix transcription factor Dec1 in the regulatory T cells. *Journal of Immunology*, *185*, 7330–7339.

Montagner, M., Enzo, E., Forcato, M., Zanconato, F., Parenti, A., Rampazzo, E., et al. (2012). SHARP1 suppresses breast cancer metastasis by promoting degradation of hypoxia-inducible factors. *Nature*, *487*, 380–384.

Morosetti, R., Mirabella, M., Gliubizzi, C., Broccolini, A., De Angelis, L., Tagliafico, E., et al. (2006). MyoD expression restores defective myogenic differentiation of human mesoangioblasts from inclusion-body myositis muscle. *Proceedings of the National Academy of Sciences of the United States of America*, *103*, 16995–17000.

Nakamura, H., Tanimoto, K., Hiyama, K., Yunokawa, M., Kawamoto, T., Kato, Y., et al. (2008). Human mismatch repair gene, MLH1, is transcriptionally repressed by the hypoxia-inducible transcription factors, DEC1 and DEC2. *Oncogene*, *27*, 4200–4209.

Noshiro, M., Usui, E., Kawamoto, T., Sato, F., Nakashima, A., Ueshima, T., et al. (2009). Liver X receptors (LXRalpha and LXRbeta) are potent regulators for hepatic Dec1 expression. *Genes to Cells*, *14*, 29–40.

Olbryt, M., Jarzab, M., Jazowiecka-Rakus, J., Simek, K., Szala, S., & Sochanik, A. (2006). Gene expression profile of B 16(F10) murine melanoma cells exposed to hypoxic conditions in vitro. *Gene Expression*, *13*, 191–203.

Ozaki, N., Noshiro, M., Kawamoto, T., Nakashima, A., Honda, K., Fukuzaki-Dohi, U., et al. (2012). Regulation of basic helix-loop-helix transcription factors Dec1 and Dec2 by RORα and their roles in adipogenesis. *Genes to Cells*, *17*, 109–121.

Park, Y.-K., & Park, H. (2012). Differentiated embryo chondrocyte 1 (DEC1) represses PPARγ2 gene through interacting with CCAAT/enhancer binding protein β (C/EBPβ). *Molecules and Cells*, *33*, 575–581.

Peng, Y., Liu, W., Xiong, J., Gui, H. Y., Feng, X. M., Chen, R. N., et al. (2012). Down regulation of differentiated embryonic chondrocytes 1 (DEC1) is involved in 8-methoxypsoralen-induced apoptosis in HepG2 cells. *Toxicology*, *301*, 58–65.

Phadke, M., Krynetskaia, N., Mishra, A., & Krynetskiy, E. (2011). Accelerated cellular senescence phenotype of GAPDH-depleted human lung carcinoma cells. *Biochemical and Biophysical Research Communications*, *411*, 409–415.

Preusser, M., Birner, P., Ambros, I. M., Ambros, P. F., Budka, H., Harris, A. L., et al. (2005). DEC1 expression in 1p-aberrant oligodendroglial neoplasms. *Histology and Histopathology*, *20*, 1173–1177.

Qian, Y., Jung, Y. S., & Chen, X. (2012). Differentiated embryo-chondrocyte expressed gene 1 regulates p53-dependent cell survival versus cell death through macrophage inhibitory cytokine-1. *Proceedings of the National Academy of Sciences of the United States of America*, *109*, 11300–11305.

Qian, Y., Zhang, J., Yan, B., & Chen, X. (2008). DEC1, a basic helix-loop-helix transcription factor and a novel target gene of the p53 family, mediates p53-dependent premature senescence. *Journal of Biological Chemistry*, *283*, 2896–2905.

Ren, J., Wen, L., Gao, X., Jin, C., Xue, Y., & Yao, X. (2009). DOG 1.0: Illustrator of protein domain structures. *Cell Research*, *19*, 271–273.

Rossner, M. J., Dörr, J., Gass, P., Schwab, M. H., & Nave, K. A. (1997). SHARPs: Mammalian enhancer-of-split- and hairy-related proteins coupled to neuronal stimulation. *Molecular and Cellular Neurosciences*, *10*, 460–475.

Rossner, M. J., Oster, H., Wichert, S. P., Reinecke, L., Wehr, M. C., Reinecke, J., et al. (2008). Disturbed clockwork resetting in Sharp-1 and Sharp-2 single and double mutant mice. *PLoS One*, *3*, e2762.

Sato, F., Kawamura, H., Wu, Y., Sato, H., Jin, D., Bhawal, U. K., et al. (2012). The basic helix-loop-helix transcription factor DEC2 inhibits TGF-β-induced tumor progression in human pancreatic cancer BxPC-3 cells. *International Journal of Molecular Medicine*, *30*, 495–501.

Seimiya, M., Bahar, R., Wang, Y., Kawamura, K., Tada, Y., Okada, S., et al. (2002). Clast5/Stra13 is a negative regulator of B lymphocyte activation. *Biochemical and Biophysical Research Communications*, *292*, 121–127.

Seimiya, M., Wada, A., Kawamura, K., Sakamoto, A., Ohkubo, Y., Okada, S., et al. (2004). Impaired lymphocyte development and function in Clast5/Stra13/DEC1-transgenic mice. *European Journal of Immunology*, *34*, 1322–1332.

Semenza, G. L. (2009). Regulation of cancer cell metabolism by hypoxia-inducible factor 1. *Seminars in Cancer Biology*, *19*, 12–16.

Shen, M., Kawamoto, T., Teramoto, M., Makihira, S., Fujimoto, K., Yan, W., et al. (2001). Induction of basic helix-loop-helix protein DEC1 (BHLHB2)/Stra13/Sharp2 in response to the cyclic adenosine monophosphate pathway. *European Journal of Cell Biology*, *80*, 329–334.

Shen, M., Kawamoto, T., Yan, W., Nakamasu, K., Tamagami, M., Koyano, Y., et al. (1997). Molecular characterization of the novel basic helix-loop-helix protein DEC1 expressed in differentiated human embryo chondrocytes. *Biochemical and Biophysical Research Communications*, *236*, 294–298.

Shen, M., Yoshida, E., Yan, W., Kawamoto, T., Suardita, K., Koyano, Y., et al. (2002). Basic helix-loop-helix protein DEC1 promotes chondrocyte differentiation at the early and terminal stages. *Journal of Biological Chemistry*, *277*, 50112–50120.

Shi, X.-H., Zheng, Y., Sun, Q., Cui, J., Liu, Q.-H., Qü, F., et al. (2011). DEC1 nuclear expression: A marker of differentiation grade in hepatocellular carcinoma. *World Journal of Gastroenterology*, *17*, 2037–2043.

St-Pierre, B., Flock, G., Zacksenhaus, E., & Egan, S. E. (2002). Stra13 homodimers repress transcription through class B E-box elements. *Journal of Biological Chemistry*, *277*, 46544–46551.

Sun, H., Li, L., Vercherat, C., Gulbagci, N. T., Acharjee, S., Li, J., et al. (2007). Stra13 regulates satellite cell activation by antagonizing Notch signaling. *Journal of Cell Biology*, *177*, 647–657.

Sun, H., Lu, B., Li, R. Q., Flavell, R. A., & Taneja, R. (2001). Defective T cell activation and autoimmune disorder in Stra13-deficient mice. *Nature Immunology*, *2*, 1040–1047.

Sun, H., & Taneja, R. (2000). Stra13 expression is associated with growth arrest and represses transcription through histone deacetylase (HDAC)-dependent and HDAC-independent mechanisms. *Proceedings of the National Academy of Sciences of the United States of America*, *97*, 4058–4063.

Thin, T. H., Li, L., Chung, T.-K., Sun, H., & Taneja, R. (2007). Stra13 is induced by genotoxic stress and regulates ionizing-radiation-induced apoptosis. *EMBO Reports*, *8*, 401–407.

Turley, H., Wykoff, C. C., Troup, S., Watson, P. H., Gatter, K. C., & Harris, A. L. (2004). The hypoxia-regulated transcription factor DEC1 (Stra13, SHARP-2) and its expression in human tissues and tumours. *Journal of Pathology*, *203*, 808–813.

Vercherat, C., Chung, T.-K., Yalcin, S., Gulbagci, N., Gopinadhan, S., Ghaffari, S., et al. (2009). Stra13 regulates oxidative stress mediated skeletal muscle degeneration. *Human Molecular Genetics*, *18*, 4304–4316.

Wang, Y., Rao, V. K., Kok, W. K., Roy, D. N., Sethi, S., Ling, B. M. T., et al. (2012). SUMO modification of Stra13 is required for repression of cyclin D1 expression and cellular growth arrest. *PLoS One*, 7, e43137.

Wang, Y., Shankar, S. R., Kher, D., Ling, B. M. T., & Taneja, R. (2013). Sumoylation of the basic helix-loop-helix transcription factor sharp-1 regulates recruitment of the histone methyltransferase G9a and function in myogenesis. *Journal of Biological Chemistry*, *288*, 17654–17662.

Wu, K., Kim, H.-T., Rodriquez, J. L., Hilsenbeck, S. G., Mohsin, S. K., Xu, X.-C., et al. (2002). Suppression of mammary tumorigenesis in transgenic mice by the RXR-selective retinoid, LGD1069. *Cancer Epidemiology, Biomarkers & Prevention*, *11*, 467–474.

Wu, Y., Sato, F., Bhawal, U. K., Kawamoto, T., Fujimoto, K., Noshiro, M., et al. (2011). Basic helix-loop-helix transcription factors DEC1 and DEC2 regulate the paclitaxel-induced apoptotic pathway of MCF-7 human breast cancer cells. *International Journal of Molecular Medicine*, *27*, 491–495.

Wu, Y., Sato, F., Bhawal, U. K., Kawamoto, T., Fujimoto, K., Noshiro, M., et al. (2012). BHLH transcription factor DEC2 regulates pro-apoptotic factor Bim in human oral cancer HSC-3 cells. *Biomedical Research*, *33*, 75–82.

Wu, Y., Sato, F., Yamada, T., Bhawal, U. K., Kawamoto, T., Fujimoto, K., et al. (2012). The BHLH transcription factor DEC1 plays an important role in the epithelial-mesenchymal transition of pancreatic cancer. *International Journal of Oncology*, *41*, 1337–1346.

Wykoff, C. C., Pugh, C. W., Maxwell, P. H., Harris, A. L., & Ratcliffe, P. J. (2000). Identification of novel hypoxia dependent and independent target genes of the von Hippel-Lindau (VHL) tumour suppressor by mRNA differential expression profiling. *Oncogene*, *19*, 6297–6305.

Xu, Q., Ma, P., Hu, C., Chen, L., Xue, L., Wang, Z., et al. (2012). Overexpression of the DEC1 protein induces senescence in vitro and is related to better survival in esophageal squamous cell carcinoma. *PLoS One*, 7, e41862.

Yamada, K., Kawata, H., Shou, Z., Mizutani, T., Noguchi, T., & Miyamoto, K. (2003). Insulin induces the expression of the SHARP-2/Stra13/DEC1 gene via a phosphoinositide 3-kinase pathway. *Journal of Biological Chemistry*, *278*, 30719–30724.

Yamada, K., Ogata-Kawata, H., Matsuura, K., & Miyamoto, K. (2005). SHARP-2/Stra13/DEC1 as a potential repressor of phosphoenolpyruvate carboxykinase gene expression. *FEBS Letters*, *579*, 1509–1514.

Yang, X. O., Angkasekwinai, P., Zhu, J., Peng, J., Liu, Z., Nurieva, R., et al. (2009). Requirement for the basic helix-loop-helix transcription factor Dec2 in initial TH2 lineage commitment. *Nature Immunology*, *10*, 1260–1266.

Yang, Y., Sun, M., Wang, L., & Jiao, B. (2013). HIFs, angiogenesis, and cancer. *Journal of Cellular Biochemistry*, *114*, 967–974.

Yao, J., Wang, L., Chen, L., Zhang, S., Zhao, Q., Jia, W., et al. (2006). Cloning and developmental expression of the DEC1 ortholog gene in zebrafish. *Gene Expression Patterns*, *6*, 919–927.

Yi, G., He, Z., Zhou, X., Xian, L., Yuan, T., Jia, X., et al. (2013). Low concentration of metformin induces a p53-dependent senescence in hepatoma cells via activation of the AMPK pathway. *International Journal of Oncology*, *43*, 1503–1510.

Yoon, D. Y., Buchler, P., Saarikoski, S. T., Hines, O. J., Reber, H. A., & Hankinson, O. (2001). Identification of genes differentially induced by hypoxia in pancreatic cancer cells. *Biochemical and Biophysical Research Communications*, *288*, 882–886.

Yuan, J., Narayanan, L., Rockwell, S., & Glazer, P. M. (2000). Diminished DNA repair and elevated mutagenesis in mammalian cells exposed to hypoxia and low pH. *Cancer Research, 60*, 4372–4376.

Yun, Z., Maecker, H. L., Johnson, R. S., & Giaccia, A. J. (2002). Inhibition of PPAR gamma 2 gene expression by the HIF-1-regulated gene DEC1/Stra13: A mechanism for regulation of adipogenesis by hypoxia. *Developmental Cell, 2*, 331–341.

Yunokawa, M., Tanimoto, K., Nakamura, H., Nagai, N., Kudo, Y., Kawamoto, T., et al. (2007). Differential regulation of DEC2 among hypoxia-inducible genes in endometrial carcinomas. *Oncology Reports, 17*, 871–878.

Zawel, L., Yu, J., Torrance, C. J., Markowitz, S., Kinzler, K. W., Vogelstein, B., et al. (2002). DEC1 is a downstream target of TGF-beta with sequence-specific transcriptional repressor activities. *Proceedings of the National Academy of Sciences of the United States of America, 99*, 2848–2853.

Zhang, L., & Li, Q. (2007). Embryo-chondrocyte expressed gene 1, downregulating hypoxia-inducible factor 1alpha, is another marker of lung tumor hypoxia. *Acta Pharmacologica Sinica, 28*, 549–558.

Zheng, Y., Jia, Y., Wang, Y., Wang, M., Li, B., Shi, X., et al. (2009). The hypoxia-regulated transcription factor DEC1 (Stra13, SHARP-2) and its expression in gastric cancer. *Omics: A Journal of Integrative Biology, 13*, 301–306.

Zheng, Y., Shi, X., Wang, M., Jia, Y., Li, B., Zhang, Y., et al. (2012). The increased expression of DEC1 gene is related to HIF-1α protein in gastric cancer cell lines. *Molecular Biology Reports, 39*, 4229–4236.

Zhu, Y., Xu, L., Zhang, J., Hu, X., Liu, Y., Yin, H., et al. (2013). Sunitinib induces cellular senescence via p53/Dec1 activation in renal cell carcinoma cells. *Cancer Science, 104*, 1052–1061.

CHAPTER TEN

DEC1/STRA13/SHARP2 and DEC2/SHARP1 Coordinate Physiological Processes, Including Circadian Rhythms in Response to Environmental Stimuli

Yukio Kato[1], Takeshi Kawamoto, Katsumi Fujimoto, Mitsuhide Noshiro

Department of Dental and Medical Biochemistry, Basic Life Sciences, Institute of Biomedical and Health Sciences, Hiroshima University, Hiroshima, Japan

[1]Corresponding author: e-mail address: ykato@hiroshima-u.ac.jp

Contents

ISSN 0070-2153
http://dx.doi.org/10.1016/B978-0-12-405943-6.00010-5
© 2014 Elsevier Inc. All rights reserved.

Abstract

Daily physiological and behavioral rhythms are regulated by endogenous circadian molecular clocks. Clock proteins DEC1 (BHLHe40) and DEC2 (BHLHe41) belong to the basic helix–loop–helix protein superfamily, which contains other clock proteins CLOCK and BMAL1. DEC1 and DEC2 are induced by CLOCK:BMAL1 heterodimer via the CACGTG E-box in the promoter and, thereafter, suppress their own expression by competing with CLOCK:BMAL1 for the DNA binding. This negative feedback DEC loop together with the PER loop involving PER and CRY, the other negative clock regulators, maintains the circadian rhythm of *Dec1* and *Dec2* expression. DEC1 is induced by light pulse and adjusts the circadian phase of the central clock in the suprachiasmatic nucleus, whereas DEC1 upregulation by TGF-β resets the circadian phase of the peripheral clocks in tissues. Furthermore, DEC1 and DEC2 modulate the clock output signals to control circadian rhythms in behavior and metabolism. In addition to the functions in the clocks, DEC1 and DEC2 are involved in hypoxia responses, immunological reactions, and carcinogenesis. These DEC actions are mediated by the direct binding to the E-box elements in target genes or by protein–protein interactions with transcription factors such as HIF-1α, RXRα, MyoD, and STAT. Notably, numerous growth factors, hormones, and cytokines, along with ionizing radiation and DNA-damaging agents, induce *Dec1* and/or *Dec2* in a tissue-specific manner. These findings suggest that DEC1 and DEC2 play a critical role in animal adaptation to various environmental stimuli.

1. INTRODUCTION

Transcription factors DEC1 and DEC2 are members of the basic helix–loop–helix (bHLH) superfamily found in various animals, including humans and rodents, which have been known under several names. Thus, DEC1 (differentiated embryonic chondrocyte 1) is also referred to as BHLHe40 (Shen et al., 1997), Stimulated with retinoic acid 13 (Stra13) (Boudjelal et al., 1997), E47-interacting protein 1 (Eip1) (Dear et al., 1997), Enhancer-of-Split (E(spl)) and Hairy-related protein-2 (Sharp-2) (Rossner, Dorr, Gass, Schwab, & Nave, 1997), cytokine response gene 8 (CR8) (Beadling, Cereseto, Fan, Naramura, & Smith, 2001), or Clast5 (Seimiya et al., 2002). DEC2 is also called BHLHe41 or Sharp-1 (Rossner et al., 1997; Fujimoto et al., 2001).

DEC1 and DEC2 may play a role in the adaptation to environmental signals, such as day–night cycle, availability of nutrients, oxygen concentration, infection, and ionizing radiation. The day–night cycle is the most prominent and universal environmental condition to influence living organisms on Earth, and most animals show circadian rhythms in behavior, and cellular and physiological processes (Albrecht, 2012). A circadian rhythm is a biological process regulated by the molecular clock system that displays an endogenous entrainable oscillation about 24 h. In animals, the circadian core oscillator is thought to be composed of autoregulatory feedback loops involving several sets of clock genes. DEC1 and DEC2 compose the DEC loop, which may participate in the generation of circadian rhythms in behavior and cell metabolism (Honma et al., 2002; Kato, Noshiro, Fujimoto, & Kawamoto, 2010; Kawamoto et al., 2004; Nakashima et al., 2008; Rossner et al., 2008).

Hypoxia is a severe environmental condition that critically affects survival, development, metabolism, and tumor progression. DEC1 and DEC2 are upregulated by hypoxia and modulate some effects of hypoxia-inducible factor-1 (HIF-1) (Ehata et al., 2007; Ivanova, Ivanov, Danilkovitch-Miagkova, & Lerman, 2001; Miyazaki et al., 2002; Montagner et al., 2012; Qian & Chen, 2008; Sato et al., 2012; Thin, Li, Chung, Sun, & Taneja, 2007). In addition, DEC1 and DEC2 are involved in responses to ionizing radiation and immunological reactions (Ivanova et al., 2004; Miyazaki et al., 2010; Seimiya et al., 2004; Sun, Lu, Li, Flavell, & Taneja, 2001; Thin et al., 2007).

At the present time, we cannot comprehensively integrate the diverse actions of DEC1 and DEC2 into a simple concept. However, the information presented in this chapter on the DEC genomic and protein structure, the role of DEC1 and DEC2 in the clock system, molecular mechanisms of DEC1 and DEC2 activity, and regulatory factors of *Dec1* and *Dec2* expression may provide insights into the mechanisms of animal adaptation to the changing environments.

2. STRUCTURES OF DEC1 AND DEC2

2.1. Primary structures of DEC1 and DEC2 proteins

DEC1 and DEC2 are bHLH domain-containing transcription factors (Fig. 10.1A) that bind to the CACGTG E-box in the promoter regions of various target genes (Honma et al., 2002; Kawamoto et al., 2004; Sato et al., 2004). DEC1 and DEC2 share 97% identity in the amino acid sequence of the bHLH domain, which contains an LXXLL motif specific for a nuclear retinoid X receptor (RXRα, NR2B1). The interaction of bHLH with RXRα can repress the ligand-dependent transactivation of nuclear receptors that heterodimerize with RXRα, such as liver

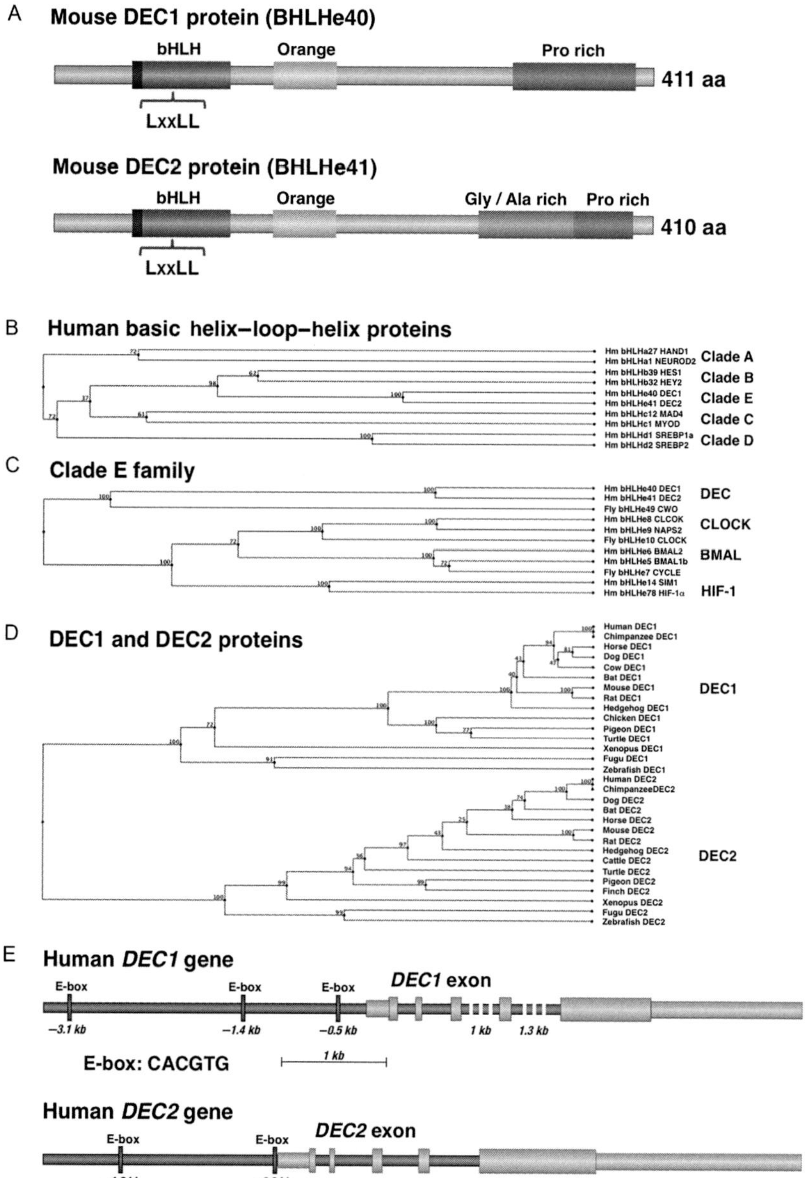

Figure 10.1 Structures and phylogenetic trees of DEC1 and DEC2. (A) Domain structures of mouse DEC1 and DEC2. A short dark bar in the basic helix–loop–helix (bHLH) indicates the basic region. (B–D) Phylogenetic trees were created with entire amino acid sequences of bHLH proteins using the UPGMA method. Two representative members of each clade are selected and used for the phylogenetic tree of bHLH proteins (B). Phylogenetic trees of Clade E family (C) and DEC proteins (D). (E) Structures of human *DEC1*

X receptor α (LXRα) (NR1H3), farnesoid X receptor (FXR) (NR1H4), retinoic acid receptor α (RARα) (NR1B4), and vitamin D receptor (VDR) (NR1I1) (Cho et al., 2009) (see Section 5.2.3 and Table 10.1). Besides the bHLH domain, DEC1 and DEC2 contain the Orange domain and a proline-rich region near the N-terminus that provides additional functionality to the transcription factors. In addition, DEC2 contains a glycine/alanine-rich region that overlaps the proline-rich region.

Based on a significant sequence homology to *Drosophila melanogaster* E(spl) and Hairy and Enhancer of split (Hes) in the bHLH and Orange domains, DEC1 and DEC2 were formerly grouped with Hes and E(spl) proteins (Group B) (Teramoto et al., 2001) and were named BHLHB2 and BHLHB3, respectively. However, DEC1 and DEC2 do not have C-terminal WRPW or YRPW tetrapeptide motifs, which are conserved in other Group B members and are critical for transcriptional repression (Fisher, Ohsako, & Caudy, 1996). Recent phylogenetic analysis based on the entire sequence comparison of the bHLH proteins indicates that the animal bHLH superfamily members can be classified into five clades (A–E) and that DEC1 and DEC2 belong to Clade E (Fig. 10.1B), which includes the bHLH–PAS proteins essential for circadian rhythm (BMAL1, CLOCK, NPAS2, and SIM1) and hypoxia response (HIF-1α and HIF-1β) (Fig. 10.1C; Skinner, Rawls, Wilson-Rawls, & Roalson, 2010).

All examined vertebrates, including fish, amphibians, reptiles, birds, and mammals, have orthologs of both DEC1 and DEC2, whereas the information on invertebrates is limited (Fig. 10.1D). Only clockwork orange (CWO) is reported as a structurally and functionally orthologous protein in the fruit fly *D. melanogaster* (Kadener, Stoleru, McDonald, Nawathean, & Rosbash, 2007; Matsumoto et al., 2007).

Interestingly, *D. melanogaster* CLOCK and CYCLE are orthologous to CLOCK and BMAL1, respectively, in vertebrates and belong to Clade E (Fig. 10.1C). In the animal evolution, the divergence of Clade E from the bHLH protein superfamily might have occurred in the process of

and *DEC2* genes. Wide bars indicate coding regions and narrower bars of the same color indicate 5′- or 3′-noncoding regions. Positions of E-boxes indicate the distances from translation initiation sites (ATG). LXXLL, leucine-x-x-leucine-leucine; CWO, clockwork orange protein; BMAL1, brain and muscle aryl hydrocarbon receptor nuclear translocator-like bHLH–PAS transcription factor; CLOCK, circadian locomotor output cycles kaput (a bHLH–PAS transcription factor); NPAS, neuronal PAS domain protein; SIM1, single-minded family bHLH transcription factor 1; HIF-1α, hypoxia-inducible factor 1α; Hm, human; kb, kilobase. (See the color plate.)

Table 10.1 DEC-interacting transcription factors

Transcription factors	Target genes	DEC functions (references)
HIF-1α	*Vegf, Angptl4, CaIX*	Inhibition of angiogenesis and cancer metastasis (Montagner et al., 2012; Sato et al., 2008)
MyoD	*Mck, M-cadherin*	Inhibition of skeletal muscle differentiation (Azmi et al., 2004, 2003; Fujimoto et al., 2007; Hsiao & Chen, 2010)
RXRα	*Srebp-1c, Abca1*	Inhibition of RXR actions (Cho et al., 2009)
Sp1	*Dec1, cyclin D1, Survivin*	Regulation of Sp1 activity (Azmi et al., 2003; Li et al., 2006, 2011)
C/EBP	*C/ebpα, Pparγ2, Ap2, Cyp2d6*	Inhibition of adipogenesis (Gulbagci et al., 2009; Matsunaga et al., 2012)
SREBP-1c	*Fasn, Srebp-1c*	Inhibition of SREBP-1c actions (Choi et al., 2008)
STAT3	*Fas*	Activation of STAT signaling pathway (Ivanova et al., 2004)
USF2	*c-Myc*	Inhibition of USF2 actions (Dhar & Taneja, 2001)
BMAL1	*Dec1, Per1*	Regulation of circadian rhythms (Honma et al., 2002; Rossner et al., 2008; Sato et al., 2004)

MyoD, myoblast determination protein; Sp1, specificity protein 1; C/EBP, CCAAT/enhancer binding protein; SREBP-1c, sterol regulatory element-binding transcription factor 1c; STAT3, signal transducer and activator of transcription 3; USF2, upstream stimulatory factor 2; *Vegf*, vascular endothelial growth factor; *Angptl4*, angiopoietin-like 4; *CaIX*, carbonic anhydrase IX; *Mck*, muscle creatine kinase; *Abca1*, ATP-binding cassette transporter A1; *Pparγ2*, peroxisome proliferator-activated receptor γ2; *Ap2*, adipocyte fatty acid-binding protein 2; *Cyp2d6*, cytochrome P-450 2D6; *Fasn*, fatty acid synthase.

adaptation to cyclic changes and development of the molecular clock system for the regulation of circadian rhythms.

2.2. Genomic structure of *Dec1* and *Dec2*

The human *DEC1* and mouse *Dec1* genes span about 4 kbp and contain five exons (Fig. 10.1E). Three E-boxes responsive to the CLOCK:BMAL1 heterodimer in circadian regulation are identified at −0.5, −1.4, and −3.1 kbp positions in the human *DEC1* promoter (Fig. 10.1E); the distal E-box (at 3.1 kbp) is missing in the mouse *Dec1* promoter (Kawamoto et al., 2004; Sato et al., 2004). At about −2.4 kbp, the mouse promoter region contains AGGTCA repeats of nuclear receptor response element

(NRE), which are recognized by several nuclear receptors as confirmed by the activation of a luciferase reporter (Noshiro et al., 2009). In addition, the nuclear receptor LXR-response element (LXRE) and ROR-response element (RORE) are identified, respectively, at −2 and −1.8 kbp in the mouse *Dec1* promoter (Noshiro et al., 2009; Ozaki et al., 2012). The VDR-response element is located at −1.7 kbp in the human *DEC1* promoter (Seuter, Pehkonen, Heikkinen, & Carlberg, 2013), which also contains the carbohydrate-response element (ChoRE), p53-response element, hypoxia-response element (HRE), and SMAD-binding element (see below).

The genomic structure of *Dec2* resembles that of *Dec1*. Both human and mouse genes span about 5 kbp and also contain five exons (Fig. 10.1E). Two E-boxes are located at −0.3 and −1.8 kbp positions in the human and mouse genes (Hamaguchi et al., 2004), and two ROREs are identified in the mouse *Dec2* promoter (Ozaki et al., 2012). In contrast to the *Dec1* promoter, the information on the identified response elements in the *Dec2* promoter is limited.

3. TISSUE DISTRIBUTION AND CIRCADIAN EXPRESSION OF *DEC1* AND *DEC2*

Dec1 is expressed in many tissues and cell types of various animals (Azmi, Ozog, & Taneja, 2004; Azmi & Taneja, 2002; Fujimoto et al., 2001; Li et al., 2002). Significant day–night variations in *Dec1* expression are observed tissue-dependently in mice and rats (Fig. 10.2A, top), which may indicate the tissue-specific regulatory mechanism (Noshiro et al., 2005).

Dec2 transcript is also ubiquitously expressed in animal tissues and its day–night variations are even more obvious than those of *Dec1*. The levels of *Dec2* transcript in the muscle, white adipose tissue, lung, and brain are higher than those in the liver, kidney, spleen, small intestine, testis, and femur, indicating tissue-dependent regulation (Fig. 10.2A, bottom; Noshiro et al., 2005).

In mammals, circadian rhythms are set by the central clock located in the suprachiasmatic nucleus (SCN) of the hypothalamus. *Dec1* and *Dec2* transcripts are highly expressed in the SCN of mice and rats in a circadian fashion with a peak at zeitgeber time (ZT) 4–8 in the subjective day of 12-h light/12-h dark cycle (Fig. 10.2B; Baba, Ono, Honma, & Honma, 2008; Butler et al., 2004; Honma et al., 2002). Constant dark conditions do not change the circadian profiles of *Dec1* and *Dec2* expression in the SCN. Various lines

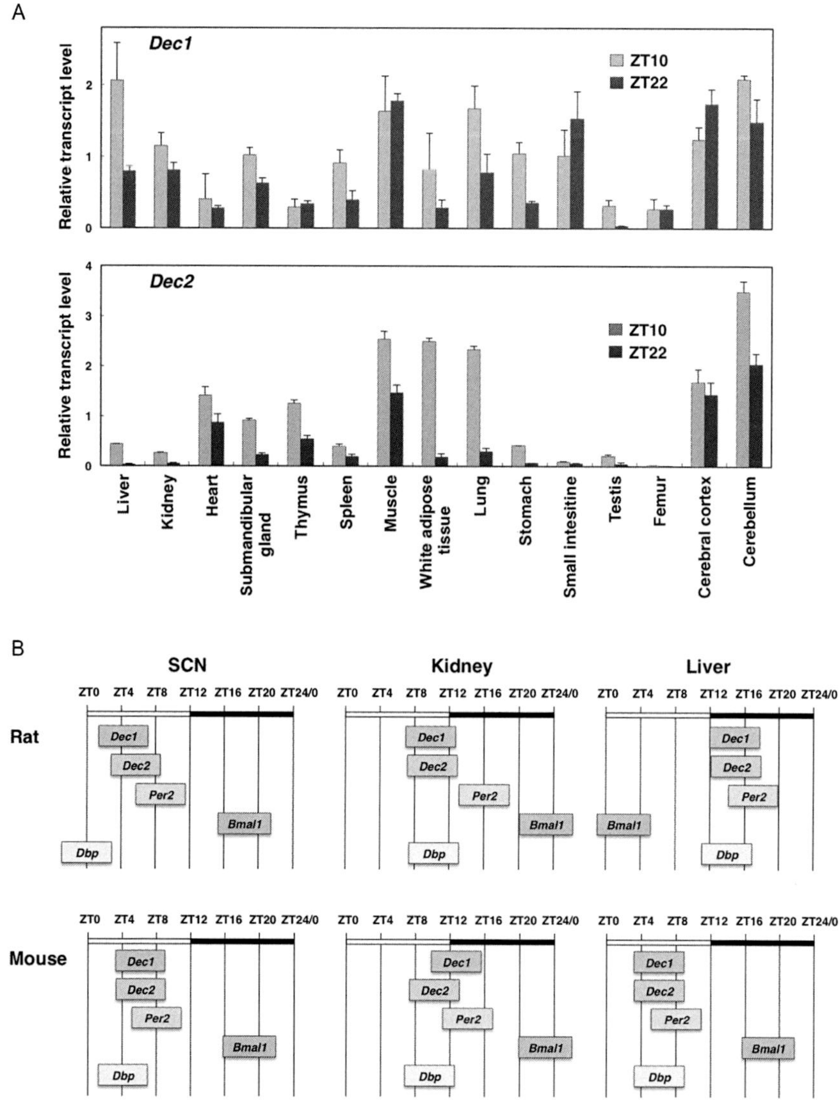

Figure 10.2 Tissue distribution and circadian expression of *Dec1* and *Dec2* (A) Tissue distribution of *Dec1* and *Dec2* transcripts. Total RNA was obtained from various tissues of C57BL/6 mice, and *Dec1* and *Dec2* transcripts were assessed by real-time PCR (Noshiro et al., 2005). (B) Circadian peak times for the mRNA expression of *Dec1*, *Dec2*, and other clock genes in the SCN, kidney, and liver of Sprague–Dawley rats and Balb/c mice. Boxes indicate peak mRNA expression. Top bars represent light (12 h) and dark (12 h) phases. DBP, D site of albumin promoter (albumin D-box)-binding protein; PER2, period circadian protein homolog 2 (a PAS transcription factor); SCN, suprachiasmatic nucleus; ZT, zeitgeber time. (See the color plate.)

of evidence indicate that DEC1 and DEC2 are negative components of the core clock loop in the SCN (Alvarez & Sehgal, 2002; Honma et al., 2002; Kawamoto et al., 2004; Nakashima et al., 2008) (see below).

The circadian expression profiles of *Dec1* and *Dec2* are different depending on the brain structure, besides the SCN. *Dec1* shows strong expression in the cerebral cortex, thalamus, and superior colliculus, whereas *Dec2* transcript levels are high in the caudate putamen, pineal gland, and granular cell layer of the cerebellum; for both genes, the expression is significant in the hippocampus, hypothalamic nuclei, piriform cortex, and olfactory bulb (Honma et al., 2002). In addition, the circadian rhythmic profiles of *Dec1* and *Dec2* expression are also reported for other tissues, including the liver, muscle, heart, kidney (Kawamoto et al., 2006; Noshiro et al., 2004, 2005; Wu, Jin, Ni, et al., 2008; Wu, Ni, Zhuge, & Fu, 2010), small intestine (Kawamoto et al., 2006), submandibular gland (Furukawa et al., 2005), cartilage (Honda et al., 2013), retina (Rohleder et al., 2006), pineal gland (Wu, Jin, Kato, & Fu, 2008), and white adipose tissue (Zvonic et al., 2006) in mice and rats.

Similar to other clock genes, the peak times of *Dec1* and *Dec2* expression are tissue specific (Fig. 10.2B). Interestingly, in the liver, the peak times of *Dec1* and *Dec2* expression show 4–8 h difference between mice and rats, whereas in the SCN and other peripheral tissues such as the kidney, the shift is insignificant (Fig. 10.2B). The relative timing of circadian peaks for most clock genes, including *Dec1*, *Dec2*, *Bmal1*, *Per2*, and *Dbp*, is similar in the respective tissues (Fig. 10.2B), possibly because of the synchronized transcriptional activation of these genes through the E-box, D-box, and RORE elements in the respective promoters (Ueda et al., 2005) (see below). The circadian rhythms of gene expression in the peripheral tissues should be tightly linked to their physiological functions; therefore, the circadian regulation of the molecular clock genes, including *Dec1* and *Dec2*, needs to be elucidated in connection with species- and tissue-specific physiological processes.

4. ROLE OF DEC1 AND DEC2 IN MOLECULAR CLOCKS

4.1. Molecular clock system

Daily physiological and behavioral rhythms such as the 24-h sleep–wake cycle are regulated by endogenous circadian molecular clocks (Albrecht, 2012). In mammals, the central clock in the SCN is entrained by day–night light cycle and controls the peripheral clocks in other tissues through neuronal and/or humoral routes. In contrast to the central clock, the peripheral clocks can be entrained by nonphotic environmental cues including food, hypoxia, and growth factors.

The central and peripheral clocks share a common molecular mechanism. The current cellular clock system comprises a number of negative feedback loops such as the DEC loop and the PER loop in which clock proteins inhibit transcription of their own genes (Fig. 10.3). In the DEC loop, *Dec1* and *Dec2* expression is induced by the CLOCK:BMAL1 heterodimer via the CACGTG E-box and suppressed by direct binding of their own products to the same E-box (Fig. 10.3 and Table 10.2; Kawamoto et al., 2004). In the PER loop, *Per1* and *Per2* expression is induced by CLOCK:BMAL1 via the CACGTG E-box and the CACGTT E′-box and suppressed by the PER:CRY heterodimer via protein–protein interactions with CLOCK:BMAL1. DEC and PER proteins can suppress each other expression via the interlocked loops (Fig. 10.3). Another level of circadian tuning is provided by DPB and E4BP4 via positive and negative, respectively, transcriptional regulation of *Per1* and *Per2* via the D-box; ROR and REV-ERB stimulate and inhibit, respectively, the expression

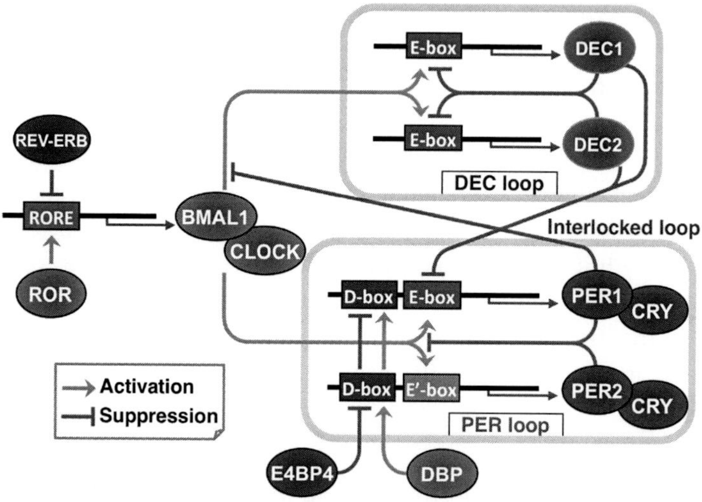

Figure 10.3 Interlocked loops of the circadian clock system. CLOCK:BMAL1 heterodimer acts as a transcriptional activator on the E-box in the *Dec* genes and on the E/E′-box in the *Per* genes. Meanwhile, DEC1 and DEC2 act as repressors to form the DEC loop. PER: CRY heterodimer also suppresses CLOCK:BMAL1-induced transcription of the *Per* genes to form the PER loop. REV-ERB, nuclear receptor NR1D1; ROR, retinoic acid receptor-related orphan receptor; E4BP4, E4 promoter-binding protein 4; PER, period circadian protein homolog; CRY, cryptochrome circadian clock. (See the color plate.)

of BMAL1 and CLOCK via RORE (Fig. 10.3; Ueda et al., 2005). This network of clock proteins is thought to finely control and stabilize robust circadian rhythms of the molecular clocks (Alvarez & Sehgal, 2002).

4.2. Molecular mechanisms involved in DEC activity in the clocks

DEC1 and DEC2 bind directly to the CACGTG E-box and powerfully suppress transactivation of *Per1*, *Dec1*, and *Dec2* by CLOCK:BMAL1 (Figs. 10.3 and 10.4; Honma et al., 2002; Hamaguchi et al., 2004; Kawamoto et al., 2004). However, unlike CLOCK:BMAL1, the affinity of DEC proteins to the CACGTT E′-box is weak, resulting in a marginal effect on the transcription induced by CLOCK:BMAL1 via the E′-box (Nakashima et al., 2008). The specificity of DEC proteins for the CACGTG E-box contributes to the DEC-based regulatory activity even in the presence of the PER loop. On the other hand, DEC proteins can directly bind to BMAL1 (Honma et al., 2002; Sato et al., 2004), but not CLOCK, which may also affect CLOCK:BMAL1-induced transcription. However, the interaction of DEC and BMAL1 proteins has a lesser effect on the transcriptional repression than DEC binding to the E-box enhancer in the promoter

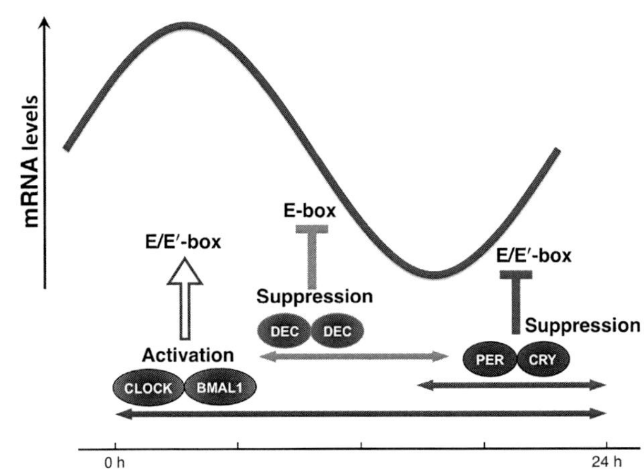

Figure 10.4 Differential suppression of circadian expression of clock and clock-controlled genes by DEC1, DEC2, and PER:CRY. CLOCK:BMAL1 heterodimer binds constitutively to the E-boxes in the target genes and activates transcription. DEC1 and DEC2 directly bind to the E-box, but not the E′-box, for the indicated time (a red bidirectional arrow) after the peak of target gene expression, whereas PER:CRY indirectly binds to the E-box and/or E′-box for the indicated time (a blue bidirectional arrow) after the trough of the expression. (See the color plate.)

region (Li et al., 2004; Sato et al., 2004), and the physiological significance of the DEC–BMAL1 complex is unknown.

The binding of DEC1 to the E-box is time dependent and shows a circadian pattern (Nakashima et al., 2008): DEC1 binding to the E-box can occur shortly after the peak of target gene expression (Fig. 10.4). In contrast, the indirect binding of PER1 to the E-box via CLOCK:BMAL1 is observed after the trough of target gene expression. DEC1 and PER1 bind to the E-box of target genes at different circadian phases, suggesting a unique role of DEC proteins. Collectively, DEC, PER, and CRY have similar, but somewhat distinct, actions in the negative feedback loops of the molecular clock. The distinct functions may result from their direct or indirect binding to the E-box and vary depending on tissue type and/or stimuli (Bode, Shahmoradi, Taneja, Rossner, & Oster, 2011; Bode, Taneja, Rossner, & Oster, 2011; Tsang et al., 2012). Further investigations are needed to understand the tissue-specific circadian actions of DEC proteins.

4.3. Resetting of the clocks by DEC

The major cue of the circadian clock is light exposure, which can reset the central clock in the SCN (Fig. 10.5). In addition, food intake and certain growth factors can act as cues in the autonomous peripheral clocks in a tissue-dependent manner (Albrecht, 2012; Kawamoto et al., 2006; Kon et al., 2008). Thus, it has been shown that food intake alters the circadian phase of peripheral clocks in the liver without affecting the central clock (Damiola et al., 2000).

In the SCN, *Dec1* expression is rapidly induced by light exposure, whereas in peripheral tissues, it is stimulated by food intake and growth factors (Honma et al., 2002; Kawamoto et al., 2006; Kon et al., 2008; Oike et al., 2011). Double knockout of *Dec1* and *Dec2* suppresses the light pulse-induced *c-Fos*, *c-Jun*, and *Per1* expression in the SCN and the resetting of the central clock (Rossner et al., 2008). *Dec1* knockout also

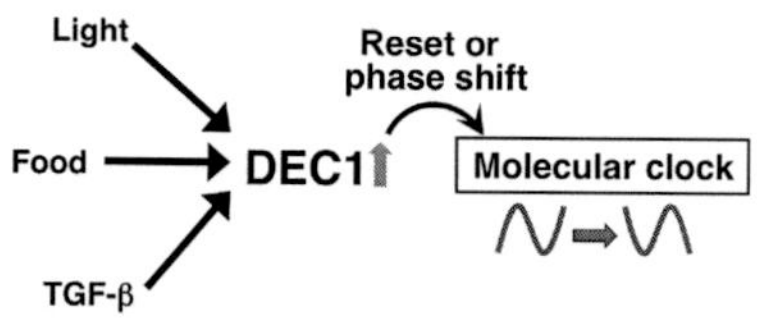

Figure 10.5 DEC1 function in the regulation of the molecular clock system. In addition to light, the major cue of the central clock, food intake, and growth factors act as cues of the peripheral clocks in a tissue-specific manner. (See the color plate.)

suppresses the transforming growth factor (TGF)-β-induced phase advance of the peripheral clocks in the kidney and adrenal gland. No phase advance of *Per1* and *Dbp* circadian expression has been observed in the *Dec1*-deficient tissues after the injection of TGF-β *in vivo* (Kon et al., 2008). These findings suggest that DEC1 is involved in the resetting of the central and peripheral clocks.

4.4. Effects of DEC deficiency on circadian rhythms

The negative circadian regulators DEC1, DEC2, PER1, PER2, PER3, CRY1, and CRY2 suppress the CLOCK:BMAL1-induced transcriptional activation, so it can be expected that the deficiency in any of these clock proteins should have similar effects on the molecular clock system. However, in CRY1- and CRY2-deficient mice, the circadian cycle becomes shorter and longer, respectively, compared to that in wild-type mice (van der Horst et al., 1999), whereas in both PER1- and PER3-deficient mice, it is shortened (Shearman, Jin, Lee, Reppert, & Weaver, 2000; Zheng et al., 2001). Furthermore, CRY1/CRY2 double-deficient and PER2-deficient mice are arrhythmic (van der Horst et al., 1999; Zheng et al., 2001). These data indicate that the function of the clock proteins is not necessarily redundant. Compared to these clock proteins, DEC deficiency has less effect on the circadian rhythms. Thus, the disruption of DEC1 extends the circadian period (24 h) of mouse locomotion (Nakashima et al., 2008) and shortens or has little effect on that of wheel-running activity (Bode, Shahmoradi, et al., 2011; Rossner et al., 2008). On the other hand, DEC1/DEC2-double deficiency has been shown to extend the circadian cycle of wheel-running activity (Rossner et al., 2008), although little effect was observed in the other study (Bode, Shahmoradi, et al., 2011). The effect of DEC1 or DEC2 deficiency on the circadian behavior in mice is thus ambiguous. In contrast, the knockdown of the *Drosophila cwo* gene markedly delays the circadian period of locomotion activity (Matsumoto et al., 2007); *cwo*-deficient flies have weaker and longer oscillations than control strains, which eventually become arrhythmic (Kadener et al., 2007). The importance of DEC in humans is indicated by the effect of a point mutation in DEC2 (P385R) found to be associated with human short sleep phenotype (He et al., 2009). Furthermore, transgenic mice carrying this mutation show a decrease in sleep time compared to wild-type mice. Consequently, these findings indicate that DEC1 and DEC2, as well as CWO, are indispensable for the circadian behavior.

5. MECHANISMS OF DEC1 AND DEC2 ACTIONS

DEC1 and DEC2 function as transcriptional repressors via two mechanisms: (i) DEC proteins bind directly to the E-box and compete with the E-box-binding transactivators, which results in the recruitment of corepressors (Section 5.1); (ii) DEC proteins inhibit the activity of transcription factors by direct protein–protein interactions independently of DNA binding (Section 5.2). However, DEC proteins can also activate the transcription of some target genes. The list of DEC-interacting transcription factors is presented in Table 10.1. We also discuss structure–function relationship in DEC proteins (Section 5.3).

5.1. Transcriptional repression by direct DNA binding

As described above, DEC1 and DEC2 bind to the CACGTG E-box sequence, as indicated by strong repression of the transcriptional activation in the luciferase reporter assays. The CACGTG E-box can be recognized by some bHLH transcriptional activators, including CLOCK:BMAL1 and cMyc:Max heterodimers, suggesting that DEC proteins compete with these factors for the binding to DNA. Through the interaction with the CACGTG E-box in the promoter region, DEC proteins cause the transcriptional repression of *Per1*, cyclin D1 (*Ccnd1*), and *c-Myc* as well as their own genes *Dec1* and *Dec2* (Azmi et al., 2003; Bhawal et al., 2011; Hamaguchi et al., 2004; Kawamoto et al., 2004; Kumaki et al., 2008; Li et al., 2003; Sun & Taneja, 2000). In the luciferase reporter assays, DEC2 shows higher repression activity than DEC1, although there is no difference in their affinity to DNA (Fujimoto et al., 2007; Honma et al., 2002).

In addition to the CACGTG E-box, DEC proteins are able to bind other E-boxes (TCAGGTGA, CACATG, and CACGCG) and the E′-box (CACGTT), although the resultant transcriptional repression is weak (Fujimoto et al., 2007; Nakashima et al., 2008; St-Pierre, Flock, Zacksenhaus, & Egan, 2002). Actually, DEC2 represses expression of the cholesterol 7α-hydroxylase gene (*Cyp7a*) via the CACATG E-box (Noshiro et al., 2004). Consequently, DEC may be able to influence the transcription of the genes carrying other E-box sequences (Kumaki et al., 2008).

DEC1 and DEC2 have been shown to recruit histone deacetylases HDAC1, HDAC2, and sirtuin 1 (SIRT1) to the E-box (Fujimoto et al., 2007; Qian, Jung, & Chen, 2011; Sun & Taneja, 2000). An HDAC inhibitor trichostatin-A partially relieves DEC-mediated transcriptional repression via the CACGTG E-box (Fujimoto et al., 2007; Garriga-Canut, Roopra, & Buckley, 2001; Sun & Taneja, 2000), suggesting that HDAC recruitment by DEC1 or DEC2 contributes to the promoter repression via E-box elements.

5.2. Transcriptional regulation by protein–protein interaction with other transcription factors

5.2.1 Hypoxia-inducible factor-1α

DEC2, but not DEC1, interacts with HIF-1α and inhibits the expression of HIF-1α targets such as the *Vegf* gene in NIH3T3 and Sarcoma 180 cells, and the *Angptl4* and *CaIX* genes in MDA-231 and MII cells (Montagner et al., 2012; Sato et al., 2008). Furthermore, DEC2 and VEGF show reciprocal circadian expression in implanted tumor cells and cartilage *in vivo*, suggesting a negative role of DEC2 in angiogenesis. In this context, DEC2 and HIF-1α should have the opposite effects on invasion, migration, and metastasis of cancer cells. Interestingly, DEC2 facilitates the degradation of HIF-1α through the association with the α4 subunit of 20S proteasome under both normoxic and hypoxic conditions (Montagner et al., 2012; Sato et al., 2008).

5.2.2 Myoblast determination protein

Myoblast determination protein (MyoD) is a member of the bHLH protein superfamily Clade C, which acts as a master regulator of skeletal muscle differentiation. MyoD forms heterodimers with the bHLH E-proteins E12 and E47 and activates transcription through binding to the CACCTG/CAGCTG E-box sequences. In contrast, DEC1 and DEC2 associate with MyoD to repress MyoD-mediated transactivation (Azmi et al., 2003; Hsiao, Huang, Chang, & Chen, 2009). DEC1 or DEC2 overexpression leads to a reduction in the levels of MyoD target genes and suppression of myogenic differentiation in C2C12 cells (Azmi et al., 2004; Fujimoto et al., 2007). Since DEC proteins do not have the affinity to the CACCTG/CAGCTG E-boxes, it is likely that DEC1 and DEC2 inhibit MyoD function by preventing the formation of MyoD:E-protein heterodimers rather than by competing with the heterodimer for the binding to the promoter.

5.2.3 Nuclear receptor retinoid X receptor

DEC1 and DEC2 directly bind to RXRα through the interaction of the LXXLL motif in the HLH region with the RXRα C-terminal ligand-binding domain (see Section 2.1) (Cho et al., 2009), which is enhanced by the treatment with an RXR ligand 9-*cis*-retinoic acid. Accordingly, DEC1 and DEC2 repress the ligand-induced transactivation by the

RXRα:RXRα homodimer or RXRα:LXRα, RXRα:FXR, RXRα:VDR, and RXRα:RARα heterodimers. Consequently, the expression of LXRα target genes, including *Srebp-1c* and *Abca1*, is downregulated by DEC1 and even more so by DEC2, at least *in vitro*.

5.2.4 Specificity protein 1

DEC1 and DEC2 have differential effects on the transcriptional regulation through the Specificity protein 1 (Sp1) binding site. DEC2 interacts with Sp1 and represses *Dec1* and *Ccnd1* expression by inhibiting Sp1 activity (Azmi et al., 2003), whereas DEC1 enhances *Survivin* expression via the Sp1 binding sites in the promoter (Li et al., 2006, 2011). The mechanism of DEC1-induced transactivation remains unknown.

5.2.5 Other factors

DEC1 and DEC2 act as negative regulators of adipogenesis by inhibiting the activity of C/EBP-α and C/EBP-β, key transcriptional mediators of the early stage of adipogenesis (Gulbagci et al., 2009; Matsunaga et al., 2012; Park & Park, 2012; Yun, Maecker, Johnson, & Giaccia, 2002). DEC1 represses *Pparγ2* promoter activity through the C/EBP binding site. The expression levels of C/EBP-β target genes such as *C/ebpα*, *Pparγ2*, and *Ap2* are increased in the DEC2-deficent mice and decreased in DEC2-overexpressing cells.

DEC1 and DEC2 directly interact with and inhibit the transcriptional activity of SREBP-1c (Choi et al., 2008) and decrease *Srebp-1c* expression by competing with SREBP-1c for the binding to the *Srebp-1c* promoter. However, only DEC2, but not DEC1, negatively affects the expression of the SREBP-1c-regulated fatty acid synthase (*Fasn*) gene.

Dec1 expression is induced by many cytokines that use the signal transducer and activator of transcription (STAT) pathways. DEC1 has been shown to activate STAT signaling through direct interaction with STAT proteins (Ivanova et al., 2004). In fact, DEC1 overexpression increases the promoter activity of the STAT-targeted Fas receptor gene and induces cell death.

DEC1 binds to the upstream stimulatory factor 2 (USF2), a Clade B bHLH protein, and inhibits USF2-mediated transactivation of the *c-Myc* promoter. In addition, DEC1 represses *c-Myc* transcription by competing with USF2 for the binding to the E-box in the *c-Myc* promoter (Dhar & Taneja, 2001).

DEC1 also binds to general transcription factors such as TATA-binding protein, transcription factor IIB (TFIIB), and TFIID (Boudjelal et al., 1997), although the physiological significance is unknown.

5.3. Structure–function relationship in DEC proteins

DEC1 and DEC2 form homodimers via their HLH regions, which is an essential step for the active repression through the E-box, because the HLH deletion abolishes DEC binding to the promoter (Fujimoto et al., 2007; Li et al., 2003). DEC mutants lacking the basic region in the bHLH domain (Fig. 10.1A) lose the DNA-binding capability and exert a dominant negative effect on DEC activity (Li et al., 2003; Sato et al., 2004). Arginine residues (R58 and R65 in human DEC1 or R57 in human DEC2) located within the basic region are essential for DEC binding to DNA (Kondo et al., 2006; Li et al., 2003; Sato et al., 2004). Because the bHLH domains of DEC1 and DEC2 show high amino acid sequence similarity, they are also able to form a heterodimer (Sato et al., 2004), although its biological relevance is unknown. DEC2 bHLH domain, together with the basic and C-terminal regions, is involved in DEC2 heterodimerization with MyoD (Fujimoto et al., 2007). In addition, the bHLH domain is required for DEC1 and DEC2 interactions with BMAL1, C/EBPα, C/EBPβ, STAT3, and SIRT1 (Fujimoto et al., 2007; Gulbagci et al., 2009; Ivanova et al., 2004; Sato et al., 2004).

Interestingly, DEC proteins can also exert transcriptional repression independently of the E-box as has been shown for DEC–GAL4DBD fusion protein interaction with GAL4-UAS reporter genes (Boudjelal et al., 1997; Fujimoto et al., 2007). Mutagenesis studies indicate that this activity is localized to DEC C-terminal domain, which is also required for DEC interaction with HDAC1 (Fujimoto et al., 2007; Sun & Taneja, 2000).

6. REGULATORY FACTORS FOR *DEC1* AND *DEC2* EXPRESSION

Dec1 and *Dec2* expression is regulated by various transcription factors, intracellular signaling molecules, growth factors/hormones, and cytokines (Table 10.2a–e) as well as environmental stimuli, including nutrients, light pulse, and hypoxia (Table 10.2f–g). The regulation is mostly exerted at the transcriptional level and, in certain cases, at the posttranslational level. *Dec1* expression can be induced by a multitude of stimuli, whereas the number of factors shown to regulate *Dec2* is more limited.

Table 10.2 Regulatory factors of *Dec1* and *Dec2* expression

Regulatory factors	Expression *Dec1*	Expression *Dec2*	Cells/tissues (references)
a. Clock proteins			
CLOCK: BMAL1/2	Up	Up	Fibroblasts, SCN (Butler et al., 2004; Hamaguchi et al., 2004; Honma et al., 2002; Kawamoto et al., 2004)
NPAS2: BMAL1	Up	–	Fibroblasts (Ueshima et al., 2012)
Dec1	Down	Down	Fibroblasts (Kawamoto et al., 2004; Li et al., 2003)
Dec2	Down	Down	Fibroblasts, 10T1/2 cells (Azmi, Sun, Ozog, & Taneja, 2003; Hamaguchi et al., 2004)
PER1, PER2	Down	Down	Fibroblasts (Hamaguchi et al., 2004; Kawamoto et al., 2004)
CRY1, CRY2	Down	Down	Fibroblasts (Grechez-Cassiau et al., 2004; Hamaguchi et al., 2004)
b. Transcription factors			
HIF-1α	Up	Up	Fibroblasts, tumors, chondrocytes (Miyazaki et al., 2002; Sato et al., 2008)
LXRα	Up	Unchanged	Liver (Noshiro et al., 2009)
RORα	Up	Up	3T3-L1 cells (Ozaki et al., 2012)
VDR	Up	–	Leukemia (Seuter et al., 2013)
ChREBP	Up	–	Hepatocytes (Iizuka & Horikawa, 2008)
p53 family	Up	Up	Tumors (Boominathan, 2010; Montagner et al., 2012; Qian, Zhang, Yan, & Chen, 2008)
Sp1	Up	–	Fibroblasts (Azmi et al., 2003)
GLI1, GLI2	–	Up	Tumors (Inaguma et al., 2013)
TCF4	Down	–	Brain (Forrest, Waite, Martin-Rendon, & Blake, 2013)
GATA3	–	Up	T cells (Yang et al., 2009)
SMAD3, SMAD4	Up	–	Tumors, fibroblasts (Ehata et al., 2007; Kon et al., 2008)

Table 10.2 Regulatory factors of *Dec1* and *Dec2* expression—cont'd

Regulatory factors	Expression *Dec1*	Expression *Dec2*	Cells/tissues (references)
c. Intracellular signaling molecules			
Cyclic AMP	Up	–	Chondrocytes, podocytes (Bek et al., 2003; Shen et al., 1997, 2001)
ALK	Up	–	Fibroblasts (Kon et al., 2008)
PI3 kinase	Up	–	Liver, gingiva (Yamada et al., 2003; Bhawal et al., 2012)
Akt	Up	–	Gingiva (Bhawal et al., 2012)
MAPK p38	–	Up	Bone marrow (Bragado et al., 2013)
d. Growth factors, hormones, and autacoids			
TGF-β	Up	Up	Keratinocytes, chondrocytes, kidney, tumors (Bragado et al., 2013; Kon et al., 2008; Sato et al., 2012; Shen et al., 2002; Zawel et al., 2002)
BMP-2	Up	–	Chondrocytes (Shen et al., 2002)
Activin	Up	–	Fibroblasts (Kon et al., 2008)
PTH/PTHrp	Up	–	Chondrocytes, mesenchymal stem cells, trophoblasts (El-Hashash, Esbrit, & Kimber, 2005; Liu, Li, et al., 2009; Shen et al., 2001, 2002)
NGF	Up	Up	PC12 cells (Rossner et al., 1997)
EGF	Up	–	Mammary glands (St-Pierre, Cooper, Jiang, Zacksenhaus, & Egan, 2004)
Insulin	Up	–	Chondrocytes, liver (Shen et al., 2002; Yamada et al., 2003)
GH	Up	–	Adipocytes (Shang & Waters, 2003)
Prostaglandin E_2	Up	–	Podocytes (Bek et al., 2003)
Glucocorticoid	Down	–	Mammary glands (St-Pierre et al., 2004)
Retinoic acid	Up	–	P19 cells, podocytes (Bek et al., 2003; Boudjelal et al., 1997)
Vitamin D	Up	–	Leukemia (Seuter et al., 2013)

Continued

Table 10.2 Regulatory factors of *Dec1* and *Dec2* expression—cont'd

Regulatory factors	Expression *Dec1*	Expression *Dec2*	Cells/tissues (references)
e. Cytokines			
IL-1, 2, 6, 12, 15	Up	–	Natural killer cells, gingiva (Bhawal et al., 2012; Ivanova et al., 2004)
IL-4, LPS	Down	–	B cells (Seimiya et al., 2002)
TNF-α, IFN-β	Up	–	Natural killer cells, fibroblasts (Ivanova et al., 2004; Petrzilka, Taraborrelli, Cavadini, Fontana, & Birchler, 2009)
SCF	Up	–	Hematopoietic cells (Pedersen et al., 2008)
f. Nutrients			
Feeding	Up	Unchanged	Liver (Kawamoto et al., 2006)
Glucose	Up	–	Hepatocytes (Iizuka & Horikawa, 2008)
Glucose + amino acids	Up	Up	Liver (Oike, Nagai, Fukushima, Ishida, & Kobori, 2011)
Serum	Up	Unchanged	Fibroblasts (Nakashima et al., 2008)
g. Environmental stimuli			
Light pulse	Up	Unchanged	SCN (Honma et al., 2002; Rossner et al., 2008)
γ-Irradiation	Up	Unchanged	Fibroblasts, tumors (Bhawal et al., 2011; Thin et al., 2007)
Hypoxia	Up	Up	Tumors, fibroblasts (Ivanova et al., 2001; Miyazaki et al., 2002)
Hyperosmolar stress	Up	–	Trophoblasts (Liu, Xu, et al., 2009)
Cell confluence	Up	–	Tumors (Li et al., 2002)
Serum starvation	Up	–	Fibroblasts (Sun & Taneja, 2000)

–, no information.

Akt, Ak-thymoma; ALK, activin receptor-like kinase; BMP2, bone morphogenetic protein 2; ChREBP, carbohydrate response element-binding protein; EGF, epidermal growth factor; ERK, extracellular signal-regulated kinase; GATA3, GATA-binding protein 3; GH, growth hormone; IL-1, interleukin-1; INF-β, interferon-α; LPS, lipopolysaccharide; NGF, nerve growth factor; p38, p38 mitogen-activated protein kinase; PI3 kinase, phosphatidylinositol 3 kinase; PTH, parathyroid hormone; PTHrp, PTH-related protein; SCF, stem cell factor; TCF4, transcription factor 4; TGF-β, transforming growth factor-β; TNF-α, tumor necrosis factor-α.

6.1. Transcription factors

Clock proteins modulate the ubiquitous expression of both *Dec1* and *Dec2*, whereas the tissue-specific expression of *Dec1* or *Dec2* is mediated separately by nonclock transcription factors, which may account for the distinct tissue distribution and functional activity of DEC1 and DEC2 proteins (Fig. 10.2A).

6.1.1 Clock proteins

CLOCK:BMAL1, CLOCK:BMAL2, and NPAS2:BMAL1 bind to the E-boxes in the *Dec1* and *Dec2* promoters and enhance their expression (Table 10.2a), whereas DEC1, DEC2, PER1, PER2, CRY1, and CRY2 suppress the transcriptional activity of CLOCK:BMAL1 via the same E-boxes (Fig. 10.3) as described above.

6.1.2 Hypoxia-inducible factor-1

Hypoxic conditions stimulate the expression of both *Dec2* and *Dec1* genes in normal and tumor cells *in vitro* and *in vivo* (Ivanova et al., 2001; Miyazaki et al., 2002; Wykoff, Pugh, Maxwell, Harris, & Ratcliffe, 2000; Yun et al., 2002; Yunokawa et al., 2007), but the induction pattern is distinct. In tumor cells subjected to hypoxia, *Dec1* mRNA levels are maintained for a long time (6–72 h) after the exposure, whereas *Dec2* mRNA increases transiently (6–24 h) (Nakamura et al., 2008).

The effect of hypoxia on *Dec1* and *Dec2* expression has been shown to be mediated by HIF-1 (Table 10.2b; Miyazaki et al., 2002). Hypoxic conditions cause the stabilization of HIF-1α:HIF-1β heterodimer, which then binds to the HRE in the mouse *Dec1* (−462 to −446 nt) and *Dec2* (−311 to −295 nt) promoters and enhances transcription.

Tumor tissues usually have defective vasculature, which creates hypoxic environment; therefore, most tumors show higher *Dec1* expression levels than normal tissues (Li et al., 2002). However, *Dec2* expression profile is considerably different from that of *Dec1* in various tumors, indicating distinct regulatory mechanisms of *Dec1* and *Dec2* (Yunokawa et al., 2007). Hypoxia decreases pH of tissue fluids because of enhanced lactate production in glycolysis, and pH changes have been shown to affect *Dec1* expression in culture (Kon et al., 2008).

6.1.3 Nuclear receptors

The *Dec1* promoter contains multiple AGGTCA motifs present in NRE (Noshiro et al., 2009). LXRα and LXRβ bind to LXRE (−2070 to

−2043 nt) of the mouse *Dec1* promoter causing its activation. Accordingly, an LXR ligand (T0901317) enhances *Dec1* expression in hepatoma cells (Noshiro et al., 2009). However, LXR has little effect on *Dec2* expression possibly because of the absence of LXRE in the *Dec2* promoter (Noshiro et al., 2009).

RORα, but not REV-ERBα, binds to RORE in the mouse *Dec1* (−1783 nt) and *Dec2* (−2902 and −672 nt) promoters and enhances the transcription without changing the circadian pattern (Ozaki et al., 2012). RORα suppresses adipogenic differentiation partly via the enhancement of *Dec1* expression (Ozaki et al., 2012), as DEC1 suppresses adipogenic differentiation (Iwata et al., 2006; Yun et al., 2002).

All-*trans* retinoic acid enhances *Dec1* expression in P19 embryonic carcinoma cells and kidney podocytes (Bek et al., 2003; Boudjelal et al., 1997; Sun & Taneja, 2000). In our studies, RARα caused 30–40% induction of the *Dec1*, but not *Dec2*, promoter in the luciferase assays (Noshiro et al., 2009). However, the position of RARE in the *Dec1* promoter has not been identified.

In leukemia cells, calcitriol (1,25(OH)$_2$ vitamin D$_3$) enhances *Dec1* expression (Seuter et al., 2013). In these cells, the VDR has been shown to bind to VDR element (−1731 to −1714 nt) in the human *DEC1* promoter and to stimulate local dynamic chromatin opening enhancing the transcription (Table 10.2b).

6.1.4 Other transcription factors

Glucose-activated carbohydrate-response element-binding protein (ChREBP) binds to the ChoRE (−160 to −143 nt) in the mouse *Dec1* promoter and enhances the transcription, as has been demonstrated by the addition of glucose to cultured hepatocytes (Iizuka & Horikawa, 2008).

Various drugs, including DNA-damaging agents, have been shown to induce *Dec1* or *Dec2* in a p53-dependent or p53-independent manner (Qian et al., 2008; Thin et al., 2007; Wu, Sato, Bhawal, et al., 2012). The p53 family proteins (p53, p63, and p73) bind to a p53-responsive element (−4211 to −4181 nt) of the human *DEC1* promoter and enhance the transcription, suggesting that DEC1 may be involved in the mediation of the p53-dependent survival, G1 arrest, and premature senescence in tumor cells (Qian & Chen, 2008; Qian, Jung, & Chen, 2012).

The Kruppel-like zinc finger transcription factors GLI are effectors of the Hedgehog signaling involved in cell proliferation and differentiation. GLI1 and GLI2 bind to the GLI-binding site (−1126 to −1117 nt) in the human

DEC2 promoter and enhance *DEC2* expression but inhibit that of MutL homolog 1 (MLH1), essential for the DNA mismatch repair and indispensable for the fidelity of DNA replication. It has been shown that in tumor cells, DEC2 mediates the GLI-induced MLH1 repression (Inaguma et al., 2013) and that the hypoxia-induced expression of DEC1 and DEC2 is followed by the reduction of MLH1 levels (Nakamura et al., 2008), suggesting that the upregulation of DEC proteins could increase the genomic instability and facilitate tumor progression through the repression of MLH1.

GATA-binding protein 3 (GATA3) induces *Dec2* expression during T_H2 differentiation of T lymphocytes (Yang et al., 2009). By a feed-forward regulatory mechanism, DEC2 directly binds to and activates the transcription of the *Gata3* and *JunB* genes and the expression of T_H2 cytokines IL-4, IL-5, IL-13, and IL-2R (Liu, Li, et al., 2009; Yang et al., 2009).

Another transcriptional regulator of *Dec1* expression is transcription factor 4 (TCF4), a bHLH protein essential for brain development. Knockdown of TCF4 upregulates *Dec1* expression in neuroblastoma cells; however, the mechanism remains unclear (Forrest et al., 2013; Table 10.2b).

6.2. Growth factors, hormones, and autacoids

The TGF-β superfamily members, including TGF-β, bone morphogenetic protein-2, and activin, induce *Dec1* expression in various cells. DEC1 partially mediates the effects of the TGF-β family proteins on apoptosis, differentiation, and metastasis (Ehata et al., 2007; Kon et al., 2008; Shen et al., 2002; Zawel et al., 2002). The TGF-β–DEC1 pathway also plays a role in the epithelial–mesenchymal transition of pancreatic cancer (Wu, Sato, Yamada, et al., 2012), whereas TGF-β-induced inhibition of cell proliferation can be observed even with DEC1-deficient keratinocytes (Zawel et al., 2002).

TGF-β binds to and activates the activin receptor-like kinase (ALK) receptor that phosphorylates SMADs. The activated SMADs induce *Dec1* expression, as demonstrated by the experiments with the knockdown of *Smad3* or *Smad4* and the forced expression of an inhibitory SMAD7, which suppresses TGF-β-induced *Dec1* expression (Ehata et al., 2007; Kon et al., 2008). However, the location of the SMAD-responsive element in the *Dec1* promoter is not yet identified.

In some tissues, *Dec2* expression is upregulated by TGF-β via activation of the mitogen-activated protein kinase (MAPK) p38 (Table 10.2c; Bragado et al., 2013; Kon et al., 2008; Sato et al., 2012); DEC2, contrary to DEC1, suppresses TGF-β-induced tumor progression (Sato et al., 2012; Wu, Sato, Yamada, et al., 2012).

Parathyroid hormone (PTH) and PTH-related protein (PTHrp) maintain calcium homeostasis and control the differentiation of chondrocytes, osteoblasts, and trophoblasts. PTH/PTHrp enhances *Dec1* expression in chondrogenic cells at the early stage (1 h), repressing it at the late stage (>24 h) (Shen et al., 2002). The early induction of *Dec1* is due to a transient increase in the intracellular cyclic AMP levels (Table 10.2c; Shen et al., 2001), whereas the repression of *Dec1* at the late stage is related to the inhibition of chondrocyte differentiation by PTH/PTHrp. In trophoblast stem cells, PTHrp enhances the expression of DEC1, which then stimulates the differentiation into trophoblast giant cells (El-Hashash et al., 2005; Liu, Xu, et al., 2009).

Nerve growth factor (NGF), which promotes the survival and differentiation of sensory and sympathetic neurons and is essential for neuronal development, has been shown to upregulate *Dec1* and *Dec2* expression in PC12 cells (Rossner et al., 1997).

In mammary epithelial cells, epidermal growth factor (EGF) increases DEC1 protein expression, which is suppressed by dexamethasone (St-Pierre et al., 2004). However, EGF and dexamethasone have little effect on *Dec1* mRNA, indicating the regulation at the posttranslational level. *Dec2* expression profile is different from that of *Dec1* during the development of the mammary gland (St-Pierre et al., 2004).

Insulin induces *Dec1* expression via the phosphoinositide 3 kinase (PI3 kinase) pathway in the liver, hepatoma cells, and chondrogenic cells (Table 10.2c; Shen et al., 2002; Yamada et al., 2003). DEC1 partially mediates the effect of insulin on the energy metabolism in the liver and on the differentiation of chondrogenic cells.

Growth hormone induces *Dec1* expression in preadipocytes during the stimulation of adipogenic differentiation (Shang & Waters, 2003). However, the physiological significance of this phenomenon is unknown because usually DEC1 suppresses adipogenic differentiation (Iwata et al., 2006; Yun et al., 2002).

Prostaglandin E_2 and forskolin induce *Dec1* expression in podocytes, possibly via the induction of cyclic AMP (Table 10.2c and d; Bek et al., 2003; Shen et al., 2001).

6.3. Cytokines

Many cytokines enhance or suppress *Dec1* expression, indicating the role of DEC1 in inflammatory and immunological reactions (Table 10.2e). Tumor necrosis factor-α (TNF-α), interleukin-2 (IL-2), and interferon-β (IFN-β) upregulate *Dec1* expression in T lymphocytes (Ivanova et al., 2004). TNF-α also upregulates *Dec1* expression in fibroblasts (Petrzilka et al., 2009). The mechanism is probably related to the activation of the putative binding sites for the cytokine-inducible STAT proteins in the *Dec1* promoter (Ivanova et al., 2004).

IL-1β enhances *Dec1* expression in gingival epithelial cells, but not gingival fibroblasts, via the PI3 kinase and Akt/protein kinase B pathway (Table 10.2c; Bhawal et al., 2012). On the other hand, various stimuli that activate B lymphocytes, including IL-4, lipopolysaccharide (LPS), CD40 ligand, and anti-IgM antibodies, downregulate *Dec1* (Seimiya et al., 2002). Other cytokines in Table 10.2e also modulate *Dec1* expression in lymphocytes and hematopoietic cells.

6.4. Nutrients

Dec1 expression levels depend upon the nutritional status (Table 10.2f). Feeding has been shown to upregulate and fasting to downregulate the expression of *Dec1* in the liver and some other tissues (Iizuka & Horikawa, 2008; Kawamoto et al., 2006; Yamada et al., 2003). In its turn, DEC1 affects energy metabolism by regulating adipogenesis and gluconeogenesis (Yamada, Ogata-Kawata, Matsuura, & Miyamoto, 2005; Yun et al., 2002). The induction of hepatic *Dec1* by feeding is rapid and prominent compared with the other clock genes (Kawamoto et al., 2006), suggesting that DEC1 may play a role in the food-entrainable circadian oscillatory mechanism of the peripheral clocks.

In fibroblast cultures, high serum concentration has been shown to induce *Dec1* expression and, subsequently, to synchronize its circadian rhythm (Nakashima et al., 2008).

The molecular components in food or serum responsible for *Dec1* induction are unknown. Carbohydrates obviously play a role because glucose at high concentrations has been shown to directly induce *Dec1* expression via ChREBP in hepatocytes (Iizuka & Horikawa, 2008). On the other hand, the intraperitoneal injection of glucose with 12 amino acids, but not alone, enhances *Dec1* expression in the liver (Oike et al., 2011). In addition to carbohydrates and amino acids, lipids may contribute to the feeding-induced

Dec1 expression because the oxysterol-activated LXR enhances *Dec1* expression (Noshiro et al., 2009).

6.5. Environmental stimuli

Light pulse at night rapidly induces *Dec1*, but not *Dec2*, in the SCN (Honma et al., 2002; Rossner et al., 2008). On the other hand, different photoperiods (light–dark 8–16 h or 16–8 h) have been shown to alter the circadian pattern of *Dec2* expression in the pineal gland (Engel, Lorenzkowski, Langer, Rohleder, & Spessert, 2005).

In thymocytes, γ-irradiation increases DEC1 protein levels in a p53-independent manner, whereas DEC1 increases p53 levels by preventing its ubiquitination and nuclear export, suggesting a close relationship between DEC1 and p53 (Thin et al., 2007).

Hypoxia induces *Dec1* and *Dec2* expression in various cells. Hyperglycemia and dehydration may also influence DEC levels by causing hyperosmolar stress, which has been shown to induce *Dec1* expression in trophoblast stem cells (Liu, Xu, et al., 2009).

In fibroblast cultures, cell confluence, serum starvation, and cell cycle blockers induce *Dec1* expression, indicating the role of DEC1 in growth arrest in some situations (Table 10.2g; Li et al., 2002; Sun & Taneja, 2000).

7. PERSPECTIVES

DEC1 and DEC2 are regulators of the mammalian molecular clock and form the DEC feedback loop of the clock system. DEC1 and DEC2 function as synchronizers of the clock outputs to clock-controlled genes in response to various environmental stimuli, although in some situations, they can also transduce external signals to molecular targets independently of the clocks. DEC proteins belong to Clade E of the bHLH protein superfamily, which includes the main clock components BMAL1, CLOCK, NPAS2, and SIM1, along with the hypoxia response factors HIF-1α and HIF-1β, suggesting that the evolution of *Dec* genes has been, at least partly, driven by the adaption to the day–night cycle and oxygen deficiency. In modern urban life, the day–night light cycle is often disturbed and the disruption of circadian rhythms causes sleep disorders, hypertension, obesity, diabetes, and tumorigenesis (Bonny & Firsov, 2013; Kato et al., 2010; Kelleher, Rao, & Maguire, 2014; Maury, Hong, & Bass, 2014); aberrant hypoxia responses further accelerate tumor progression. DEC proteins as components of the circadian clocks and as environmental sensors may be implicated in the pathology of these diseases. We hope that studies on DEC functional mechanisms will open new avenues for targeted therapies of hypertension, diabetes, autoimmune disorders, and cancer.

ACKNOWLEDGMENTS

The authors were supported by grants from the Japanese Ministry of Education, Culture, Sport, Science and Technology.

REFERENCES

Albrecht, U. (2012). Timing to perfection: The biology of central and peripheral circadian clocks. *Neuron, 74*, 246–260.

Alvarez, J. D., & Sehgal, A. (2002). Circadian rhythms: Finer clock control. *Nature, 419*, 798–799.

Azmi, S., Ozog, A., & Taneja, R. (2004). Sharp-1/DEC2 inhibits skeletal muscle differentiation through repression of myogenic transcription factors. *The Journal of Biological Chemistry, 279*, 52643–52652.

Azmi, S., Sun, H., Ozog, A., & Taneja, R. (2003). mSharp-1/DEC2, a basic helix-loop-helix protein functions as a transcriptional repressor of E box activity and Stra13 expression. *The Journal of Biological Chemistry, 278*, 20098–20109.

Azmi, S., & Taneja, R. (2002). Embryonic expression of mSharp-1/mDEC2, which encodes a basic helix-loop-helix transcription factor. *Mechanisms of Development, 114*, 181–185.

Baba, K., Ono, D., Honma, S., & Honma, K. (2008). A TTX-sensitive local circuit is involved in the expression of PK2 and BDNF circadian rhythms in the mouse suprachiasmatic nucleus. *The European Journal of Neuroscience, 27*, 909–916.

Beadling, C., Cereseto, A., Fan, W., Naramura, M., & Smith, K. A. (2001). Cytokine response gene 8 (CR8) regulates the cell cycle G1-S phase transition and promotes cellular survival. *Oncogene, 20*, 1771–1783.

Bek, M. J., Wahle, S., Muller, B., Benzing, T., Huber, T. B., Kretzler, M., et al. (2003). Stra13, a prostaglandin E2-induced gene, regulates the cellular redox state of podocytes. *FASEB Journal, 17*, 682–684.

Bhawal, U. K., Ito, Y., Tanimoto, K., Sato, F., Fujimoto, K., Kawamoto, T., et al. (2012). IL-1beta-mediated up-regulation of DEC1 in human gingiva cells via the Akt pathway. *Journal of Cellular Biochemistry, 113*, 3246–3253.

Bhawal, U. K., Sato, F., Arakawa, Y., Fujimoto, K., Kawamoto, T., Tanimoto, K., et al. (2011). Basic helix-loop-helix transcription factor DEC1 negatively regulates cyclin D1. *The Journal of Pathology, 224*, 420–429.

Bode, B., Shahmoradi, A., Taneja, R., Rossner, M. J., & Oster, H. (2011). Genetic interaction of Per1 and Dec1/2 in the regulation of circadian locomotor activity. *Journal of Biological Rhythms, 26*, 530–540.

Bode, B., Taneja, R., Rossner, M. J., & Oster, H. (2011). Advanced light-entrained activity onsets and restored free-running suprachiasmatic nucleus circadian rhythms in per2/dec mutant mice. *Chronobiology International, 28*, 737–750.

Bonny, O., & Firsov, D. (2013). Circadian regulation of renal function and potential role in hypertension. *Current Opinion in Nephrology and Hypertension, 22*, 439–444.

Boominathan, L. (2010). The guardians of the genome (p53, TA-p73, and TA-p63) are regulators of tumor suppressor miRNAs network. *Cancer Metastasis Reviews, 29*, 613–639.

Boudjelal, M., Taneja, R., Matsubara, S., Bouillet, P., Dolle, P., & Chambon, P. (1997). Overexpression of Stra13, a novel retinoic acid-inducible gene of the basic helix-loop-helix family, inhibits mesodermal and promotes neuronal differentiation of P19 cells. *Genes & Development*, *11*, 2052–2065.

Bragado, P., Estrada, Y., Parikh, F., Krause, S., Capobianco, C., Farina, H. G., et al. (2013). TGF-beta2 dictates disseminated tumour cell fate in target organs through TGF-beta-RIII and p38alpha/beta signalling. *Nature Cell Biology*, *15*, 1351–1361.

Butler, M. P., Honma, S., Fukumoto, T., Kawamoto, T., Fujimoto, K., Noshiro, M., et al. (2004). Dec1 and Dec2 expression is disrupted in the suprachiasmatic nuclei of Clock mutant mice. *Journal of Biological Rhythms*, *19*, 126–134.

Cho, Y., Noshiro, M., Choi, M., Morita, K., Kawamoto, T., Fujimoto, K., et al. (2009). The basic helix-loop-helix proteins differentiated embryo chondrocyte (DEC) 1 and DEC2 function as corepressors of retinoid X receptors. *Molecular Pharmacology*, *76*, 1360–1369.

Choi, S. M., Cho, H. J., Cho, H., Kim, K. H., Kim, J. B., & Park, H. (2008). Stra13/DEC1 and DEC2 inhibit sterol regulatory element binding protein-1c in a hypoxia-inducible factor-dependent mechanism. *Nucleic Acids Research*, *36*, 6372–6385.

Damiola, F., Le Minh, N., Preitner, N., Kornmann, B., Fleury-Olela, F., & Schibler, U. (2000). Restricted feeding uncouples circadian oscillators in peripheral tissues from the central pacemaker in the suprachiasmatic nucleus. *Genes & Development*, *14*, 2950–2961.

Dear, T. N., Hainzl, T., Follo, M., Nehls, M., Wilmore, H., Matena, K., et al. (1997). Identification of interaction partners for the basic-helix-loop-helix protein E47. *Oncogene*, *14*, 891–898.

Dhar, M., & Taneja, R. (2001). Cross-regulatory interaction between Stra13 and USF results in functional antagonism. *Oncogene*, *20*, 4750–4756.

Ehata, S., Hanyu, A., Hayashi, M., Aburatani, H., Kato, Y., Fujime, M., et al. (2007). Transforming growth factor-beta promotes survival of mammary carcinoma cells through induction of antiapoptotic transcription factor DEC1. *Cancer Research*, *67*, 9694–9703.

El-Hashash, A. H., Esbrit, P., & Kimber, S. J. (2005). PTHrP promotes murine secondary trophoblast giant cell differentiation through induction of endocycle, upregulation of giant-cell-promoting transcription factors and suppression of other trophoblast cell types. *Differentiation; Research in Biological Diversity*, *73*, 154–174.

Engel, L., Lorenzkowski, V., Langer, C., Rohleder, N., & Spessert, R. (2005). The photoperiod entrains the molecular clock of the rat pineal. *The European Journal of Neuroscience*, *21*, 2297–2304.

Fisher, A. L., Ohsako, S., & Caudy, M. (1996). The WRPW motif of the hairy-related basic helix-loop-helix repressor proteins acts as a 4-amino-acid transcription repression and protein-protein interaction domain. *Molecular and Cellular Biology*, *16*, 2670–2677.

Forrest, M. P., Waite, A. J., Martin-Rendon, E., & Blake, D. J. (2013). Knockdown of human TCF4 affects multiple signaling pathways involved in cell survival, epithelial to mesenchymal transition and neuronal differentiation. *PLoS One*, *8*, e73169.

Fujimoto, K., Hamaguchi, H., Hashiba, T., Nakamura, T., Kawamoto, T., Sato, F., et al. (2007). Transcriptional repression by the basic helix-loop-helix protein Dec2: Multiple mechanisms through E-box elements. *International Journal of Molecular Medicine*, *19*, 925–932.

Fujimoto, K., Shen, M., Noshiro, M., Matsubara, K., Shingu, S., Honda, K., et al. (2001). Molecular cloning and characterization of DEC2, a new member of basic helix-loop-helix proteins. *Biochemical and Biophysical Research Communications*, *280*, 164–171.

Furukawa, M., Kawamoto, T., Noshiro, M., Honda, K. K., Sakai, M., Fujimoto, K., et al. (2005). Clock gene expression in the submandibular glands. *Journal of Dental Research*, *84*, 1193–1197.

Garriga-Canut, M., Roopra, A., & Buckley, N. J. (2001). The basic helix-loop-helix protein, sharp-1, represses transcription by a histone deacetylase-dependent and histone deacetylase-independent mechanism. *The Journal of Biological Chemistry*, *276*, 14821–14828.

Grechez-Cassiau, A., Panda, S., Lacoche, S., Teboul, M., Azmi, S., Laudet, V., et al. (2004). The transcriptional repressor STRA13 regulates a subset of peripheral circadian outputs. *The Journal of Biological Chemistry*, *279*, 1141–1150.

Gulbagci, N. T., Li, L., Ling, B., Gopinadhan, S., Walsh, M., Rossner, M., et al. (2009). SHARP1/DEC2 inhibits adipogenic differentiation by regulating the activity of C/EBP. *EMBO Reports*, *10*, 79–86.

Hamaguchi, H., Fujimoto, K., Kawamoto, T., Noshiro, M., Maemura, K., Takeda, N., et al. (2004). Expression of the gene for Dec2, a basic helix-loop-helix transcription factor, is regulated by a molecular clock system. *The Biochemical Journal*, *382*, 43–50.

He, Y., Jones, C. R., Fujiki, N., Xu, Y., Guo, B., Holder, J. L., Jr., et al. (2009). The transcriptional repressor DEC2 regulates sleep length in mammals. *Science*, *325*, 866–870.

Honda, K. K., Kawamoto, T., Ueda, H. R., Nakashima, A., Ueshima, T., Yamada, R. G., et al. (2013). Different circadian expression of major matrix-related genes in various types of cartilage: Modulation by light-dark conditions. *Journal of Biochemistry*, *154*, 373–381.

Honma, S., Kawamoto, T., Takagi, Y., Fujimoto, K., Sato, F., Noshiro, M., et al. (2002). Dec1 and Dec2 are regulators of the mammalian molecular clock. *Nature*, *419*, 841–844.

Hsiao, S. P., & Chen, S. L. (2010). Myogenic regulatory factors regulate M-cadherin expression by targeting its proximal promoter elements. *The Biochemical Journal*, *428*, 223–233.

Hsiao, S. P., Huang, K. M., Chang, H. Y., & Chen, S. L. (2009). P/CAF rescues the Bhlhe40-mediated repression of MyoD transactivation. *The Biochemical Journal*, *422*, 343–352.

Iizuka, K., & Horikawa, Y. (2008). Regulation of lipogenesis via BHLHB2/DEC1 and ChREBP feedback looping. *Biochemical and Biophysical Research Communications*, *374*, 95–100.

Inaguma, S., Riku, M., Hashimoto, M., Murakami, H., Saga, S., Ikeda, H., et al. (2013). GLI1 interferes with the DNA mismatch repair system in pancreatic cancer through BHLHE41-mediated suppression of MLH1. *Cancer Research*, *73*, 7313–7323.

Ivanova, A. V., Ivanov, S. V., Danilkovitch-Miagkova, A., & Lerman, M. I. (2001). Regulation of STRA13 by the von Hippel-Lindau tumor suppressor protein, hypoxia, and the UBC9/ubiquitin proteasome degradation pathway. *The Journal of Biological Chemistry*, *276*, 15306–15315.

Ivanova, A. V., Ivanov, S. V., Zhang, X., Ivanov, V. N., Timofeeva, O. A., & Lerman, M. I. (2004). STRA13 interacts with STAT3 and modulates transcription of STAT3-dependent targets. *Journal of Molecular Biology*, *340*, 641–653.

Iwata, T., Kawamoto, T., Sasabe, E., Miyazaki, K., Fujimoto, K., Noshiro, M., et al. (2006). Effects of overexpression of basic helix-loop-helix transcription factor Dec1 on osteogenic and adipogenic differentiation of mesenchymal stem cells. *European Journal of Cell Biology*, *85*, 423–431.

Kadener, S., Stoleru, D., McDonald, M., Nawathean, P., & Rosbash, M. (2007). Clockwork Orange is a transcriptional repressor and a new Drosophila circadian pacemaker component. *Genes & Development*, *21*, 1675–1686.

Kato, Y., Noshiro, M., Fujimoto, K., & Kawamoto, T. (2010). Roles of DEC1 and DEC2 in the core loop of the circadian clock, and clock outputs to metabolism. *Hirosaki Medical Journal*, *60*, S34–S42.

Kawamoto, T., Noshiro, M., Furukawa, M., Honda, K. K., Nakashima, A., Ueshima, T., et al. (2006). Effects of fasting and re-feeding on the expression of Dec1, Per1, and other clock-related genes. *Journal of Biochemistry*, *140*, 401–408.

Kawamoto, T., Noshiro, M., Sato, F., Maemura, K., Takeda, N., Nagai, R., et al. (2004). A novel autofeedback loop of Dec1 transcription involved in circadian rhythm regulation. *Biochemical and Biophysical Research Communications*, *313*, 117–124.

Kelleher, F. C., Rao, A., & Maguire, A. (2014). Circadian molecular clocks and cancer. *Cancer Letters*, *342*, 9–18.

Kon, N., Hirota, T., Kawamoto, T., Kato, Y., Tsubota, T., & Fukada, Y. (2008). Activation of TGF-beta/activin signalling resets the circadian clock through rapid induction of Dec1 transcripts. *Nature Cell Biology*, *10*, 1463–1469.

Kondo, J., Sato, F., Fujimoto, K., Kusumi, T., Imanaka, T., Kawamoto, T., et al. (2006). 57Arg in the bHLH transcription factor DEC2 is essential for the suppression of CLOCK/BMAL2-mediated transactivation. *International Journal of Molecular Medicine*, *17*, 1053–1056.

Kumaki, Y., Ukai-Tadenuma, M., Uno, K. D., Nishio, J., Masumoto, K. H., Nagano, M., et al. (2008). Analysis and synthesis of high-amplitude Cis-elements in the mammalian circadian clock. *Proceedings of the National Academy of Sciences of the United States of America*, *105*, 14946–14951.

Li, Y., Shen, Q., Kim, H. T., Bissonnette, R. P., Lamph, W. W., Yan, B., et al. (2011). The rexinoid bexarotene represses cyclin D1 transcription by inducing the DEC2 transcriptional repressor. *Breast Cancer Research and Treatment*, *128*, 667–677.

Li, Y., Song, X., Ma, Y., Liu, J., Yang, D., & Yan, B. (2004). DNA binding, but not interaction with Bmal1, is responsible for DEC1-mediated transcription regulation of the circadian gene mPer1. *The Biochemical Journal*, *382*, 895–904.

Li, Y., Xie, M., Song, X., Gragen, S., Sachdeva, K., Wan, Y., et al. (2003). DEC1 negatively regulates the expression of DEC2 through binding to the E-box in the proximal promoter. *The Journal of Biological Chemistry*, *278*, 16899–16907.

Li, Y., Xie, M., Yang, J., Yang, D., Deng, R., Wan, Y., et al. (2006). The expression of antiapoptotic protein survivin is transcriptionally upregulated by DEC1 primarily through multiple sp1 binding sites in the proximal promoter. *Oncogene*, *25*, 3296–3306.

Li, Y., Zhang, H., Xie, M., Hu, M., Ge, S., Yang, D., et al. (2002). Abundant expression of Dec1/stra13/sharp2 in colon carcinoma: Its antagonizing role in serum deprivation-induced apoptosis and selective inhibition of procaspase activation. *The Biochemical Journal*, *367*, 413–422.

Liu, Z., Li, Z., Mao, K., Zou, J., Wang, Y., Tao, Z., et al. (2009). Dec2 promotes Th2 cell differentiation by enhancing IL-2R signaling. *The Journal of Immunology*, *183*, 6320–6329.

Liu, J., Xu, W., Sun, T., Wang, F., Puscheck, E., Brigstock, D., et al. (2009). Hyperosmolar stress induces global mRNA responses in placental trophoblast stem cells that emulate early post-implantation differentiation. *Placenta*, *30*, 66–73.

Matsumoto, A., Ukai-Tadenuma, M., Yamada, R. G., Houl, J., Uno, K. D., Kasukawa, T., et al. (2007). A functional genomics strategy reveals clockwork orange as a transcriptional regulator in the Drosophila circadian clock. *Genes & Development*, *21*, 1687–1700.

Matsunaga, N., Inoue, M., Kusunose, N., Kakimoto, K., Hamamura, K., Hanada, Y., et al. (2012). Time-dependent interaction between differentiated embryo chondrocyte-2 and CCAAT/enhancer-binding protein alpha underlies the circadian expression of CYP2D6 in serum-shocked HepG2 cells. *Molecular Pharmacology*, *81*, 739–747.

Maury, E., Hong, H. K., & Bass, J. (2014). Circadian disruption in the pathogenesis of metabolic syndrome. *Diabetes & Metabolism*. pii: S1262-3636(13)00235-8. http://dx.doi.org/10.1016/j.diabet.2013.12.005. [Epub ahead of print].

Miyazaki, K., Kawamoto, T., Tanimoto, K., Nishiyama, M., Honda, H., & Kato, Y. (2002). Identification of functional hypoxia response elements in the promoter region of the DEC1 and DEC2 genes. *The Journal of Biological Chemistry*, *277*, 47014–47021.

Miyazaki, K., Miyazaki, M., Guo, Y., Yamasaki, N., Kanno, M., Honda, Z., et al. (2010). The role of the basic helix-loop-helix transcription factor Dec1 in the regulatory T cells. *The Journal of Immunology*, *185*, 7330–7339.

Montagner, M., Enzo, E., Forcato, M., Zanconato, F., Parenti, A., Rampazzo, E., et al. (2012). SHARP1 suppresses breast cancer metastasis by promoting degradation of hypoxia-inducible factors. *Nature*, *487*, 380–384.

Nakamura, H., Tanimoto, K., Hiyama, K., Yunokawa, M., Kawamoto, T., Kato, Y., et al. (2008). Human mismatch repair gene, MLH1, is transcriptionally repressed by the hypoxia-inducible transcription factors, DEC1 and DEC2. *Oncogene*, *27*, 4200–4209.

Nakashima, A., Kawamoto, T., Honda, K. K., Ueshima, T., Noshiro, M., Iwata, T., et al. (2008). DEC1 modulates the circadian phase of clock gene expression. *Molecular and Cellular Biology*, *28*, 4080–4092.

Noshiro, M., Furukawa, M., Honma, S., Kawamoto, T., Hamada, T., Honma, K., et al. (2005). Tissue-specific disruption of rhythmic expression of Dec1 and Dec2 in clock mutant mice. *Journal of Biological Rhythms*, *20*, 404–418.

Noshiro, M., Kawamoto, T., Furukawa, M., Fujimoto, K., Yoshida, Y., Sasabe, E., et al. (2004). Rhythmic expression ofDEC1 and DEC2 in peripheral tissues: DEC2 is a potent suppressor for hepatic cytochrome P450s opposing DBP. *Genes to Cells: Devoted to Molecular & Cellular Mechanisms*, *9*, 317–329.

Noshiro, M., Usui, E., Kawamoto, T., Sato, F., Nakashima, A., Ueshima, T., et al. (2009). Liver X receptors (LXRalpha and LXRbeta) are potent regulators for hepatic Dec1 expression. *Genes to Cells: Devoted to Molecular & Cellular Mechanisms*, *14*, 29–40.

Oike, H., Nagai, K., Fukushima, T., Ishida, N., & Kobori, M. (2011). Feeding cues and injected nutrients induce acute expression of multiple clock genes in the mouse liver. *PLoS One*, *6*, e23709.

Ozaki, N., Noshiro, M., Kawamoto, T., Nakashima, A., Honda, K., Fukuzaki-Dohi, U., et al. (2012). Regulation of basic helix-loop-helix transcription factors Dec1 and Dec2 by RORalpha and their roles in adipogenesis. *Genes to Cells: Devoted to Molecular & Cellular Mechanisms*, *17*, 109–121.

Park, Y. K., & Park, H. (2012). Differentiated embryo chondrocyte 1 (DEC1) represses PPARgamma2 gene through interacting with CCAAT/enhancer binding protein beta (C/EBPbeta). *Molecules and Cells*, *33*, 575–581.

Pedersen, M., Lofstedt, T., Sun, J., Holmquist-Mengelbier, L., Pahlman, S., & Ronnstrand, L. (2008). Stem cell factor induces HIF-1alpha at normoxia in hematopoietic cells. *Biochemical and Biophysical Research Communications*, *377*, 98–103.

Petrzilka, S., Taraborrelli, C., Cavadini, G., Fontana, A., & Birchler, T. (2009). Clock gene modulation by TNF-alpha depends on calcium and p38 MAP kinase signaling. *Journal of Biological Rhythms*, *24*, 283–294.

Qian, Y., & Chen, X. (2008). ID1, inhibitor of differentiation/DNA binding, is an effector of the p53-dependent DNA damage response pathway. *The Journal of Biological Chemistry*, *283*, 22410–22416.

Qian, Y., Jung, Y. S., & Chen, X. (2011). DeltaNp63, a target of DEC1 and histone deacetylase 2, modulates the efficacy of histone deacetylase inhibitors in growth suppression and keratinocyte differentiation. *The Journal of Biological Chemistry*, *286*, 12033–12041.

Qian, Y., Jung, Y. S., & Chen, X. (2012). Differentiated embryo-chondrocyte expressed gene 1 regulates p53-dependent cell survival versus cell death through macrophage inhibitory cytokine-1. *Proceedings of the National Academy of Sciences of the United States of America*, *109*, 11300–11305.

Qian, Y., Zhang, J., Yan, B., & Chen, X. (2008). DEC1, a basic helix-loop-helix transcription factor and a novel target gene of the p53 family, mediates p53-dependent premature senescence. *The Journal of Biological Chemistry*, *283*, 2896–2905.

Rohleder, N., Langer, C., Maus, C., Spiwoks-Becker, I., Emser, A., Engel, L., et al. (2006). Influence of photoperiodic history on clock genes and the circadian pacemaker in the rat retina. *The European Journal of Neuroscience*, *23*, 105–111.

Rossner, M. J., Dorr, J., Gass, P., Schwab, M. H., & Nave, K. A. (1997). SHARPs: Mammalian enhancer-of-split- and hairy-related proteins coupled to neuronal stimulation. *Molecular and Cellular Neurosciences*, *10*, 460–475.

Rossner, M. J., Oster, H., Wichert, S. P., Reinecke, L., Wehr, M. C., Reinecke, J., et al. (2008). Disturbed clockwork resetting in Sharp-1 and Sharp-2 single and double mutant mice. *PLoS One*, *3*, e2762.

Sato, F., Bhawal, U. K., Kawamoto, T., Fujimoto, K., Imaizumi, T., Imanaka, T., et al. (2008). Basic-helix-loop-helix (bHLH) transcription factor DEC2 negatively regulates vascular endothelial growth factor expression. *Genes to Cells: Devoted to Molecular & Cellular Mechanisms*, *13*, 131–144.

Sato, F., Kawamoto, T., Fujimoto, K., Noshiro, M., Honda, K. K., Honma, S., et al. (2004). Functional analysis of the basic helix-loop-helix transcription factor DEC1 in circadian regulation. Interaction with BMAL1. *European Journal of Biochemistry/FEBS*, *271*, 4409–4419.

Sato, F., Kawamura, H., Wu, Y., Sato, H., Jin, D., Bhawal, U. K., et al. (2012). The basic helix-loop-helix transcription factor DEC2 inhibits TGF-beta-induced tumor progression in human pancreatic cancer BxPC-3 cells. *International Journal of Molecular Medicine*, *30*, 495–501.

Seimiya, M., Bahar, R., Wang, Y., Kawamura, K., Tada, Y., Okada, S., et al. (2002). Clast5/Stra13 is a negative regulator of B lymphocyte activation. *Biochemical and Biophysical Research Communications*, *292*, 121–127.

Seimiya, M., Wada, A., Kawamura, K., Sakamoto, A., Ohkubo, Y., Okada, S., et al. (2004). Impaired lymphocyte development and function in Clast5/Stra13/DEC1-transgenic mice. *European Journal of Immunology*, *34*, 1322–1332.

Seuter, S., Pehkonen, P., Heikkinen, S., & Carlberg, C. (2013). The gene for the transcription factor BHLHE40/DEC1/stra13 is a dynamically regulated primary target of the vitamin D receptor. *The Journal of Steroid Biochemistry and Molecular Biology*, *136*, 62–67.

Shang, C. A., & Waters, M. J. (2003). Constitutively active signal transducer and activator of transcription 5 can replace the requirement for growth hormone in adipogenesis of 3T3-F442A preadipocytes. *Molecular Endocrinology*, *17*, 2494–2508.

Shearman, L. P., Jin, X., Lee, C., Reppert, S. M., & Weaver, D. R. (2000). Targeted disruption of the mPer3 gene: Subtle effects on circadian clock function. *Molecular and Cellular Biology*, *20*, 6269–6275.

Shen, M., Kawamoto, T., Teramoto, M., Makihira, S., Fujimoto, K., Yan, W., et al. (2001). Induction of basic helix-loop-helix protein DEC1 (BHLHB2)/Stra13/Sharp2 in response to the cyclic adenosine monophosphate pathway. *European Journal of Cell Biology*, *80*, 329–334.

Shen, M., Kawamoto, T., Yan, W., Nakamasu, K., Tamagami, M., Koyano, Y., et al. (1997). Molecular characterization of the novel basic helix-loop-helix protein DEC1 expressed in differentiated human embryo chondrocytes. *Biochemical and Biophysical Research Communications*, *236*, 294–298.

Shen, M., Yoshida, E., Yan, W., Kawamoto, T., Suardita, K., Koyano, Y., et al. (2002). Basic helix-loop-helix protein DEC1 promotes chondrocyte differentiation at the early and terminal stages. *The Journal of Biological Chemistry*, *277*, 50112–50120.

Skinner, M. K., Rawls, A., Wilson-Rawls, J., & Roalson, E. H. (2010). Basic helix-loop-helix transcription factor gene family phylogenetics and nomenclature. *Differentiation; Research in Biological Diversity*, *80*, 1–8.

St-Pierre, B., Cooper, M., Jiang, Z., Zacksenhaus, E., & Egan, S. E. (2004). Dynamic regulation of the Stra13/Sharp/Dec bHLH repressors in mammary epithelium. *Developmental Dynamics*, *230*, 124–130.

St-Pierre, B., Flock, G., Zacksenhaus, E., & Egan, S. E. (2002). Stra13 homodimers repress transcription through class B E-box elements. *The Journal of Biological Chemistry*, *277*, 46544–46551.

Sun, H., Lu, B., Li, R. Q., Flavell, R. A., & Taneja, R. (2001). Defective T cell activation and autoimmune disorder in Stra13-deficient mice. *Nature Immunology*, *2*, 1040–1047.

Sun, H., & Taneja, R. (2000). Stra13 expression is associated with growth arrest and represses transcription through histone deacetylase (HDAC)-dependent and HDAC-independent mechanisms. *Proceedings of the National Academy of Sciences of the United States of America*, *97*, 4058–4063.

Teramoto, M., Nakamasu, K., Noshiro, M., Matsuda, Y., Gotoh, O., Shen, M., et al. (2001). Gene structure and chromosomal location of a human bHLH transcriptional factor DEC1 x Stra13 x SHARP-2/BHLHB2. *Journal of Biochemistry*, *129*, 391–396.

Thin, T. H., Li, L., Chung, T. K., Sun, H., & Taneja, R. (2007). Stra13 is induced by genotoxic stress and regulates ionizing-radiation-induced apoptosis. *EMBO Reports*, *8*, 401–407.

Tsang, A. H., Sanchez-Moreno, C., Bode, B., Rossner, M. J., Garaulet, M., & Oster, H. (2012). Tissue-specific interaction of Per1/2 and Dec2 in the regulation of fibroblast circadian rhythms. *Journal of Biological Rhythms*, *27*, 478–489.

Ueda, H. R., Hayashi, S., Chen, W., Sano, M., Machida, M., Shigeyoshi, Y., et al. (2005). System-level identification of transcriptional circuits underlying mammalian circadian clocks. *Nature Genetics*, *37*, 187–192.

Ueshima, T., Kawamoto, T., Honda, K. K., Noshiro, M., Fujimoto, K., Nakao, S., et al. (2012). Identification of a new clock-related element EL-box involved in circadian regulation by BMAL1/CLOCK and HES1. *Gene*, *510*, 118–125.

van der Horst, G. T., Muijtjens, M., Kobayashi, K., Takano, R., Kanno, S., Takao, M., et al. (1999). Mammalian Cry1 and Cry2 are essential for maintenance of circadian rhythms. *Nature*, *398*, 627–630.

Wu, T., Jin, Y., Kato, H., & Fu, Z. (2008). Light and food signals cooperate to entrain the rat pineal circadian system. *Journal of Neuroscience Research*, *86*, 3246–3255.

Wu, T., Jin, Y., Ni, Y., Zhang, D., Kato, H., & Fu, Z. (2008). Effects of light cues on re-entrainment of the food-dominated peripheral clocks in mammals. *Gene*, *419*, 27–34.

Wu, T., Ni, Y., Zhuge, F., & Fu, Z. (2010). Resetting process of peripheral circadian gene expression after the combined reversal of feeding schedule and light/dark cycle via a 24-h light period transition in rats. *Physiological Research*, *59*, 581–590.

Wu, Y., Sato, F., Bhawal, U. K., Kawamoto, T., Fujimoto, K., Noshiro, M., et al. (2012). BHLH transcription factor DEC2 regulates pro-apoptotic factor Bim in human oral cancer HSC-3 cells. *Biomedical Research*, *33*, 75–82.

Wu, Y., Sato, F., Yamada, T., Bhawal, U. K., Kawamoto, T., Fujimoto, K., et al. (2012). The BHLH transcription factor DEC1 plays an important role in the epithelial-mesenchymal transition of pancreatic cancer. *International Journal of Oncology*, *41*, 1337–1346.

Wykoff, C. C., Pugh, C. W., Maxwell, P. H., Harris, A. L., & Ratcliffe, P. J. (2000). Identification of novel hypoxia dependent and independent target genes of the von Hippel-Lindau (VHL) tumour suppressor by mRNA differential expression profiling. *Oncogene*, *19*, 6297–6305.

Yamada, K., Kawata, H., Shou, Z., Mizutani, T., Noguchi, T., & Miyamoto, K. (2003). Insulin induces the expression of the SHARP-2/Stra13/DEC1 gene via a phosphoinositide 3-kinase pathway. *The Journal of Biological Chemistry*, *278*, 30719–30724.

Yamada, K., Ogata-Kawata, H., Matsuura, K., & Miyamoto, K. (2005). SHARP-2/Stra13/DEC1 as a potential repressor of phosphoenolpyruvate carboxykinase gene expression. *FEBS Letters*, *579*, 1509–1514.

Yang, X. O., Angkasekwinai, P., Zhu, J., Peng, J., Liu, Z., Nurieva, R., et al. (2009). Requirement for the basic helix-loop-helix transcription factor Dec2 in initial TH2 lineage commitment. *Nature Immunology*, *10*, 1260–1266.

Yun, Z., Maecker, H. L., Johnson, R. S., & Giaccia, A. J. (2002). Inhibition of PPAR gamma 2 gene expression by the HIF-1-regulated gene DEC1/Stra13: A mechanism for regulation of adipogenesis by hypoxia. *Developmental Cell*, *2*, 331–341.

Yunokawa, M., Tanimoto, K., Nakamura, H., Nagai, N., Kudo, Y., Kawamoto, T., et al. (2007). Differential regulation of DEC2 among hypoxia-inducible genes in endometrial carcinomas. *Oncology Reports*, *17*, 871–878.

Zawel, L., Yu, J., Torrance, C. J., Markowitz, S., Kinzler, K. W., Vogelstein, B., et al. (2002). DEC1 is a downstream target of TGF-beta with sequence-specific transcriptional repressor activities. *Proceedings of the National Academy of Sciences of the United States of America*, *99*, 2848–2853.

Zheng, B., Albrecht, U., Kaasik, K., Sage, M., Lu, W., Vaishnav, S., et al. (2001). Nonredundant roles of the mPer1 and mPer2 genes in the mammalian circadian clock. *Cell*, *105*, 683–694.

Zvonic, S., Ptitsyn, A. A., Conrad, S. A., Scott, L. K., Floyd, Z. E., Kilroy, G., et al. (2006). Characterization of peripheral circadian clocks in adipose tissues. *Diabetes*, *55*, 962–970.

INDEX

Note: Page numbers followed by "*f*" indicate figures and "*t*" indicate tables.

F

G

H

I

U

V

Z

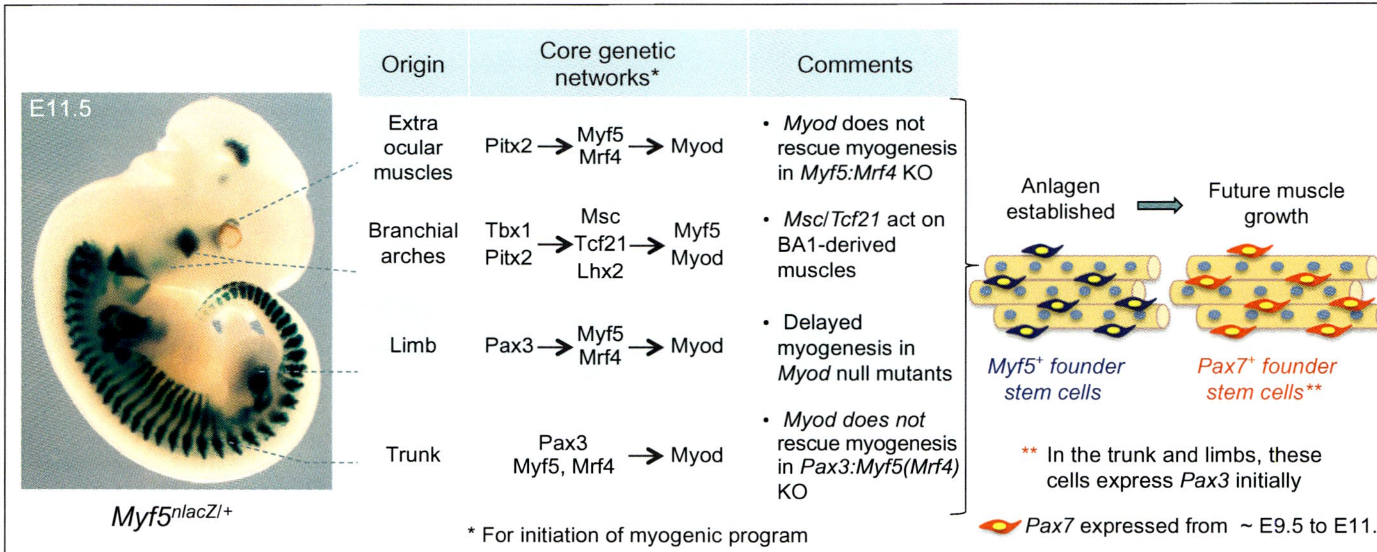

Glenda Comai and Shahragim Tajbakhsh, Figure 1.1 Tissuegenesis is initiated with an anlagen prior to the emergence of the self-renewing stem cell population. At anatomical locations throughout the embryo (*Myf5*^{nlacZ} E11.5, Tajbakhsh et al., 1996), an anlagen of differentiated cells is established before the long-term myogenic stem cell (Pax7+) population emerges. In the head and limb, myogenic cells of the anlagen are specified by the upstream regulators*Tbx1/Pitx2* or *Pax3*, respectively, prior to activation of the core set of myogenic factors *Myf5/Mrf4/Myod*. In the trunk, *Pax3* and also *Myf5/Mrf4* act as upstream regulators, given that in *Pax3* null mice, muscle is still specified by these factors in the epaxial domain of the somite.

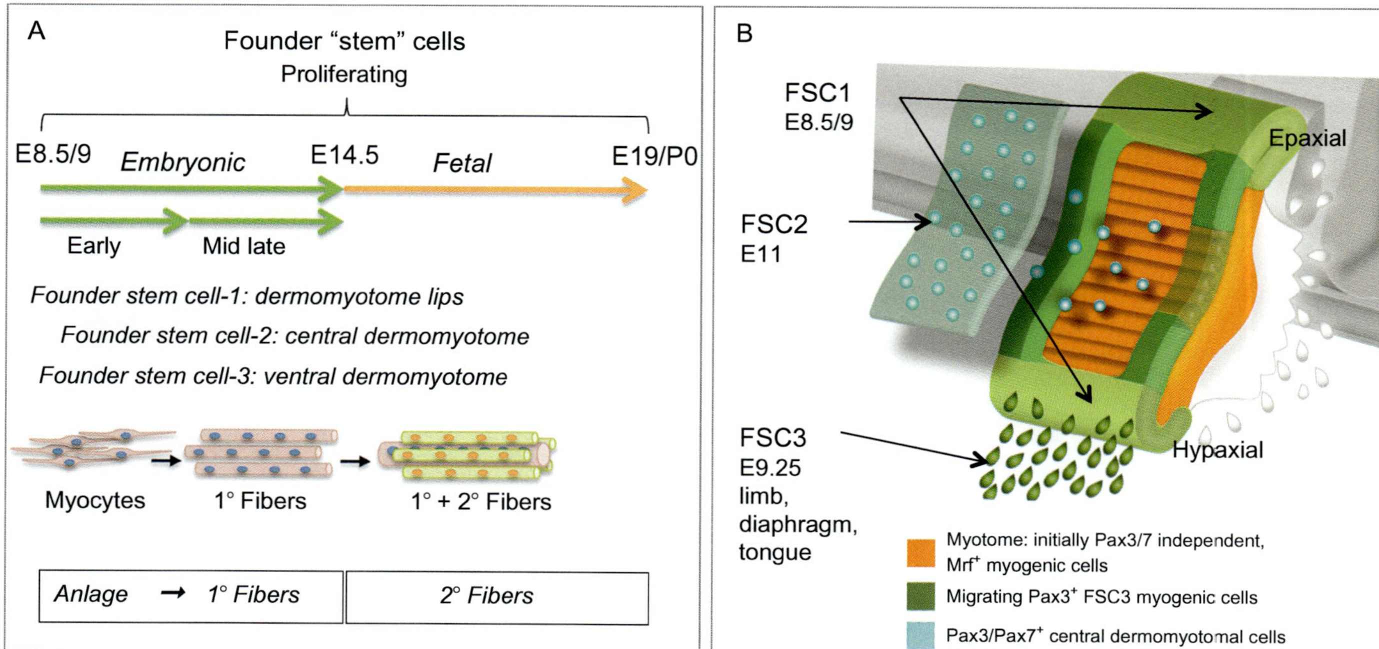
A
Founder "stem" cells
Proliferating
E8.5/9
Embryonic
E14.5
Fetal
E19/P0
Early
Mid late
Founder stem cell-1: dermomyotome lips
Founder stem cell-2: central dermomyotome
Founder stem cell-3: ventral dermomyotome
Myocytes
1° Fibers
1° + 2° Fibers
Anlage → 1° Fibers
2° Fibers
B
FSC1
E8.5/9
FSC2
E11
FSC3
E9.25
limb,
diaphragm,
tongue
Epaxial
Hypaxial
Myotome: initially Pax3/7 independent, Mrf+ myogenic cells
Migrating Pax3+ FSC3 myogenic cells
Pax3/Pax7+ central dermomyotomal cells

Glenda Comai and Shahragim Tajbakhsh, Figure 1.2 Emergence of distinct founder stem cell populations from the somite with different genetic requirements. (A) Embryonic development can be divided into early and late phases. During the early phase, progenitors are specified and determined as myoblasts and an anlage is established in the somite consisting of a primary myotome with differentiated/postmitotic mononuleated myocytes. Myocytes fuse to form multinucleated myofibers (primary fibers). Fetal myogenesis is characterized by growth and maturation of each muscle anlage and by the onset of innervation. This second wave of myogenesis involves fusion of fetal myoblasts to form secondary fibers (initially smaller and surrounding primary fibers). Cells that establish muscles prenatally are referred to as founder stem cells (FSC1, FSC2, FSC3 in somites) to distinguish them from the juvenile and adult satellite (stem) cell populations. FSC1 is largely exhausted in the embryo, while FSC2/FSC3 will give rise to the majority of the adult skeletal muscle stem cells. (B) Localization of the founder stem cell populations highlighted in (A). FSC1 (founder stem cell population-1) emerges first (E8.5/E9.0) in the dorsomedial (epaxial) and later in the ventrolateral (hypaxial) dermomyotome (DM). These are initially Pax3+ then Myf5/Mrf4+. They are reduced in number but not lost in *Pax3* mutants. FSC3 cells, that emerge from $\approx$E9.25 onwards, are Pax3+ and this migratory population (from the somites to limbs, diaphragm, and tongue) is lost in *Pax3* null embryos. Note, *Pax7* expression is initiated from E9.5, therefore it is not implicated in the establishment of FSC1 and FSC3. FSC2 cells emerge from the central dermomyotome between E10.5 and E11.5 starting from anteriormost somites. FSC2 cells are initially *Mrf*-negative, present in the *Pax3* null, but lost in the *Pax3:Pax7* double mutant (Kassar-Duchossoy et al., 2005; Relaix, Rocancourt, Mansouri, & Buckingham, 2005). Each of these populations can be distinguished by their distinct genetic requirements, as they are spared, or lost in *Pax3* or *Pax3:Pax7* mutants.

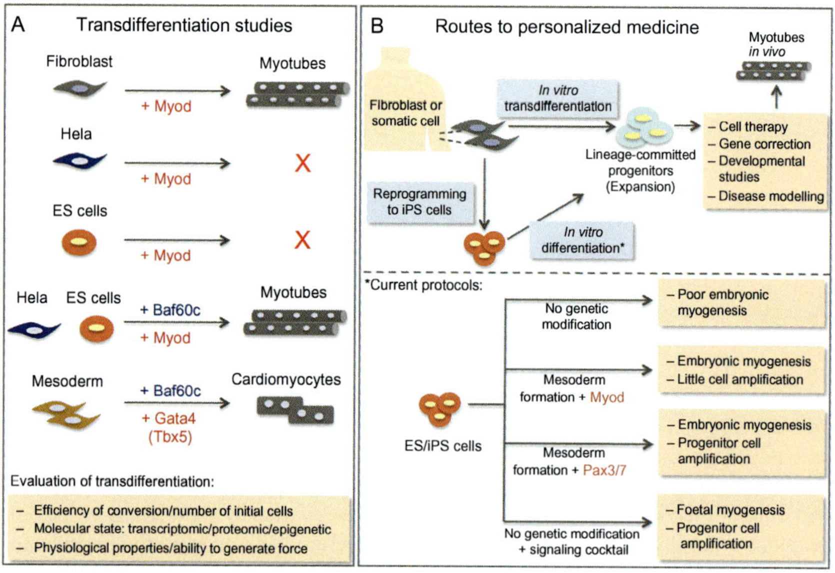

Glenda Comai and Shahragim Tajbakhsh, Figure 1.3 Myogenic conversion and routes to personalized medicine. (A) The hallmark of the reprogramming era began with the transdifferentiation of multiple cell types to skeletal myogenesis and cell-fate conversion with *Myod* (Davis et al., 1987). Some cells remained refractory to myogenic conversion (HeLa, ES cells) (Albini et al., 2013; Dekel, Magal, Pearson-White, Emerson, & Shani, 1992; Weintraub et al., 1989). More recently, chromatin modifiers (e.g., Baf60c) were found to render some of these cells permissive to conversion (Albini et al., 2013; Forcales et al., 2012; Takeuchi & Bruneau, 2009). This underscores the notion that rendering the chromatin accessible should be considered in concert with the choice of transcription factor(s) for directing the desired cell fate. In addition, a combination of parameters should be routinely evaluated to assure full cell-fate conversion. (B) Schematic representation of some of the approaches under investigation to produce new muscle tissue for therapeutic use (upper panel). Patient fibroblasts can be directed to transdifferentiate *in vitro* into lineage-committed myogenic cells directly or via a reprogramming step through iPS cells. The generation of lineage-committed progenitors has a number of advantages: reduced chance of teratoma formation, minimization of alternative cell fates and ideally, the potential to specifically generate the desired type of muscle progenitors (location specific, slow vs. fast fibers). The conversion of pluripotent ES or iPS cells to myogenesis has been a major obstacle (lower panel). Although different strategies have now made it possible to amplify skeletal muscle stem/progenitor cells, this is achieved by genetic manipulation of the ES/iPS cells. Furthermore, fetal/adult myogenesis has not been systematically generated in these *in vitro* differentiation assays (see Section 4.3 for reference).

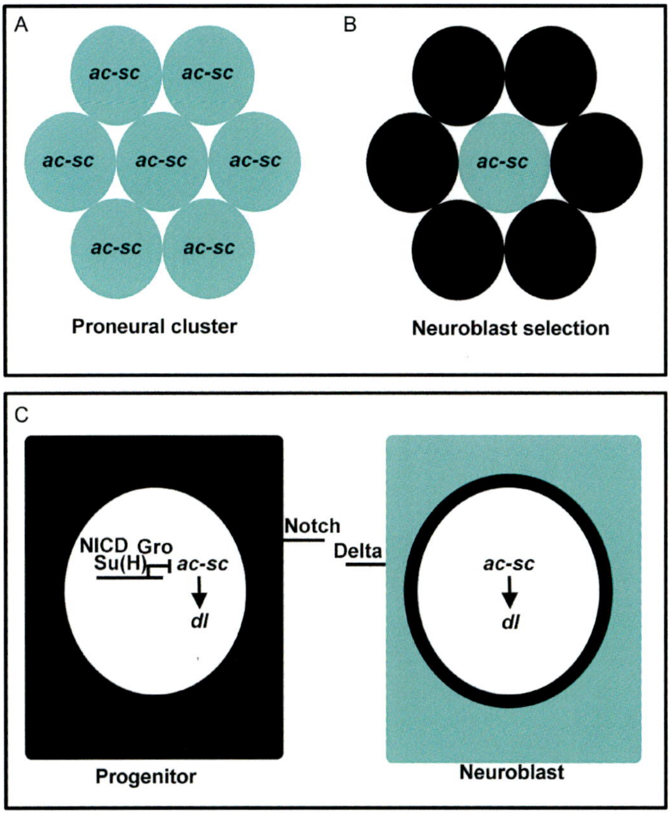

Carol Huang *et al.*, Figure 2.1 Model of lateral inhibition. (A) Within the unspecified ectoderm, cells within each proneural cluster initially express similar levels of the proneural genes (e.g., *ac-sc*). (B) One cell begins expressing *ac-sc* at a higher level, becoming the presumptive neuroblast, while lateral inhibition shuts down the proneural program in neighboring cells, which retain a progenitor phenotype. (C) Higher delta expression on the surface of the presumptive neuroblast signals to neighboring cells through their Notch receptors. Notch signals through the Notch intracellular domain, *gro*, and *Su*(*H*) to suppress *ac-sc* genes, which then cannot transactivate *dl*, keeping neighboring cells in a progenitor state.

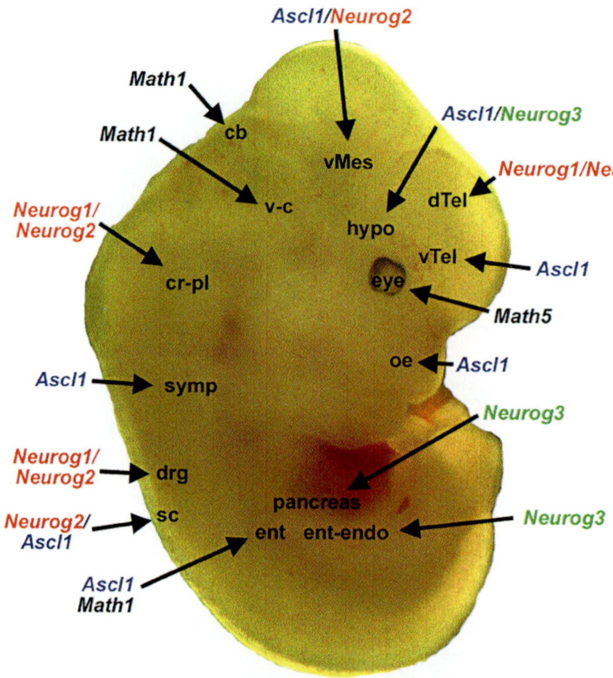

Carol Huang *et al.*, Figure 2.2 Major anatomic sites of proneural gene function in the developing embryo. *Neurog1* and *Neurog2* are active in the dorsal telencephalon (dTel), ventral mesencephalon (vMes), cranial placodes (cr-pl), dorsal root ganglia (drg), and spinal cord (sc). *Neurog3* is active in the developing hypothalamus (hypo), pancreas, and enteroendocrine (ent-endo) system. *Math1/Atoh1* is critical for cerebellar (cb) and enteric (ent) secretory cell development, whereas *Math5/Atoh7* predominantly functions in the eye. *Ascl1* is active in the ventral telencephalon (vTel), ventral mesencephalon (vMes), olfactory epithelium (OE), autonomic system (symp), spinal cord (sc), and enteric autonomic system (ent).

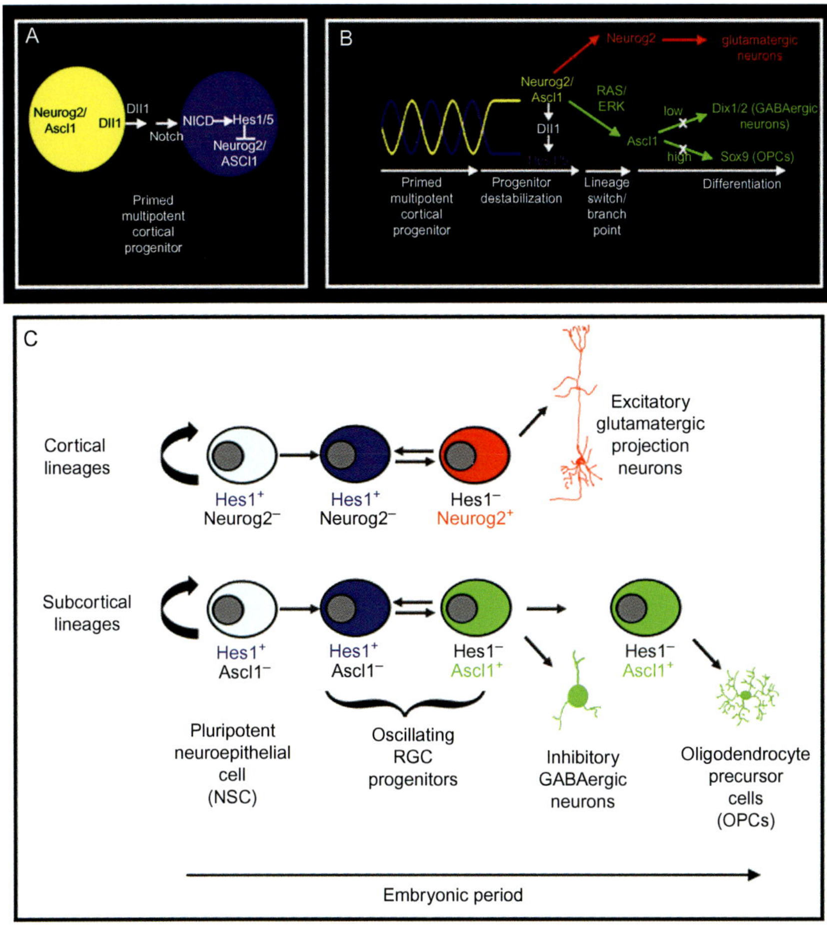

Carol Huang *et al.*, Figure 2.3 Oscillatory cycles of proneural and neurogenic bHLH proteins in cortical progenitors. (A) Both Neurog2 and Ascl1 initiate expression of the Dll1 ligand, which binds to a Notch receptor on a neighboring cell, leading to the translocation of the Notch intracellular domain (NICD) into the nucleus. NICD induces the transcription of *Hes1/5*. Hes1/5 in turn represses *Neurog2* and *Ascl1* transcription, setting up oscillatory cycles of Neurog2 and Hes1/5 proteins. (B) Schematic illustration of predicted $Neurog2^{+}Ascl1^{+}/Hes1/5^{+}$ oscillatory cycles in cortical progenitors. These cycles must be destabilized for lineage commitment to occur. RAS/ERK signaling promotes stable Ascl1 expression (while blocking Neurog2 expression). (C) Lineage selection in cortical and subcortical progenitors. After Neurog2 is stably expressed in cortical lineages, it promotes a glutamatergic neuronal phenotype in cortical cells. After Ascl1 is stably expressed in subcortical lineages, it promotes a GABAergic neuronal or OPC cell fate.

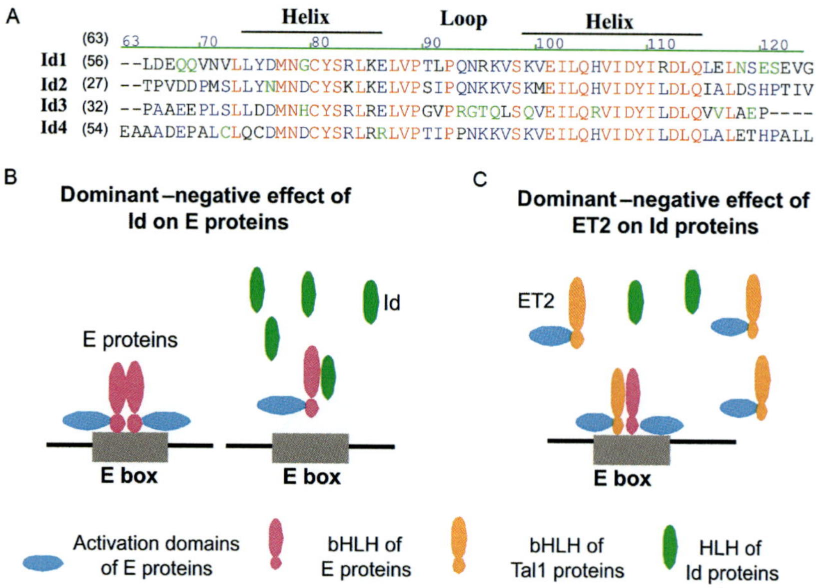

Flora Ling *et al.*, Figure 5.1 Structural similarity of Id proteins and their mode of action. (A) Sequence alignment of the HLH motif of the four human Id proteins. Identical amino acids are shown in red; conserved changes in blue and weak homologies in green. Schematic diagrams of the effect of Id proteins on the DNA binding activity of all E proteins (B) and the ability of ET2 to compete with all Id proteins to dimerize with E proteins and activate transcription of E protein target genes (C). ET2 is an artificial hybrid molecule between E47 and Tal1. The symbols for each functional domain of Id, Tal1 and E proteins are indicated by the legend included in the figure.

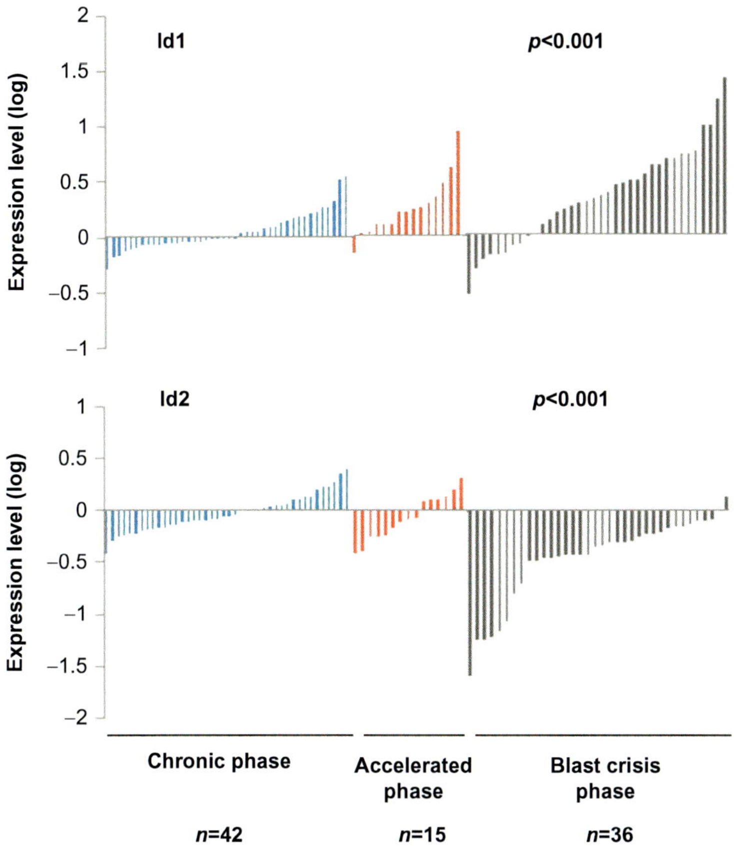

Flora Ling *et al.*, Figure 5.2 Id1 and Id2 expression in chronic myelogenous leukemia. Microarray analysis of Id1 and Id2 expression in $CD34^+$ cells in the peripheral blood of CML patients at the indicated disease stage. The number of samples in each group is as indicated. Statistical significance was determined using multigroup comparison using ANOVA F test.

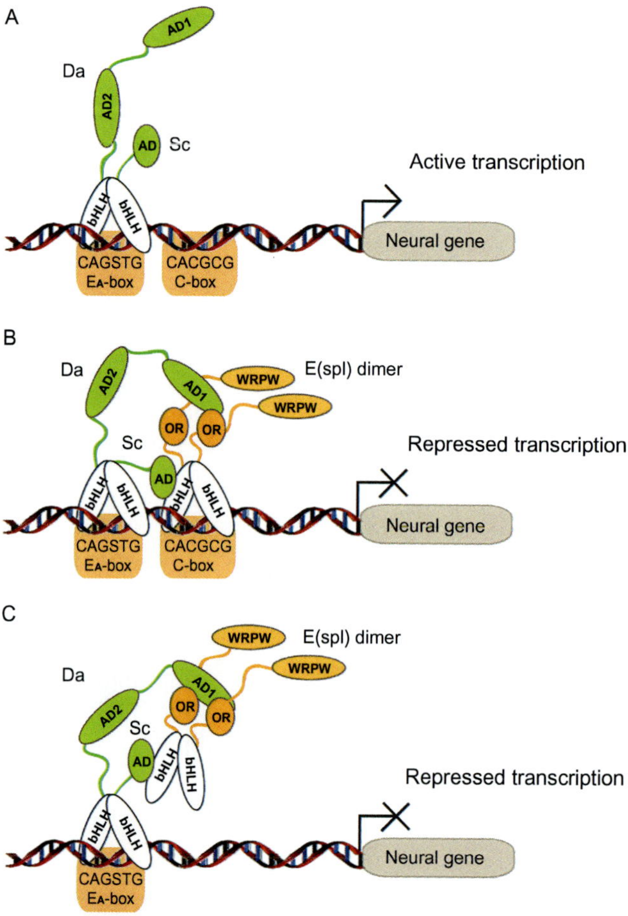

Christos Delidakis *et al.*, Figure 6.1 Graphic representation of the antagonism between proneural and E(spl) proteins. Proneural activators, such as the Da/Sc heterodimer depicted here, bind to E_A-boxes, whereas E(spl) repressors bind to E_B-, C-, and N-boxes. (A) In the absence of repressors, proneural proteins activate their target genes. Activation is mediated via two transactivation domains on Da, AD1 and AD2, as well as one transactivation domain on Sc. (B) In cells where E(spl) proteins are expressed, they are strongly bound by target enhancers thanks to DNA binding at the C-box as well as protein–protein interactions with two transactivation domains of the proneural heterodimer, the Da AD1 and the Sc AD. (C) If a cognate DNA target site is not present, E(spl) proteins can still be recruited onto proneural target enhancers via protein–protein interactions with Da and Sc. In this case, we hypothesize that higher levels of E(spl) will be needed to ensure recruitment and target gene repression, than in the case depicted in (B). *Adapted from a figure originally published in Zarifi et al. (2012), Copyright © American Society for Microbiology.*

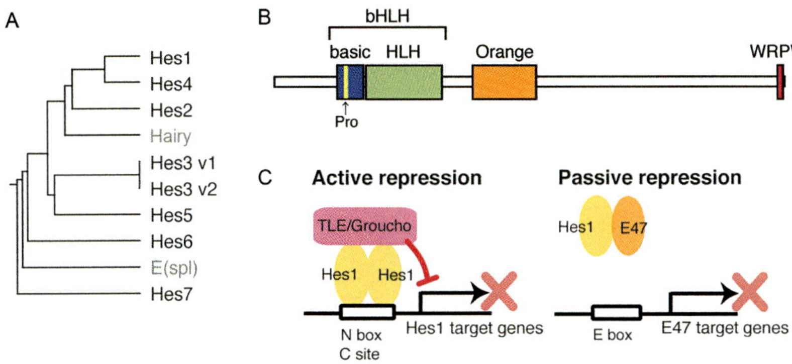

Taeko Kobayashi and Ryoichiro Kageyama, Figure 7.1 Features of Hes factors and their functions. (A) Comparison of protein homology in human Hes factors (Hes1–7, black) and their Drosophila homologues (hairy and E(spl), gray). Phylogenetic tree was generated by ClustalW software from the amino acid sequence alignment. (B) Structure of Hes1 protein. (C) Dual repressor functions of Hes factors: active repression and passive repression. During active repression, Hes factors bind to the N-box or the class C site in the promoter regions of target genes and repress their transcription. During passive repression, Hes factors form heterodimers with bHLH activators and inhibit their binding to the E box in the promoter regions of target genes.

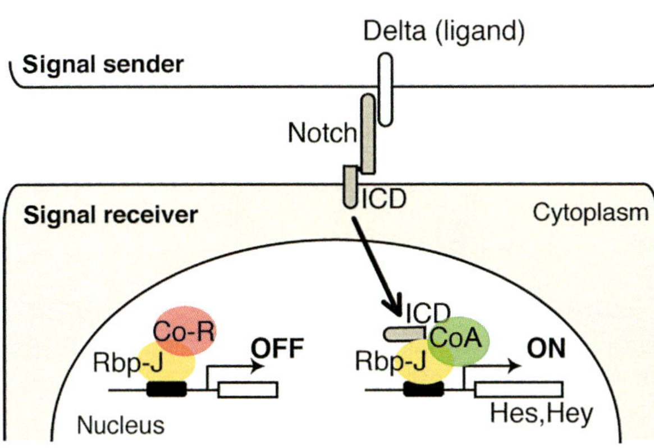

Taeko Kobayashi and Ryoichiro Kageyama, Figure 7.2 Notch signaling pathway. The Notch receptor is a transmembrane protein activated by the binding of Notch ligands expressed on the surface of neighboring cells (signal sender). Upon ligand binding, the Notch receptor on the signal receiver cell is sequentially cleaved by the ADAM protease and the gamma-secretase, releasing the intracellular domain (ICD) of Notch from the plasma membrane. ICD translocates into the nucleus and forms a complex with the DNA-binding transcription factor RBP-J. Without the ICD, RBP-J represses target gene expression, but is converted to a transcriptional activator when complexed with the ICD and the coactivators (CoA). Finally, Notch signaling activation induces the expression of target genes such as *Hes* and *Hey*.

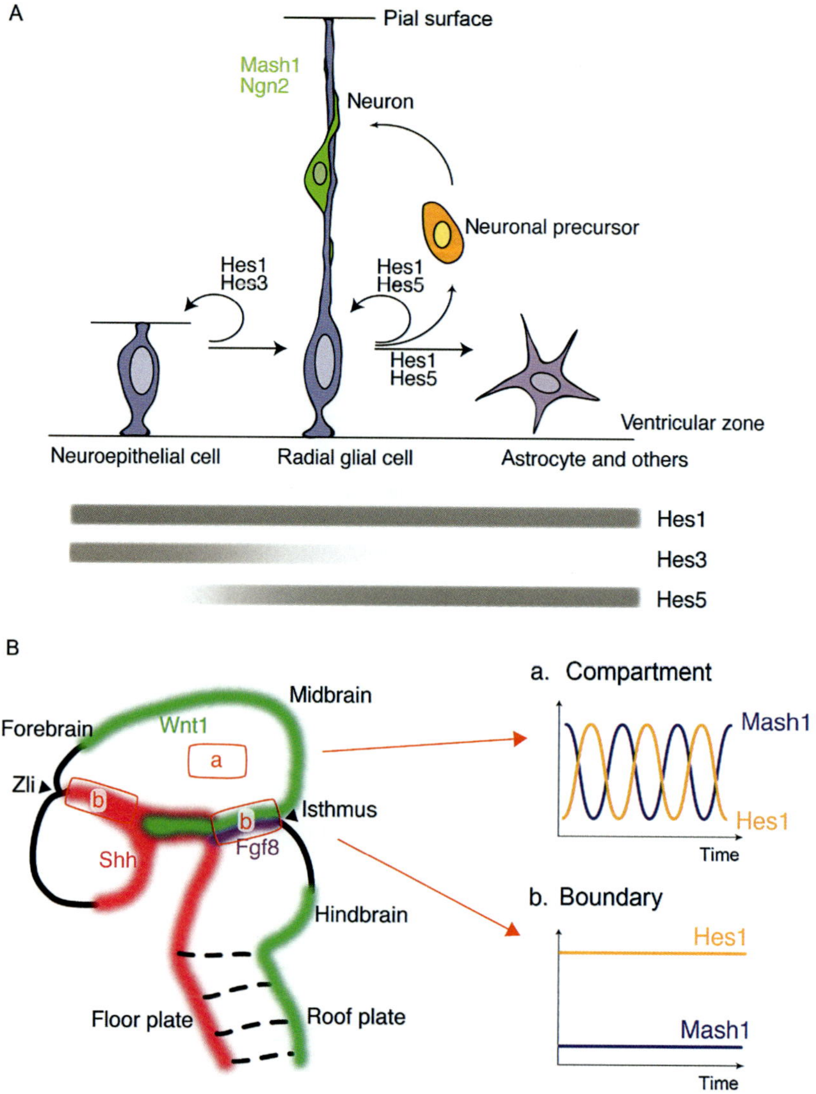

Taeko Kobayashi and Ryoichiro Kageyama, Figure 7.3 Hes functions in the central nervous system. (A) Neural stem cells and their differentiation during neural development (upper panel). Neuroepithelial cells initially proliferate by symmetric cell division and then become radial glial cells, which sequentially generate neurons by asymmetric cell division. Both neuroepithelial cells and radial glial cells are considered to be neural stem (NS) cells. Radial glial cells eventually differentiate into astrocytes, oligodendrocytes, and ependymal cells. Hes factors, expressed in these NS cells, maintain these NS cells and then promote astrocyte differentiation. Proneural bHLH genes, such as *Mash1* and *Ngn2*, are activated in differentiating neurons, leading to the promotion of neurogenesis. Expression patterns of Hes genes in the ventricular zone are indicated (lower panel). (B) The structure of the developing central nervous system (CNS) and differential expression of Hes1 gene. The CNS in the mouse at E10.5 (left) is partitioned into two regions, compartments (a) and boundaries (b). Boundary cells function as organizing centers by expressing several signaling ligands, including Shh, Wnt1, or Fgf8. The temporal expression patterns of Hes1 and Mash1 (right panels) are different in compartment and boundary cells. Hes1 expression is variable and oscillating in NS cells in compartments, but is sustained in boundary cells. *Adapted from Kageyama, Ohtsuka, & Kobayashi, 2007.*

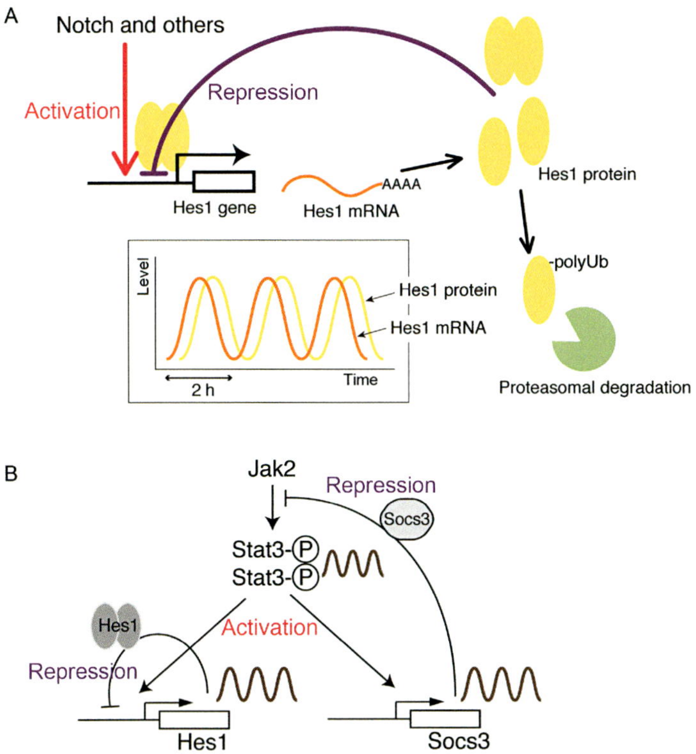

Taeko Kobayashi and Ryoichiro Kageyama, Figure 7.4 Oscillatory expression of Hes1. (A) Hes1 oscillation is regulated by negative feedback: Hes1 protein represses its own expression by directly binding to N-box sequences in its promoter. Both Hes1 mRNA and protein are unstable and disappear quickly when the expression is repressed. By this mechanism, Hes1 autonomously initiates oscillatory expression with a periodicity of about 2 h. (B) Jak–Stat signaling regulates Hes1 oscillation. Jak2 activity is regulated via negative feedback by *Socs3*, a downstream gene of Jak–Stat signaling. The levels of phosphorylated Stat3 (Stat3-P) and Socs3 are oscillating, which regulate Hes1 oscillation and cell proliferation.

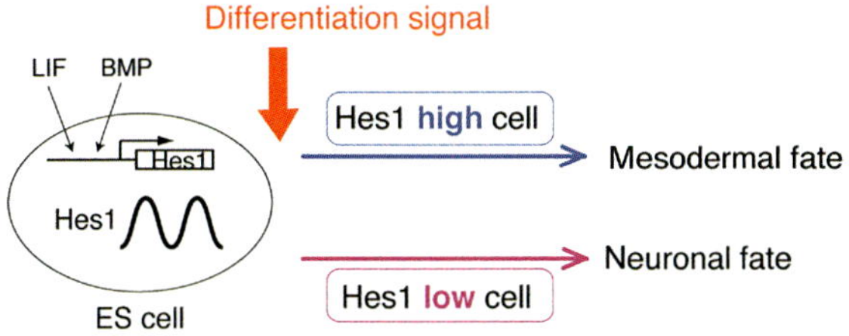

Taeko Kobayashi and Ryoichiro Kageyama, Figure 7.5 Hes1 oscillation in ES cells. Hes1 expression oscillates in embryonic stem (ES) cells via LIF and BMP stimulation. When differentiation signals activate ES cell differentiation, cells low in Hes1 protein levels differentiate into neuroectodermal cells (neuronal fate), and those high in Hes1 protein levels tend to differentiate into early mesodermal cells.

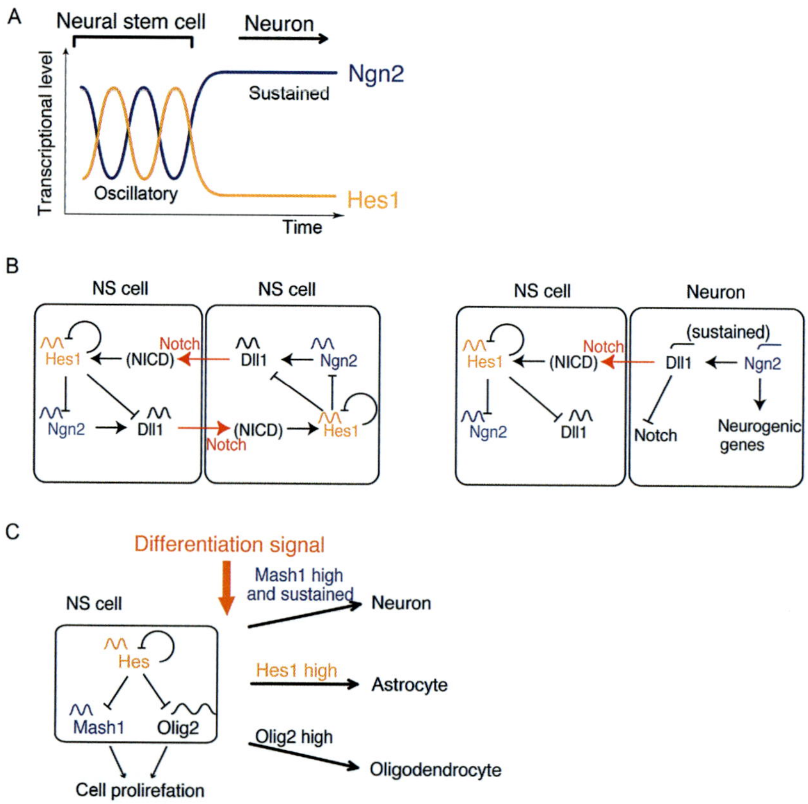

Taeko Kobayashi and Ryoichiro Kageyama, Figure 7.6 Hes1 oscillation in NS cells. (A) The levels of both Hes1 and *Ngn2*, a downstream gene of Hes1, oscillate in neural stem (NS) cells. When Hes1 expression is repressed in differentiating neurons, *Ngn2* expression becomes sustained. (B) Hes1 oscillation in NS cells induces the oscillatory expression of *Ngn2* and *Dll1* by periodic repression. *Dll1* oscillation mutually activates Notch signaling between neighboring NS cells, which maintain their proliferation without any help from neurons (left). When Hes1 is repressed, *Ngn2* and *Dll1* expression becomes sustained in differentiating neurons (right). Notch signaling is unidirectional from differentiating neurons to NS cells. Sustained Dll1 expression might prevent neurons from receiving Notch signaling by the *cis*-inhibitory function. (C) In NS cells, oscillatory expression of Hes1, Mash1, and Olig2 is important for cell proliferation and stem cell maintenance: oscillatory expression of several fate-determination factors contributes to the multipotent state of NS cells. When differentiation signals arrive, the expression of a single factor becomes sustained or high and determines the cell fate.

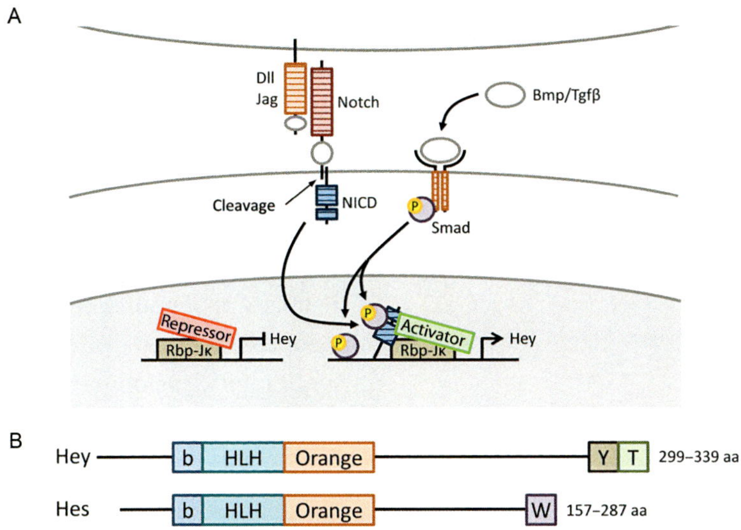

David Weber *et al.*, Figure 8.1 Hey genes are targets of canonical Notch signaling. (A) Ligands of the Delta (Dll) or Jagged (Jag) family induce intramembrane cleavage of the Notch receptor. The Notch intracellular domain (NICD) replaces transcriptional corepressors with activators in the Rbp-Jκ complex enabling transcription of Hey genes. Hey transcription can also be induced by Bmp/Tgfβ signaling: Bmp or Tgfβ ligands bind to their receptors leading to a phosphorylation of Smad proteins, which either directly bind to the Hey promoter or interact with the NICD. (B) Domain organization of Hey and Hes proteins. Amino acid (aa) numbers indicate the length of the proteins.

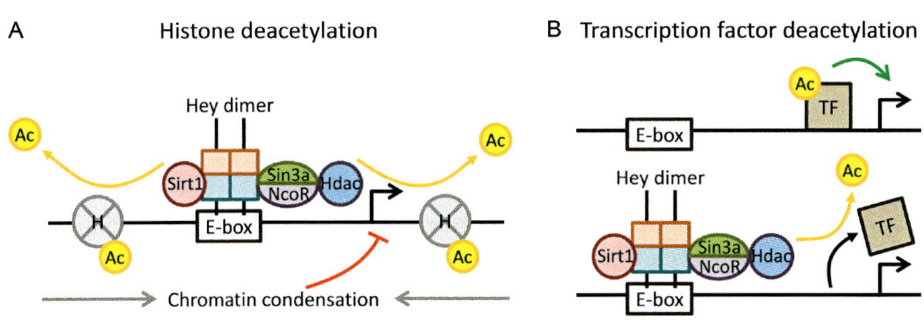

David Weber *et al.*, Figure 8.2 Model for Hey target gene regulation: (A) Hey dimers bound to target sequences (E-box) interact with histone deacetylases either directly (Sirt1) or indirectly (Hdac) through the Sin3a/NcoR complex. This leads to deacetylation of histones (H) inducing chromatin condensation and a reduction of transcription. (B) Hey proteins with attached Sirt1/Hdac may deacetylate other transcriptional activators either bound in proximity to Hey or via direct protein–protein interaction and abolish their activating capabilities, thereby reducing transcription. Alternatively, inactive heteromeric complexes may be formed.

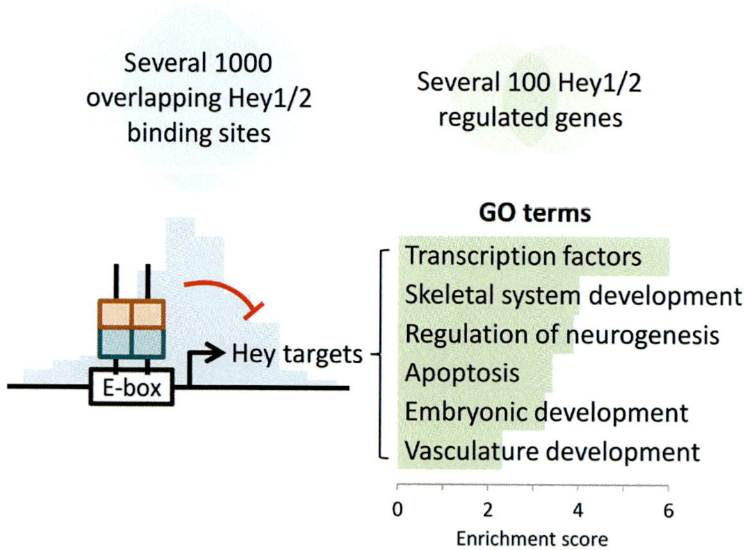

David Weber *et al.*, Figure 8.3 Gene regulation by Hey proteins. Hey proteins preferentially bind to E-box motifs in close proximity to transcription start sites of several 1000 promoters. In most screens, the number of regulated genes was limited to several hundred. These are mainly classified as other transcription factors involved in various developmental processes. *Taken from Heisig et al. (2012).*

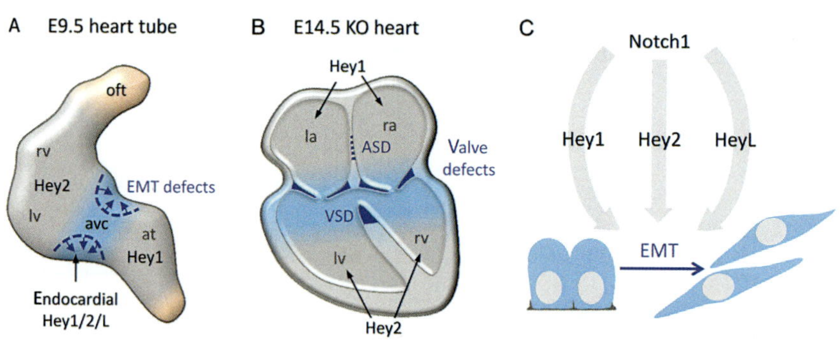

David Weber *et al.*, Figure 8.4 Hey expression in the developing heart and defects in knockout mice: (A) At embryonic day E9.5, Hey1 is expressed in cardiomyocytes of the developing atrium (at), while Hey2 is expressed in cardiomyocytes of the future left and right ventricle (rv, lv). All three Hey proteins are expressed in the endocardium. Hey2 and Hey1/L knockout mice exhibit epithelia–mesenchymal transition (EMT) defects in the endocardium of the atrioventricular canal (avc). (B) At embryonic day E14.5, Hey1 and Hey2 are expressed in atrial and ventricular cardiomyocytes, respectively. *Hey2* knockout mice have ventricular septum defects (VSDs), valve defects, and atrial septum defects (ASDs). These are the result of the earlier EMT defect, as these structures develop from mesenchymal cells, which have undergone EMT. (C) *Notch1* as well as *Hey2* and *Hey1/L* knockout mice exhibit similar cardiac EMT defects suggesting a cascade with Notch1 activating *Hey* genes in a combinatorial manner with partial redundancy to facilitate EMT.

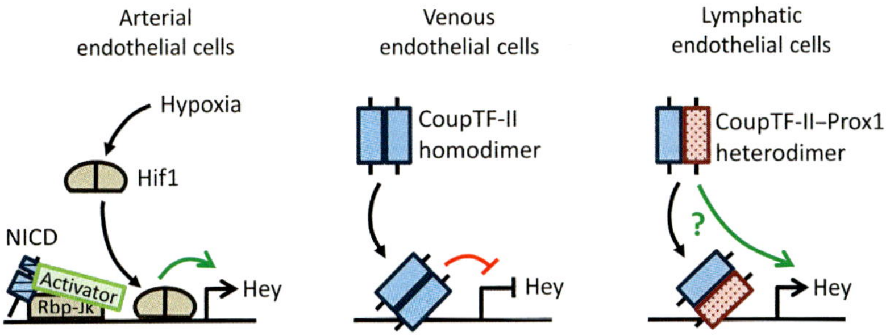

David Weber *et al.*, Figure 8.5 Regulation of Hey expression in endothelial cells: In arterial endothelial cells, Hey is directly induced by hypoxia-sensing Hif1 and the Delta–Notch pathway. In venous endothelial cells, transcription of Hey is repressed by CoupTF-II homodimers that directly bind to the Hey promoter. In lymphatic endothelial cells, repression by CoupTF-II is abolished by Prox1, which forms heterodimers with CoupTF-II and blocks its repressive capacity toward Hey genes.

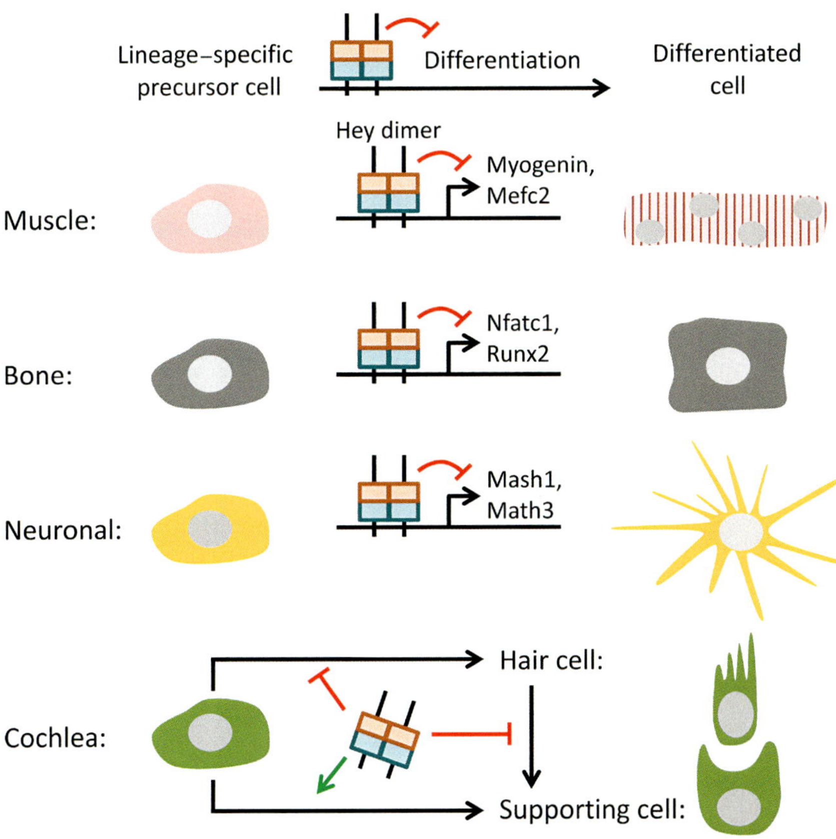

David Weber *et al.*, Figure 8.6 Hey proteins play a critical role in differentiation processes. They repress target genes, which drive differentiation within certain lineage and thereby affect myogenic, osteogenic, and neuronal differentiation.

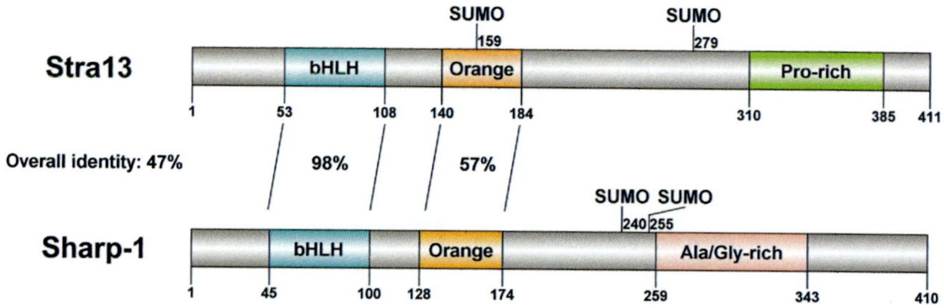

Jin Rong Ow *et al.*, Figure 9.1 Domain structure and posttranslational modifications of mouse Stra13 and Sharp-1. The conserved bHLH and Orange domains as well as sumoylation sites (SUMO) in both proteins are shown. Numbers reflect amino acid residues. The image was made using the domain visualization software D.O.G. 2.0 (Ren et al., 2009).

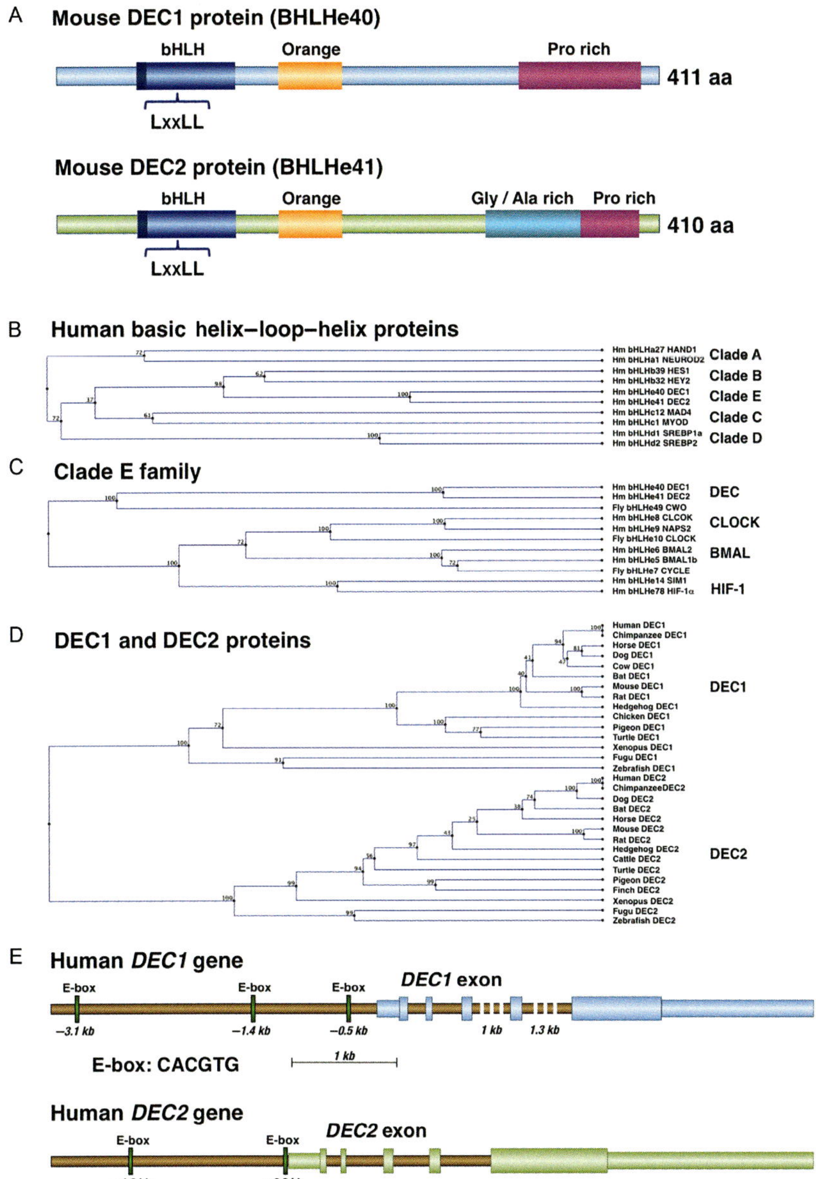

Yukio Kato *et al.*, Figure 10.1 Structures and phylogenetic trees of DEC1 and DEC2. (A) Domain structures of mouse DEC1 and DEC2. A short dark bar in the basic helix–loop–helix (bHLH) indicates the basic region. (B–D) Phylogenetic trees were created with entire amino acid sequences of bHLH proteins using the UPGMA method. Two representative members of each clade are selected and used for the phylogenetic tree of bHLH proteins (B). Phylogenetic trees of Clade E family (C) and DEC proteins (D). (E) Structures of human *DEC1* and *DEC2* genes. Wide bars indicate coding regions and narrower bars of the same color indicate 5′- or 3′-noncoding regions. Positions of E-boxes indicate the distances from translation initiation sites (ATG). LXXLL, leucine-x-x-leucine-leucine; CWO, clockwork orange protein; BMAL1, brain and muscle aryl hydrocarbon receptor nuclear translocator-like bHLH–PAS transcription factor; CLOCK, circadian locomotor output cycles kaput (a bHLH–PAS transcription factor); NPAS, neuronal PAS domain protein; SIM1, single-minded family bHLH transcription factor 1; HIF-1α, hypoxia-inducible factor 1α; Hm, human; kb, kilobase.

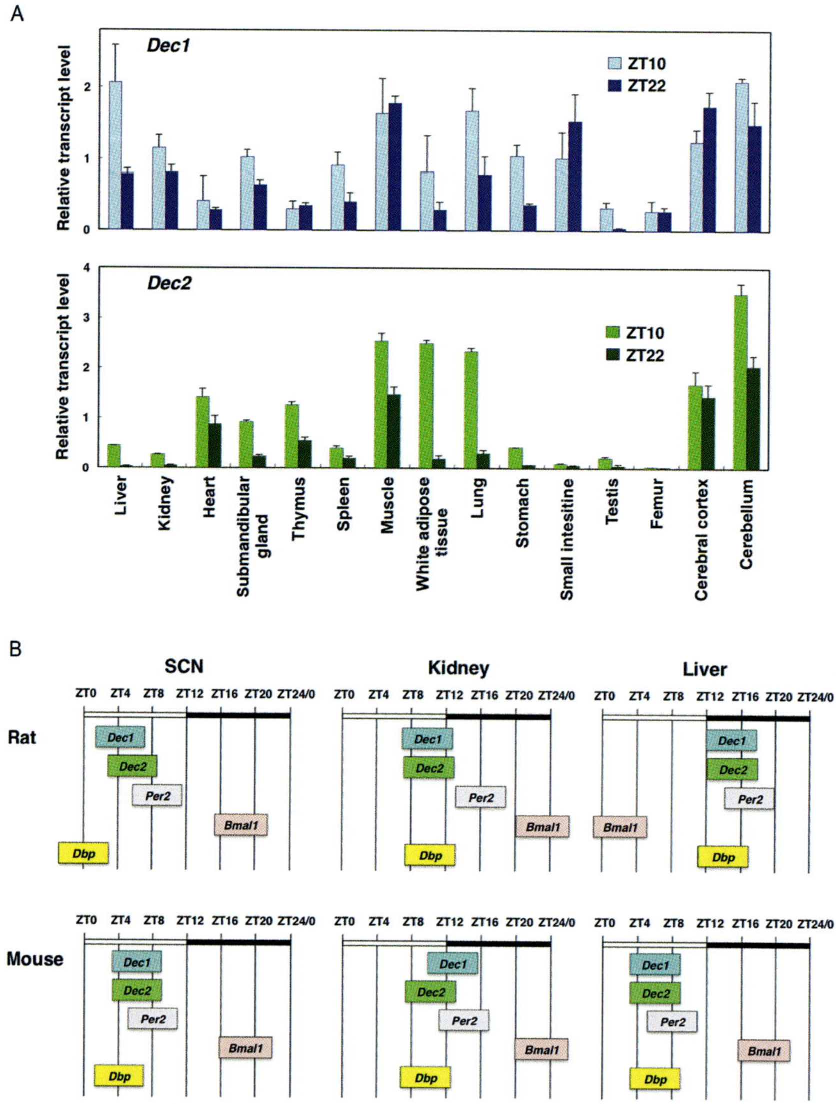

Yukio Kato *et al.*, Figure 10.2 Tissue distribution and circadian expression of *Dec1* and *Dec2* (A) Tissue distribution of *Dec1* and *Dec2* transcripts. Total RNA was obtained from various tissues of C57BL/6 mice, and *Dec1* and *Dec2* transcripts were assessed by real-time PCR (Noshiro et al., 2005). (B) Circadian peak times for the mRNA expression of *Dec1*, *Dec2*, and other clock genes in the SCN, kidney, and liver of Sprague–Dawley rats and Balb/c mice. Boxes indicate peak mRNA expression. Top bars represent light (12 h) and dark (12 h) phases. DBP, D site of albumin promoter (albumin D-box)-binding protein; PER2, period circadian protein homolog 2 (a PAS transcription factor); SCN, suprachiasmatic nucleus; ZT, zeitgeber time.

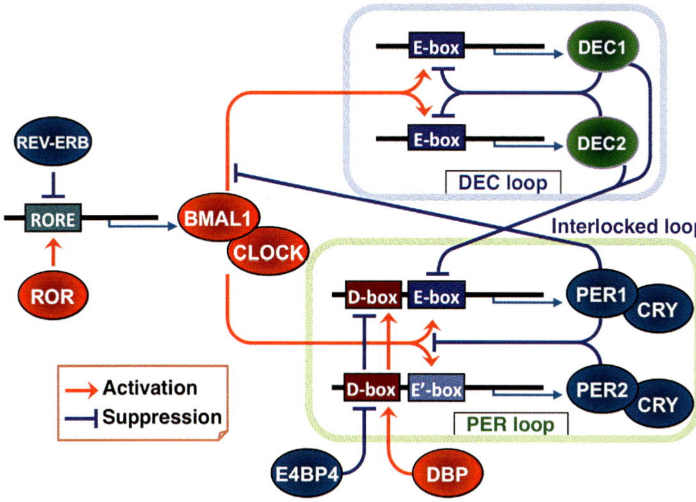

Yukio Kato *et al.*, Figure 10.3 Interlocked loops of the circadian clock system. CLOCK: BMAL1 heterodimer acts as a transcriptional activator on the E-box in the *Dec* genes and on the E/E′-box in the *Per* genes. Meanwhile, DEC1 and DEC2 act as repressors to form the DEC loop. PER:CRY heterodimer also suppresses CLOCK:BMAL1-induced transcription of the *Per* genes to form the PER loop. REV-ERB, nuclear receptor NR1D1; ROR, retinoic acid receptor-related orphan receptor; E4BP4, E4 promoter-binding protein 4; PER, period circadian protein homolog; CRY, cryptochrome circadian clock.

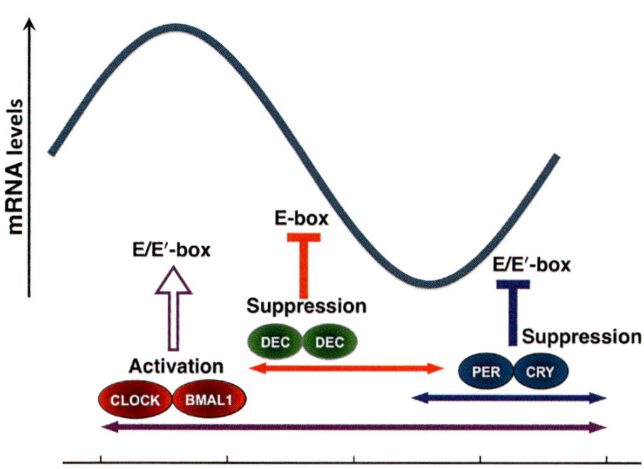

Yukio Kato *et al.*, Figure 10.4 Differential suppression of circadian expression of clock and clock-controlled genes by DEC1, DEC2, and PER:CRY. CLOCK:BMAL1 heterodimer binds constitutively to the E-boxes in the target genes and activates transcription. DEC1 and DEC2 directly bind to the E-box, but not the E′-box, for the indicated time (a red bidirectional arrow) after the peak of target gene expression, whereas PER:CRY indirectly binds to the E-box and/or E′-box for the indicated time (a blue bidirectional arrow) after the trough of the expression.

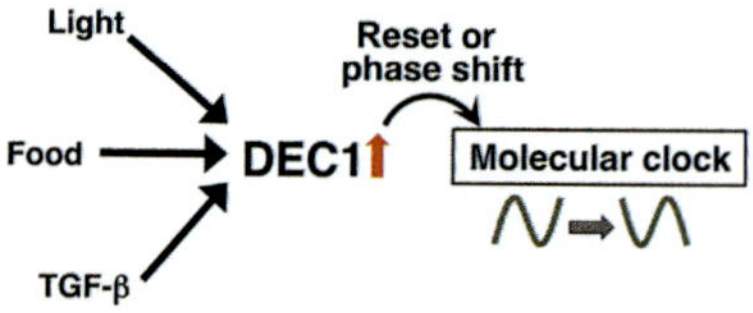

Yukio Kato *et al.*, Figure 10.5 DEC1 function in the regulation of the molecular clock system. In addition to light, the major cue of the central clock, food intake, and growth factors act as cues of the peripheral clocks in a tissue-specific manner.

Edwards Brothers, Inc.
Thorofare, NJ USA
October 16, 2014